III All-Russian Scientific and Practical Conference on Innovations in Non-Destructive Testing (SibTest 2015)

Journal of Physics: Conference Series Volume 671

Altai, Russia
27 - 31 July 2015

ISBN: 978-1-5108-1962-7
ISSN: 1742-6588

Printed from e-media with permission by:

Curran Associates, Inc.
57 Morehouse Lane
Red Hook, NY 12571

proceedings.com

Some format issues inherent in the e-media version may also appear in this print version.

Printed by Curran Associates, Inc. (2016)

For permission requests, please contact the Institute of Physics
at the address below.

Institute of Physics
Dirac House, Temple Back
Bristol BS1 6BE UK

Phone: 44 1 17 929 7481
Fax: 44 1 17 920 0979

techtracking@iop.org

Additional copies of this publication are available from:

Curran Associates, Inc.
57 Morehouse Lane
Red Hook, NY 12571 USA
Phone: 845-758-0400
Fax: 845-758-2634
Email: curran@proceedings.com
Web: www.proceedings.com

Table of contents

Volume 671

III All-Russian Scientific and Practical Conference on Innovations in Non-Destructive Testing (SibTest 2015)
27–31 July 2015, Altai, Russia

Accepted papers received: 27 November 2015
Published online: 18 January 2016

Preface

Papers

Journal of Physics: Conference Series **671** (2016) 011001 doi:10.1088/1742-6596/671/1/011001

1. Preface

This issue of the journal is devoted to the research and studies presented at the III All-Russian Scientific and Practical Conference on Innovations in Non-Destructive Testing SibTest. The conference was held in Altai, Russia, on 27–31 July 2015. The conference brought together experts from different countries and organizations who had a great opportunity to share knowledge during oral and poster presentations and to initiate discussions on topics that are of interest to the conference attendees. The conference aimed to discuss innovative methods and the application of advanced technologies in non-destructive testing. The conference also attempted to bring together university, academic and industrial science, to expand the co-operation of scientists from different countries in research and development and the commercialization of innovative technologies in non-destructive testing. The key themes of the conference were: ultrasonic and acoustic testing; electromagnetic and thermal testing; various types of radiation non-destructive testing; passive and active testing techniques. The conference organizers are the Institute of Non-Destructive Testing, Tomsk Polytechnic University, with the assistance of the Russian Society for Non-Destructive Testing and Technical Diagnostics, Institute of Strength Physics and Materials Science, Siberian Branch of the Russian Academy of Sciences, National Research Tomsk State University, Moscow State Institute of Radio Engineering, Electronics and Automation.

2. List of delegates

No.	Article	
1.	A V Yurchenko, N I Gorlov, A D Alkina, A D Mekhtiev, A A Kovtun	E-mail: barton.kz@mail.ru
2.	A Aristov, K Timchenko, A Novoseltseva, D Kustov, I A Larioshina	E-mail:tina_tim7@mail.ru, aristov@tpu.ru
3.	E Barbin, A Koleda, T Nesterenko, S Vtorushin	E-mail: bar_es@mail.ru
4.	S G Stuchebrov, A V Batranin and I A Miloichikova	E-mail: stuchebrov@tpu.ru
5.	S Bekher	E-mail: behers@mail.ru
6.	A Ubaichin, A Bespalko, A Filatov, E Alexeev, G Zhuk	E-mail: dualog@mail.ru
7.	F Yu Blynskiy	E-mail: blynskiyf@tpu.ru
8.	A Smokotin, M Bogatyreva and I Protasova	E-mail: mar289@rambler.ru
9.	S V Bordunov, O V Galtseva	E-mail: bvv@academ.tsc.ru
10.	S Osipov, S Chakhlov, O Osipov, A Shtein and J Van	E-mail: osip1809@rambler.ru
11.	A O Chulkov and V P Vavilov	E-mail:chulkovao@tpu.ru
12.	S G Stuchebrov, I A Miloichikova and I B Danilova	E-mail: stuchebrov@tpu.ru
13.	T Fursa, D Dann, A Surzhikov, M Petrov	E-mail: fursa@tpu.ru
14.	B N Borikov, O V Galtseva and G A Filippov	E-mail: piano@tpu.ru
15.	Rejection of Welded Connections According to EN ISO 10675: a New Approach to Increasing Objectivity N P Aleshin, D I Galkin, A S Sorokin and A V Khodzhaeva	E-mail: aleshin@bmstu.ru, digalkin@sertink.ru, assorokin@mesi.ru, avkhodzhaeva@cniitmash.com
16.	V Gelever, E Usachev and A Manushkin	E-mail:gelgan@yandex.ru
17.	S Gerasimov, T Sych and V Kuleshov	E-mail: 912267@gmail.com
18.	A E Goldshtein, V Y Belyankov	E-mail: algol@tpu.ru
19.	F A Gubarev, M S Klenovskii	E-mail: gubarevfa@tpu.ru
20.	A N Golikov and A A Ignatovskaya	E-mail: nastena@tpu.ru
21.	D Ivashkov, A Batranin and T Mamyrbayev	E-mail: dvi2@tpu.ru
22.	A V Kamenev, R N Satarov, V P Yakubov, S E Shipilov	E-mail: s.shipilov@gmail.com
23.	A A Filimonov, S V Kasyanov, V A Kasyanov	E-mail: kassianov2011@yandex.ru
24.	S V Chakhlov, S V Kasyanov, V A Kasyanov, S P Osipov, M M Stein, A M Stein, Sun Xiaoming	E-mail: copperman@rambler.ru
25.	A Khammatov, D Belkin and N Barbina	E-mail: hammatov@tpu.ru
26.	A Sh Bureev, E Yu Kiseleva, M S Kutsov and D S Zhdanov	E-mail: eka.kiselyova@gmail.com
27.	V Klimenov, A Ovchinnikov, S Osipov, A Shtein5, A Ustinov and A Danilson	E-mail: nauka@tsuab.ru
28.	A V Klokov, A S Zapasnoy, A S Mironchev, V P Yakubov, S S Shipilova	E-mail: zas_rff@sibmail.com
29.	A Kolomeets, S Bekher	E-mail: andreykolomeec@yandex.ru
30.	A Sechin, O Kyrmakova, S Osipenko	E-mail: olia_917@mail.ru
31.	A L Ovchinnikov, B M Lapshin	E-mail: ovchinnikov_al@inbox.ru
32.	A A Uvarov, I A Lezhnina, K V Overchuk, A S Starchak, Sh D Akhmedov, I A Larioshina	E-mail: inna84-08@mail.ru

33	N P Kalinichenko, A N Kalinichenko, I S Lobanova, A A Zaitseva, E L Loboda	E-mail: konarevai007@tpu.ru
34	E L Malikov, A P Shestak, M M Rychkov, K V Sukharnikov	E-mail: malikov@tpu.ru
35	T Mamyrbayev, A Batranin and D Ivashkov	E-mail: talgat.mamyrbayev@yandex.ru
36	S G Stuchebrov, I A Miloichikova, A L Melnikov and M A Pereverzeva	E-mail: stuchebrov@tpu.ru
37	A I Nizhegorodov, A N Gavrilin, B B Moyzes	E-mail: nastromo_irkutsk@mail.ru, tom-gawral@list.ru, mbb@tpu.ru
38	B I Kapranov and M A Mudrov	E-mail: mudrov@tpu.ru
39	Yu M Fomichev, S V Silushkin, T S Mylnikova	E-mail: slavasv@mail.ru
40	O B Nazarenko, T V Melnikova and P M Visakh	E-mail: obnaz@mail.ru
41	O B Nazarenko, P B Bukhareva, T V Melnikova and P M Visakh	E-mail: obnaz@mail.ru
42	I G Odnokopylov and V V Gneushev and I A Larioshina	E-mail: odivan@yandex.ru
43	S Osipov, S Chakhlov, O Osipov , S Badin and A Abashkin	E-mail: osip1809@rambler.ru
44	L A Payuk , N A Voronina and O V Galtseva	E-mail: lubapa81@mail.ru
45	V M Agranat, A M Goudov, V A Perminov	E-mail: perminov@tpu.ru
46	R M Mustafina, I A Plotnikov, I V Plotnikova, O N Tchaikovskaya	E-mail: inna@tpu.ru
47	Artem Popkov	E-mail: zabagy@gmail.com
48	A Sechin, A Popov and O Antonevich	E-mail: LandBio@mail.ru
49	V V Red'ko, A P Leonov, L A Red'ko and V A Bolgova	E-mail: sales@redwill.ru
50	J Fedorchuk , M Sadenova and O Rusina	E-mail: olgarusina@tpu.ru
51	S G Stuchebrov, I A Miloichikova and X O Shilova	E-mail: stuchebrov@tpu.ru
52	S E Shipilov, A I Eremeev, V P Yakubov	E-mail: s.shipilov@gmail.com
53	Yu M Fomichev, S V Silushkin, I A Larioshina	E-mail: slavasv@mail.ru
54	A I Soldatov, A A Soldatov, M A Kostina, V F Tolkatchev	E-mail: asoldatof@mail.ru
55	Y D Lugovtsova, A I Soldatov	E-mail: yevgeniya.lugovtsova@bam.de
56	N S Starikova, V V Redko, G V Vavilova	E-mail: nadushasns@tpu.ru
57	S G Stuchebrov, I A Miloichikova and I B Danilova	E-mail: stuchebrov@tpu.ru
58	K V Sukharnikov, M M Rychkov, V G Gentselman	E-mail: k.sukharnikov@gmail.com
59	V Udod , J Van, S Osipov, S Chakhlov and A Temnik	E-mail: temnik_ak@mail.ru
60	S N Torgaev, M V Trigub, G S Evtushenko and T G Evtushenko	E-mail:torgaev@tpu.ru
61	S Vavilov, G Koshkin, V Udod, O Fofanof	E-mail: baron@tpu.ru
62	A E Goldshtein, G V Vavilova and S V Mazikov	E-mail: wgw@tpu.ru
63	O V Zakharov, S V Novichkov, A V Kochetkov	E-mail: zov17@mail.ru
64	L A Khokhlova, A I Seleznev, D S Zhdanov, I Yu Zemlyakov and E Yu Kiseleva	E-mail: hohlova@diagnostic.tom.ru
65	D O Zyatkov, A V Yurchenko, V B Balashov, V I Yurchenko	E-mail: zyatkov_do@niipp.ru

3. Members of the Organizing Committee

V.N. Borikov, Chairman of the Committee, DSc, Professor, Director of the Institute of Non-Destructive Testing, Tomsk Polytechnic University.
A.P. Surzhikov, DSc, Professor, Deputy Director for Scientific Activity in the Institute of Non-Destructive Testing, Tomsk Polytechnic University.
V.P. Vavilov, DSc, Professor, Vice-President of the Russian Society for Non-Destructive Testing and Technical Diagnostics.
E.S. Gorkunov, Academician of RAS, DSc, Professor, Director of the Institute of Engineering Science, UB RAS.
A.S. Sigov, Academician of RAS, DSc, Professor, Rector of Moscow State Institute of Radiotechnics, Electronics and Automation (Technical University).
V.I. Syryamkin, DSc, Professor, Head of the Department of Quality Management, Tomsk State University
A.E. Goldstein, DSc, Professor of the Institute of Non-Destructive Testing, Tomsk Polytechnic University
Uwe Ewert, Professor, Federal Institute for Materials Research and Testing.
V.V. Kluev, Academician of RAS, DSc, Professor.
Edward Siemens, Professor, Anhalt University of Applied Sciences (Germany)
V.P. Yakubov, DSc, Professor, Head of Radiophysics Department, Tomsk State University
Xavier Maldag, Professor, Laval University, Quebec, Canada
A.V. Yurchenko, DSc, Professor, Scientific Secretary, Tomsk Polytechnic University
A.G. Yakunin, DSc, Head of the Department of Computing Systems and Information Security, Polzunov Altai State Technical University
S.E. Shipilov, DSc (Phys.-Math), Associate Professor of the Department of Radiophysics, Tomsk State University

4 Tribute to colleagues

The Organizing Committee expresses its sincere gratitude to all the participants, organizers and sponsors of the conference. This conference is not just a scientific event to learn the prospects for the research, new technologies and innovative ideas, but it is another meeting of the colleagues who have become trusted friends, partners and allies in the common cause.

Peer review statement

All papers published in this volume of *Journal of Physics: Conference Series* have been peer reviewed through processes administered by the proceedings Editors. Reviews were conducted by expert referees to the professional and scientific standards expected of a proceedings journal published by IOP Publishing.

Research of the Additional Losses Occurring in Optical Fiber at its Multiple Bends in the Range Waves 1310nm, 1550nm and 1625nm Long

A V Yurchenko [1], N I Gorlov [2], A D Alkina [3], A D Mekhtiev [3], A A Kovtun [4]

[1] Professor, Tomsk Polytechnic University, Tomsk, Russia
[2] Professor, Head of "Communication lines" Department in Siberian state university of telecommunications and information sciences, Novosibirsk, Russia
[3] TPU graduate student, master's degree in "Information Technologies and communication systems"
[4] Head of "The technology and communication systems" Department in Karaganda State Technical University, Karaganda, Kazakhstan
[5] Undergraduate in Siberian state university of telecommunications and information sciences, Novosibirsk, Russia
E-mail: barton.kz@mail.ru

Abstract. Article is devoted to research of the additional losses occurring in the optical fiber at its multiple bends in the range waves of 1310 nanometers, 1550 nanometers and 1625 nanometers long. Article is directed on creation of the external factors methods which allow to estimate and eliminate negative influence. The automated way of calculation of losses at a bend is developed. Results of scientific researches are used by engineers of "Kazaktelekom" AS for practical definition of losses service conditions. For modeling the Wolfram|Alpha environment — the knowledge base and a set of computing algorithms was chosen. The greatest losses are noted on wavelength 1310nm and 1625nm. All dependences are nonlinear. Losses with each following excess are multiplicative.

1. Introduction

Further development of communication and telecommunication is already impossible without digital fiber-optical systems using, in view of their indisputable advantages before the directing systems with copper conductors. Despite of positive results and the moments of using fiber – optical transmission lines (FOTL) there are shortcomings connected with the feature of information signal distribution in the optical fiber, its design connected with its criticality to different external influences and loadings. Distinctive feature of the optical cable (OC) in comparison with electrical cabels is availability of quartz fiber instead of copper, which is subject to aging, degradation and can be the reason of the cable exit out of the operation. Therefore it is necessary to consider separately the optical fiber (OF) and the optical cable OC in general.

The introduction features of optical technologies in Kazakhstan are the considerable extent of highways, because of the huge territory taking the 9th place in the world on the area of the terrestrial land with rather low population density falling on one inhabitant. It is necessary to create extended main and intra-zone lines with large volume OC various types and other technical equipment.

Economic consequences of that are relative expenses 2,3 times more in comparison with the average world expenses. It must be kept in mind feature that Kazakhstan in considerable part is in a zone of a severe temperate climate with sharp temperature differences. An important point in ensuring standard reliability and efficiency of the optical directing systems operation throughout the long term of OC operation is the question of elimination of factors of negative external mechanical loads influence of the OF and a cable and also on the transfer system of transfer in general [1]. The creation of methods which allows us to estimate and eliminate negative influence, in particular a bend which reduce reliability of FOTL and serve as the reason of increase in losses service conditions of "Kazaktelekom" AS based on the hierarchical approach, is a novelty and solves an actual problem of the applied direction. FOTL is one of conservative telecommunication parts, which has to be operated based on considerable capital and labor OC expenditure [2].

While laying of the fiber-optical communication lines there are questions of external impacts influence on the optical cable characteristics of an optical cable. The most important is the influence of optical fibers bends on additional losses of radiation power in an optical cable. In case of deformation, namely a bend of optical fiber there are additional losses of energy. The size of these losses depends on many reasons, but the main can be at achievement of a certain critical bend radius. In case when the bend radius makes only some millimeters at fibers with a high numerical aperture and when radius about tens centimeters for fibers in the single-mode mode with a big cross mode area [2].

Losses are observed at macroscopic bends of fiber, which arise at operation owing to emergence of roughnesses (violations of structure) in fiber, which are explained by imperfection of manufacturing techniques or installation. The bends arising in fiber can be divided into two types [2] conditionally. Microbends are the small local violations of fiber straightforwardness caused by constructive and technological discontinuity, which can arise at a fiber production, and also laying and production of a cable [2]. Fiber macrobends appear as a result of their twist on the cable length and winding on a drum. Losses are caused by an effluence or radiation of the directed modes and become inadmissibly big at the radius reduction of the bend curvature to critical values. Critical radius of a fiber bend approximately pays off on a formula (1):

$$R_{cr} \approx \frac{3n_1^2\lambda}{4\pi\left(n_1^2-n_2^2\right)^{3/2}} \qquad (1)$$

where n_1, n_2– refraction indicators of a core and cover; λ – the wavelength of the applied radiation.

We executed the influence researches on external impacts on OC in particular at a bend leading to growth of additional losses dB on which results basis the radius calculation method the optical cable (OC) at mechanical influence is developed. The obtained data will allow to create automated calculation method for the solution of an actual problem of the "Kazakhtelekom" AS FOTL operation efficiency. In particular the task of additional losses definition in dB at various bend angles of a cable in 45, 90, 135 degrees was set, depending on wavelength 1310nm, 1550nm and 1625nm, the quantity of fiber bends is also considered. As object of research single-mode G-652 fiber 10 meters long was used. The dead zone of the reflectometer makes 0.75 meters (according to the passport).

At the decision of the theory of the directing systems, theories of transfers lines, calculus mathematics, computer processing of experiment tabular data, programming in the environment of DBMS, methods of computer and imitating modeling with the using of Wolfram|Alpha were used. The approximations executed by us will allow to receive value of additional losses in dB at various combinations of bend angles and their quantities.

During the first experiment 70 measurements were taken. On ten measurements for one, two, three, four, five, six and seven excesses, with a radius of curve not less than 5 mm, at an angle in 45 degrees for one site of FOTL. Also the other experiments are similarly made. For researches the special stand representing the panel with a grid at a slow pace in 1sm with fixing bolts on each centimeter, for giving it any form to a fiber-optical cable without possibility of its mechanical damage was manufactured (figure 1).

During research nine groups of experiments will be carried out. On three types of a corner for each frequency separately. All experiments are made by means of the high-precision reflectometer of

Journal of Physics: Conference Series **671** (2016) 012001 doi:10.1088/1742-6596/671/1/012001

Yokogawa AQ1200E firm. A series of optical Yokogawa AQ1200E reflectometers are developed especially for work on city or local networks when it isn't necessary to measure long distances, and compactness when moving is necessary on foot between objects. Yokogawa AQ1200E is a multipurpose tester, it can be completed with the certain port of the measuring instrument of power for measurement of losses, a source of visible radiation for search of damages, the video probe for optical sockets check.

For losses definition on fiber at bends the stand of five probe zones in size 0.5m×0.5m for a cable laying was made. Each probe zone represents an array 11×11 with a step in 1sm. While laying a cable various variations – with a corner 45, 90,135 degrees are possible. Radius of a cable excess makes 1.5–1.7sm that corresponds to critical angles of a bend which most often turn out at the wrong cable laying or at other situations when the cable is bent under critical angles.

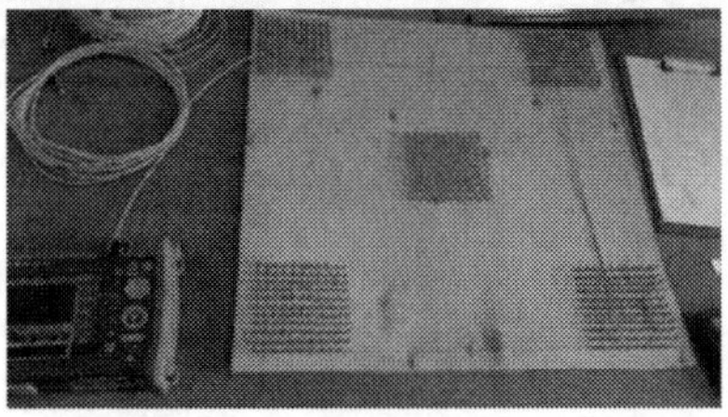

Figure 1. Laboratory stand.

The important point is that the automated method development capable to give out values of losses, when using various lengths of waves and at various laying of a cable, for further planning of laying FOTL within laboratories, buildings, server rooms, offices, etc. PO has to apply various dependences of losses calculation, to obtaining the most exact data on short sites of a fiber-optical cable or to definition of losses on a long site of FOTL at a smaller quantity of excesses.

During the first experiment the fiber-optical cable was fixed at the stand at an angle in 45 degrees in quantity from one to seven. For more exact statistical data it was removed on 10 indications for each corner.

The schedule have a nonlinear dependence, about increase in quantity of bends losses increase. During the following experiments made similarly dependences for all other cases were revealed. Schedules are compared and revealed the main regularities. Nonlinearity of growth of losses. Multiplication of losses for each corner.

For modeling the Wolfram|Alpha environment was chosen the knowledge base and a set of computing algorithms [3]. Wolfram Alpha is the system intended for storage, processing and delivery to users of the structured data on inquiries in natural English. Wolfram Alpha isn't a search engine. It is caused by that it isn't intended for automatic processing of unstructured texts. For its work it is necessary to enter manually previously factual information into a database, and also to develop and realize algorithms of its processing. These procedures are carried out manually by community of developers and experts of Wolfram Alpha.

The Wolfram|Alpha project is created by the well-known mathematician Stephen Volfram on the basis of the theory about a computability. Actually, its sphere all is built on IT. Stephen devoted to idea of calculations more than thirty years and created two significant projects: Mathematica and Wolfram Alpha. The first represents system of computer algebra, and the second — the knowledge base and a set of computing algorithms on the basis of the first

Wolfram Mathematica (WM) is a package of symbolical mathematics. The huge number of the functions put by developers, and also the open environment allowing to supplement a package with own expansions makes its opportunitues truly boundless. Mathematica has high speed and almost not limited accuracy of calculations that allows it to work as at very powerful computers, and not really strong personal computers. On the basis of a kernel of a package there is a Web server which allows to use its opportunities to unlimited number of people.

General competitors of a package are called Maple, MathCAD and MatLab. If with the first it is difficult to argue, about MathCAD and MatLab it is possible. The matter is that these two packages occupy absolutely other niche, than Mathematica. Both use numerical algorithms at calculation but not symbolical. Symbolical calculations are poorly developed (in comparison with packages of symbolical calculations) additions. Much more similar product is free of charge extended Maxima package.

Results of experiments were processed with use of Wolphram Alpha, the most plausible models of approximation are: polinomal approximation of the fourth degree, cubic approximation, square approximation.

For wavelength 1310nm, for corners 45 °.

Experiment studying results are presented on the figure 2.

According to the handler: polinomal approximation (2) of the fourth degree (quartic):

$$-0.0000751894\,x^4 + 0.0011976\,x^3 - 0.00660133\,x^2 + 0.0168207x \quad + 0.0363341 \qquad (2)$$

cubic approximation (cubic) (3):

$$-0.000144949\,x^3 - 0.00203626\,x^2 + 0.0106552\,x + 0.0372364 \qquad (3)$$

square approximation (quadratic) (4):

$$-0.000514286\,x^2 + 0.00666905\,x + 0.0387583 \qquad (4)$$

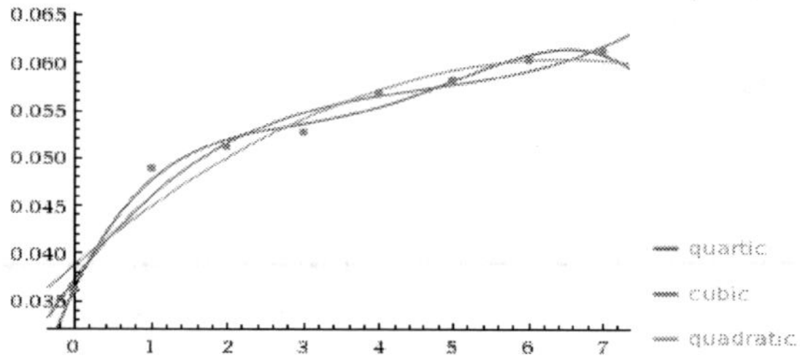

Figure 2. Schedules of accuracy comparison.

Here it is well noticeable that the most satisfying result is yielded by polinomal approximation of the 4th degree and a cubic approximation.

For wavelength 1550nm for corners 45 °.

Experiment studying results are presented on the figure 3.

Polinomal approximation (5) of the fourth degree (quartic):

$$0.0000105114\,x^4 + 0.000213068\,x^3 - 0.000987973\,x^2 + 0.00292265\,x \quad + 0.048486 \quad (5)$$

Cubic approximation (6):

$$0.0000659091\,x^3 - 0.000349784\,x^2 + 0.00206071\,x \quad + 0.0486121 \qquad (6)$$

Square approximation (quadratic) (7):

$$0.000342262\,x^2 + 0.000248214x \quad + 0.0493042 \qquad (7)$$

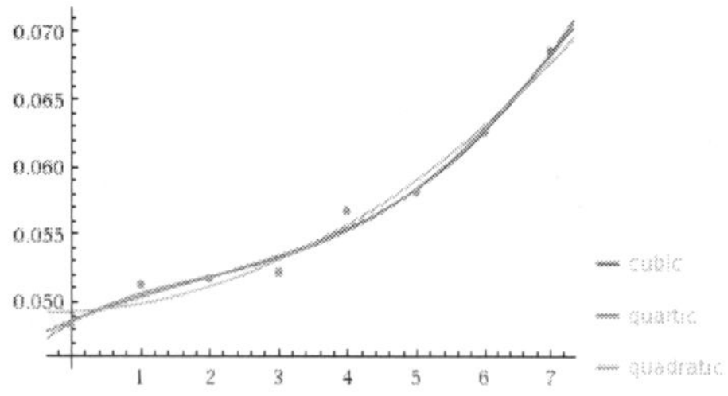

Figure 3. Schedules of accuracy comparison.

Here it is well noticeable that the most satisfying result is yielded by polinomal approximation of the 4th degree and cubic approximation.

For wavelength 1625nm for corners 45 °.

Experiment studying results are presented on the figure 4.

Polinomal approximation of the fourth degree (quartic) (8):

$$0.000104072\,x^4 + 0.00167948\;x^3 - 0.00868381\,x^2 + 0.019243\,x \quad + 0.0362693 \qquad (8)$$

cubic approximation (cubic) (9):

$$-0.000222475\,x^3 - 0.00236515\,x^2 + 0.0107091\,x \quad + 0.03751820 \qquad (9)$$

square approximation (quadratic) (10):

$$-0.000514286\,x^2 + 0.00666905\,x + 0.0387583 \quad (10)$$

Schedules of accuracy comparison:

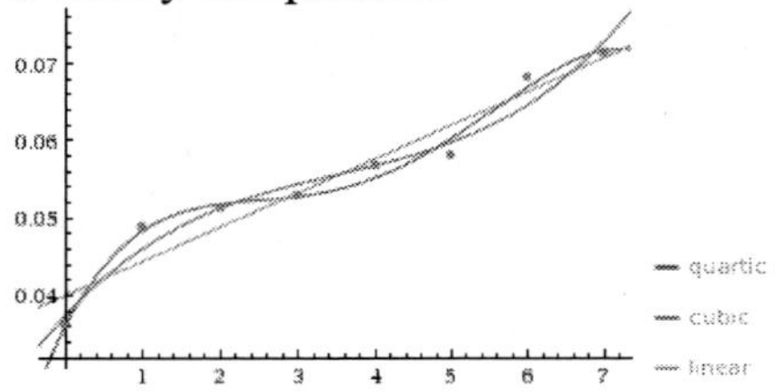

Figure 4. Schedules of accuracy comparison.

Experiment general results are presented on the figure 5.

Here it is well noticeable that the most satisfying result is yielded by polinomal approximation of the 4th degree and cubic approximation.

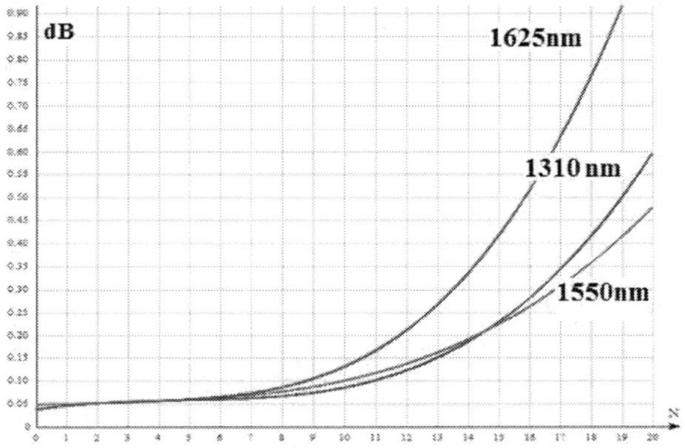

Figure 5. Schedules of comparison of losses for the wavelength of 1310 nanometers, 1550 nanometers, 1625 nanometers, with a bend angle 45 °.

We made experiments of measurement of additional losses in wavelengths 1310nm, 1550nm, 1625nm for corners 90 ° on the figure 6. Results of experiments are processed by means of the Wolphram Alpha program. We made experiments of measurement of additional losses in wavelengths 1310nm, 1550nm, 1625nm for corners 135 ° on the figure 7.

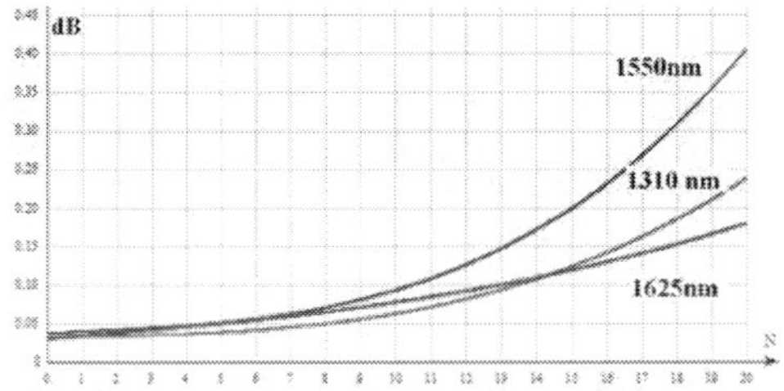

Figure 6. Schedules of comparison of losses for the wavelength of 1310 nanometers, 1550 nanometers, 1625 nanometers, with a bend angle 90 °.

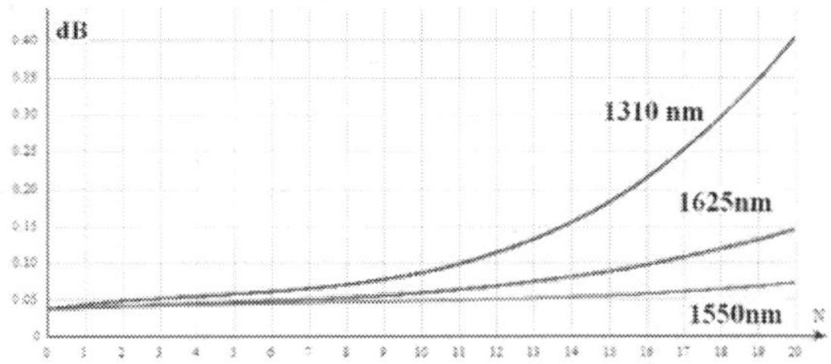

Figure 7. Schedules of comparison of losses for the wavelength of 1310 nanometers, 1550 nanometers, 1625 nanometers with a bend angle 135 °.

Drawing up the program for calculations of losses on different lengths of waves at different corners. In the environment of Delphi 7 development of the program with which it is possible to estimate losses in optical fiber at multiple bends of critical radius was developed. A window of the program for the automated calculation of additional losses of OV with its multiple bends in the range waves 1310nm, 1550nm and 1625nm long is shown on the figure 8.

The program has intuitively cleared interface and allowed to carry out visualization of bends losses. For creation of the schedule the program uses a method of cubic approximation which showed on the highest precision.

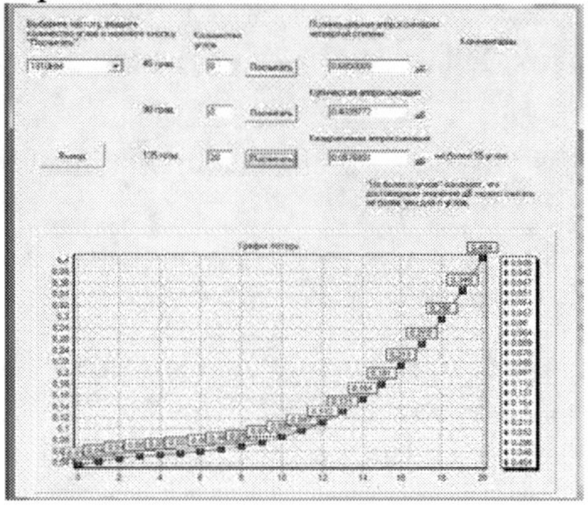

Figure 8. A window of the program for the automated calculation of additional losses of OV with its multiple bends in the range waves 1310nm, 1550nm and 1625nm long.

4. Conclusion

During research work experiments with various corners of optical fiber excesses and also loss of FOTL were made for each of standards of work: 1310nm, 1550nm, 1625nm. The greatest losses are noted for waves 1310nm and 1625nm long. All dependences are nonlinear. Losses with each following excess are multiplicative.

In most cases losses at a bend strongly increase with a length of waves 1625nm because they are strongly defined by existence of an interference of light reflected from cover/border of a covering and/or from an external surface of a covering. The increase in losses at a fiber bend on big lengths of waves limits ranges a transmission of single-mode fibers. The greatest losses are caused by an excess in 45 degrees. It is possible to see that with increase in a corner of loss fall as well as with increase in wavelength.

During research data according to which mathematical models of dependences of losses on quantity a catch, types of corners and wavelength were multiply obtained. The software prepared by them allowing to calculate automatically losses of OC service conditions of "Kazaktelekom" AS at various quantities of bends in the waves range 1310nm, 1550nm and 1625nm long.

References

[1] Gorlov N, Mekhtiyev A, Airikh V, Aldoshina O, Kshalova A 2014 *Methods and gages of optical fiber communication lines parameters* (Karaganda: Publishing house of KSTU)

[2] Gambling W (2000) *The Rise and Rise of Optical Fiber* IEEE Journal on Selected Topics in Quantum Electronics **6** (6) 1084–1093 doi: 10.1109/2944.902157

[3] Mitschke F 2009 *Fiber Optics – Physics and Technology* (Berlin; Heidelberg: Springer)

[4] Ramaswami R, Sivarajan K, Sasaki G 2009 *Optical Networks: A Practical Perspective.*(Morgan Kaufmann; 3rd edition)

[5] http://www.wolframalpha-ru.com

Designing of Phantom Head Used in Optical Diagnostics of Brain Injury

A Aristov[1], K Timchenko[2], A Novoseltseva[3], D Kustov[4], I A Larioshina[5]

[1]Associate professor, Institute of Non-Destructive Testing, Tomsk Polytechnic University, Tomsk, Russia
[2]Undergraduate, Institute of Non-Destructive Testing, Tomsk Polytechnic University, Tomsk, Russia
[3]Undergraduate, Institute of Non-Destructive Testing,, Tomsk Polytechnic University, Tomsk, Russia
[4]Bachelor, Institute of international education and language communication, Tomsk Polytechnic University, Tomsk, Russia
[5]Assistant, Tomsk State University, Tomsk, Russia
E-mail:tina_tim7@mail.ru, aristov@tpu.ru

Abstract. This article shows the results of an experimental research on properties of the materials chosen for designing of a phantom head, which is to be used in testing of a brain hematoma diagnostics device. We have conducted a comparative research of the optical properties of model materials and real head tissues

1. Introduction and terms of reference

It is important to detect intracraneal hematoma when facing brain injury. Nowadays magnetic resonance tomography and computed tomography is the basic precise instrumental method for detecting such injuries, however a possibility of undergoing tomographic screening for a patient right after obtaining an injury is practically non-existent. As a result, development of a device that would be able to immediately detect intracraneal injuries is important today. Unfortunately, sound field visualization method [1] used for gathering information on brain structure doesn't provide good results due to the intense diffusion of ultrasound in skullbones. To solve this problem there has been an approach suggested, which is based on spectroscopy in optical and near infrared wave range. The method is based on facts that molecules absorbance spectra are feature-based for a substance given, and absorbing power is tied to content of absorbing constituent in irradiated sample. The wave range used is harmless both for an operator and a sample being analyzed. Along with relevant hardware it allows to explore sites of animals and human beings without any damage to integration and bodies [2]. InfraScan has developed a device called Infrascanner model 3000 [3], this is a mobile intracraneal hematoma detector, which works in a near infrared wave range (808 nm). Published experimental research carried out on models of intracerebral bleeding, as well as clinical investigations have shows quite high diagnostic sensitivity of the device. The scannes is able to detect hematomas with a probability of 91% (in case a hematoma's maximum depth is 2.5 cm from brain cortex and with size more than 3.5 cm^3) [4, 5].

However, the device is not able to detect properties of a hematoma (dimension, depth of location, density, exact location), which are of a great diagnostic importance for further resuscitation

Journal of Physics: Conference Series **671** (2016) 012002 doi:10.1088/1742-6596/671/1/012002

procedures. Since diagnostics are conducted with a differential method when optical properties of symmetrical locus in exact set points on a head surface, there is a high probability to miss small or deep-seated hematomas out of sections scanned. Therefore, further development of method and device is needed, for the purpose of obtaining more perfect medical and technical characteristics.

2. Theoretical aspects of phantom head designing

We have designed a prototype of a device which allows to conduct laboratory studies aiming to search possible ways of improving diagnostic capabilities of optical method of hematoma detection [6]. The device will help to determine follows things: optimal structure of primary measuring transducer, a method to process analogue signal from photon detector, digital processing and visualization of data received from the device.

In order to model different variants of pathologies (size, location and depth of hematomas) and also minimize or take into account interfering factors on first stages of research (inhomogenous and different thickness of skull bones, different optical properties of tissues), researches need to be conducted on a model sample with known properties. This artificial research object (phantom) must have properties close to the real object. Since we use optical radiation for gathering information on an object's structure optical properties of phantom's materials have to be similar or close to the properties of the real tissues that are being simulated.

During transmission of a head, bone and brain tissues absorb the bigger part of optical radiation. Therefore, within the framework of our project we will simulate three types of tissues: bone tissue, brain tissue and hematoma. Brain tissue is a complex of big number of neurons, which are the nerve fibers surrounded by myelin sheath. Myelin sheath makes a major contribution in brain tissue's optical properties. Since myelin sheath consist of lipoprotein complexes, we suggest replacing this tissue with oil emulsion-based viscous dispersed liquid. Particularly, during the first stage of the research we used mayonnaise. Phase state of this sample in a liquid state is essential, because this way it is possible to move the subject of research (simulated "hematoma") inside the sample. Thereby, we can set its location towards optical part. A bone is one of the most important phantom environments which have quite strong modificative impact on transmitted radiation. For this reason the bone has to best meet the properties of a real object. According to comparative visual estimation of transmitted radiation intensity, and also according to the spot size of scattered radiation during transmitting of trephined human skull and samples of muddy plastics, we prefer to use fluoroplastic. Hematoma is clotted blood, that is why it is possible to simulate it with real human blood placed in a silicone clear container. By changing size of the container we can simulate size of the simulated hematoma.

3. Experimental research

We have conducted an estimation of optical properties of the chosen model materials compared to the real samples of the simulated tissues. The following bio-tissues have been conducted research on: cortical substance of a pig brain, a sample of trephined human skull, clotted and incoagulated (with addition of heparin) human blood. Polychromatic radiation source (halogen lamp) was used in the research. Intensity of transmitted and backscatter radiation that went through the study samples of materials and bio-tissues has been estimated.

Measurements have been conducted with a help of photometric sphere. Transmitted or scattered radiation from the corresponding outputs of optical sphere was recorded and fixed with Ocean Optics USB4000-IVS-NIR-ES spectrograph. Liquid and viscous samples were put into single-use thin-walled cuvets. Thickness of a sample tested in such cuvet was 2 mm. Examined fluoroplastic sheet was as thick as an average human skull bone. Below you can see comparative spectral dependence of examined samples available from experiments.

Figure 1 shows the spectra of radiation transmitted through the brain and the environment which was selected for simulating the brain in a phantom.

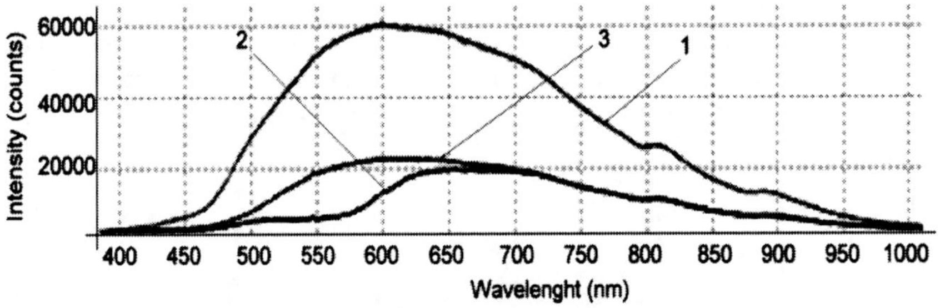

Figure 1. The spectra of radiation transmitted through objects: 1 - the spectrum of the radiation source without an object, 2 - after the radiation transmitted through the brain, 3 - after the radiation transmitted through the mayonnaise.

The graphs show that absorbance spectra of the examined environments best match at wavelengths of 650 nm to 900nm. This is proved by the graph of the intensity of the transmitted radiation (figure 2). Both samples of the same thickness transmit about 30% of radiation in this area. Unfortunately, spectral characteristics of the spectrometer we used did not allow us to examine longer waves.

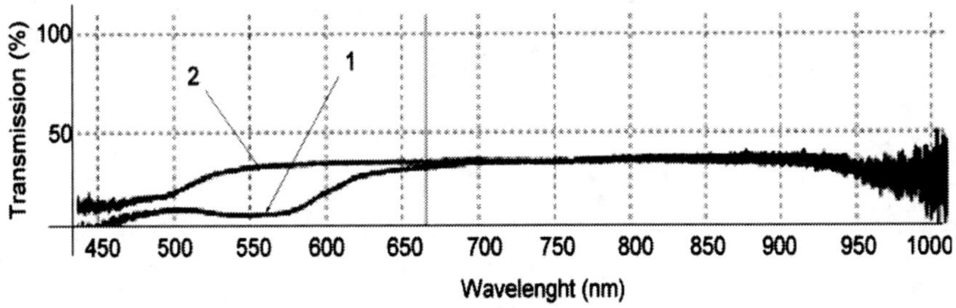

Figure 2. Graphs of the intensity of radiation transmitted through objects: 1 - brain, 2 - mayonnaise. Thickness of a layer transmitted is 2 mm.

A similar comparative research was conducted with a bone tissue and fluoroplast (figure 3). Examination of a skull bone is a specific in some way as a skull has significant differences in thickness and structure all over its surface, so the absorption intensity for the object is changed largely for each segment. We used fluoroplastic plate as thick as an average skull bone. Measurements were held on a trephined parietal bone (60×60 mm). Intensity of radiation transmitted through the bone was being changed by 2 times depending on a certain part of the bone. Therefore, we chose a part of it that would be approximately of the same thickness as fluoroplastic plate.

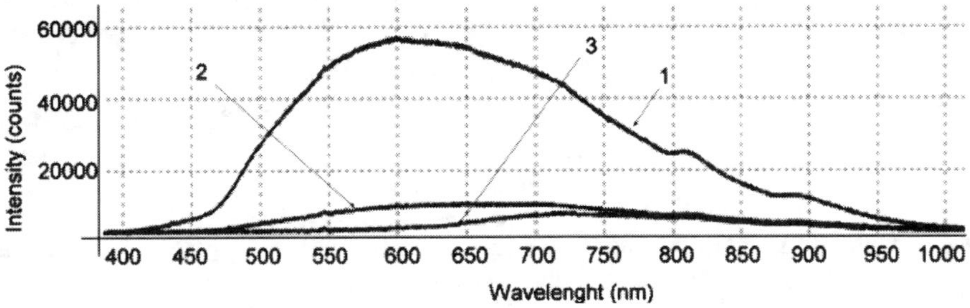

Figure 3. The spectra of radiation transmitted through objects: 1 - the spectrum of the radiation source without an object, 2 - after transmitting through a fluoroplastic plate, 3 - after transmitting through a bone.

The graph shows that spectral characteristics of these samples best match at wavelengths of 750 nm to 900nm. This is the most noticeable on the intensity curves of the transmitted radiation (figure 4). As we can see from the graphs, dispersion of parameters of the bone and fluoroplastic plate is about 5% in the range of 750 nm to 900 nm.

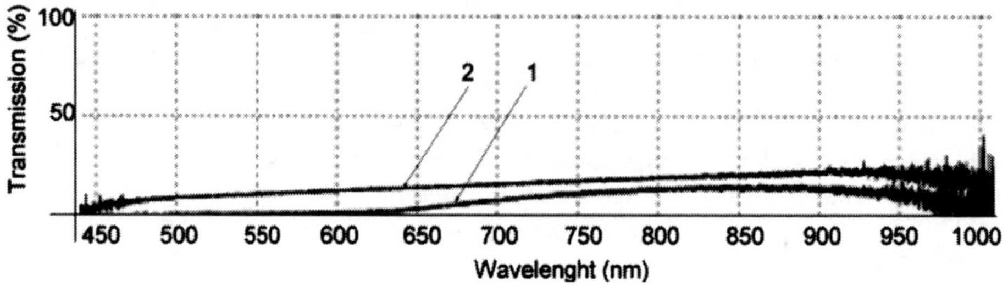

Figure 4. Graphs of the intensity of radiation transmitted through samples: 1 - bone, 2 - fluoroplastic.

During the examination of samples of clotted and incoagulated blood (figure 5) we discovered that waves shorter than 600 nm are absorbed by both clotted and incoagulated blood. On 650 nm spectral field we can see the biggest difference in the intensity of absorption by samples of clotted and incoagulated blood (figure 6). On this range the difference is 15-18%. On the range of 800-900 nm we can see the smalls difference in the spectral characteristics of these samples.

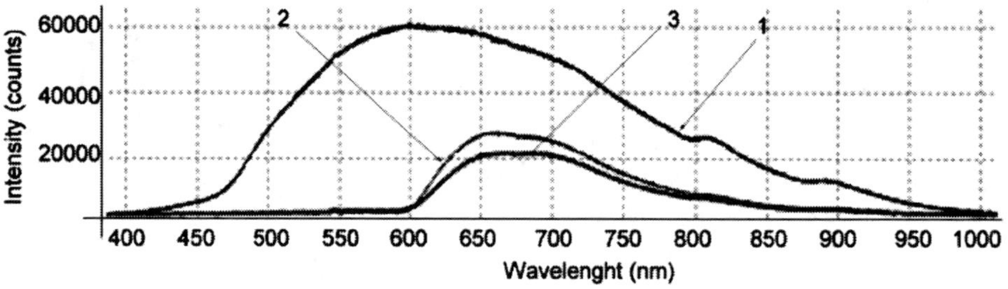

Figure5. The spectra of radiation transmitted through objects. 1 - without objects, 2 - incoagulated blood, 3 - clotted blood.

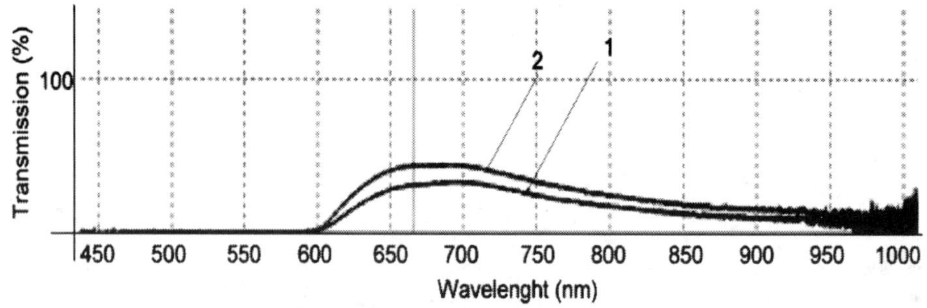

Figure 6. Graphs of intensity of radiation transmitted through samples. 1 - clotted blood, 2 - incoagulated blood. Thickness of a layer transmitted is 2 mm.

4. Results and conclusions

We can draw some conclusions based on the results of our experimental investigations. Thus, due to the large difference of spectral characteristics of clotted and incoagulated blood, two-wave spectroscopy in red (650 nm) and in near infrared (800 nm) spectral range is the most specific for detecting a hematoma. It is possible to detect a brain region with a hematoma near it more precisely by conducting a comparative analysis of the two wave ranges mention above. Estimation of optical characteristics of a pig's cortical substance and 70% oil emulsion (mayonnaise) has proved almost full match of their spectral characteristics in 650-850 nm spectral range. Thus, oil emulsion of a certain concentration can fully simulate brain tissue during optical investigations in this wave range. Notably, emulsion has certain goods on solid phantoms due to the possibility to easily move diagnosable object in liquid fluid, and, as a result, to obtain a phantom with new properties. Examination of optical properties of fluoroplastic plate as a skull bone in a head phantom has shown similarities of their spectral characteristics in 750-850 nm wave range. Significant reduction of transmission through skull bone in the range lower than 630 nm can be explained by the presence of tissues filled with blood in the skull bone. This property of a material can be created only with a help of composite materials with addition of proper absorption filler.

Altogether, the results of experimental investigations we conducted have proved the possibility of using artificial materials we selected for designing phantom head for future optical researches.

References

[1] Soldatov A I, Makarov V *et al.* (2012) Dynamic imaging acoustic fields in research practice *Proceedings - 2012 7th International Forum on Strategic Technology, IFOST 2012* 6357734 doi: 10.1109/IFOST.2012.6357734

[2] ZimnyakovD A, Tuchin V V (2002) Optical tomography of tissues *Quantum Electronics* **32**(10) 849-867

[3] The Infrascanner Model 2000. Electronic resource - access mode: http://www.infrascanner.com/ models/ (date of treatment 22/03/2014)

[4] Bressan S, Daverio M, Martinolli F, Dona' D, Mario F, Steiner I P, Da Dalt L (2013) The use of handheld near-infrared device (Infrascanner) for detecting intracranial hamorrhages in children with minor head injury *Child's Nervous System* **30**(3) 477-484

[5] Robertson C S, Zager E L, Narayan R K *et al* (2010) Clinical evaluation of a portable near-infrared device for detection of traumatic intracranial hematomas *Journal of Neurotrauma* **27**(9) 1597-1604

[6] Timchenko K A, Aristov A A, Musorov I S, Evtushenko T G. (2014) Development of optoelectronic system for subdural hematoma diagnostics *International Conference of Young Specialists on Micro/Nanotechnologies and Electron Devices, EDM* 319-322 doi: 10.1109/EDM.2014.6882538

Journal of Physics: Conference Series **671** (2016) 012003
doi:10.1088/1742-6596/671/1/012003

Three-axis MEMS Accelerometer for Structural Inspection

E Barbin[1], A Koleda[1], T Nesterenko[2], S Vtorushin[3]

[1]Engineer, Tomsk Polytechnic University, Tomsk, Russia
[2]Associate professor, Tomsk Polytechnic University, Tomsk, Russia
[3]Ph.d. student, Tomsk Polytechnic University, Tomsk, Russia
E-mail: bar_es@mail.ru

Abstract. Microelectromechanical system accelerometers are widely used for metrological measurements of acceleration, tilt, vibration, and shock in moving objects. The paper presents the analysis of MEMS accelerometer that can be used for the structural inspection. ANSYS Multiphysics platform is used to simulate the behavior of MEMS accelerometer by employing a finite element model and MATLAB/Simulink tools for modeling nonlinear dynamic systems.

1. Introduction

The modern instrumentation engineering opens new vistas for the development of low-weight and small-size devices of low cost, energy consumption, and high reliability. Microelectromechanical systems (MEMS) comply with these properties due to their micron size, advanced functionality, low energy consumption and cost, and mass-scale application. Nowadays, the MEMS market rapidly grows: the annual increase of the market demand comes to 14–20 % as shown in figure 1 [1].

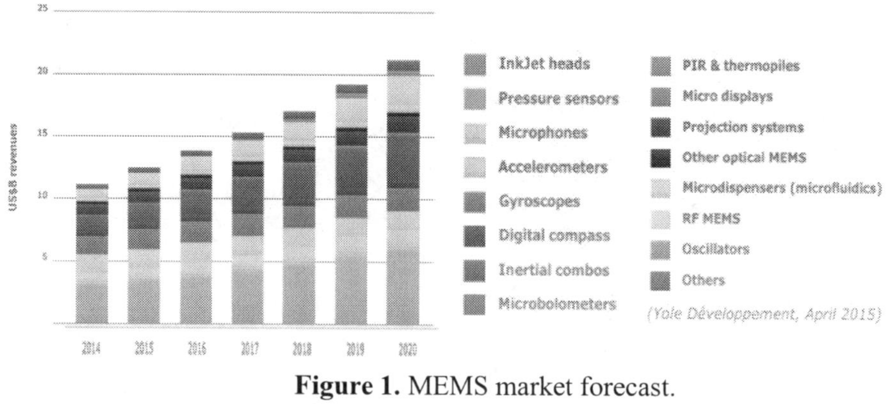

Figure 1. MEMS market forecast.

The increase of the buildings service life as well as the improvement of occupational safety conditions is the worldwide tendency of today. The development of safety systems and continuous monitoring of buildings, bridges, railways is very important to achieve these purposes. MEMS seismic sensors are used in the capacity of railway safety systems. Conventional patrolling is being replaced by the intelligent inspection systems. If somebody tries to do a mine hole under a railroad bed, the alarm is generated on the operator's desk [2].

In connection with the new national standard of the structural inspection accepted in Russia in 2011, it is necessary to perform the building construction monitoring. This becomes possible with the use of the wireless sensor network (WSN) that allows inspecting such characteristics as vibration, deformation, deviation, etc. with the periodicity convenient for a user. A wireless sensor network consists of spatially distributed autonomous sensor nodes and execution units unified by a radio channel to monitor physical or/and environmental conditions. WSN coverage area can achieve several kilometers due to the messages forwarding from one element to another. The monitoring system includes different types of sensors installed on structural elements of a building with a view to detect both the physical (humidity and temperature) and force (static and dynamic loads) effects on their strength and deformation properties [3]. Therefore, the market of building construction monitoring is rather a promising direction of the WSN implementation in Russia. Moreover, the appropriate control for the execution of diagnosis and industrial safety acts affects the WSN market that can be estimated at billions of dollars [2].

2. Description and architecture of accelerometer

A microelectromechanical system accelerometer is one of the wireless sensors both embedded in inclinometer (an instrument for measuring angles of borehole, elevation or depression of an object with respect to the Earth's gravitational field and independent 1-, 2-, and 3-axis units intended for measuring vibrations along the selected axes of an object.

The purpose of this work is to design a three-axis MEMS accelerometer. The function scheme of this accelerometer sensor is shown in figure 2. The frame 3 is mounted on the bottom silicon wafer 12 by the elastic suspension 9 and anchors 10. Sensor is covered with the top silicon wafer 11. Planar electrodes 15, 16 are sprayed on top and bottom silicon wafers. The elastic suspension provides the frame movement along Z axis together with the moving masses 1, 6 moving along X and Y axes, respectively.

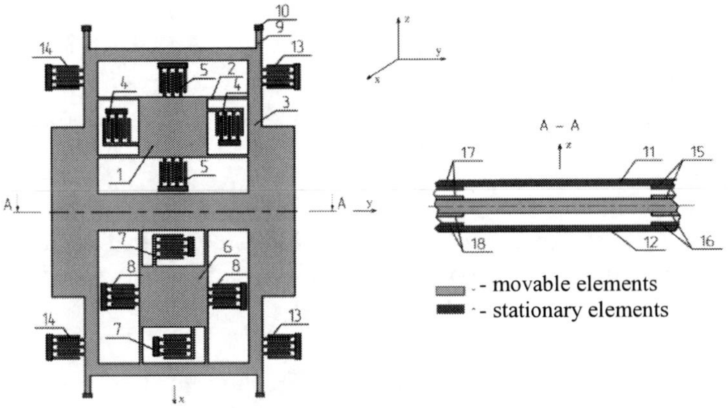

Figure 2. Function scheme of accelerometer sensor: *1, 6* – moving masses; *2, 9* – elastic suspension; *3* – frame; *4, 5, 7, 8 13, 14* – comb electrodes; *10* – anchor; *11*- top silicon wafer, *12* – bottom silicon wafer; *15, 16, 17, 18* – planar electrodes.

Comb electrodes 4 are incorporated into the mechanical movement transducer of the moving mass 1 along X axis. This mechanical movement occurs due to the acceleration along X axis. Comb electrodes 5 serve as the execution units that form the correction forces and compensating feedback in X-direction. Comb electrodes 7 are incorporated into the mechanical movement transducer of the moving mass 6 along Y axis. This mechanical movement occurs due to the acceleration along Y axis. Comb electrodes 8 serve as the execution units that form the correction forces and compensating feedback in Y-direction. Planar electrodes 15, 16, 17, 18 are incorporated into the mechanical movement transducer to detect the motion of frame 3 along Z axis. This mechanical movement occurs in case of the acceleration along Z axis. Comb electrodes 13, 14 serve as the execution units that form the correction forces and compensating feedback in Z-direction.

Let us generate the equation for the movement of the accelerometer sensor the equivalent scheme of which is presented in figure 3. It is assumed that stiffness G_z, G_x, G_y of elastic suspensions is much lesser on the sensor sensitive axes than that of the same suspensions in crossing directions. This assumption is ground and very important for the design of accelerometer sensor. Sensor coordinates *OXYZ* shown in figure 4 are associated with the frame which performs a translational motion with the acceleration having *XYZ*-components, namely *ax, ay, az* and rotates with Ω_x, Ω_y, Ω_z angular velocities. Coordinates $O_1X_1Y_1Z_1$ are connected with the moving mass 1, coordinates $O_2X_2Y_2Z_2$ – with the moving mass 2, while coordinates $O_3X_3Y_3Z_3$ are connected with the frame.

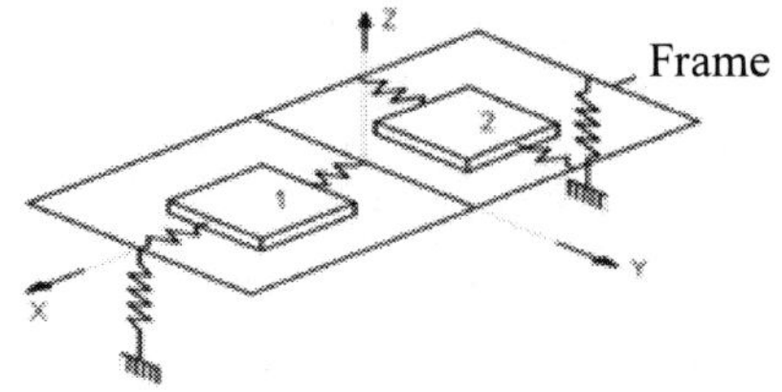

Figure 3. Equivalent scheme of accelerometer sensor.

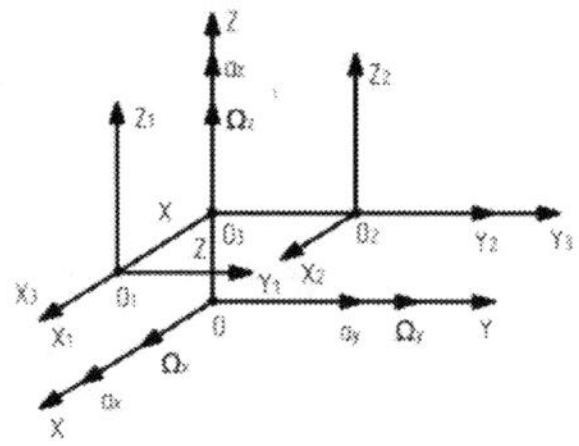

Figure 4. Sensor coordinates.

The motion equations for the accelerometer sensor can be obtained from Lagrange equations of the second order:

$$(m_1 + m_2 + m_3)\ddot{z} + z\left[G_z - (m_1 + m_2 + m_3)\left(\Omega_y^2 + \Omega_x^2\right)\right] + \mu_z\dot{z} =$$
$$= (m_1 + m_2 + m_3)a_z - 2m_2\Omega_x\dot{y} + 2m_1\Omega_y\dot{x} - m_2y\cdot\Omega_z\Omega_y - m_1x\Omega_x\Omega_z$$
$$m_1\ddot{x} + \mu_x\dot{x} + x\left[G_x - m_1(\Omega_z^2 + \Omega_y^2)\right] = m_1a_x - 2m_1\Omega_y\dot{z} + m_1z\Omega_x\Omega_z$$
$$m_2\ddot{y} + \mu_y\dot{y} + y\left[G_y - m_2(\Omega_x^2 + \Omega_z^2)\right] = m_2a_y + 2m_2\Omega_x\dot{z} - m_2z\Omega_y\Omega_z,$$

where m_1, m_2, m_3 are masses of bodies 1, 5 and 4 respectively; G_z, G_x, G_y are stiffness coefficients of suspensions; μ_x, μ_y, μ_z are the coefficients of viscous friction force in the direction of the respective generalized coordinates; a_x, a_y, a_z are accelerations of the translational motion of the object.

3. Research and results

Eigenfrequencies and eigenmodes of the accelerometer are presented in figure 5. Its behavior is simulated using the ANSYS Multiphysics platform. The geometry of MEMS accelerometer is designed such that the eigenfrequencies are close to each other along all three axes.

Displacement in *X*-direction (f_x=3021.2 Hz) | Displacement in *Y*-direction (f_y=3021.8 Hz) | Displacement in *Z*-direction (f_z=3075 Hz)

Figure 5. Eigenfrequencies and eigenmodes of MEMS accelerometer.

The modal analysis allows obtaining the coefficients of stiffness along the respective axes:

$$G_i = \frac{m_i}{\omega_i^2}.$$

Journal of Physics: Conference Series **671** (2016) 012003 doi:10.1088/1742-6596/671/1/012003

MEMS accelerometers should operate at the temperature range of –40°C to 85°C guaranteeing their functional properties. The internal stresses occurred due to the extension of the sensor structure is the main factor that affects eigenfrequencies. The temperature effect induces structural changes in geometry, Young's modulus of silicon and, thereby, internal stresses [4–6].

The temperature effect is investigated within the range from –40°C to 105°C. A change of eigenfrequencies results in a change of the mechanical scale factor of accelerometer sensor. Figures 6 and 7 contain the plots of the temperature dependencies for eigenfrequencies and the scale factor.

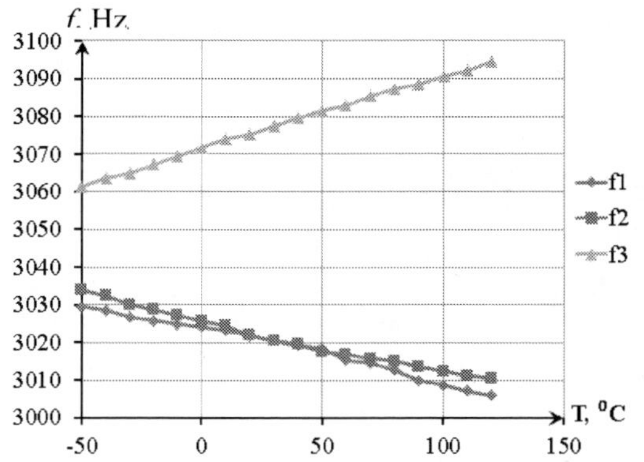

Figure 6. Eigenfrequencies *vs*. temperature.

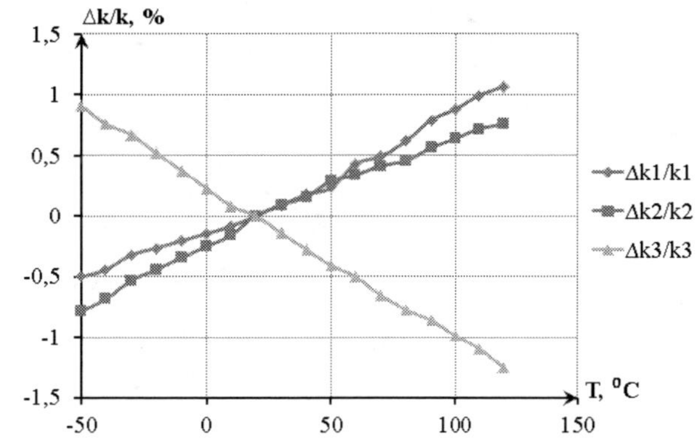

Figure 7. Scale factor *vs*. temperature.

The temperature effect on the accelerometer geometry modifies the capacitance of electrode structures, and an error is introduced in the output signal of the sensor. The accelerometer capacitances are presented in figure 8. The range of motion of the movable sensor element is ±1 μm. The meshes generated using the finite element method (FEM) for the dependencies between capacitances of the comb structure and the temperature are presented in figure 9.

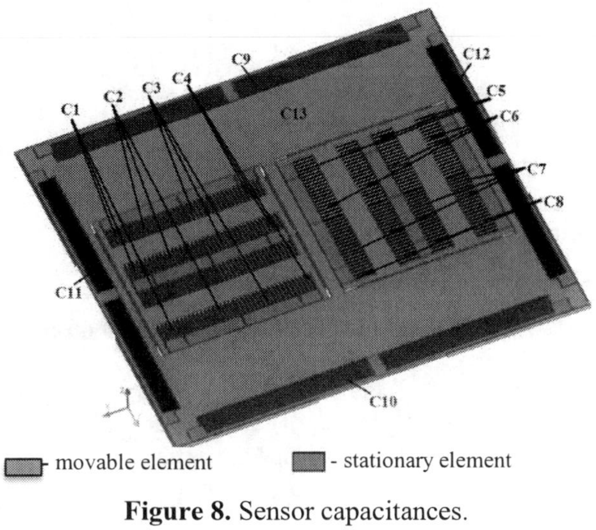

Figure 8. Sensor capacitances.

Thus, the maximum change in scale factors of the accelerometer sensor due to temperature changes within the selected temperature range is less than 1%. Errors produced by changes in capacitances of comb structures are not over 2%. Methods of compensation of temperature errors are described in works [4–6].

The results of the modal, thermal, and electrostatic analyses allow the design of the system model of accelerometer shown in figure 10.

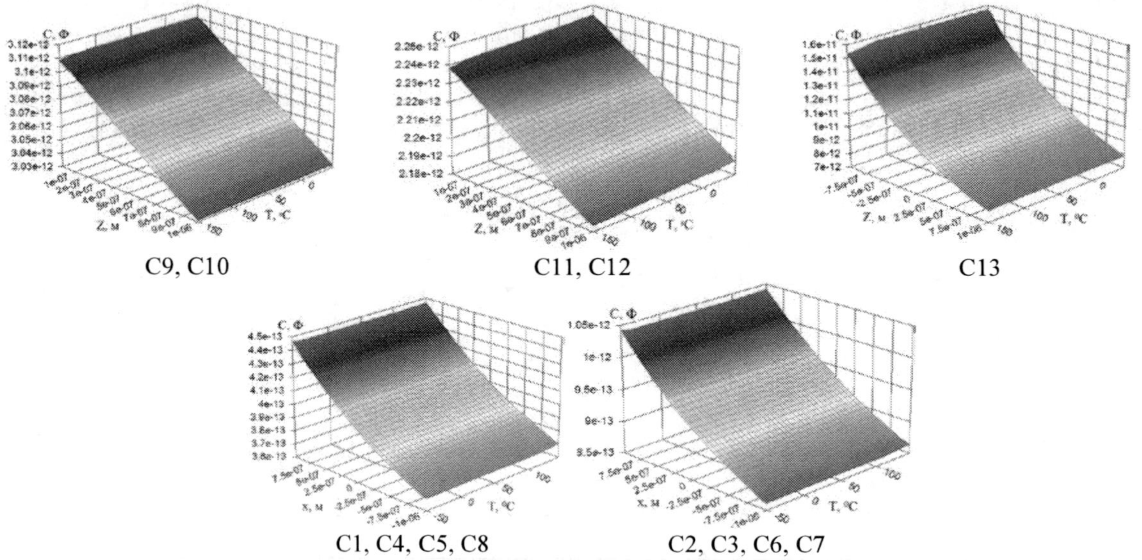

Figure 9. Dependences between comb structure capacitances and temperature. Motion range ±1 μm.

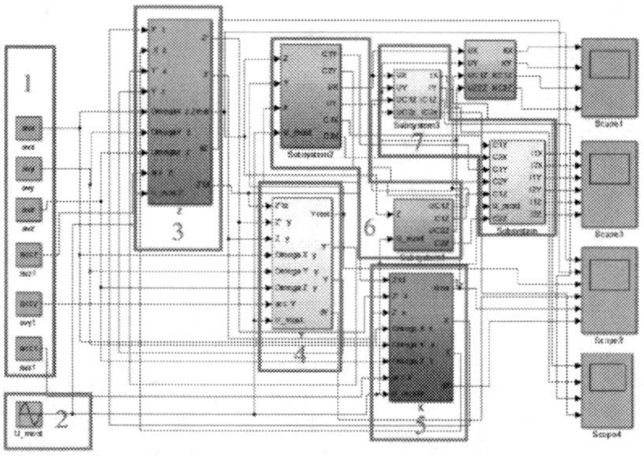

Figure 10. Layout of microaccelerometer modeled using MATLAB/Simulink software:
1 – sources of disturbance; *2* – the source of voltage for data retrieval form measuring bridge;
3 – acceleration measurement channel on the *X*-axis; *4* – acceleration measurement channel on the *Y*-axis; *5* - acceleration measurement channel on the *Z*-axis; *6* – subsystem of measuring bridge;
7 – subsystem for calculation of output currents.

The elastic force vector is calculated at each time interval. This model has been developed in several stages, beginning from separate parts to the whole system and including modeling of the ideal system. Figure 11 contains a plot of the dependence between the output voltage and acceleration, and figure 12 describes the motion of frames at sinusoidal behavior of acceleration.

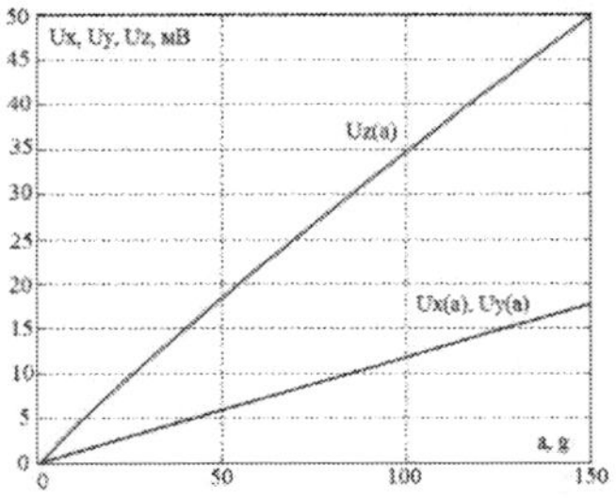

Figure 11. Output voltage/acceleration dependence.

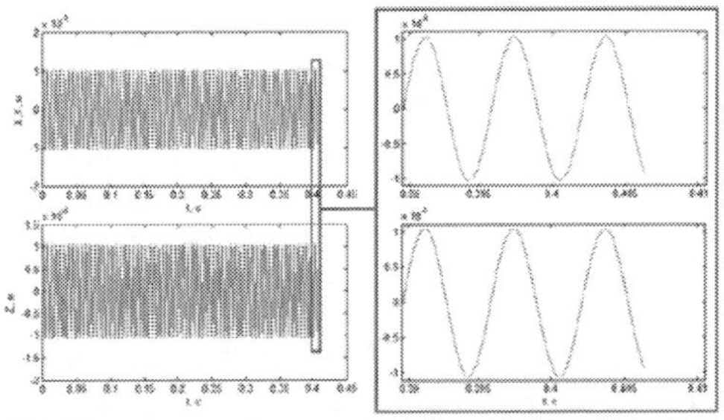

Figure 12. Frame motion at sinusoidal behavior of acceleration.

Cross-coupling errors are then modeled subject to angular velocities Ωy, Ωx, Ωz, and the output signal error is determined. The dependence between the variability of the cross-coupling error and angular velocities is given in figure 13.

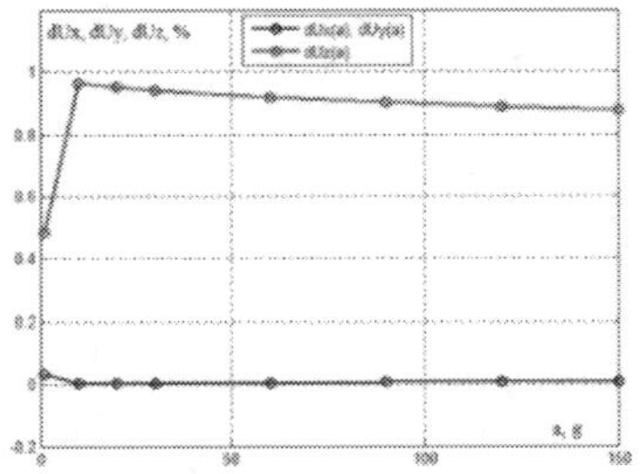

Figure13. Cross-coupling error and angular velocity dependence on *X* and *Y* axes.

4. Conclusion

As a result of this research, the three-axis MEMS accelerometer was designed. It can be applied in the field of building construction monitoring.

Acknowledgements

The authors acknowledge the financial support from the Ministry of Education and Science of the Russian Federation (agreement N 14.575.21.0068, unique identifier RFMEFI57514X0068) of this research that was carried out at Tomsk Polytechnic University within the Federal Target Program on research and development according to priority research fields of the Russian science and technology sector for 2014-2020.

References

[1] Yole Development //www.yole.fr

[2] Russian MEMS association // www.mems-russia.ru/

[3] Sistema monitoringa napryazhennogo sostoyaniya zdanii i konstruktsii. Institut tochnoi mekhaniki i vychislitel'noi tekhniki im. S A Lebedeva RAN - nauchno-issledovatel'skii institut v oblasti informatsionnykh tekhnologii, vychislitel'noi tekhniki i mikroelektroniki [Stress-strain state monitoring systems. Lebedev Institute of Precision Mechanics and Computer Engineering] // www.ipmce.ru/custom/sensornetworks/products/smbuilding/

[4] Nesterenko T G, Koleda A N, Barbin E S, Uchaikin SV (2014) *IEEE Transactions on components, packaging, and manufacturing technology* **4**(10) 1598–1605

[5] Koleda A N, Barbin E S, Nesterenko T G (2015) Three-component microelectromechanical accelerometer *22nd Saint Petersburg International Conference on Integrated Navigation Systems, ICINS 2015 – Proceedings* 338–342

[6] Nesterenko T G, Barbin E S, Koleda A N, Arshinova A A (2015) Metrological performance of integrated multiple axis MEMS accelerometers under thermal effect *XXI IMEKO World Congress 2015*

Modernization of the X-Ray Tomographic Scanner Based on Gas-Discharge Linear Detector

S G Stuchebrov[1], A V Batranin[2] and I A Miloichikova[3]

[1] Senior Researcher, Tomsk Polytechnic University, Tomsk, Russia
[2] Junior Researcher, Tomsk Polytechnic University, Tomsk, Russia
[3] Engineer, Tomsk Polytechnic University, Tomsk, Russia
E-mail: stuchebrov@tpu.ru

Abstract. In this paper, we describe the modernization of the tomographic scanner based on multi-channel linear gas-discharge detector. We have changed the principle of acquisition the projection data, which allowed to receive a bulk three-dimensional tomographic data instead of single slices of the studied samples. Modified scanner has shown increasing contrast and spatial resolution of single slices. The volume of interest in studied objects has been significantly increased and are as high as 25 000 cubic cm, which is determined by 1536 pixels in high

1. Introduction

Nowadays, multi-channel gas-discharge X-ray detectors are almost superseded by semiconductor detectors, because of their advantages such as speed, large dynamic range, small size of achievable detection step and others. However, the potential of multi-channel gas-discharge detectors has not been exhausted. Most studies of internal structures of objects, including medical studies, do not require high spatial resolution, while test objects often are characterized by relatively large geometrical dimensions. Higher cost of semiconductor detectors limits their usage in radiography and tomography for studying large objects. Here, multi-channel gas-discharge X-ray systems seem to be the effective solution due to their lower costs and larger sensitive area.

In Tomsk Polytechnic University tomographic setup based on a multi-channel gas-discharge detector PRIZ 1536 was developed to study the imaging performance in biological and medical application [1]. Obtained results have shown the potential of this type of detectors. The setup demonstrated necessary spatial resolution on standard test objects, required for medical studies. We used a pulsed radiation source and the setup was operating in synchronization mode to coordinate the X-ray tube and the detector. The pulsed X-ray tube produces not sufficient dose rate to obtain required contrast resolution on the test object with tissue-equivalent density and 20 cm in diameter, which are typical to the human body [2]. Attempts to deliver the required dose rate in a few cycles of exposure did not lead to success, since the noise in the gas-discharge detector increased proportionally. To overcome this obstacle we replaced the pulsed X-ray source by continuous source with higher brightness. For this task industrial X-ray apparatus Comet MXR-451HP/11 [3] was chosen.

The first developed setup used fan beam scanning geometry which provides tomographic data in the form of sinogram. After reconstruction, we could obtain information of internal structure of an object in one tomographic slice. The scanning geometry was changed and then detector moved across the cone beam providing full shadow projections. Thus, necessary data for volumetric reconstruction were acquired.

2. Materials and methods

2.1. Emitting source

We used an industrial X-ray apparatus MXR-451HP/11 (manufactured by Comet Group, Switzerland). The source is equipped by bipolar oil-cooled tube with tungsten target and it has dual focal spot: 0.4 and 1.0 mm. Nominal tube voltage is 450 kV; the highest power is 500 or 1500 W depending on the focal spot size. Radiation coverage is 40°×30° [3].

2.2. Experiment geometry

In the original geometry, the detector and the source were fixed and the tested object rotates along an axis perpendicular to the detector. This scanning geometry is known as fan beam scheme (see figure 1). A sinogram is the result image (see figure 2). Reconstruction reveals the internal structure of one slice with one pixel in height. Thus, to obtain a volumetric representation of an object one should transfer the detector along vertical direction step by step.

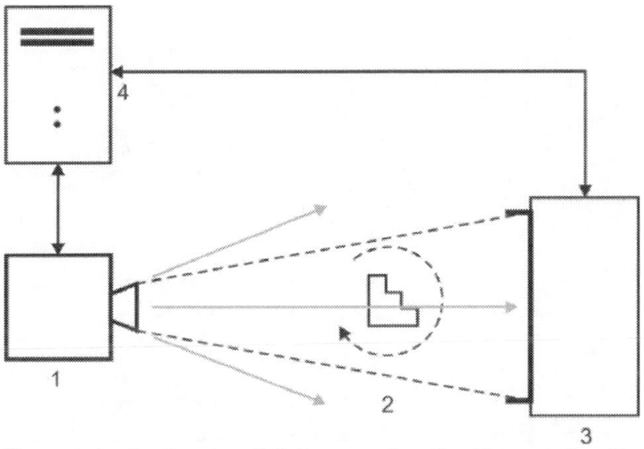

Figure 1. The original scheme of the setup: 1 – the X-ray tube; 2 – an object; 3 – the multi-channel gas-discharge X-ray detector; 4 – a computer.

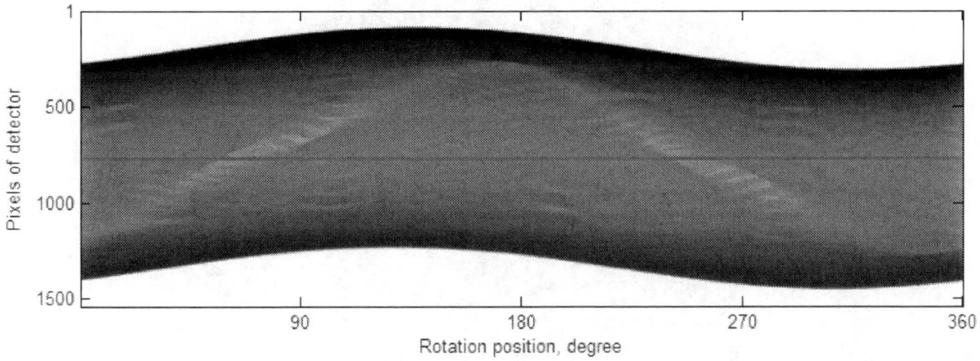

Figure 2. Sinogram of the test module obtained from the multi-channel gas-discharge X-ray detector PRIZ 1536.

In the new geometry scheme, only X-ray source was fixed. Rotary stage with the test object is adjusted between X-ray source and detector to obtain necessary magnification and voxel size. The detector is set up vertically and moves across X-ray beam by means of its own manipulator. The new geometry scheme is shown on the figure 3. While the detector moves it collects full X-ray projection of the test object. Then object rotates and the next projection is acquired. As a result, necessary X-ray data to volumetric reconstruction are collected.

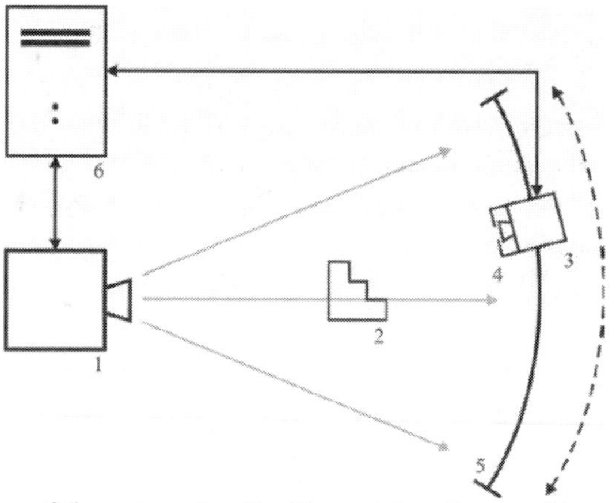

Figure 3. The new scheme of the setup: 1 – the X-ray tube; 2 – an object; 3 – scanning path of the detector; 4 – the multi-channel gas-discharge X-ray detector; 5 – the way of detector manipulator; 6 – a computer.

2.3. Test object

As the test object, we used AAPM CT Performance Phantom 610 for performance evaluation and quality assurance of clinical CT-scanners [4]. The phantom has overall dimensions of 21.6 cm in diameter and 39.4 cm in height. The exterior of the phantom is shown on figure 4.

Figure 4. The exterior of AAPM CT Performance Phantom, model 610.

The phantom includes several inserts to measure the key parameters of CT scanners such as spatial and contrast resolution, CT numbers linearity, slice thickness and some others. Thus, the phantom is a proper test object to evaluate the performance of the developed setup due to its size and different included inserts.

2.4. Image processing

The updated geometry scheme is close to cone-beam geometry widely used in tomographic scanners. For reconstruction of volumetric data, we applied the Nrecon software (made by Bruker microCT, Belgium). This software employs FDK algorithm for cone-beam geometry [5] and allows different enhancement techniques for image processing such as beam hardening correction, ring artifact reduction and smoothing [6]. For volumetric visualization we applied the CTvox software made by Bruker microCT as well. The software allows to operate with the full slice stack, to apply virtual lighting, to change point of view and other actions. To analyze slices and to make measurements on images, we used the "ISee!" software developed in Federal Institute for Materials Research and Testing (Berlin, Germany) [6].

3. Results and discussions

We have carried out two experiments to evaluate possibilities of volumetric tomography by means of the multi-channel gas-discharge X-ray detector PRIZ 1536. The scanning geometry in both experiments was the same. The distance between the source and the test object was 2440 mm and the distance between the source and the detector was 3490 mm, thus the geometry magnification was 1.43. The pitch of detecting elements is 0.261 mm, consequently, the voxel size is 0.183 mm.

The voltage was 140 kV, the current was 5 mA and filters were not applied in both experiments. The first test scan was made with 5 degree angle step and consists of 40 projections. The second scan was performed with 1 degree angle step to obtain 200 projections. The result of volumetric reconstruction after these experiments is shown on the figure 5.

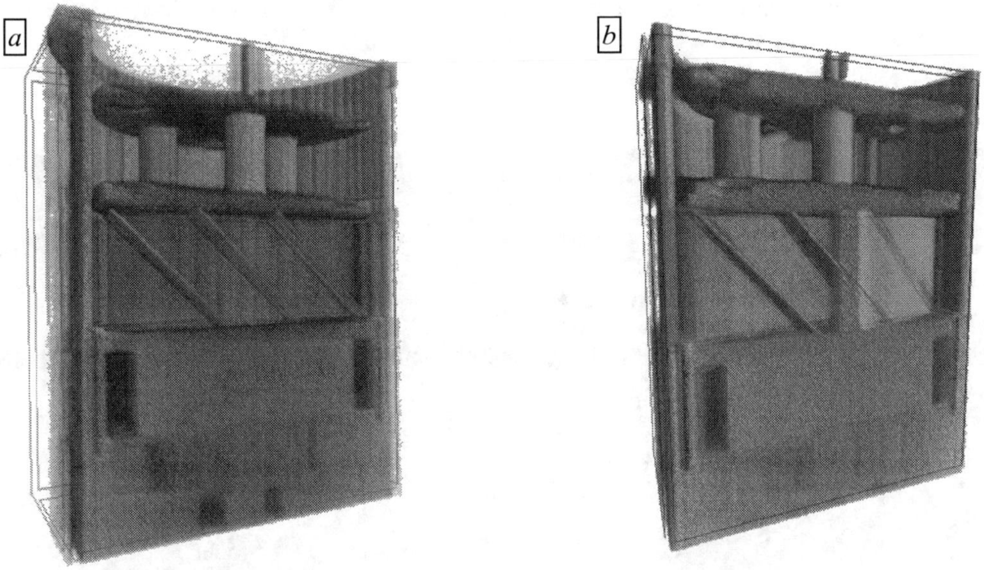

Figure 5. Volumetric representation of the CT phantom after reconstruction: a – from 40 projections; b – from 200 projections.

As seen on figure 5a, the reconstructed test object suffers from artificial ripples which are caused by undersampling due to large angle step. Reconstruction from 200 projections is almost free from such artifacts. However, in both cases, the reconstructed phantom looks realistic.

To evaluate contrast resolution we used the "ISee!" software. We chose one projection from the obtained data set where all cylinders in CT linearity insert are seen separately. Then, by drawing profiler line on the specific area we have got values of attenuated radiation. Figure 6 shows that all cylinders have different attenuation.

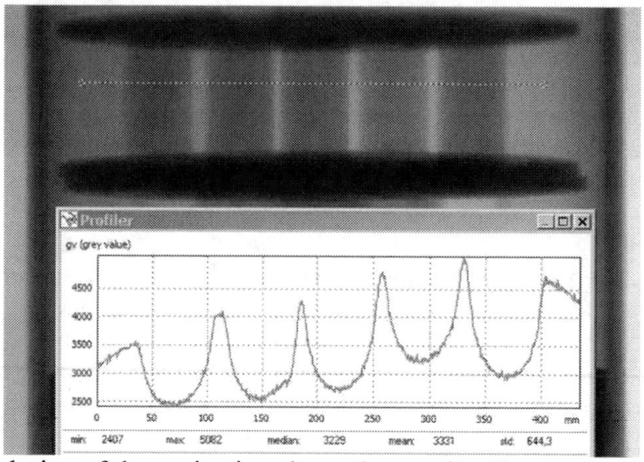

Figure 6. Contrast resolution of the projection shown by profiler line on the CT linearity insert of the phantom.

After the reconstruction procedure we analyzed contrast and spatial resolution of tomographic slices. As shown on the figure 7a, calibration cylinders have different gray values and that in agreement with their attenuation seen on the projection on figure 6.

On the figure 7b four rows of holes are seen. In addition, calibration wire is seen on the center of the image, closer to the bottom.

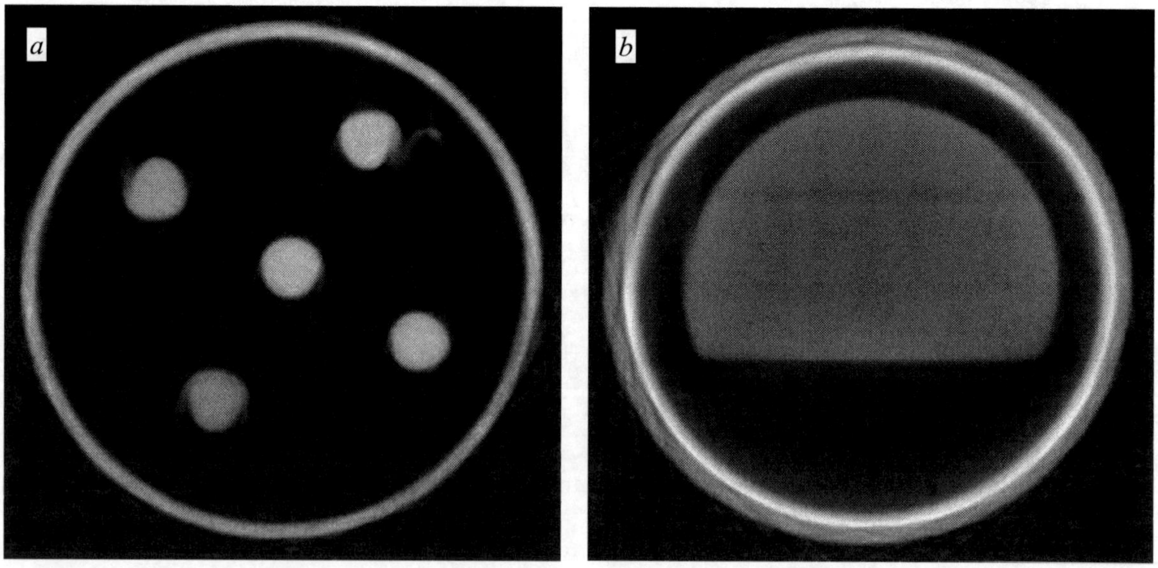

Figure 7. The slice of the CT: a – the linearity insert; b – the spatial resolution insert.

The analysis has shown that contrast resolution is a linear function, which means the tested detector demonstrate a proper sensitivity to X-ray attenuation. Spatial resolution is estimated as 0.28 mm or 1.8 lp/mm. The resulted resolution is lower than achievable value, which is equal to the voxel size or 0.182 mm. The main reason is movement of the detector across to the beam was not enough smooth and regular. Thus, different projections were not adjusted to the center of rotation. We believe that improvement of mechanical part would make possible to obtain better spatial resolution up to the value of the voxel size.

4. Summary

In the modernized tomographic setup based on the multi-channel gas-discharge detector PRIZ-1536 a pulsed radiation source was replaced by the continuous X-ray tube Comet MXR-451HP/11. In addition, fan-beam geometry of scanning was changed into cone-beam one. These improvement made possible volumetric reconstruction and the proper contrast resolution was shown. So far, achieved spatial resolution is estimated as 0.280 mm or 1.8 lp/mm. Improving of resolution requires some minor changes in mechanical part of the setup. The field of view for the developed setup as big as 400×400 mm approximately and limited by the length of the detector and the scanning path of the mechanical system. Thus, the modernized setup is suitable for volumetric reconstruction of large object with high contrast resolution and submillimetric spatial resolution.

Acknowledgements

This work was financially supported by The Ministry of Education and Science of the Russian Federation in part of the science program (N 5.1485.2015).

References

[1] Operation manual 2006 *Multichannel ionization chamber for the scanning type roentgenographic device* (Novosibirsk: IJF the Siberian Branch of the Russian Academy of Science) [in Russian]
[2] Stuchebrov S, Batranin A, Verigin D and Wagner A (2014) Setups for tomographic imaging with submillimeter spatial resolution *J. of Phys.: Conf. Ser.* **517**(1) Article number 012046
[3] Information on http://www.comet-xray.com/Products-en/X-Ray-Tubes/MXR-451HP-11
[4] Information on http://www.cirsinc.com/products/all/31/aapm-ct-performance-phantom/
[5] Feldkamp L A, Davis L C and Kress J W (1984) Practical cone-beam algorithm *Journal of the Optical Society of America A: Optics and Image Science, and Vision* **1**(6) 612–619
[6] Information on http://www.skyscan.be/products/downloads.htm
[7] Information on http://dir.bam.de/ic/
[8] Stuchebrov S, Verigin D, Lukyanenko Y, Siniagina M and Wagner A (2014) Digital X-ray apparatus based on the scanning r-ray gas-discharge detector for studying of interior structure of biological objects *Advanced Materials Research* **880** 168–173

Regularities of Acoustic Emission in the Freight Car Solebar Materials

S Bekher

Department of non-destructive testing, Siberian Transport University, Novosibirsk, Russia
E-mail: behers@mail.ru

Abstract. Acoustic emission results which were obtained during tests of the samples, which were made from foundry solebars with the developing fatigue crack, are presented. The dependences of the acoustic emission event count, the force critical value during the stationary acoustic emission process, and the growth rate of the event count from the cycles number are determined. The amplitude signal distributions relating to the crack growth were received. It is offered to use the force critical value and the amplitude threshold in the rejection criteria.

1. Introduction

Eddy current and magnetic methods of nondestructive testing are applied to diagnostics freight car solebars and bolster beams during planned types of repair on the Russia railroad. Bogies cast details, which resource is carried out, are testing by acoustic emission [1] method. The existing acoustic emission testing technique is directed on prevention refusals connected with old suitable details using. The rejection criteria are the most rigid and directed to the selection of faultless details from obviously unsuitable to operation on service life.

The acoustic emission control during planned types repair demands other approach in which it is necessary to minimize not only defects admission probabilities, but also false rejection probabilities. Necessary dependences of diagnostic signals critical values with errors probabilities can be received only in the course of acoustic emission testing of the samples with the real defects.

2. Experiment description

The samples for mechanical testing were prepared from solebar material (steel 20ГЛ in GOST 3240). Samples thickness varied in the range from 8 to 13 mm as samples were made from the real details with the foundry defects. The sources of the cracks development were the longitudinal slots 45 mm long.

The tests of the samples were carried out according to the similar scheme to the operation and repair process of the freight cars. The samples were loaded in the MTS electrohydraulic machine with the stretching cyclic force (figure 1) with amplitude 25 kN. Loading were stopped through out 5...20 thousand cycles and the samples were testing by acoustic emission method using monotonously increasing force to the maximum value 30 kN. The sufficient cycle quantity for samples destruction was varied from 60 to 250 thousand and was depended on thickness, foundry defects number and their extent. The detailed metallographic analysis result of the destroyed fragments is provided in [2].

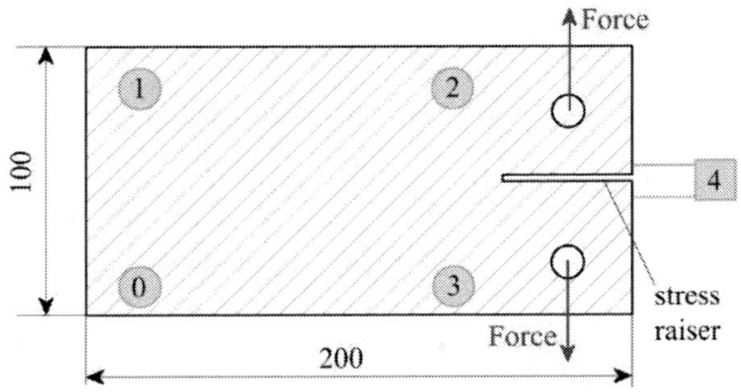

Figure 1. The sample for the test.

The cracks development was controlled by the regular MTS disclosure sensor (position 4 on the figure 1), which time of measurements was synchronized with acoustic emission system. Acoustic emission signals were registered by the acoustic emission transducers with a pass-band from 0.1 to 0.7 MHz and the discrimination threshold equal 20 μV. The transducers were installed on the one side of the samples in the rectangle corners (position 1, 2, 3, 4 on the figure 1). The acoustic emission system SDAD 16.02/03 [3] was used for signals registration, filtration and preprocessing. Final signals processing was carried out in the specialized developed software. A plane source location was realized by the difference time of the signal arrival to the transducers. The transducer coordinates were used for definition and location of a crack top and signal filtration, which are arising in the places of the interaction between the samples and the loading device captures.

3. Experimental results

The increase crack disclosure was fixed already during the initial testing stage and the disclosure speed was monotonously increased for all testing samples. Similar regularities are typical for the low-cyclic fatigue testing and are caused with repeated-plastic material deformation.

During the testing of the samples No. 5 and No. 6 the sharp increase of the crack disclosure (more than twice) before the destruction are fixed. But such regularity was not found in a sample No. 2. The crack development unevenness is caused by the different foundry defects in the samples.

In the conditions of the low-cyclic fatigue the basic process of the acoustic emission signals generation is the plastic deformation of the vicinity of the stress raiser. Any obvious regularities which would allow to establish the dependence between the event count and the cycles number or the crack disclosure were not revealed. For example, for the sample No. 2 the considerable increase of the event count is characteristic not long before the destruction, for a sample No. 6 the existence of the event count maximum at 80 thousand cycles is characteristic, and for the sample No. 5 event count is fixed at the constant level from 20 to 40 acoustic emission impulses. First of all the acoustic emission event count depends from the plastic deformation value and material volume [4]. The lack of the regularity depends with the composite crack path and the existence in the sample material defects which influence on crack development. Therefore the plastic current in the crack mouth change in a random way in the course of the crack moving. It is defines the event count dependence from the cycles number.

The graphics of the dependences the test force and the acoustic emission activity during the static test on the time are submitted on the figure 2 for the sample No. 6 after 83000 cycles. These dependences are typical for all samples. Acoustic emission signals are registered in the course of the test force value. It testifies that plastic deformation for both directions test forces change. The acoustic emission signals arising during loading and unloading was analyzed separately.

Acoustic emission signals were registered in the certain force range. On the figure 2 it is shown the behavior of acoustic emission activity. The acoustic emission signals arise when the force increase above the critical value F_i and when the force decrease below F_d [5]. Critical values can be

Journal of Physics: Conference Series **671** (2016) 012005 doi:10.1088/1742-6596/671/1/012005

characterized by dimensionless coefficient K_F, which is equal to the relation critical value and cyclic force amplitude.

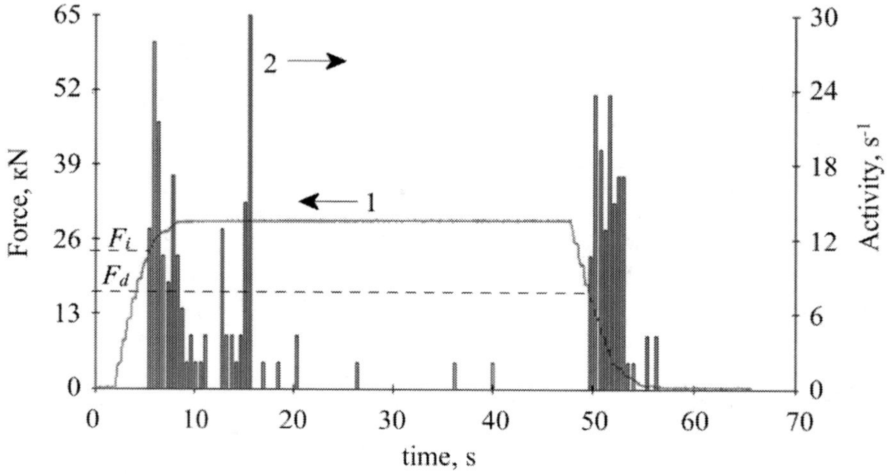

Figure 2. The graphics of the dependences the test force (1) and the acoustic emission activity (2) during the static test on the time, the sample No. 6 after 83000 cycles, where F_i, F_d – critical force when increase and decrease, respectively.

The graphic dependence of the coefficient K_F from cycles number are shown on the figure 3. Sharp decrease force critical value F_i is observed after 5...10 thousand cycles (5...10% total of cycles) destruction in all tests. The distinctive feature of the sample No. 2 is the existence of the tendency to decrease F_i at all stages of crack development. The critical value decrease to 90% is observed for 30 thousand cycles (20%) before the destruction. The force value F_d has considerable fluctuations, as shown on the figure 3, therefore it can't be effectively applied for using as single rejection criterion.

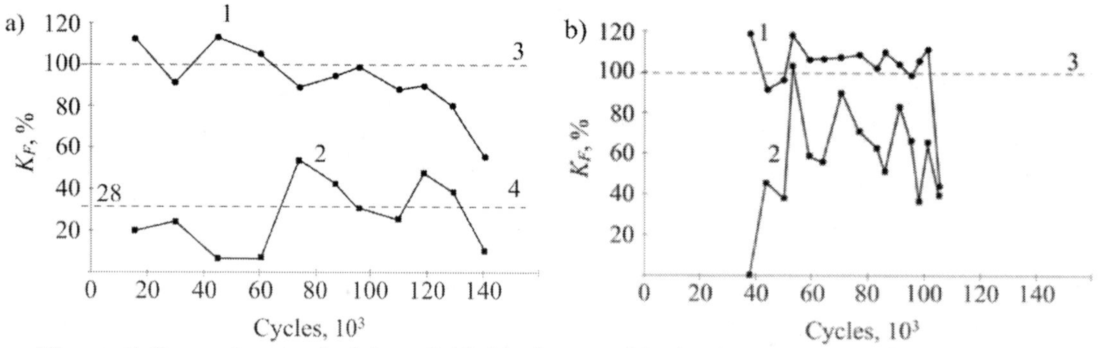

Figure 3. Dependence schedules of relative forces critical values F_i (1) and F_d (2) from cycles number for samples No. 2 (a) and No. 6 (b); 3 – amplitude of cyclic loading force; 4 – the average value of F_d.

The critical value F_i divides the dependence of acoustic emission event count from increasing force on two parts (figure 4). In the force field less than F_i (about 20 kN) separate impulses are registered, acoustic emission has non-stationary character. The non-stationary acoustic emission event count didn't exceed 20 for all tests. Loading with a force more than 20 kN causes emergence stationary acoustic emission and event count linearly depends from force:

$$N = a \cdot (F - F_i) + N_0, \quad (1)$$

a – the coefficient determined by least-square analysis, kN^{-1}; F – the force value, kN; F_i – the force

Journal of Physics: Conference Series **671** (2016) 012005 doi:10.1088/1742-6596/671/1/012005

critical value, kN; N_0 –non-stationary acoustic emission event count.

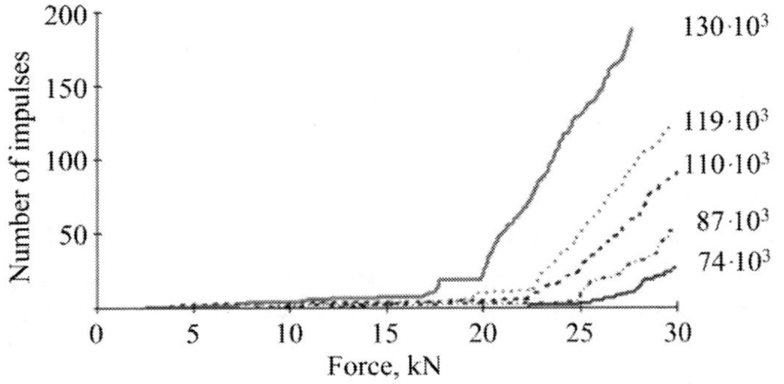

Figure 4. Schedules of event count dependence from force for various cycles number.

Coefficient a in the equation (1) describes event count growth rate. Stationary acoustic emission appears at force above the amplitude cyclic loading before cycle quantity less 87 thousand. The crack length increasing leads to reduction force critical value F_i, which equal 40% force amplitude before destruction. Emergence of stationary acoustic emission before cyclic amplitude force testifies about Kaiser effect violation and fatigue defects develop actively. In the course of crack growing the coefficient a from 12 to 21 kN^{-1} in the cycles range from 110 to 130 for the sample No.2.

Acoustic emission signal amplitude distribution [6] for various cycles is presented on the figure 5. The high-amplitude signals number with an amplitude more than 300 mV increase before 10...20 thousand cycles (for a sample No. 2 before 40...50 thousand) destruction for all samples increase. It testifies about plasticity stock exhaustion and fragile destruction beginning. In the sample No. 2 relative quantity of high-amplitude signals doesn't exceed 5%, thus the average amplitude during tests monotonously increases from 40 to 75 μV (figure 6).

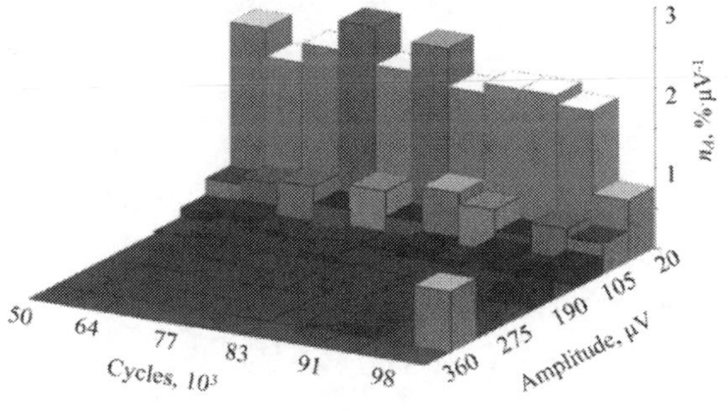

Figure 5. The dynamic of the amplitude distribution at static tests.

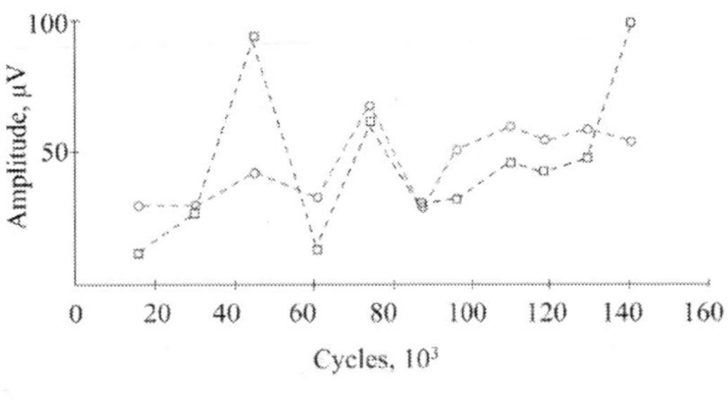

Figure 6. Average amplitude at static tests.

In the course of fatigue crack growing the concentration coefficient increases and, therefore, a stress increases close to crack top. Thus the average signal amplitudes increase. For example, in the sample No. 2 the growth rate of acoustic emission signals amplitude equal 23 μV divide on 103 thousand cycles (figure 6). Amplitude threshold equal 200 μV will allow finding fatigue cracks 10... 20 thousand cycles before destruction in the course of control.

4. Conclusion

Solebars rejection is based on the event count analysis in the course of tests in the specialized test desk for loading. At the same time weak sites, in which plastic deformation develops, are found in details. It leads to the high level rejection (5–10%) of new details. It is offered to use critical force value, event count growth rate coefficient, amplitude threshold in acoustic emission rejections for increasing test reliability. The recommended relative force threshold equal 80% of average working force and amplitude threshold equal 200 μV.

The executed tests of the samples, which was made from solebar material, showed that composite processes crack development in foundry molded material defines weak dependence between event count and crack disclosure. Monotonous growth event count in crack development is observed only in one of the samples. Force critical value allows observing a crack in the sample (5–20) % cycles total number before destruction. Using for rejection high-amplitude signals allows observing transition from viscous destruction to fragile destruction (10–40) % cycles total number before destruction.

References

[1] Kanji Ono (2011) Acoustic emission in materials research – a review *Journal of acoustic emission* **29** 284–308

[2] Ser'eznov A N, Murav'ev V V, Stepanova L N, Kabanov S I, Kozhemyakin V L, El'tsov A E and Lebedev E YU (1998) A high-speed acoustic-emission diagnostic system *Russian Journal of Nondestructive Testing* **34**(7) 481–485

[3] Stepanova L N, Bekher S A, Boyarkin E V, Moreva N A (2013) A study of specimens from solebar material with boxlike cross sections using the acoustic-emission method *Russian Journal of Nondestructive Testing* **49**(4) 215–224

[4] Christian U G and Masayasu O 2008 *Acoustic emission testing* (Berlin: Springer)

[5] Tomoki Shiotani, Xiu Luo and Hiroshi Haya (2006) Damage diagnosis of railway concrete structures by means of one-dimensional AE sources *Journal of acoustic emission* **24** 205–214

[6] Pollock A A 1981 Acoustic emission amplitude distribution *International advances in nondestractive testing* (ed.: Warren J. McGonnagle) **7** pp 215–240

Journal of Physics: Conference Series **671** (2016) 012006 doi:10.1088/1742-6596/671/1/012006

Patch Antenna for Measuring the Internal Temperature of Biological Objects Using the Near-Field Microwave Radiometric Method

A Ubaichin[1,2], A Bespalko[3], A Filatov[4], E Alexeev[5], G Zhuk[5]

[1] Researcher, Institute of Non-destructive Testing, TPU, Tomsk, Russia
[2] Researcher, Special Design Bureau "Smena", TUSUR, Tomsk, Russia
[3] Leading researcher, Institute of Non-destructive Testing, TPU, Tomsk, Russia
[4] Professor, Special Design Bureau "Smena", TUSUR, Tomsk, Russia
[5] Engineer, Special Design Bureau "Smena", TUSUR, Tomsk, Russia
E-mail: dualog@mail.ru

Abstract. The near-field microwave antenna with central frequency of 2.23 GHz has been designed and manufactured to be used as a part of the medical microwave radiometric system. Experimental studies of the reflection coefficient in different parts of the human body were conducted using the developed antenna. The experimental studies were carried out in a group of volunteers with normal somatic growth. The results of the experiments were used to perform the analysis of the potential errors in the measurements obtained via the developed antenna.

1. Introduction

One of the most important tasks in early medical diagnosis is to determine an internal temperature of the biological tissue. Invasive methods of temperature measuring, not to mention the additional pain, are not effective and cause significant measurement errors. Widely used non-invasive methods are infrared and microwave imaging [1]. Infrared imaging is commonly used in medical practice; however, this method allows measuring only near-surface temperature. Therefore, it is impossible to fully determine the internal thermal processes. The internal temperature distribution with the accuracy level adequate for medical practice can be measured by methods based on microwave imaging [2].

The most widely used technique is the near-field microwave radiometry. In this case (figure 1), antenna *A* is in a direct contact with the human body, as a result, resolution increases, the reflection coefficient of the antenna-medium complex can be estimated, and the influence of the radio interference decreases. The accuracy in comparison with remote radiometric methods increases as well.

2. Antenna designing

The accuracy of the results obtained via the near-field microwave imaging depends both on the biological tissue structure under study and the absolute accuracy (and sensitivity) of the measuring device – microwave radiometer. Among different types of microwave radiometers [3–7] widely-spread in medical diagnosis, the microwave radiometer based on the modified zero measuring method is most advantageous [5]. This microwave radiometer allows measuring the internal temperature of biological objects and the reflection coefficient with no effect of the main destabilizing factors inherent in other types of radiometers. The main feature of the near-field microwave radiometric method is the need of impedance matching the antenna with the biological tissue.

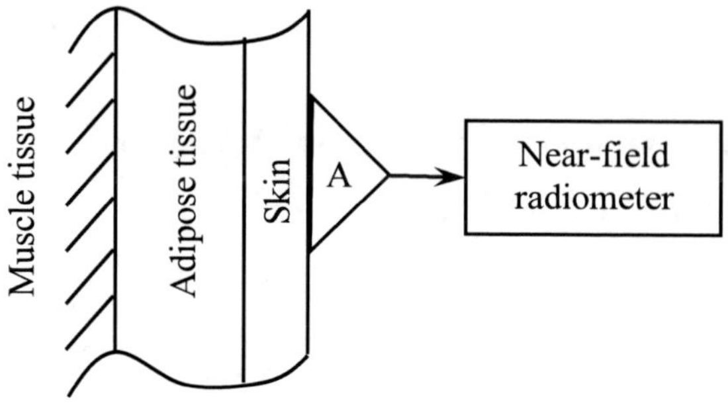

Figure 1. Biological tissue structure and the near-field radiometer for internal temperature measurement.

In [9], a comparative analysis of microwave antennas for the near-field radiometric method is provided. According to [9], waveguide antennas with regard to their main technical characteristics have a number of advantages (operating frequency range, matching with a layered medium), as compared to patch antennas. Technologically, patch antennas are most simple for manufacturing.

In [10–15], the authors consider various configurations of the radiating elements to achieve the required parameters: matching at the operating frequency range (reflection coefficient module is less than 0.5) [8], operating frequency width is at least 100 MHz.

Near-field patch antennas with circular radiating element are widespread. The design of these near-field antennas provides shielding, small size and ease of implementation by using a subtractive manufacturing technology in comparison with waveguide antennas. The technique of designing an antenna with a circular radiating element for the far zone is presented in [16]. A brief description of the methodology for development of the near-field antenna with a circular radiating element is as follows: dielectric substrates with low dielectric permittivity $\varepsilon = 2...3.8$ are used, thickness of dielectric substrates is at least 1 mm, location of the feeder in the radiating element is determined by the required impedance value.

A near-field patch antenna with a circular radiating element has been designed. The development of the mathematical model and structural modeling of the antenna were carried out to meet the requirements of the impedance matching of the antenna-feeder path with the biological tissue of the "skin - adipose tissue - muscle tissue" type.

The developed antennas are made of microwave dielectric material “FLAN” $\varepsilon = 2.2$, the substrate thickness is about 1 mm, the transverse dimension is about 13×13 mm, the central frequency is 2.23 GHz, and the 3 dB level bandwidth is 500 MHz.

Figure 2. Designed models of the near-field antennas and the scalar network analyzer R2M-3200.

Variations in the biological tissue structure in the near-field measurements lead to antenna impedance mismatch. The influence of the biological tissue parameter variation (thicknesses of the skin, adipose and muscle tissue) on antenna matching has been evaluated at the design stage by using the computer-aided design CST Microwave Studio 2010. Some of the simulation results are shown in figure 3.

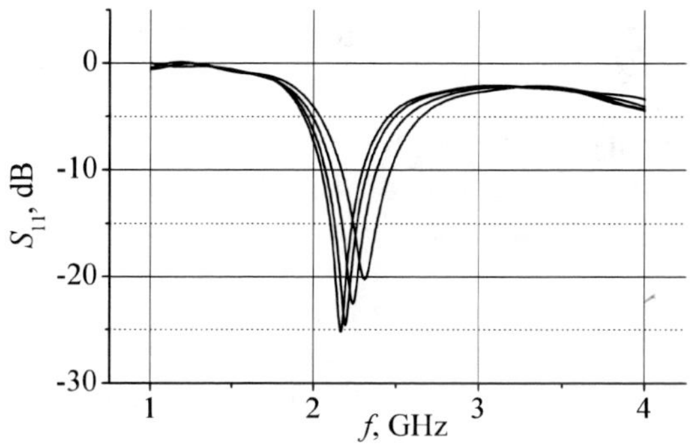

Figure 3. Effect of the changes in the layered medium thickness of the "skin-adipose tissue-muscle tissue" components on the S_{11} module.

The simulation results are obtained by varying the thicknesses of the biological medium layers within the following ranges: the thickness of the skin is from 0.5 to 2 mm, the thickness of the adipose tissue is from 0 to 20 mm, and the remaining space in the model has the dielectric parameters of the muscle tissue.

The experiments to study the reflection coefficient module of the biological medium were performed by using the designed antenna. The measurements were carried out in a group of volunteers with a normal somatic growth on particular parts of the body. The measurement results obtained using the scalar network analyzer R2M-3200 produced by JSC "Micran" are presented in Table 1.

Table 1. The results of the reflection coefficient module measurements in the frequency range from 2.18 to 2.28 GHz

Body area	The module of the reflection coefficient in the operating frequency band for the correlation number of the subjects, not more than								
	No.1	No.2	No.3	No.4	No.5	No.6	No.7	No.8	No.9
The inner side of forearm	0.107	0.075	0.184	0.110	0.200	0.104	0.110	0.095	0.101
The shoulder (outside)	0.158	0.127	0.135	0.127	0.170	0.115	0.124	0.117	0.130
The shoulder (inside)	0.193	0.188	0.190	0.178	0.202	0.098	0.175	0.198	0.180
The front surface of abdomen	0.316	0.124	0.191	0.210	0.305	0.256	0.180	0.230	0.185
The side surface of abdomen	0.101	0.03	0.120	0.130	0.280	0.090	0.129	0.189	0.123
The inner surface of thigh	0.203	0.185	0.212	0.199	0.253	0.150	0.207	0.200	0,175
The rear surface of lower leg	0.050	0.048	0.098	0,026	0,150	0.033	0.037	0.058	0.035
The larynx	0.079	0.084	0.070	0.090	0.124	0.050	0.048	0.073	0.034
The left half of chest	0.141	0.090	0.134	0.152	0.203	0.112	0.127	0.112	0.120
The right half of chest	0.125	0.085	0.147	0.167	0.234	0.090	0.130	0.124	0.110
The cheek area	0.251	0.199	0.208	0.203	0.280	0,199	0.153	0.235	0.205
The top of head area	0.177	0.162	0.180	0.170	0.180	0.165	0.179	0.150	0.139

There are two major factors affecting the accuracy of the biological tissue temperature profile measuring: the first factor is the efficiency of algorithms for inverse problem solving and the second one is absolute accuracy of the microwave radiometer used for measurements [17].

A mathematical model of the zero-type analog radiometer considered in [3] is limited to an ideal case with no insertion loss and with full reflection of the energy in the microwave switch. This model does not consider the impact of losses in the antenna-feeding path. In practice, the development of the microwave components with the parameters close to ideal ones is difficult.

The analysis shows that the influence of these factors decreases the accuracy of the measurement results, and at the level of the reflection coefficient module $R \geq 0.1$ (see Table 1) it causes significant errors.

To evaluate the efficiency of the developed antenna, the analysis of the errors in the noise temperature measurements of the biological tissue using the radiometer [5] based on the modified zero method measurements was performed. The input unit of the microwave radiometer [5] implemented in the block diagram is shown in figure 4.

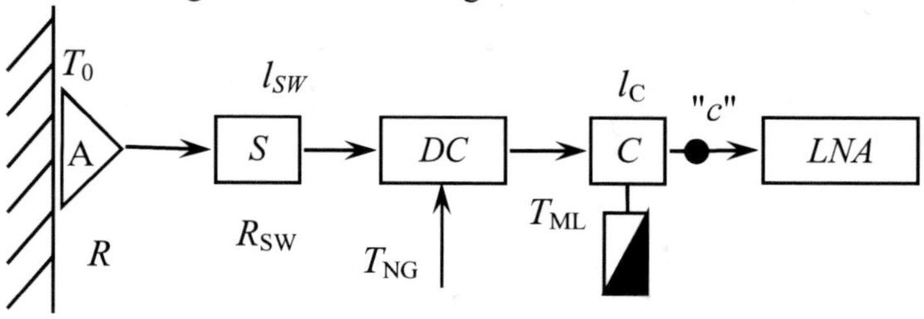

Figure 4. Block diagram of the input unit of the microwave radiometer: the tested object with temperature T_0, antenna A, microwave reflective switch S, directional coupler DC, circulator C and low noise amplifier LNA.

Figure 5 shows the timing diagram of the radiometer [18]. The pulse amplitude modulation signal t_{APM} (see figure 5) controls the switch (see figure 4) – in the time domain of the logical high signal t_{APM}, the switch is opened and the antenna is connected to the input of the LNA, in the time domain of the logical low signal t_{APM}, the switch is closed and the LNA receives signals T_{ML} and T_{NG} considering reflections in the closed switch [19]. The power transmission T_{NG} control is carried out by the pulse-width modulation signal t_{PWM}. The transfer of the signal T_{NG} occurs in the time domain of the logical high pulse-width modulation signal in the antenna waveguide towards the receiver, in the time domain of the logical zero, the signal T_{NG} is travels to the reflective switch. The pulse-width duration is changing until the condition of the equality of volt-second areas is fulfilled [5, 20, 21].

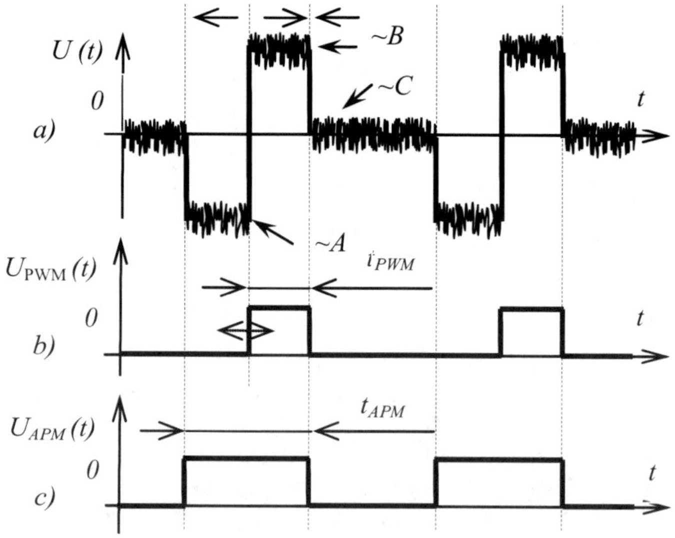

Figure 5. Timing diagrams of the zero-type radiometer: a) timing diagram of the logical signal at the input of the low noise amplifier at "c" point (see figure 4), the direction of the T_{NG} signal is indicated by arrows,
b) timing diagram of the amplitude-pulse modulation signal,
c) timing diagram of the pulse-width modulation signal.

In the following analysis, the efficiency of the near-field antenna is not considered. The near-field antenna together with the biological medium is characterized by the module of the power reflection coefficient R, the reflective switch is characterized by the transmission coefficient l_{SW} and by the module of the power reflection coefficient in a closed state R_{SW}, the transmission of the signal T_{NG} to the antenna waveguide is carried out by the directional coupler, and the matching load of the circulator (isolator operating mode) generates the T_{ML} signal.

In the above block diagram (see figure 4), the case of complete removal of the effect of antenna mismatching with the tested medium is realized only with full reflection from the closed switch disregarding the insertion loss in it. In practice, it is impossible to realize lossless switch and\or with full reflection of the signal from the closed input.

In the timing diagram a) (see figure 5), the levels of the signals are shown taking into account the imperfections of the reflective switch:

$$A \sim T_0 \cdot (1-R) \cdot l_{SW} + R \cdot l_{SW}^2 \cdot T_{NG} + R \cdot l_{SW}^2 \cdot T_{ML} + (1-l_{SW}) \cdot T_{ML}$$

$$B \sim T_0 \cdot (1-R) \cdot l_{SW} + T_{NG} + R \cdot l_{SW}^2 \cdot T_{ML} + (1-l_{SW}) \cdot T_{ML}$$

$$C \sim R_{SW} \cdot l_{SW} \cdot (T_{NG} + T_{ML}),$$

$$A \sim T_0 \cdot (1-R) \cdot l_{SW} + R \cdot l_{SW}^2 \cdot T_{NG} + R \cdot l_{SW}^2 \cdot T_{ML} + (1-l_{SW}) \cdot T_{ML}$$

$$B \sim T_0 \cdot (1-R) \cdot l_{SW} + T_{NG} + R \cdot l_{NG}^2 \cdot T_{ML} + (1-l_{SW}) \cdot T_{ML}$$

$$C \sim R_{ML} \cdot l_{ML} \cdot (T_{NG} + T_{ML})$$

where l_{SW} is the insertion loss in the reflective switch, R is reflection from the "object-antenna" border, T_{NG} is the noise temperature of the noise generator added to the antenna noise temperature T_A, T_{ML} is the noise temperature of the matching load (see figure 1 and figure 4).

Let us write the expression to find the dependence of the pulse-width modulation signal on the medium noise temperature considering the data and the values of *A, B, C* levels provided in [7]

$$T_0 = \frac{(1+R) \cdot T_{ML} \cdot (l_{SW}-1) - (R-R_{SW}) \cdot l_{SW}^2 \cdot T_{ML} + R_{SW} \cdot T_{NG} \cdot l_{SW}^2}{(R-1) \cdot l_{SW}} - \frac{t_{PWM}}{t_{APM}} \cdot \frac{T_{NG} \cdot (1+R \cdot l_{SW}^2)}{(R-1) \cdot l_{SW}}. \quad (1)$$

For the case of an ideal reflective switch (R_{SW}=1, l_{SW}=1), the expression (1) takes the form

$$T_0 = T_{ML} + T_{NG} - \frac{t_{PWM}}{t_{APM}} \cdot T_{NG}. \quad (2)$$

Expression (2) is consistent with the data in [5] for idealized elements.

The modern microwave reflection switch has an insertion loss level of 0.01...0.1 and the reflection coefficient of 0.9...0.99 at the shortwave region of the *S*-range. Taking into account the data range of the reflection coefficient module (see Table 1 – the maximum antenna reflection coefficient R_{MAX}=0.316) and the range of the input module imperfections (insertion loss and not full reflection in the switch), determine the possible measurement error by using expression (1). In solving the inequality (3) obtained by (1) taking into account the range of values $0.9 \le l_{SW} \le 0.99$, $0.9 \le R_{SW} \le 0.99$ и $0 \le R_{MAX} \le 0.316$ for the dynamic measurement range of 10 K at the range boundaries l_{SW}, R_{SW}, R_{MAX}, the errors exceed the desired accuracy by more than one order of magnitude.

$$\Delta T \ge \frac{(1+R_{MAX}) \cdot T_{ML} \cdot (l_{SW}-1) - (R_{MAX}-R_{SW}) \cdot l_{SW}^2 \cdot T_{ML} + R_{SW} \cdot T_{NG} \cdot l_{SW}^2}{(R_{MAX}-1) \cdot l_{SW}} - \frac{t_{PWM}}{t_{APM}} \cdot \frac{T_{NG} \cdot (1+R \cdot l_{SW}^2)}{(R_{MAX}-1) \cdot l_{SW}}, \quad (3)$$

The measurement accuracy ΔT=0.05 K is provided for the range of values of the antenna reflection coefficient module within the limits $0 \le R \le 0.316$ under the condition of $0.98 \le l_{SW} \le 0.99$, $0.98 \le R_{SW} \le 0.99$.

3. Conclusion

Thus, the developed antenna can be used for testing the biological tissue by the near-field microwave radiometer in case of using the microwave input unit with the reflection switch characteristics close to ideal parameters. If the input module characteristics do not provide the desired accuracy, it is necessary to use antennas that comply with the requirements for matching of the tested tissue (body area).

A single antenna can also be used if both the noise temperature and the reflection coefficient module are measured simultaneously with subsequent correction of the measurement results taking into account the influence of the reflection coefficient module.

This work was supported by The Ministry of Education and Science of the Russian Federation in part of the Science program.

References

[1] Hand J W, Van Leeuwen G M J, Mizushina S, Van De Kamer J B, Maruyama K, Suiura T, Azzopardi D V and Edwards AD (2001) Monitoring of deep brain temperature in infants using multi-frequency microwave radiometry and thermal modelling *Phys. Med. Biol.* **46**(7) 1885–1903

[2] Rosen A, Stuchly M A and Vander Vorst A (2002) Applications of RF/microwaves in medicine *IEEE Transactions on Microwave Theory and Techniques* **50**(3) 963–974

[3] Oikonomou A, Karanasiou I S and Uzunoglu N K (2010) Phased-array near field radiometry for brain intracranial applications *Progress in Electromagnetics Research* **109** 345–360

[4] Karanasiou I S, Uzunoglu N K and Papageorgiou C C (2004) Towards functional noninvasive imaging of excitable tissues inside the human body using focused microwave radiometry *IEEE Trans. Microwave Theory Tech.* **52**(8 II) 1898–1908

[5] Yablonskiy D A, Ackerman J J H and Raichle M E (2000) *Proceedings of the National Academy of Sciences of the United States of America* **97**(17) 9819

[6] Filatov A V, Lekhanov A G and Ubaichin A V (2015) A Microwave radio thermometer for measuring the depth temperatures of biological objects by a noninvasive method *Measurement Techniques* (Article in press) doi: 10.1007/s11018-015-0731-1

[7] Guy A W (1971) Analyses of electromagnetic fields induced in biological tissues by thermographic studies on equivalent phantom models *IEEE Trans. Microwave Theory and Tech.* **19**(2) 205–214

[8] Lin C *et al* (2010) A novel three-fed microstrip antenna for circular polarization application *Journal of Electromagnetic Waves and Applications* **24**(11–12) 1511–1520

[9] Asimakis N P *et al* (2011) Non-invasive microwave radiometric system for intracranial applications: A study using the conformal L-notch microstrip patch antenna *Progress in Electromagnetics Research* **117** 83–101

[10] O'Halloran M, Glavin M and Jones E (2010) *Prog. Electromag. Res.* **107** 203–217

[11] Conceicao R C, O'Halloran M, Glavin M and Jones E (2009) Comparison of planar and circular antenna configurations for breast cancer detection using microwave imaging *Prog. Electromag. Res.* **99** 1–20

[12] Lee J N and Park J K (2010) *Jour. Electromag. Waves Appl.* **24**(2–3,1) 179–188

[13] Corbett R, Laptook A and Weatherall P (1997) *J. Cereb. Blood Flow Metab.* **17**(4) 363–369

[14] Jacobsen S and Klemetsen O (2008) *IEEE Trans. Biomed. Eng.* **55**(12) 2778–2785

[15] Garg R 2001 *Microstrip antennas design handbook* (Norwood, MA 02062)

[16] Kraus J D 1984 *Radio astronomy* (Michigan, Cygnus-Quasar Books)

[17] Filatov A V and Ubaichin A V (2015) *Inst. and exp. tech.* **58**(1) 81–85

[18] Filatov A V, Ubaichin AV and Bombizov A A (2013) *Meas. Tech.* **55**(11) 1281–1286

[19] Filatov A V, Ubaichin A V and Paraev D E (2012) *Inst. Exp. Tech.* **55**(1) 59–64

[20] Filatov AV and Ubaichin A V (2012) *Meas. Tech.* **54** 1–6 doi: 10.1007/s11018-012-9865-6

[21] Ubaichin A V and Filatov A V (2014) *CriMiCo 2014 - 2014 24th International Crimean Conference Microwave and Telecommunication Technology, Conference Proceedings* 928–929

Method for Automated Bone Shape Correction within Bone Distraction Procedure

F Yu Blynskiy[1]

[1]Postgraduate student, Department of Industrial and Medical Electronics, Tomsk Polytechnic University, Tomsk, Russia
E-mail: blynskiyf@tpu.ru

Abstract. The method for automated bone shape correction within bone distraction procedure is presented. High precision deformation angle measurement is provided by the software for X-Ray images processing. Special BDC v.1.0.1. application is designed. The purpose of the BDC is modeling of the bone geometry structure to calculate the appropriate distraction forces. The correction procedure control is realized by the hardware of the distraction system.

1. Introduction

Bone distraction and compression osteogenesis (DCO) is an effective orthopedics procedure to heal both inborn and acquired injures of the human musculoskeletal system. The history of the DCO started at the beginning of 20th century when an Italian surgeon, Alessandro Codivilla, performed the first bone lengthening procedure [1]. However, the Ilizarov system, developed in 1950s, became a leader in clinical practice. According to G.A. Ilizarov, the bone distraction rhythm is be more than 60 distraction steps per day to obtain the natural bone growth temp [2]. This rhythm is provided by automated bone lengthening systems with adjusted distraction force values [3].

Apart from the linear bone lengthening issues, there are other groups of injuries related to bone shape deformation. They are marked out as varus and valgus deformations (figure 1 [4]). The causes of these injuries are childhood diseases (e.g. rachitis, metabolic disorders, etc.), teenage diseases (e.g. Ca and vitamin D deficiency, excessive limb load, etc.), adult diseases (e.g. traumas and joint pathology processes).

The conventional method for bone deformation correction is distraction osteogenesis (DO). Further, the studies [5,6] provide the advantages of continuous DO for bone regeneration acceleration. For complex deformations, surgical methods are applied. Increased painfulness for patients and a subjective component within the distraction force adjustment are major shortcomings of the bone shape correction procedure.

The new bone shape correction method was designed to eliminate the existing drawbacks. It is based on the limb X-Ray image processing to create a bone geometry model. According to this model, the special software will provide the total control of the bone correction procedure.

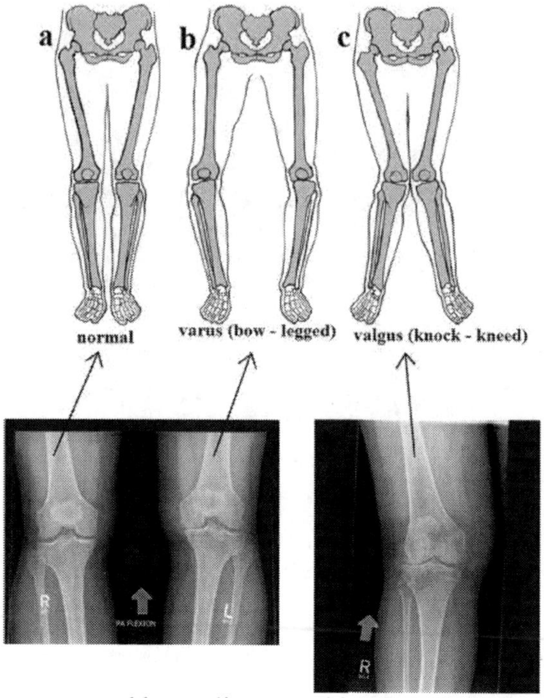

Figure 1. The images show a normal knee alignment (a), varus (bowed legs) (b) and valgus (knock knee) (c) alignment.

2. Materials and methods

Modern X-Ray procession techniques provide high-precision determination of the bone deformation angle. The investigation was performed using the Agfa NX v. 8800 SU1 software for digital X-Ray images processing. We identified the deformation angle for 10 patients from the Emergency Hospital in Tomsk, Russia. Figure 2 shows an NX image of one of the patients. The built-in instruments of this software are able to determine and process the region of interests.

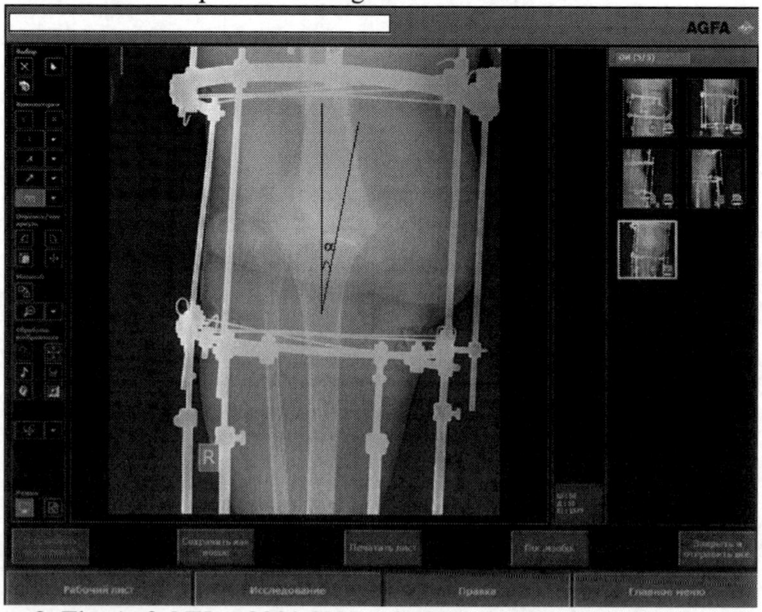

Figure 2. The Agfa NX v. 8800 SU1 software with pelvic limb X-Ray image.

Before the geometry model design, we set the external fixation system ring diameter *l* equal to 160 mm and the threaded nail length *r* equal to 200 mm to simplify further calculations. The example of the bone geometry model is shown in figure 3. The main issue is to identify the deformation value of *x*. After that, the appropriate gain to decrease the deformation can be set. For distraction force setup and calculation of the variable *x*, a special software BDC (Bone Distraction Control) v. 1.0.1 was designed.

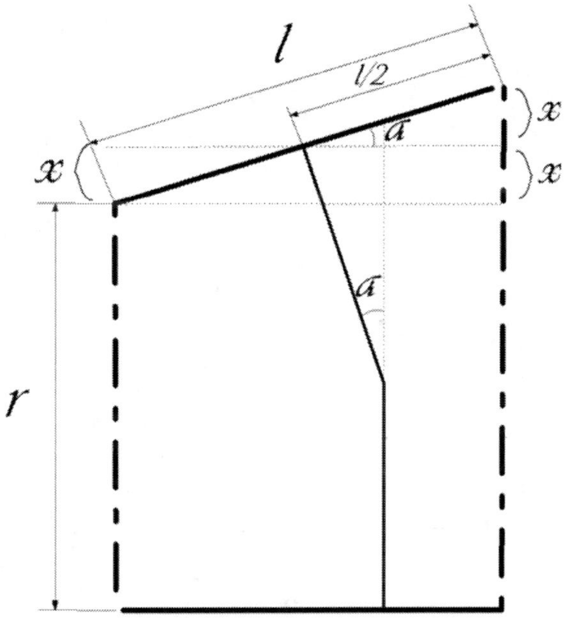

Figure 3. Defected bone geometry model.

3. Results

As can be seen in figure 3, the defected bone is fastened in the upper ring center (point *l/2*). The variable *x* represents the difference between the upper ring horizontal axis and the actual ring position. The algorithm to determine the motor output force value is as follows:

1. The deformation angle is measured with NX software;
2. The *X* value is calculated by BDC v.1.0.1. software;
3. The *ratio* between the motor ganes is defined.
4. In compliance with this ratio, the output motor force is set.

The *X* and *ratio* values are calculated by the formulas:

$$x = \frac{l \cdot \sin(\alpha)}{2} \quad (1)$$

$$ratio = \frac{r + 2x}{2} \quad (2)$$

The basic motor gain value is set as 50% of the maximum motor power level. For example, if ratio=1.3, then the first motor gain level is 50%, and the second one is 65% ($50 \cdot 1.3$). The system model is shown in figure 4, where G1 and G2 are the output gains of the motors.

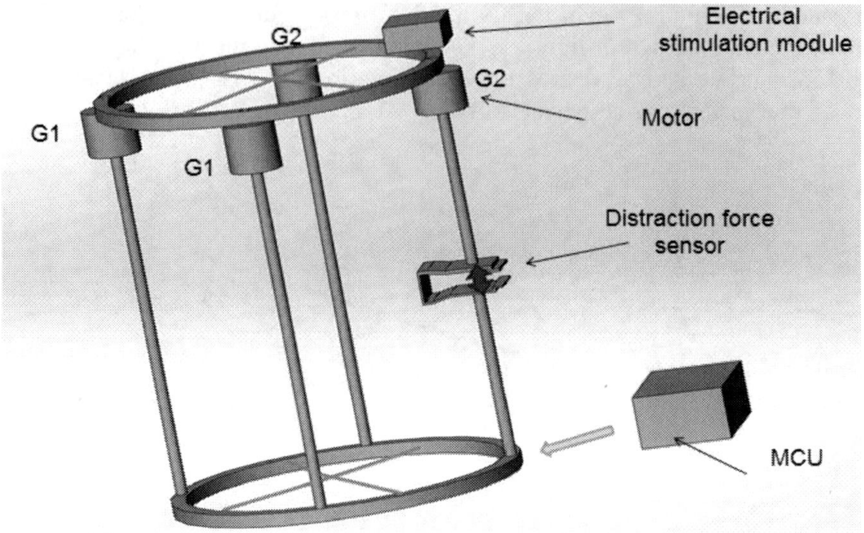

Figure 4. 3D model of distraction system for limb correction.

Software BDC v.1.0.1. was created on the basis of the IDE Qt v. 4.8.4. The main window screenshot is shown in figure 5. The measured deformation angle value is set in the field "Angle" of the software. The gains of the motors are set in accordance with the calculated ratio value. Communication with hardware is realized via the Bluetooth interface.

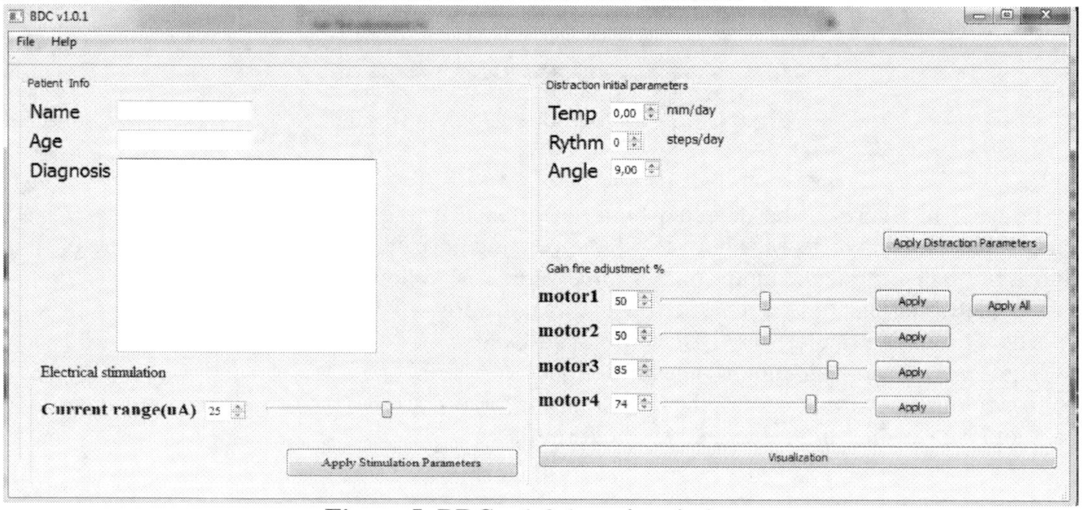

Figure 5. BDC v.1.0.1. main window.

4. Conclusion

The designed method provides opportunities for remote control of the bone shape correction. The major advantage of the method is high precision deformation angle calculation and automatic deformation removal. The physician is able to adjust the motor gain parameters to adapt the method for an individual patient.

References

[1] Reggie C H, Juan S R and Maryam T 2012 *Distraction Osteogenesis and Its Challenges in Bone Regeneration* (INTECH Open Access Publisher)

[2] Shevtsov V I, Popkov A V 1998 *Operativnoe udlineniye nizhnih konechnostey* (M.: Medicina)

[3] Shevtsov V I, Popkov A V (2003) Automated mode full-time bone lengthening *Regenerativnnaya hirurgiya journal* **1** 1–16
[4] Dr. John Cameron (2013) *Knee Conditions and Disability* Available from: http://www.wsiat.on.ca/english/mlo/knee.htm
[5] Djasim U M, Mathot B J, Wolvius E B, van Neck J W, van der Wal K G H (2009) Histomorphometric comparison between continuous and discontinuous distraction osteogenesis *Journal of Cranio-Maxillofacial Surgery* **37** 398–404
[6] Pat. USA №US20080051779 A1. Force-controlled autodistraction/ William Mackenzie, Tariq Rahman, Robert Akins, Alastair Younger. Aug. 2, 2006

The Quality Evaluation of Concrete Structures with the Ultrasonic Low-Frequency Tomograph

A Smokotin[1], M Bogatyreva[2] and I Protasova[2]

[1] The Head of "Non-Destructive Testing and Diagnostics" Laboratory, Foundations and Testing of Constructions, Tomsk State University of Architecture and Building, Tomsk, Russia
[2] Research Assistant, Foundations and Testing of Constructions, Tomsk State University of Architecture and Building, Tomsk, Russia
E-mail: mar289@rambler.ru

Abstract. The paper presents the research findings obtained as a result of the ultrasonic nondestructive testing of flaw detection in one-sided access confined concrete. The ultrasonic nondestructive testing is applied to detect the size and location of internal defects in confined concrete that allows the accurate taking cores and their mechanical testing. Tomographic images of the scanned internal structure are given herein as well as the results of compressive strength testing of cores taken from defective portions of the concrete one-sided access construction.

1. Introduction

Nowadays, the ultrasonic nondestructive testing is widely used along with other techniques for the quality control and inspection of in-situ reinforced concrete constructions. The ultrasonic nondestructive testing allows the investigation of one-sided access constructions, field applications, non-hazardous and eco-friendly applications, and provides the parameter informativeness. Ultrasonic waves is the only type of radiation which responds to various mechanical defects of the material structure [1–4].

In 2014, the authors conducted research into the flaw detection, namely cracks, voids, cavities, etc. occurred in the reinforced concrete load-bearing structures, that have a negative effect on mechanical conditions of test object. Examination was carried out on ultrasonic nondestructive testing tools.

2. Experimental

The nondestructive testing was carried out on the ultrasonic low-frequency tomograph A1040 POLYGON that performs the echo method for testing, does not require a coupling medium to promote the transfer of sound energy into the test object, and provides the structure visualization of concrete and masonry products and constructions with the one-sided access [5–6].
The following problems should be solved to accomplish the aim of this research.

- Selection of the appropriate portions of the concrete construction and their preparation for the inspection.
- Representation of experimental findings by tomographic imaging B-, C-, and D-scans.
- Laboratory processing of obtained experimental findings.

- Analysis of obtained experimental findings and making conclusions in accordance with the aim of research.

Tomographic imaging is a two-dimensional reconstruction of the material structure represented by its three mutually perpendicular cross-sections as shown in figure 1. The screening and evaluation method is used to generate tomographic images of these cross-sections. The screening and evaluation method comprises a plurality of scanning steps in which the scanner assembly performs a successive scanning of an object plane in order to capture a series of tomographic images which can be evaluated. The direction and step of scanning are pre-selected and not changed during the whole procedure as a result of which only preliminary conclusions can be made about test object. We draw parallel lines on the object plane at regular intervals [7] prior to the scanning procedure.

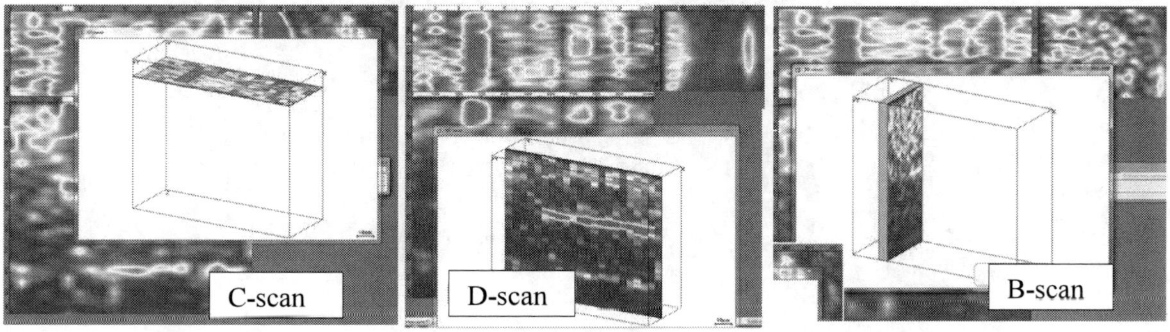

Figure 1. Scans of tomographic images.

Figures 2–4 contain the fragments of tomographic images of scanned concrete structures. The red colour shows that there are regions (reflecting surfaces) with acoustic properties differing from that of the material of the structure, i.e. concrete [8]. This reflecting surface is the non-homogeneous concrete having voids, cavities, channels, and steel bars.

Journal of Physics: Conference Series **671** (2016) 012008 doi:10.1088/1742-6596/671/1/012008

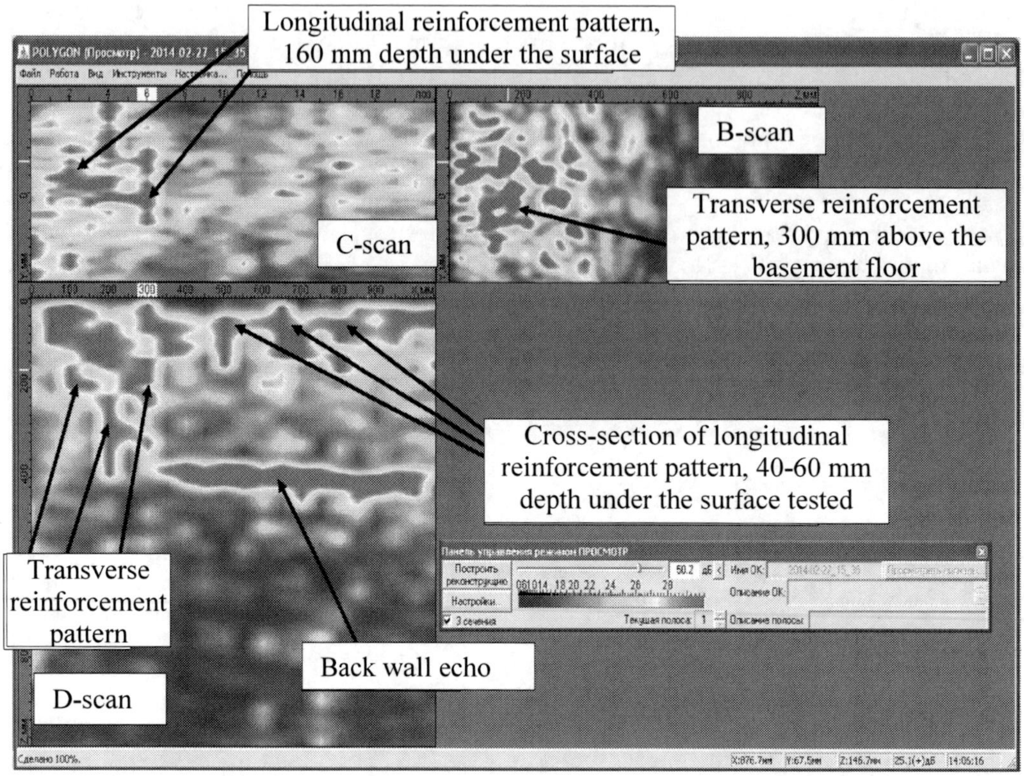

Figure 2. Scans of tomographic images. Test object – confined concrete in strip foundation and basement floor, 400 mm thick.

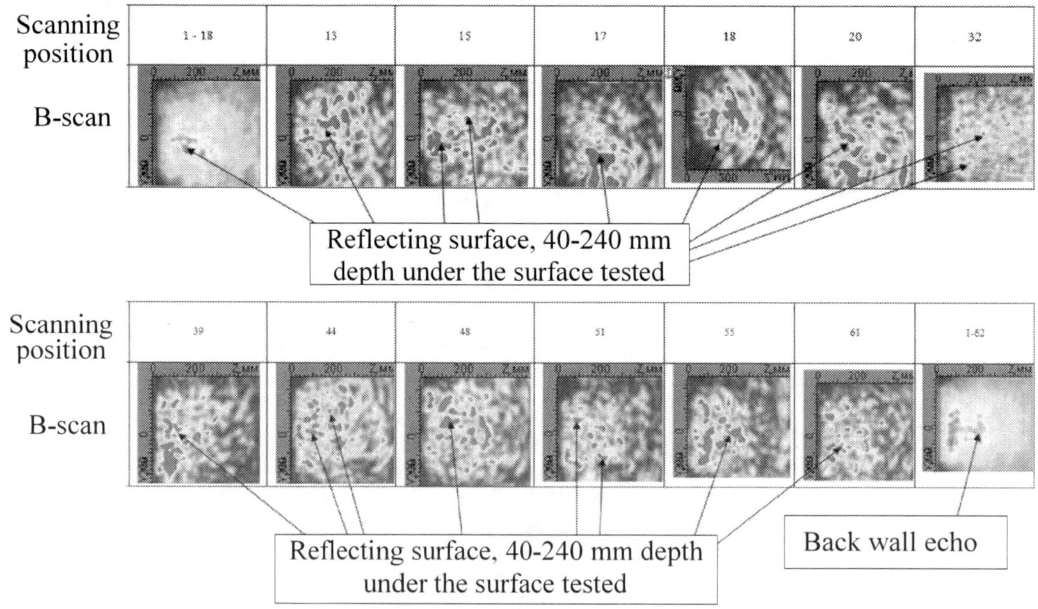

Figure 3. B-scans of tomographic images obtained at different scanning positions. Test object – in-situ concrete floor slab, 250 mm thick.

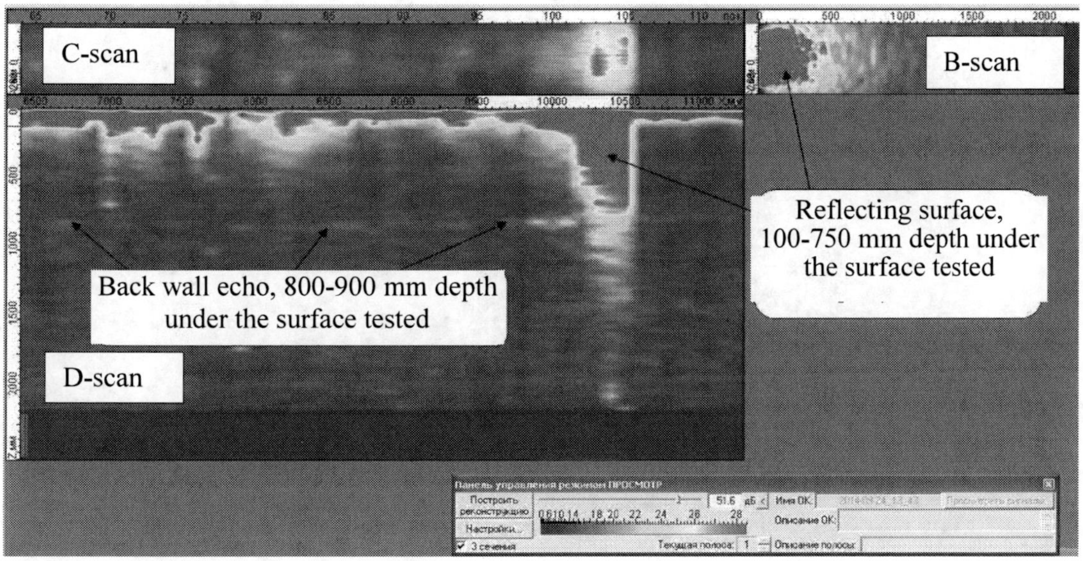

Figure 4. Scans of tomographic images. Test object – in-situ base slab, 900 mm thick.

B- and D-scans of tomographic images presented in figure 4, show the back wall echo the distance from which to the surface comes to 800–900 mm. The thickness of the inspected construction is detected by the location of the back wall echo.

The in-depth analysis of these tomographic images allows the detection of test object regions having supposedly a non-homogeneous structure [9–10]. Cores are then taken from the detected region in order to determine the strength properties of concrete using direct methods.

Thus, two cores are taken from the in-situ base slab during the scanning procedure. The length of the core taken from the first region is 780 mm, while that of the second comes to 930 mm. The quality control and inspection of test object includes the visual examination and compressive strength testing of cores on a test machine.

The visual examination of cores shows that some of them have small single voids and pores distributed over the concrete core as shown in figure 5. Also, single clay inclusions are observed in the core presented in figure 6. The strength testing of the cores was carried out at Tomsk State University of Architecture and Building, Tomsk, Russia in the Laboratory 'Stromtest'.

Figure 5. Small voids in core N 2.

Figure 6. Single clay inclusions in core N 1(right) and N 2 (left).

Strength properties of concrete obtained during the compressive strength testing are given in the table below.

Table. Mechanical and physical properties of cores

Cores	Specimen number	Ultimate compressive strength (MPa)	Concrete type by compressive strength	Concrete type by design documents
Core 1	1.1/1	30.0	B22.5	B25
	1.1/6	18.3	B12.5	
	1.2/1	29.0	B20	
	1.2/4	16.8	B12.5	
Core 2	2.1/1	29.7	B22.5	
	2.1/5	31.9	B22.5	
	2.2/1	35.2	B25	

3. Conclusions

This research resulted in the generation of tomographic images of the material structure of inspected concrete constructions. Scan imaging was carried out in the real time mode. The flaw detection technique allowed obtaining information about the sizes and locations of defects in the confined concrete structures.

However, the following disadvantages were detected during the application of A1040 POLYGON:

- A relative complexity of data interpretation.
- A rather high degree of the human factor in the quality evaluation of concrete structures.

4. References

[1] Klyuev V V, Sosnin F R 2008 *Nerazrushayushchii kontrol'. Spravochnik* (M.: Mashinostroenie)

[2] Klevtsov V A, Korevitskaya M G, Matveev Y K (1991) Primenenie nerazrushajushhih metodov ispytanij pri obsledovanii monolitnyh konstrukcij *Beton i zhelezobeton* **7** 1920

[3] Hoegh K, Khazanovich L, Yu H T (2011) Ultrasonic tomography for evaluation of concrete pavements *Transportation Research Board* **1** 85–94

[4] Mayer K *et al* (2008) Characterization of reflector types by phase-sensitive ultrasonic data processing and imaging *Nondestructive Evaluation* **27**(1–3)35–45

[5] Bishko A, Samokrutov A A, Shevaldykin V G (2008) Ultrasonic echo-pulse tomography of concrete using shear waves low-frequency phased antenna arrays *17th World Conference on Nondestructive Testing, Shanghai* 1–9

[6] Samokrutov A A, Kozlov V N, Shevaldykin V G (2006) *The 5th International Conference on Non-Destructive Testing and Technical Diagnostics in Industry, Proceedings* 152

[7] *Opisanie tomografa A1040M Poligon* [Tomograph A1040 Polygon] Information on: www.ndtpribor.ru/product/tomograf-a1040m-poligon (last visited Jul 2014)

[8] *Ul'trazvukovoi tomograf A1040M polygon* [Tomograph A1040 Polygon] Information on: www.ncontrol.ru/catalog/Ultrazvukovoj-kontrol/Ultrazvukovye-defektoskopy/Ul-trazvukovoj-tomograf-A1040M-POLIGON (last visited Jul 2014)

[9] Nedavnii O I, Smokotin A V, Protasova I B, Bogatyreva M M (2015) Pulse-echo technique used at non-destructive testing of concrete supporting constructions *Vestnik of Tomsk state university of architecture and building. English version appendix* **1** 140-147

[10] Shvabovich K, Suvorov V A (2014) Nondestructive testing of a bottom surface and construction of its profile by ultrasonic tomography *Russian Journal of Nondestructive Testing* **50(2)** 109–119

About Prospects of Enrichment of Mineral Raw Materials and Chemical Activation of Water Suspensions by Electroexplosive Method

S V Bordunov[1], O V Galtseva[2]

[1]Superviser, Public Corporation Scientific - Innovation Enterprise “Echteh”, Tomsk, Russia
[2]Associate Professor, NDT Institute, Tomsk Polytechnic University, Tomsk
E-mail: bvv@academ.tsc.ru

Abstract. A series of experiments is conducted on laboratory and pilot-production installation for the application of high-voltage pulsed discharge in the processes of enrichment of gold-bearing mineral raw materials with a relatively low energy electrical discharge in the pulse of 120 *J*. At the same time the conglomerates of clay components are breaking down to micron and submicron sizes completely. Solid minerals such as quartz are simultaneously destroyed by defects and grains and withdrawn from the installation with water, thus increasing degree of enrichment of concentrate. The results of the processing by high-voltage pulsed discharges of building sand from the Ob riverbed are given. Discharge energy is 45 *J*. Relations of fractions in the original sand and in the sand after electroexplosive process show the uniform crushing of sand particles with splitting off particles with size of less than 0.045 *mm*. A content of chipped particles is 18 %. The results of particle-size and *X*-ray structural analyses are shown.

1. Introduction

According to scientific literature 70 % of the gold in the lithosphere is represented by particles with size of 100 microns or less. Thin and fine - grained gold contained in the technogenic placers - dumps of current and long - closed mines. In [1] is reported that the large and the very large gold with size more than 100 microns is produced in Russia. In [2] are shown the data about the gold deposits in the Magadan region, only 5.1 % of this gold with particle size of 50 microns or less is recovered by gravity methods. In [3] is reported about the necessity of disintegration of clay scatterings of gold, and thin gold can be extracted only after removal of clay from these scatterings.

In [4] is presented the method of disintegration of clay, based on the dispersion of clay lumps by rotating water jets with pressure about 150 atmospheres. As a result, the raw material is separated into sludge (clay) fraction with thin gold and crystal fraction with approximately equal gold content. Gold from fractions leached separately. This method didn't allow removing the clay from raw material without loss of gold, as the traditional method of washing out of raw materials from clay in the scrubber-trommels. Therefore, the most important problems in the process of enrichment of raw material are the removal of clay components from raw material and the extraction of thin gold (platinum) with particle size less than 50 microns into concentrate.

According to [5] the clays are composed of clay particles, ultramicrounits, units, grains, inclusions of microfauna and flora, microcrystals and ore minerals. The individual particles form ultramicrounits and microunits. Ultramicrounits are the strong formations by length from shares of micron to several microns; they are not destroyed by the granulometric analysis. Microunits (size up to several tens of microns) are the associations of clay particles and ultramicrounits, which not disintegrate in water independently at absence of mechanical dispersers.

Interaction of clay particles with clastic grains (minerals and particles of gold) leads to formation of a dense shell-like “coat” on grains; thus, these grains are always submersed in clay mass. The contact between grains takes place only through “coat”. The strength of such phase contacts is 10 - 6 *N*, such “coats” do not lose the strength when wet, and a particle of gold in “coat” not only inaccessible to aqueous leaching solutions, but also loses its essential dignity, used at the gravitational enrichment – high specific weight. Extracting of fine - grained gold into concentrates is mainly by gravity methods based on the difference in the speed of sedimentation of particles with different sizes and density.

The deposition rates of equal-sized particles of gold, pyrite and quartz calculated for laminar mode correlate as 14:3:1; and for turbulent mode correlate as 5:2:1, which is used in most modern devices for obtaining gravity concentrates, and a particle of gold in “coat” is deposited not faster than particles of pyrite. In the literature there are reports about the use of effects of secondary cavitation of ultrasound for dispersing clays, but the effectiveness of this method is small, as well as the efficiency coefficient of energy use.

The most effective tool for dispersing clays with simultaneous destruction of the mineral particle on adhesions, boundaries of section in particles and other defects in the mineral particles, is an electrical discharge in water suspensions. The electric discharge ($E=CU^2/2$) generates the shock waves of compression and extension (up to 60% energy level), ionizes of water with the formation of such reactive products, such as O_2, H_2, OH^-, H_2^+, O_2^+, O^-, H^+, secondary volumetric cavitation on particles of suspension, ultraviolet and electromagnetic radiation.

Physical and chemical factors of effect of electric discharge on solid phase of water suspensions are known for a long time. The most publications based on the use of high-voltage pulsed discharges with voltage higher than 110 *kV* and energy of 10 - 400 *kJ*, comparable to the explosion energy of industrial explosives. Probably why the industrial electroexplosive installations, the walls of which should be thick enough and durable, were not created.

We were focused on the voltage to 30 *kV* and the energy in the discharge of up to 120 *J*, which is allowed to create the technological equipment of sheet steel with a wall thickness of 3 to 4 *mm*.

2. Experimental

The working laboratory installation is shown schematically in figure 1; it is designed for the primary research of raw materials and determination of the necessary technological modes of processing.

The following tasks are solved with the help of this stand:

- the optimization of energy consumption (*kW•h/t*);
- the speed optimization of water consumption (*l/min*) required for the maximum removal of "empty" rock from the reactor with the flow of water and increase of gold content in the concentrate.

Further the experiments are carried out at the experimental industrial electroexplosive installation (figure 2) with four electrode pairs [6], which arranged in pairs on opposite walls of the reactor.

The installation operates as follows. The reactors, tank of concentrate and settlement pond are filled with water, which flows through the reactor and is discharged from the settlement pond through the drain with a certain predetermined speed.

The linear speed of the water and suspension "water and raw material" in the reactor is defined as the quotient of the division of the preset productivity of installation by suspension (dm^3) on the cross-sectional area of the reactor (dm^3). The power supply is switched on after filling of the system with water, the frequency of pulse repetition is set and the conveyor of giving of raw materials into reactor turns on. The serve of the initial solid raw materials with natural moisture is carried from the top down to meet the upward flow of water, in which the air can be introduced through the ejector as necessary.

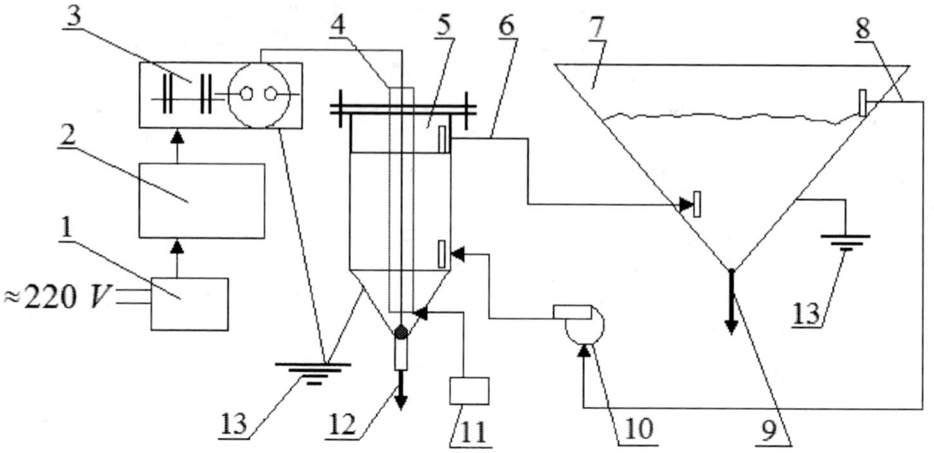

Figure 1. Research electroexplosive installation: 1 – control panel; 2 – high-voltage transformer; 3 – the pulse generator (capacitor, ballast resistors, spark gaps); 4 – insulated electrode; 5 – electro-hydraulic reactor; 6 – line with removal of water from fine-grained rock; 7 – settlement pond; 8 – withdrawal line of clarified water; 9 – line of withdrawal of slurry tailings; 10 – pump; 11 – compressor; 12 – output line of concentrates; 13 – grounding.

At movement of suspension on reactor length the each particle of mineral raw materials should be at least 20 times in the area of the shock waves of four electrodes, sequentially installed in the insulators on the lid of the reactor along its length. Grounded electrodes are the shelves which are installed inside of reactor at an angle of 30 to 45 ^{0} to the vertical axis of the reactor. There are spring-loaded horizontal supports of shelves.

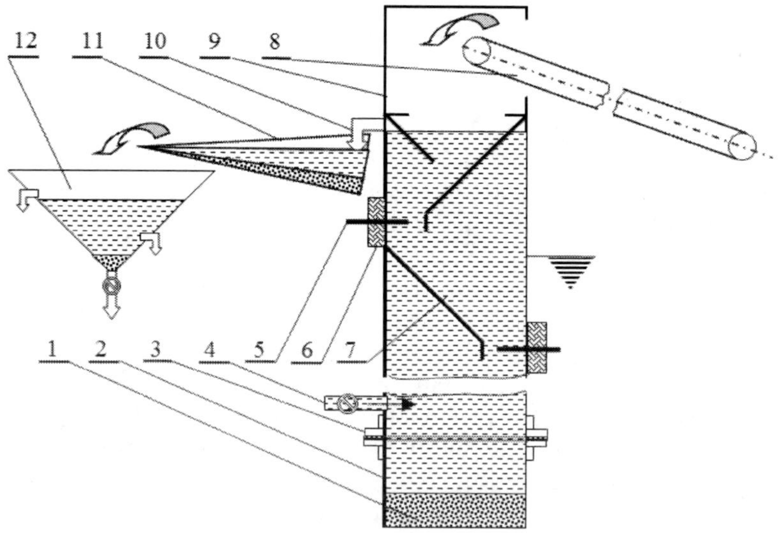

Figure 2. Experimental industrial electroexplosive installation with four electrode pairs: 1 – heavy concentrate; 2 – heavy concentrate bunker; 3 – flanged connection; 4 – water inlet pipe sleeve; 5 – high-voltage electrode; 6 – high-voltage electrode isolator; 7 – landing - ground electrode; 8 – initial raw material transporter; 9 – reactor housing, square in section; 10 – fine slurry outlet pipe sleeve; 11 – fine concentrate trap; 12 – settlement pond.

Thus, in the experiments it is necessary to set the frequency of pulse repetition from power source based on the speed of movement of suspension and the length of zone of action of discharges from electrodes. In reactor the rock breaking occurs by action of discharges, the crushed clay components entrained with the flow of water into the settlement pond. The removal of particles which are not deposited in the upward flow of water (tailings) is carried out through nozzles in the upper part of the reactor.

The residence time in the reactor increases due to the sliding of solid phase on shelves, the slipping on the shelves of the smallest particles is provided by the vibration of shelves under the action of shock waves and springs. Heavy particles with speed of sedimentation higher than the speed of the upward flow of water fall into tank of concentrate.

The electroexplosive processing of mineral raw materials with size up to 100 *mm* and the classification of particles of solid phases by size in the tailings and concentrate by the speed regulation of upward flow of water are possible in this reactor. The design of shelves in the upper part of reactor provides the removal of thin suspension of tailings into special chamber without pulsation in the mode close to laminar.

The batcher is switched off after recycling of the batch of raw materials, the water for washing out of the reactor volume from the residue of a solid phase in suspension is served within 5 minutes through the reactor, the power supply is switched off, the voltage of the high - voltage elements of installation is switched off, the water is discharged from the reactor, the concentrate and the tailings from tanks are discharged.

As the clays on the *Prof.* Protodyakonov's scale of hardness rocks is among the least hard rocks, they are almost evenly dispersed to ultramicrounits (3 – 6 μm) at processing by high-voltage pulsed discharge up to 30 *kV* on installations (figures 1 and 2), the speed of sedimentation of this particles at the laminar mode less than that of gold particles with size of 1 micron.

3. Results and considerations

We studied the possibility of dispersion the clays when exposed to an electric discharge on pieces with size of 10 *mm*. Samples were taken from the ilmenite placer of Tisulsky district of the Kemerovo region. The experiment showed that pieces of clay are dispersed almost to the colloidal state after 20 discharges with energies up to 120 *J*. Speed of clarification of column of such clay suspension and accumulation of sediment at the bottom of the cylinder with height of 20 *cm* is less than 1 *cm*/*day*.

The equality of the speed of clarification and sediment shows that there are no harmful factors (the conglutination and the consolidation of clay particles) for the extraction process of fine - grained gold from clay raw material. Stronger rocks are also subject to crushing, mainly, by defects in the structure of the grains.

The results of experiment of electroexplosive process of building sand from the riverbed of the Ob with discharge energy 45 *J* are shown on figure 3. Changes of relations of fractions in the original sand and in the sand after electroexplosive process show the uniform crushing of sand particles with splitting off particles with size of less than 0.045 *mm*. A content of chipped particles is 18 %.

X–ray analysis showed that the amount of the crystal phase of SiO_2 increased to 94.9 % after processing by the high-voltage pulsed discharge (90.2 % in the initial sand) in the sand fractions 1 – 7; the amount of tailings decreased to 83 % in the sand fraction 8, but 17 % of *Si** phase appeared. The strength of concrete on compression with filler of sand of fractions 1 – 7 after processing by the high-voltage pulsed discharge on 40 % higher than strength of concrete with filler of initial sand [7].

The effect of grinding of strong rock SiO_2 by discharges with energy of 45 *J* was tested on tailings of three gold processing plants. The gold content in leaching solution increases by 3 to 6 % after leaching of gold in the Aqua Regia for 24 hours in the sample after processing by the high-voltage pulsed discharge. This is the gold screened within grains of accompanying rock or clay "coats".

High efficiency of processing by the high-voltage pulsed discharge of clay gold-bearing raw material with clay content up to 90 %, which is almost completely removed from the process, the concentrate with weight 8 – 2 % of weight of raw material contains more than 95 % of the initial gold [8].

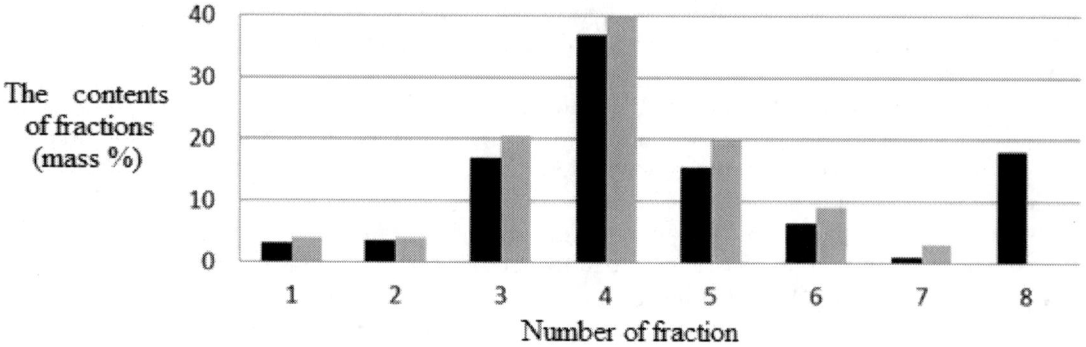

Figure 3. The result of processing of sand from the river Ob by 10 discharges at 45 *J*, where are ■ – the sand after processing by the high-voltage pulsed discharge and □ – the original sand. Number of sand fraction corresponds to the following particle size:
1÷ **5**: – 2.5 *mm*; **2**: 2.5 – 2 *mm*; **3**: 2 – 1 *mm*; **4**: 1 – 0.315 *mm*; **5**: 0.315 – 0.2 *mm*; **6**: 0.2 – 0.1 *mm*; **7**: 0.1 – 0.045 *mm*; **8**: 0.045 *mm*.

4. Summary

Three types of industrial electroexplosive installations with productivity of thin suspension up to 100 m^3/h and of solid raw material up to 10 *t/h* [6, 8–10] are developed by us on the basis of more than fifteen years of research. The developed technology is characterized by low consumption of electric energy. Depending on the type of processed raw materials the consumption of installation at enrichment of mineral raw materials at processing by high-voltage pulsed discharge is 0.05 – 0.1 $kW{\bullet}h/m^3$ of suspension (not more than 1.5 *kW·h/t* of solid raw materials).

Experimental industrial tests showed that the wall thickness of electroexplosive reactors for processing of suspensions with size of particles of the solid phase is less than 0.63 *mm*, productivity is 100 m^3/h and not more than 4 *mm*, and the dimensions of the reactor do not exceed 400×800×1200 *mm*. Technology can be used in utilities in the purification of drinking and waste water [11], hydrometallurgy, enrichment of ores of nonferrous and precious metals, in the construction of buildings and structures made of concrete with high impact strength and compressive strength.

References

[1] Melamud V S (1999) *Proceedings of 1st International Symposium "Gold of Siberia"* 115–117
[2] Melnikov M S *et al* (1981) *Proceedings of TsNIGRI* **158** 38–41
[3] Generalov M E, Naumov V A (1998) *Uralian Geological Journal* **4** 19–56
[4] Metodi K *et al* (2014) *Gold and Technology* **1** 28–32
[5] Galtseva O V *et al* 2013 *Electroexplosive equipment for dispersing, chemical activation and mineral processing* (Tomsk, TPU Publishing House)
[6] Bordunov S V *et al* 2013 RF Patent 2477173 *Byull. Izobret.* **7**
[7] Bordunov S V and Kulaga I G (2012) *Industrial and Civil Engineering* **4** 58–61
[8] Bordunov S V *et al* (2008) *Rus. J. Non-Fer. Met.* **9** 24–28
[9] Bordunov S V *et al* 2012 RF Patent 2470875 *Byull. Izobret.* **36**
[10] Bordunov S V *et al* (2008) *Rus. J. Non-Fer. Met.* **2** 65–68
[11] Petrova A Y *et al* (2015) *Technical Physics* **60 (4)** 592–594

Journal of Physics: Conference Series **671** (2016) 012010 doi:10.1088/1742-6596/671/1/012010

Adequacy Criteria of Models of the Cargo Inspection System with Material Discrimination Option

S Osipov[1], S Chakhlov[2], O Osipov[3], A Shtein[4] and J Van[5]

[1]Leading reseacher, NDT Institute, Tomsk Polytechnic University, Russia
[2]Head laboratory of technical tomography and introscopy, NDT institute, Tomsk Polytechnic University, Russia
[3]Engineer-researcher, NDT Institute, Tomsk Polytechnic University, Russia
[4]Engineer, NDT Institute, Tomsk Polytechnic University, Russia
[5]Postgraduate student, NDT Institute, Tomsk Polytechnic University, China
E-mail: osip1809@rambler.ru

Abstract. Generalized adequacy criteria for mathematical models in order to discriminate materials in X-ray inspection systems by the dual-energy method were developed. Two main approaches of the examination systems to produce the adequacy criteria by the final and the intermediate parameters of the dual-energy method were analyzed. The criteria were specified in respect to the discrimination by the effective atomic number and by the method of level functions. Experimental and theoretical estimates of the discrimination parameters of the test object constituents scanned by fan beams of X-ray radiation with the maximal energies of 4.5 and 9 MeV are given.

1. Introduction

Digital radiography is widely used to examine luggage, vehicles, and containers [1–4]. In order to discriminate various materials in the inspected objects, the dual-energy method (DEM) is often used [4–6]. The discrimination is based on the correlation of X-ray parameters relevant for the object materials with those for known materials. An effective atomic number or a related parameter of a material is often used as a discrimination parameter [5, 7–9]. There are several approaches to generate primary radiographs based on the DEM, as well as its transformation to final DEM images or to images of the discrimination parameters as described in detail in [7]. Different physical factors affect the quality of the discrimination both for low [10] and high X-ray energies [11]. Until now, there is a lack of adequate mathematical models of digital radiography systems dealing with the discrimination of the test object materials and their constituents. Here, we show that adequate criteria for discrimination parameter can be obtained for the DEM method.

2. Generalised adequacy criteria for systems with discrimination by the dual energy method

Let A be a finite set, each element $a \in \mathbf{A}$ match one and only one group of the materials. It is most convenient for matching to use the effective atomic number which is the average atomic number for a mixture of materials in a given compound. In a radiographic inspection, it must be possible to identify the object's constituents independent of their size, their mass density ρ and the transmission length H of the X-rays within the object. As an integral characteristic one may use the product ρH. From an application point of view, it is of interest to identify ρH within $[\rho H_{min}, \rho H_{max}]$, in which the material from the group $a \in \mathbf{A}$ is recognized with a given confidence probability. The object to be inspected consists of a finite number of constituents. Therefore it is logical to define a set

$$\mathbf{R}(a) = \{\rho H_i(a),\ i = 1...n(a),\ a \in \mathbf{A}\}, \tag{1}$$

where $n(a)$ is the number of constituents with ρH as the indentifying parameter for a group of materials a. Furthermore, let us introduce a set $\mathbf{M}_I$, which defines a region of interest in the following way:

$$M_I = \{(\rho H, a):\ \rho H \in \mathbf{R}(a);\ a \in \mathbf{A}\}, \tag{2}$$

In processing the radiographs obtained by DEM for each test object constituent, a pair of intermediate parameters, a pair of DEM parameters, and a final identifying parameter may be computed. Any specified groups of parameters can be used to create adequacy criteria. The criteria parameters are represented in vectors $\mathbf{p} = (p_1, p_2, \ldots, p_k)$. To define the criteria two vectors are used: the theoretical vector $\mathbf{p}$ and the experimental vector $\tilde{\mathbf{p}}$. The required criteria reduce in the most generalized form to

$$M_I \subset M^* = \{(\rho H, a):\ |p_i(\rho H, a) - \tilde{p}_i(\rho H, a)| \le \Delta p_i(\rho H, a);\ i = 1, k\}. \tag{3}$$

where $\Delta p_i(\rho H, a)$ are the maximal deviations of the theoretical and the experimental parameters.

3. Discrimination by effective atomic number

The primary radiographs are obtained by scanning the test object with X-ray radiation at two energies E_1 and E_2. The primary radiographs are transformed in a first process into DEM images and in second stage to a final identity image, which represents the distribution of the effective atomic number Z overlaid on the primary radiograph. The algorithm to identify the test object material by DEM is described in detail in Ref. [7–13]. To estimate the adequacy of the corresponding mathematical model, it is necessary to compare the simulated images with the measured images of the test object. Ref. [7] describes a test object whose constituents have an equal quadratic cross-sections. Let us assign a set of elements $a \in \mathbf{A}$, as described in the first chapter, a value of the effective atomic number Z. The test object constituents are characterized by a pair of numbers (ρH, Z), where ρH is the ray thickness of the constituent in g/cm^2 and Z is the effective atomic number.

As a result of the primary processing of the original radiographs of the test objects $\boldsymbol{I}_1$ and $\boldsymbol{I}_2$, corresponding to the X-ray energies E_1 and E_2, the mean values of ray-thicknesses of all test object constituents can be estimated for energies E_1 and E_2. The constituent with the characteristics (ρH,Z) is associated with its ray-thicknesses $Y_1(\rho H,Z)=Y(E_1,\rho H,Z)$ and $Y_2(\rho H,Z)=Y(E_2,\rho H,Z)$, as measured by the absorption length of the X-ray.

As a next step the DEM parameters $A(\rho H,Z) \approx \rho H$ and $B(\rho H,Z) \approx \rho HF(Z)$ must be found for all constituents (ρH,Z), $\rho H \in \mathbf{R}(a)$, $a \in \mathbf{A}$. For X-ray sources with energies up to 150 keV, the function $F(Z) \approx Z^{3.8}$ [14], but for high-energy X-rays with energies above 1.022 MeV, the function $F(Z) \approx Z$ [1]. The last stage of the algorithm reduces to the estimation of the effective atomic numbers of materials for all test object constituents, that is the set $\tilde{Z}(\rho H, Z)$, $\rho H \in \mathbf{R}(Z)$, $Z \in \mathbf{A}$. The corresponding formula has the form

$$\tilde{Z}(\rho H, Z) = F^{-1}\left(\frac{B(\rho H, Z)}{A(\rho H, Z)}\right), \tag{4}$$

where F^{-1} is the inverse function of F. The concretization of the adequacy criterion (3) in respect to the discrimination by the effective atomic number is in a formalized form given by

$$M_I \subset M^* = \{(\rho H, Z): |\tilde{Z}(\rho H, Z) - Z_t(\rho H, Z)| \le \Delta Z(\rho H, Z)\} \tag{5}$$

Here, in a first approximation we set $Z_t(H,Z) = Z$.

The presented procedure calls for an adequate criterion of discrimination by an effective atomic number (finite parameter of DEM). Like in chapter 1, one can use an approach which is based on the analysis of the pairs of experimental primary or intermediate DEM images with the pairs of theoretical images. This approach is valid because the equality of theoretical and the experimental images cause the equality of the final DEM images.

Let us call the corresponding criterion the adequacy criterion for intermediate DEM parameters $Y_1(\rho H,Z)$ and $Y_2(\rho H,Z)$. Furthermore, we formalise the introduced criterion in respect to the test object. This formalization is implemented by analogy with the approach based on expression (3). The adequacy criterion with the intermediate DEM parameters has the form

$$M_I \subset M^* = \{(\rho H,Z):\ |p_i(\rho H,Z)-\tilde{p}_i(\rho H,Z)| \le \Delta p_i(\rho H,Z);\ i=1,2\}. \quad (6)$$

As pairs of theoretical and experimental parameters p_1, p_2 and $\tilde{p}_1, \tilde{p}_2$ in expression (6), one can use corresponding values of ray thicknesses $Y(E_1,\rho H,a)$ and $Y(E_2,\rho H,a)$, or the final DEM parameters – $A(\rho H,Z)$ and $B(\rho H,Z)$.

4. Discrimination by the method of level functions

The procedure for the discrimination of test-object materials by level functions is described in detail in Ref. [7]. The method is based on the calculation of the discrimination parameter q for each point (x,y) of the original radiographs. The value of the discrimination parameter q for point with coordinates (x,y) is calculated by expression

$$q(x,y) = \frac{Y_2(x,y)}{Y_1(x,y)} = Q(Y_1(x,y)). \quad (7)$$

The ratio q is a function of the ray thickness for an X-ray radiation source with the maximal energy E_1. Here, for maximal X-ray energies less than 1 MeV, we select $E_2 < E_1$, but for maximal energies larger than 1 MeV we select $E_2 > E_1$. The functions $Q(Y_1(x,y))$ are generated by special test objects, described in chapters 1 and 2, for all classes a, $a \in \mathbf{A}$ of identifiable materials and serve to build a set of level functions $U_-(Y_1(x,y))$ and $U_+(Y_1(x,y))$.

The test object material along the beam connecting the source and the point (x,y), refers to class a, $a \in \mathbf{A}$, if the following relation is satisfied

$$U_-(Y_1(x,y),a) \le q(x,y) < U_+(Y_1(x,y),a). \quad (8)$$

For the concretization of the adequacy criterion (3) for material discrimination by level functions, we make use of the fact that the set of level functions is defined uniquely by the set of the functions $Q(Y_1(x,y),a)$, $a \in A$. Here, a material class a is selected as the material most applicable for this class. We shall use polyethylene, aluminium, and iron as materials.

The first version of the criterion can be named the criterion of identifying functions. Using the approach outlined in the first chapter, we define the adequacy criterion

$$M_I \subset M^* = \{(\rho H,a):\ |Q(Y_1(\rho H,a)) - \tilde{Q}(Y_1(\rho H,a))| \le \Delta Q(Y_1(\rho H,a))\} \quad (9)$$

The adequacy criterion based on expression (9) is much less dependent on the change of mutual geometric positions of the test object and the inspection system than the adequacy criterion by the final DEM parameters mentioned in the first chapter. The second concretization of the criterion (3) is similar to the adequacy criterion of the intermediate DEM parameters described in the previous chapter.

5. Computational formulae

5.1. Discrimination by effective atomic number

The theoretical values of the ray thicknesses $(\rho H,Z)$ of the constituents for X-rays with the maximal energies E_1, E_2 are computed from the expressions [7]:

$$Y_1(\rho H,Z) = -\ln\left\{\frac{\mathrm{int}\left(N_{01}\int_0^{E_1} E_{ab}(E) f(E,E_1)\varepsilon(E,h)e^{-m(E,Z)\rho H}dE \Big/ \Delta I\right)}{\mathrm{int}\left(N_{01}\int_0^{E_1} E_{ab}(E) f(E,E_1)\varepsilon(E,h)dE \Big/ \Delta I\right)}\right\},$$

$$Y_2(\rho H,Z) = -\ln\left\{\frac{\mathrm{int}\left(N_{02}\int_0^{E_2} E_{ab}(E) f(E,E_2)\varepsilon(E,h)e^{-m(E,Z)\rho H}dE \Big/ \Delta I\right)}{\mathrm{int}\left(N_{02}\int_0^{E_2} E_{ab}(E) f(E,E_2)\varepsilon(E,h)dE \Big/ \Delta I\right)}\right\}. \quad (10)$$

Here, N_{01} and N_{02} are the number of the X-ray quanta with energies E_1 and E_2, which are recorded by the detector in the absence of a test object. Furthermore, $E_{ab}(E)$ is the mean value of the absorbed energy of the recorded photons with the energy E, $f(E,E_i)$ is the energy spectrum of the X-ray radiation with maximal energy E_i, $\varepsilon(E,h)$ is the detection efficiency of the scintillator of thickness h for photons with energy E, $m(E,Z)$ is the mass attenuation coefficient of the photon radiation with energy E, int(x) represents the integer part of number x, and ΔI is the sampling rate.

In references [7, 13] the integral equations to find the DEM parameters A and B are given. In order to continue with the purpose of our work, we shall use this system of equations to find A and B, similar to Eq. (10):

$$-\ln\left\{\frac{\int_0^{E_1} E_{ab}(E) f(E,E_1)\varepsilon(E,h) e^{-g_1(E)A-g_2(E)B} dE}{\int_0^{E_1} E_{ab}(E) f(E,E_1)\varepsilon(E,h) dE}\right\} = Y_1(\rho H, Z),$$

$$-\ln\left\{\frac{\int_0^{E_2} E_{ab}(E) f(E,E_2)\varepsilon(E,h) e^{-g_1(E)A-g_2(E)B} dE}{\int_0^{E_2} E_{ab}(E) f(E,E_2)\varepsilon(E,h) dE}\right\} = Y_2(\rho H, Z). \tag{11}$$

Here, $g_1(E)$, $g_2(E)$ describe the energy dependencies of the x-ray interaction with the material constituents. Index «1» corresponds to the Compton effect. Index «2» corresponds to the photo-effect for X-ray energies less than 1.022 MeV, and to the effect of pair creation for energies larger than 1.022 MeV. It should be mentioned that Eqs. (11) were derived under the condition $\Delta I \approx 0$, that is for an analog-to-digital converter (ADC) with an infinite number of bits. In Ref. [10], the influence of the ADC capacity on the accuracy of the effective atomic number definition was examined in detail. As a result of the solution of the Eqs. (11), there are defined DEM parameters $A(\rho H,Z)$ and $B(\rho H,Z)$ for each constituent $(\rho H,Z)$ of the test object.

It was emphasized above that in the final stage of the algorithm for the constituent $(\rho H,Z)$, an estimation of the effective atomic number can be found:

$$Z_t = F^{-1}\left(\frac{B}{A}\right). \tag{12}$$

The function F^{-1} has a different form for low-energy and high-energy realizations of the DEM. For the low-energies $F^{-1}\left(\frac{B}{A}\right) = \sqrt[3.8]{\frac{B}{A}}$, but for the high-energies $F^{-1}\left(\frac{B}{A}\right) = \frac{B}{A}$. Evidently, there is a difference between the estimation of the effective atomic number Z_t and its true value Z. The deviation is caused by some inaccuracies and computational errors.

Each element of primary radiographs of the test objects $\boldsymbol{I}_1$ and $\boldsymbol{I}_2$ presents itself an integer number ranging from 0 to 2^m -1, where m is the ADC capacity. The primary images are calibrated by use «black level» and «white level» images, and are then transformed by the logarithm operator. For the test object constituents with the characteristics $(\rho H,Z)$, there are computed experimental estimations of the ray thicknesses $\tilde{Y}_1(\rho H,Z)$ and $\tilde{Y}_2(\rho H,Z)$. The values $\tilde{Y}_1(\rho H,Z)$ and $\tilde{Y}_2(\rho H,Z)$ are used as the right part of the system (11) to get the experimental estimations of the DEM parameters $\tilde{A}(\rho H,Z)$ and $\tilde{B}(\rho H,Z)$, and then the experimental estimation of the effective atomic number $\tilde{Z}$, which is obtained by Eq. (12).

5.2. Discrimination by the method of level functions

The theoretical value of the parameter Q for a constituent with the initial characteristics $(\rho H,Z)$ and the corresponding ray thicknesses $Y_1(\rho H,Z)$ and $Y_2(\rho H,Z)$ are founded on the equation

$$Q(Y_1(\rho H,Z)) = \frac{Y_2(\rho H,Z)}{Y_1(\rho H,Z)}. \tag{13}$$

To estimate the ray thicknesses $Y_1(\rho H,Z)$ and $Y_2(\rho H,Z)$ the expressions (10) are used. The corresponding experimental values of the discrimination parameter $\tilde{Q}$ for a constituent with the parameters ($\rho H,Z$) are defined by substituting in Eq. (13) the experimental values of the ray thicknesses $Y_1(\rho H,Z)$ and $Y_2(\rho H,Z)$.

6. Experimental validation of the adequacy of discrimination system-models by the dual energy method

The correspondence of the theoretical and the experimental discrimination parameters was carried out on the cargo inspection system of the National Research Tomsk Polytechnic University (TPU). The region of the consumer interest was limited by a set

$$M_I = \{(\rho H, a): \ \rho H \in \mathbf{R}(a) = \{20, 40, 60, 80, 100\}; \ a \in \mathbf{A} = \{6, 13, 26\}\}. \quad (14)$$

Any element a from the set **A** is associated with the material's effective atomic number Z. In the high-energy realization of the DEM, the primary radiographs were produced for a pair of maximal X-ray energies E_1 = 4.5 MeV, E_2 = 9 MeV. To describe the energy spectrum of X-ray radiation of betatron Shiff's formula [15] was used.

The dependencies of the mass coefficients of the radiation attenuation of energy $m(E)$ were obtained from data libraries [16, 17] together with the energy dependencies $g_1(E)$, $g_2(E)$ for the Compton effect and for the pair production [16, 17]. To record the X-ray radiation in the TPU cargo inspection system, $CdWO_4$ scintillator detectors were used having a length of 35 mm. The ADC capacity was m = 16. Two adequacy criteria of the DEM-discrimination system models were verified experimentally. Figure 1 shows the theoretical, the computational, and the experimental dependencies $Z_{eff}(\rho H)$. From the analysis of the data shown in figure 1, one can verify the adequacy of the discrimination model by an effective atomic number for the considered region of interest (14), and the maximal deviation $\Delta Z_{eff} \approx 5$ of the effective atomic number.

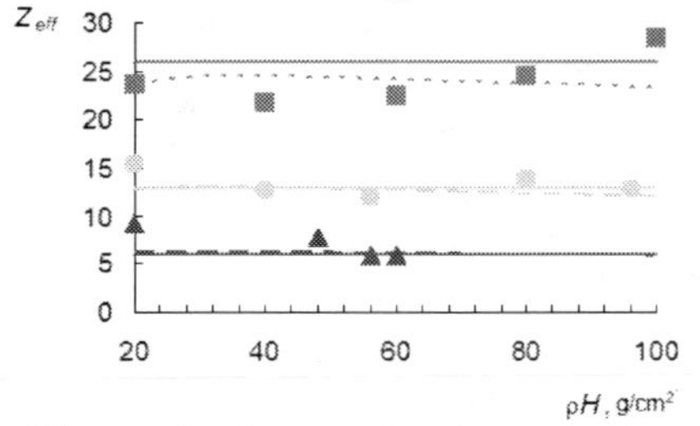

Figure 1. Theoretical and experimental dependencies $Z_{eff}(\rho H)$ as a function of ρH. The solid lines indicate the theoretical expectation, the dashed lines the computational result and the symbols represent the experimental result: full squares for Fe; dots for Al, and triangles for $(CH_2)_n$.

Figure 2 shows the computational and the experimental dependencies of the discrimination parameters $Q(Y_1)$. From the analysis of the data given in figure 2, one can conclude that for maximal deviations $\Delta Q \approx 0.012$ of the discrimination parameter, the adequacy of the discrimination model by the method of level functions for the considered region of interest (14) is warranted.

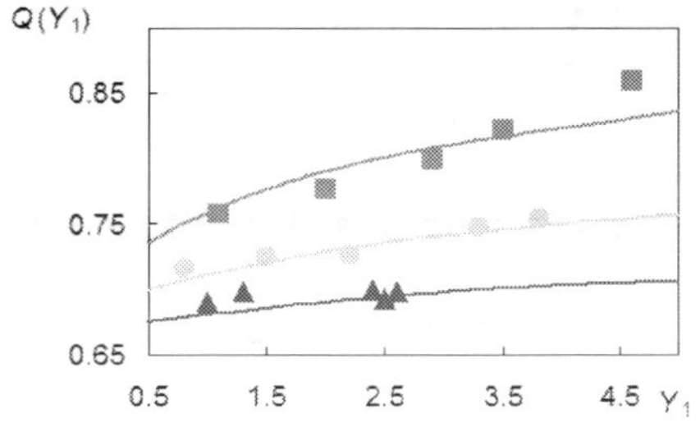

Figure 2. Theoretical and experimental dependencies $Q(Y_1)$. Again as in Figure 1, the solid lines indicate the theoretical expectation, the dashed lines the computational result and the symbols the experimental results for Fe (full squares), for Al (dots) and for $(CH_2)_n$ (triangles).

7. Summary

The proposed adequacy criteria of the examination system models with the material discrimination option for the test objects and their constituents by the dual energy method can be used to design and to compare quantitatively different cargo inspection systems. Two main approaches to build the adequacy criteria by the final and the intermediate DEM parameters were analyzed. The criteria are selected in order to discriminate by the effective atomic number and by the method of level functions. The results of the experimental and the theoretical estimations of the discrimination parameters of the test object constituents, scanned by an X-ray fan-beam with the maximal energies 4.5 and 9 MeV, confirm the efficiency of the proposed adequacy criteria.

Acknowledgement
The authors acknowledge the financial support from The Ministry of Education and Science of the Russian Federation in part of the science program and also the Russian Foundation for Basic Research (Grant 13–08–98027).

References
[1] Runkle R C, White T A, Miller E A, Caggiano J A, Collins B A (2009) Photon and neutron interrogation techniques for chemical explosives detection in air cargo: A critical review *Nuclear Instruments and Methods in Physics Research, Section (A)* **603** (3) 510–528
[2] Reed W 2012 *Non-destructive Testing and Inspection Using Electron Linacs* in: Hamm R W, Hamm M E (Eds.) Industrial Accelerators and Their Applications (Singapore, World Scientific Publishing Co. Pte. Ltd.)
[3] Mery D (2014) Computer vision technology for X-ray testing *Insight: Non-Destructive Testing and Condition Monitoring* **56** (3) 147–155
[4] Kolkoori S, Wrobel N, Osterloh K, Redmer B, Deresch A, Ewert U (2013) High-Energy Radiography for Detecting Details in Highly Complex Packings *Materials Testing* **55**(9) 683–688
[5] Rebuffel V, Dinten J (2007) Dual-energy x-ray imaging: benefits and limits *Insight: Non-Destructive Testing and Condition Monitoring* **49** (10) 589–594
[6] Fuchs T, Keßling P, Firsching M, Nachtrab F, Scholz G (2012) Industrial Applications of Dual X-ray Energy Computed Tomography (2X-CT) *RILEM Booseries* **6** 97–103
[7] Chakhlov S, Osipov S (2013) High-energy digital X-ray imaging method for material identification *Testing. Diagnostics* **183** (9) 9–17
[8] Mery D (2013) X-ray Testing: The State of the Art *The e-Journal of Nondestructive Testing & Ultrasonics* **18** (9) 1–12
[9] Park J S, Kim J K (2013) Calculation of effective atomic number and normal density using a source weighting method in a dual energy x-ray inspection system *Journal of the Korean Physical Society* **59** (4) 2709–2713
[10] Klimenov V, Osipov S, Temnik A (2013) Identification of the substance of a test object using the dual-energy method *Russian Journal of Nondestructive Testing* **49** (11) 642–649
[11] Osipov S, Temnik A, Chakhlov S (2014) The effects of physical factors on the quality of the dual high energy identification of the material of an inspected object *Russian Journal of Nondestructive Testing* **50** (8) 491–498
[12] Xing Y, Zhang L, Duan X, Cheng J, Chen Z (2011) A reconstruction method for dual high-energy CT with MeV X-rays *IEEE Transactions on Nuclear Science* **58** (2) 537–546
[13] Nedavnii O, Osipov S, Sidulenko O (2002) Computational aspects of dual energy digital X-ray photography *Russian Journal of Nondestructive Testing* **38** (3) 170–175
[14] Vinegar H, Wellington S (1987) Tomographic imaging of three-phase flow experiments *Review of Scientific Instruments* **58** (1) 96–107
[15] Shiff L (1951) Energy-Angle Distribution of Thin Target Bremsstrahlung *Physical Review* **83** (2) 252–253
[16] Department of nuclear power plant, ABBN laboratory http:\www.ippe.ru/podr/abbn/libr/groupkon.php
[17] Chadwick M B, Obložinský P, Herman M, Greene N M *et al* (2006) ENDF/B-VII.0: Next generation evaluated nuclear data library for nuclear science and technology *Nuclear Data Sheets* **107** (12) 2931–3060

Hardware and Software for Thermal Nondestructive Testing of Metallic and Composite Materials

A O Chulkov[1] and V P Vavilov[2]

[1]Engeneer of Lab №34, Institute of Non-destructive Testing, National Research Tomsk Polytechnic University,Tomsk, Russia
[2]Head of Lab №34, Institute of Non-destructive Testing, National Research Tomsk Polytechnic University,Tomsk, Russia
E-mail:chulkovao@tpu.ru

Abstract. Modeling and processing software, as well as experimental units, developed at Tomsk Polytechnic University for the last decade in the field of thermal/infrared nondestructive testing, are shortly described in this paper along with some illustrations of using this technique in the detection of impact damage in composites and corrosion in metals.

1. Introduction

A strong demand for high quality of technical installations and responsible parts in high-tech industries, such as military, aero space, automotive, ship building and power production, continues to stimulate development of novel and combined techniques of nondestructive testing (NDT). For the last decades, the so-called "big five" NDT methods have been: X-ray, ultrasonic, eddy current, magnetic particle and liquid penetrants [1]. Presently, thermal nondestructive testing (TNDT) is often included in this list as an indispensable method due to the following [2, 3]: 1) TNDT can be applied to all types of solids even if particular materials pose a certain challenge for efficient inspection, 2) TNDT is very fast, thus being considered as a screening technique operating in combination with others, 3) TNDT is illustrative, and human being's skills in interpreting 'images' can be useful in treatment of infrared (IR) images, and 4) TNDT is safe and often portable that is important when conducting NDT in public places.

Tomsk Polytechnic University has long traditions in TNDT. In the Soviet period, the research emphasis was done on development of specialized NDT hardware based on Russian-made components, because foreign IR imagers used to be practically unavailable in the former USSR in that period of time. Since the 1990s, the research focus was shifted onto software, both for modeling and data processing. Two major software packages have been developed and continuously modernized for the last years: ThermoCalc-6L and its modifications (simulation of TNDT problems) and ThermoFit Pro (processing experimental and synthetic IR image sequences).

Presently, certain efforts are being inputted in the development of a portable unit for performing active thermal NDT in both laboratory and outdoor conditions.

First of all, the research target is inspection of composite materials, mainly used in aviation, such as carbon and glass fiber reinforced plastics (CFRP and GFRP) and honeycomb structures. Another research area is the detection of hidden corrosion in above-ground steel tanks, piping, tanks, etc. In both areas, the above-mentioned advantages of TNDT are fully realized.

Journal of Physics: Conference Series **671** (2016) 012011 doi:10.1088/1742-6596/671/1/012011

2. TNDT setups

The concept is to develop both experimental laboratory systems and portable units for in-field inspection. A flexible laboratory system for active TNDT is shown in figure 1a in its implementation to corrosion detection in steel containers intended for keeping low-active radioactive wastes. A cylindrical test object is rotated being stopped for about 20 seconds needed for inspection in each position. An IR imager captures from 20 to 100 IR thermograms which are processed, and results are exhibited as raw or binary images. Images taken at each check position are composed to present circular evolution of temperature (figure 1b). Several IR imagers can be used in this system (FLIR P-65, Nec Avio-9100, FLIR-A325 SC and FLIR SC 7700M IR cameras are currently used at Tomsk Polytechnic University). There are a couple of heat sources which are used in the system, for instance: 1) Xenon flash lamps, and 2) halogen lamps (shown in figure 1a).

a)

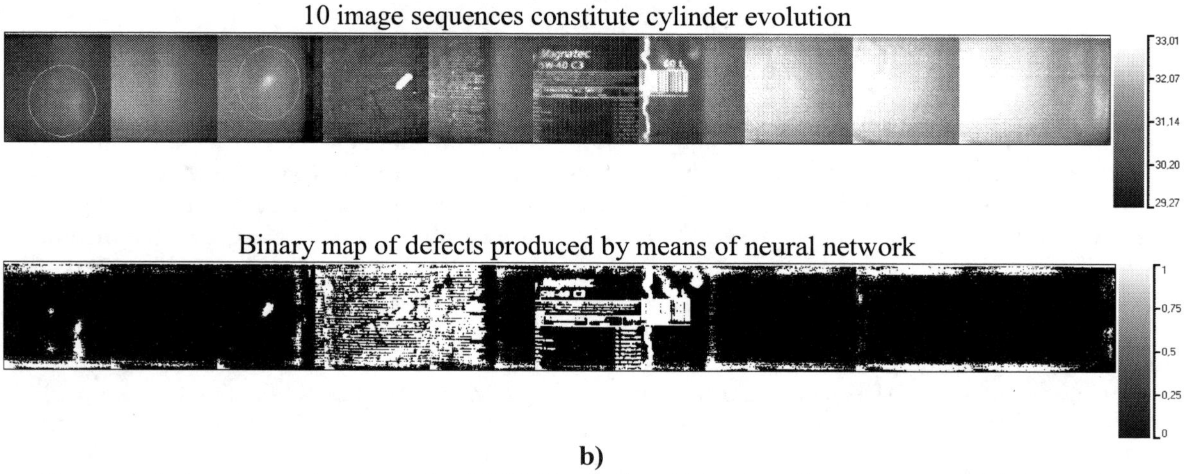

b)

Figure 1. Laboratory setup for active TNDT including NEC Avio IR camera and 2 halogen lamp heaters (a) and example of test results (b).

A portable IR thermographic unit implements the same principle of active TNDT but allows out-door operation by one or two thermographers. The current unit is presented in figure 2a. Four halogen lamps 500 W each illuminate an object to be tested, such an airplane panel, for a controllable time (typically, 3-15 s). Since the heat source is turned off, its residual thermal radiation is cut off by means of a 4-lamella mechanical shutter thus preventing reflected radiation that is a notorious source of noise in one-sided TNDT.

The illustrations of using the portable unit in NDT are shown in figure 2b, c. When checking airplane panels made of honeycombs, there is an inspection problem of distinguishing between water and epoxy glue hidden in honeycomb cells. Water represents a serious exploitation problem because its presence leads to honeycombs damage [4]. However, water surface thermal 'footprints' can be easily confused with those of epoxy glue due to close amplitudes of differential temperature signals

ΔT (figure 2b, left). However, specialized data treatment based on the analysis of the $T(\tau)\cdot\tau^n$ synthetic temperature function as a novel processing algorithm proposed in [5] allows fairly good discrimination between two defective sites (figure 2b, right).

Another example of using TNDT in the inspection of impact damage in a 4 mm-thick CFRP sample (low-velocity 59 J impact) is presented in figure 2c. The raw temperature image bears no detailed information about structure of the induced defect. Better results have been obtained by applying the algorithms of dynamic thermal tomography proposed at Tomsk Polytechnic University in the 1980s, see more details the recent paper [6]). The tomograms in figure 2c exhibit sample structure within three different layers and clearly show that cracks (delaminations) induced by the impact are minor on the impacted surface but become larger with material depth, This is in accordance with the known fact that impact damage can be invisible on the airplane outer surface but cause significant damage deeper in the composite, often closely to the panel rear surface.

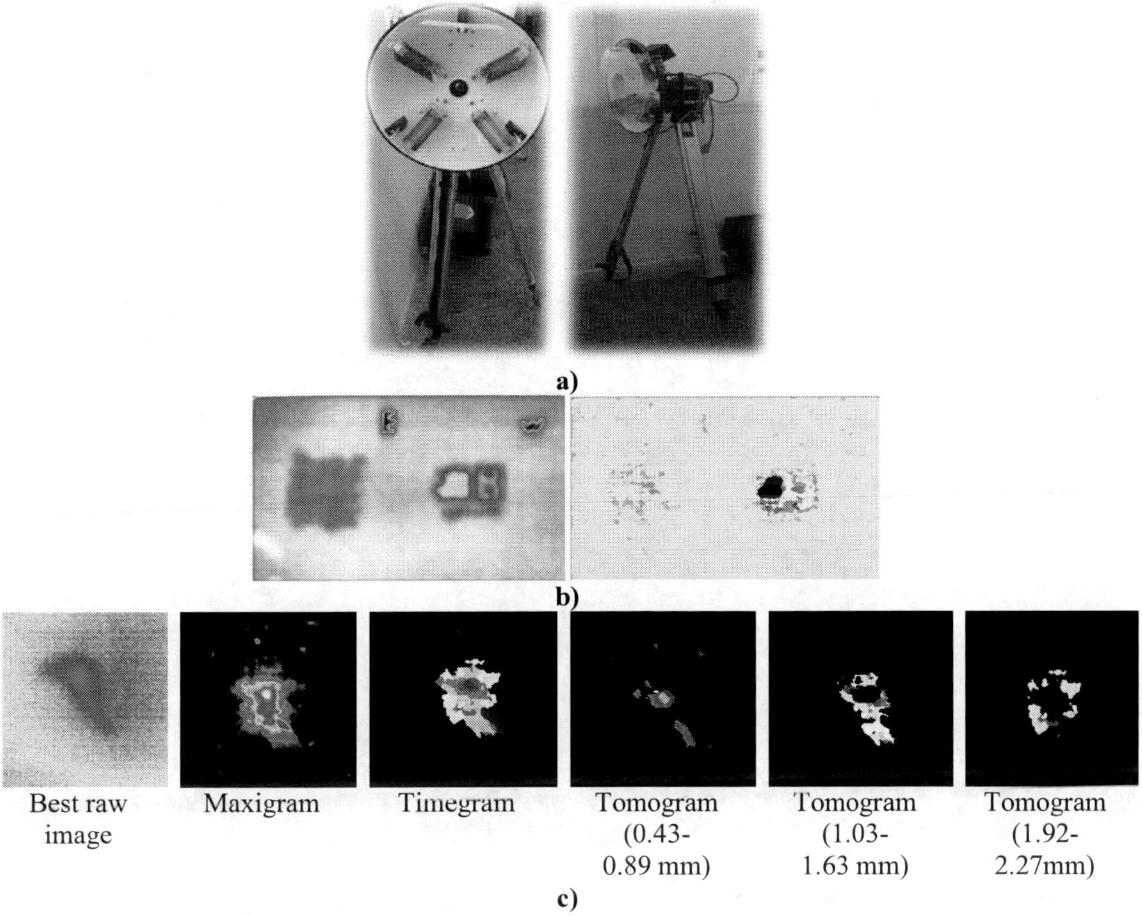

Figure 2. Portable IR thermographic NDT unit and examples of test results: a – unit photo; b – distinguishing between hidden water and epoxy glue;in honeycomb panel (left – best raw image, right – result of processing by using the analysis of $T\cdot\tau^{0.43}$ synthetic function [5]); c – thermal tomography of impact damage in 4 mm-thick CFRP sample (one-sided inspection, 59 J impact energy).

Some technical characteristics of the portable unit from figure 2a are given in table 1.

Journal of Physics: Conference Series **671** (2016) 012011 doi:10.1088/1742-6596/671/1/012011

Table 1. Technical characteristics of portable IR thermographic NDT unit (figure 2a).

Parameter	Value
Test objects and their thickness	Steel, titanium, aluminum (up to 6 mm*) CFRP and other composites (up to 4 mm*)
Test area	0.04 m^2
Test productivity	4.7 $m^2 \cdot hr^{-1}$
Minimum detectable materials loss in steel **	10 %
Minimum lateral size of detectable defects in composites **	10 mm
Temperature sensitivity	0.02-0.06 °C ***
Power supply	220 V AC
Mass	5–10 kg

* Depending on a used heat source

** Depending on sample thickness and some test parameters

***Depending on a used IR imager

Another concept of a portable IR thermographic NDT unit is shown in figure 3a. It differs from that in figure 2a by the use of two LED heat sources, 500 W each. Such heaters only recently started to be used in TNDT. Their advantage in comparison with lamps is the monochromatic operation mode involving radiation in the visual band from 0.3 to 0.7 μm. Since IR imagers normally operate at wavelengths from 3 to 5 or from 7–13 μm, LED heaters produce no reflected radiation sensed by IR imagers. Unfortunately, because of high reflectivity of many materials at these wavelengths, the LED radiation may be inefficient to warm a test sample up to a required temperature. The corresponding illustration is given in figure 3b in the detection of hidden corrosion 25 % material loss in a 1 mm-thick steel sample) by using both halogen lamps and LED heaters. This example shows that LEDs provide worse detection results although their power is much less than in the case of 30 kW halogen lamps.

a) b)

Figure 3. Concept of a portable IR thermographic unit using LED heat source (a) and corrosion detection in 1 mm-thick steel sample by applying 30 kW halogen lamps (left) and 2×500 W LEDs (right) heat sources; the sample contained 5 sites of rear-surface 25 % material loss of different size.

3. Determining TNDT detection limits

It is very typical for end-users to ask about detection limits of NDT methods and particular equipment. To answer this question is particularly difficult in TNDT because test results depend on many factors, such as defect size by all coordinates, defect depth, type of material and surface conditions, type of a used heat source, wavelength band, IR imager temperature resolution, etc.

The corresponding approach allowing the evaluation of minimum detected defects has been proposed by combining results of modeling and experimental estimates of the noise adhering to each particular test case.

Modeling is done by using a software package developed at Tomsk Polytechnic University and allowing calculation of surface temperature signals in 1D, 2D and 3D cases. The Layer-3 Analytic, ThermoCalc-2D and ThermoCalc-36L programs with their many modifications are simple and robust enabling calculation of temperature distributions over hidden defects. The main calculated parameters are differential temperature signals ΔT_m and temperature contrasts C_m, as well as times of their appearance τ_m.

A detailed description of modeling is beyond the scope of this paper, see the handbook [2]. Let us consider the detection of delaminations in CFRP as an example. The temperature signals that can be achieved in this test case depend on defect depth and size and duration of heating, as demonstrated by figure 4, where the curve family has been calculated by using the ThermoCalc-36L software for solving the corresponding 3D heat conduction model (see details in [3]). The contrast $C=\Delta T/T$ is a dimensionless parameter which shows a relative temperature signal in regard to sample excess temperature T. Each material, while being heated, is characterized by the temperature noise contrast conditioned by uneven heating and surface clutter. It is important that such noise increases proportionally with excess temperature, therefore, it can be expressed in percent to characterize each particular material. For example, it has been experimentally found that CFRP composite has noise contrast from 1-2 %, hence, its level is often assumed to be 5 % (at the level of 2-3 standard deviations). If to draw a horizontal line in figure 4 corresponding to 5 %, its crossings with family curves will specify minimum detectable defects. It follows from figure 4 that, in one-sided TNDT of a 5 mm-thick CFRP samples, delaminations of various lateral size and thickness can be detected to the depths of 3.3-4 mm that matches well earlier published experimental results [2].

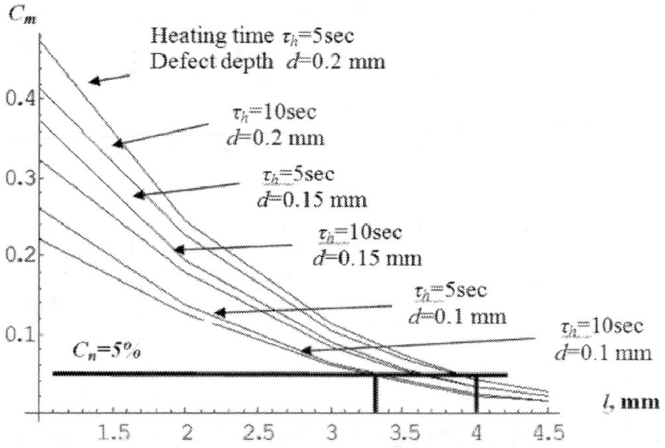

Figure 4. Determining defect detection limit by depth in TNDT of 5 mm-thick CFRP.

The same approach applied to the detection of corrosion in steel is shown in figure 5. The thickness of steel samples heated with a thermal pulse was 1, 5 and 10 mm and the size of a corroded area was 10×10 m. The noise level was assumed 10 % that corresponds to many painted metallic surface. It is seen that, in this case, the detection limit is 10-40% by material loss depending on sample thickness.

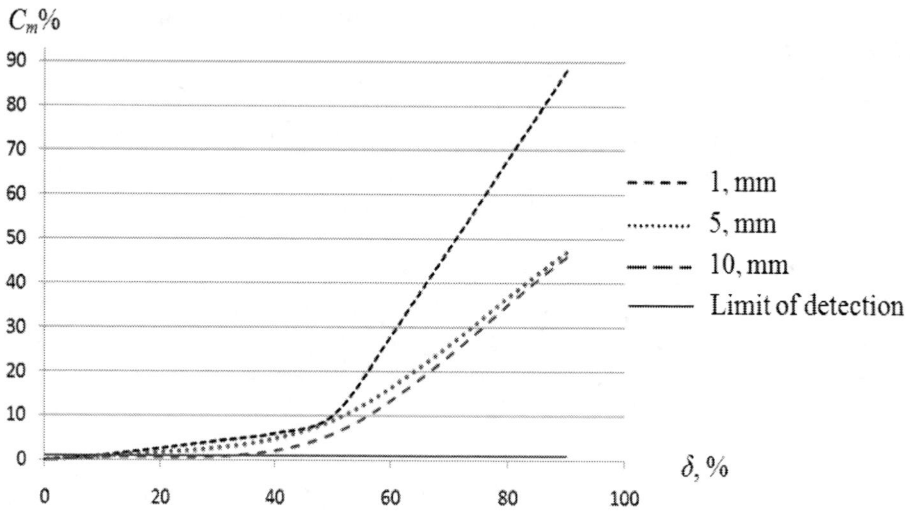

Figure 5. Determining corrosion detection limit by material loss in steel.

4. Conclusion

A universal character of thermal/infrared NDT allows developing universal inspection units which may be used for testing a wide range of materials and objects. This feature, along with illustrativeness and high productivity of the thermal methodб makes it attractive as a screening inspection technique which can be used solely or in combination with other NDT methods. Composites constitute the most successful application area for thermal NDT, particularly, in aviation where large areas subject to NDT and safety requirements make difficult using other NDT methods. Another perspective application area is the detection of corrosion in metallic shells. The thermal/infrared NDT research being conducted at Tomsk Polytechnic University involves the development of both software and hardware, as well as inspection guidelines and approaches for determining NDT limits.

References

[1] Carpenter J L, Stuhrke W F (1976) NDE: An effective approach to improved reliability and safety. A technology survey *NASA Technical Reports Server* **NASA-CR-134963** 163
[2] Vavilov V P (2006) *Thermal/Infrared Testing NDT Handbook* Series **5** (Moscow, Russia Spektrum Publisher)
[3] Vavilov V P (2014) Modeling thermal NDT problems *International Journal of Heat and Mass Transfer* **72** 75–86
[4] Nesteruk D A 2012 Infrared thermographic detection of water in aviation honeycomb structures (Lambert Academic Publishing)
[5] Vavilov V P, Pawar S (2015) A novel approach for one-sided thermal nondestructive testing of composites by using infrared thermography *Polymer testing* **44** 224–233
[6] Vavilov V P (2015) Dynamic thermal tomography: Recent improvements and applications. *NDT & E International* **71** 23–32

The X-ray Beam Passage through the Collimator Made of Different Materials: Numerical Simulation

S G Stuchebrov[1], I A Miloichikova[2] and I B Danilova[3]

[1] Senior Researcher, Tomsk Polytechnic University, Tomsk, Russia
[2] Engineer, Tomsk Polytechnic University, Tomsk, Russia
[3] Bachelor student, Tomsk Polytechnic University, Tomsk, Russia
E-mail: stuchebrov@tpu.ru

Abstract. The X-ray beam application grows in the research investigations, in the medical diagnosis and treatment, in the industry. In this paper the theoretical model of the actual pulsed X-ray generator RAP-160-5 beam developed in the program "Computer Laboratory (PCLab)" is shown. The simulation data of the X-ray beam profile and shape collimated by different materials (gypsum, corund-zirconia ceramic, lead) are illustrated.

1. Introduction

The X-ray sources have a wide range of application, for example for the radiographic analysis in the nondestructive and medical examinations [1–6]. In this regard, the possibility to manage of the X-ray beam parameters in accordance with specific purposes is an actual task. As it is known that the numerical simulation allows estimating the X-ray beam parameters, making it easier and faster to obtain these characteristics than the actual measurements. Nowadays, the theoretical model development of the X-ray beam is a contemporary task.

In this research the theoretical analysis of the pulsed X-ray generator RAP-160-5 beam was carried out. The first objective of this investigation was to create the actual X-ray beam model correspond to the experimental result obtained previously in the paper [7]. The models of the X-ray beam shape modulation were developed in the program "Computer Laboratory (PCLab)". The next objective of this research was to analyze the X-ray beam collimation by different materials (gypsum, corund-zirconia ceramic, lead).

2. Materials and methods

2.1. Emitting source

In this research the following parameters of the pulsed X-ray generator RAP-160-5 were used as the emitting source: the anode material is copper; the focal spot size is 1.2 ×1.2 mm; the angular divergence of a beam is 40°; the anode voltage varies from 40 to 160 kV; the anode current varies from 0.4 to 5 mA [8].

2.2. Simulation program

The program "Computer laboratory (PCLab)" version 9.5 was used for the generator RAP-160-5 X-ray beam model creation. Numerical simulation is carried out by applying the Monte Carlo method.

The software package allows calculating the propagation process of photons, electrons, protons and positrons in substance with specified characteristics [9].

2.3. Experiment geometry

In this investigation the simulation results of the copper anode X-ray continuous spectrum was used. The normal plane disc (diameter – 1.2 mm) monoenergetic X-ray source with energy of 70 kV, corresponding to the actual pulsed X-ray generator RAP-160-5 beam, was used in the numerical simulation. The source was located in front of the plexiglas output window (thickness – 2 mm; diameter – 40 mm).

In the first part of this research the geometry of experiment is correspond to the experimental setup in the paper [7]. At the distance from the output window equal to 35 mm the aluminum filter was installed (thickness – 1.1 mm) to reduce the contribution of soft X-rays of the spectrum.

In the second part of this research the collimated X-ray beam shape was analyzed. In the simulation with a collimated X-ray beam the output window was overlapped by the plates from different materials (collimator channel length – 5 mm and 30 mm) with a cylindrical hole (taper diameter – 30 mm). The collimator materials are gypsum, corund-zirconia ceramic and lead. The beam shape analysis was carried out in the air.

The figure 1 (a, b) illustrates the calculation geometry with noncollimated and collimated X-ray beams, correspondingly.

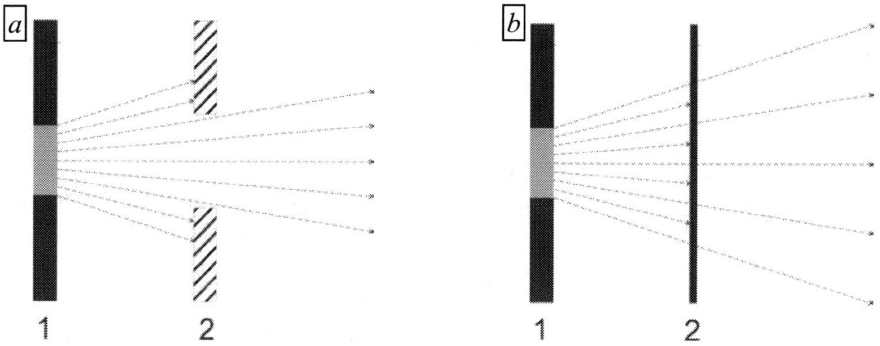

Figure 1. The simulation geometry:
a – noncollimated X-ray beam (1 – output window; 2 – an aluminum filter);
b – collimated X-ray beam (1 – output window; 2 – collimator).

3. Results and discussions

In the first part of this investigation the experimental and simulation results of the X-ray beam profile and shape were compared [7]. The figure 2 illustrates the measurement and simulation X-ray beam profile at the 30 cm distance from the X-ray tube focus.

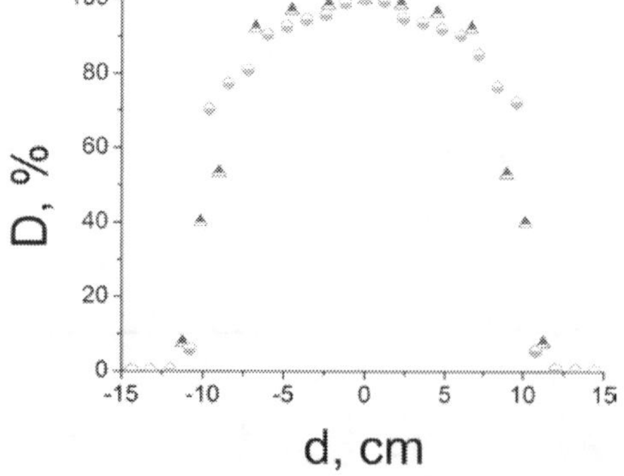

Figure 2. The X-ray beam profile at the 30 cm distance from the X-ray tube focus (the energy was 70 kV): ▲ – experimental results; ● – simulation results.

The figure 2 shows that the model of the pulsed X-ray generator RAP-160-5 beam and the experimental results are in a good agreement. These points to the fact that this model can be used for the real X-ray beams analyze. The insignificant divergences between the obtained X-ray beam profiles in the figure 2 can be explained by the following factors: the size of the plane-parallel ionization chamber type 23342 effective area; an inaccuracy of the experiment geometry reproduction; the ideal numerical simulation area.

In the second part of this investigation the X-ray beam collimation by different materials are analyzed. In the figure 3 (a, b, c) the simulation data of the pulsed X-ray generator RAP-160-5 beam shape with the gypsum, corund-zirconia ceramic and lead collimator (channel length – 5 mm, taper diameter – 30 mm) at 30 cm distance from the X-ray tube focus are shown, correspondingly. The dose results were normalized to the maximum simulation dose in the layer.

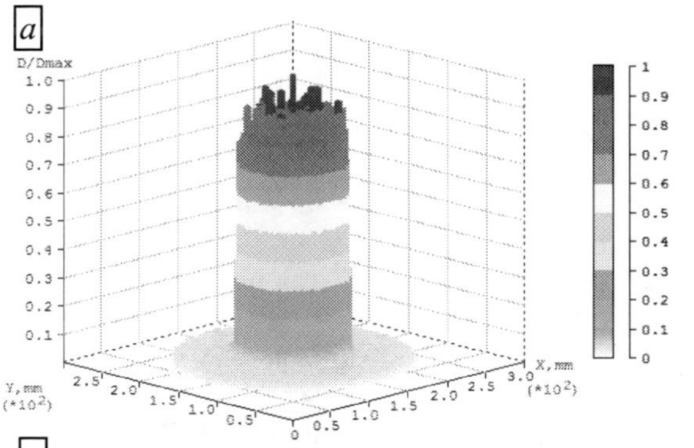

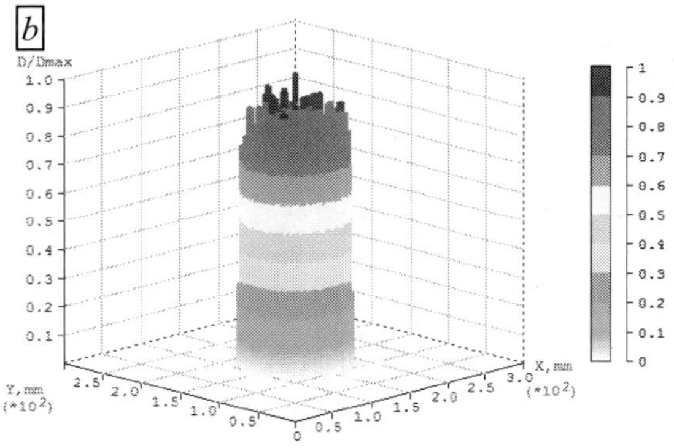

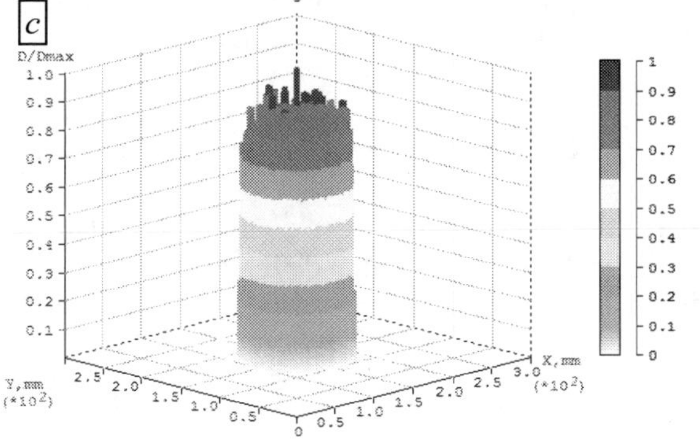

Figure 3. The collimated X-ray beam shape at the 30 cm distance from the X-ray tube focus (collimator channel length – 5 mm, taper diameter – 30 mm, the X-ray energy – 70 kV): a – gypsum collimator; b – corund-zirconia ceramic collimator; c – lead collimator.

In the figure 4 (a, b, c) the simulation data of the pulsed X-ray generator RAP-160-5 beam shape with the gypsum, corund-zirconia ceramic and lead collimator (channel length – 30 mm, taper diameter – 30 mm) at 30 cm distance from the X-ray tube focus are shown, correspondingly. The dose results were normalized to the maximum simulation dose in the layer.

The simulation model (figures 3, 4) shows that the gypsum collimator can be used with the collimator channel length is equal to or greater than 30 mm, the corund-zirconia ceramic and lead collimators can be used with the collimator channel length is equal to or greater than 5 mm, this material thickness is enough to absorb the X-ray beam with the 70 kV energy. The obtained results illustrates that the dose in the layer has the regularly spaced distribution in the collimation area.

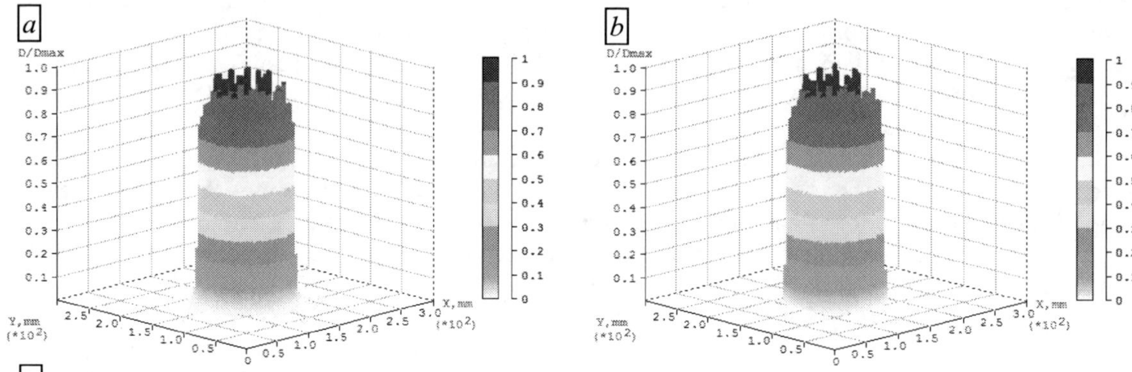

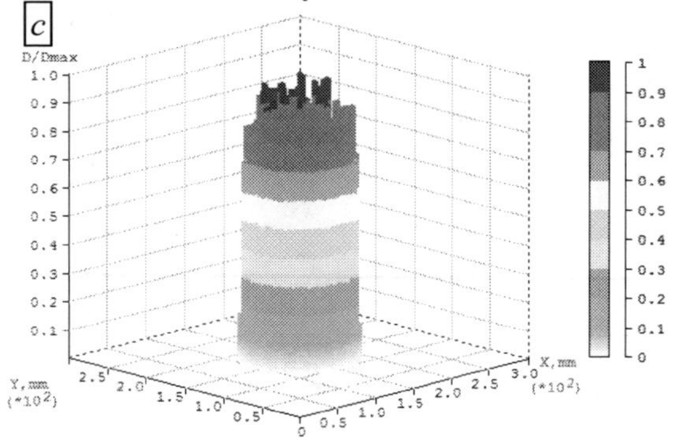

Figure 4. The collimated X-ray beam shape at the 30 cm distance from the X-ray tube focus (collimator channel length – 30 mm, taper diameter – 30 mm, the X-ray energy – 70 kV): a – gypsum collimator; b – corund-zirconia ceramic collimator; c – lead collimator.

4. Summary

In this paper the theoretical model of the actual pulsed X-ray generator RAP-160-5 beam was developed in the simulation program "Computer laboratory (PCLab)". The calculation data show the suitability of this program for the real X-ray beams analyzing. The obtained results allow modulating the X-ray beam profile and shape for the specific practical task using the different type of the collimation devices. The next step of this research is an experimental approbation of the obtained simulation data of the collimated X-ray beam.

Acknowledgements

This work was financially supported by the Ministry of Education and Science of the Russian Federation in part of the science program (N 5.1485.2015).

References

[1] Chair C-M Ma, Coffey C W, DeWerd L A, Liu C, Nath R, Seltzer S M and Seuntjens J P (2001) AAPM protocol for 40–300 kV X-ray beam dosimetry in radiotherapy and radiobiology *Med. Phys.* **28** (6) 868–893

[2] Cherepennikov Y and Gogolev A (2014) Device for X-ray spectral absorption analysis with use of acoustic monochromator *J. Phys.: Conf. Ser.* **517**(1) Article number 012037

[3] Information on http://www.ndt.net/article/aero2013/content/papers/14_Lifton_Rev2.pdf

[4] Cherepennikov Y and Gogolev A (2012) Method to reduce radiation exposure in the medical X-ray diagnostics *Proceedings - 2012 7th International Forum on Strategic Technology, IFOST 2012* Article number 6357737

[5] Information on http://www.icdd.com/resources/axa/vol49/v49_02.pdf

[6] Gogolev A, Stuchebrov S, Wagner A, Cherepennikov Yu and Potylitsyn A (201)2 Acoustic "pumping effect" for quartz monochromators *J. Phys.: Conf. Ser.* **357**(1) Article number 012031

[7] Miloichikova I, Stuchebrov S, Krasnykh A and Wagner A (2015) Dose rate spatial distribution produced by the pulsed X-ray source in the radiographic examination *Advanced Materials Research* **1084** 598–602
[8] Operations manual 2008 *Portable X-ray device for the industrial utilization* (Tomsk: Photon) [in Russian]
[9] Bespalov V I 2015 *Computer laboratory (Version 9.5)* (TPU: Tomsk) [in Russian]

The Relationship Between the Parameters of the Electric and the Acoustic Signal with the Destruction of Concrete Under Cyclic Freeze-Thaw

T Fursa[1], D Dann[2], A Surzhikov[3], M Petrov[4]

[1]Leading researcher, NDT Institute, Tomsk Polytechnic University, Russia
[2]Reseacher, NDT Institute, Tomsk Polytechnic University, Russia
[3]Professor, NDT Institute, Tomsk Polytechnic University, Russia
[4]Undergraduate, NDT institute, Tomsk Polytechnic University, Russia
E-mail: fursa@tpu.ru

Abstract. The paper presents the research results of the effect of formation crack process on the parameters of the electric and acoustic response to impact excitation. The physical basis of mechanoelectric transformations is described. It was found that with increasing number of freeze-thaw cycles observed increase of the attenuation coefficient of energy of the electric and acoustic response by a linear relationship. Differences in the dynamics of change of attenuation coefficient of energy of the electric and acoustic response associated with differences in formation and registration of electric and acoustic response.

1. Introduction

Building structures made of concrete are often exploited in significant seasonal temperature fluctuations, which lead to cracking. Concrete damage caused by freeze-thawing is the main problem in cold climate. These damages caused by freeze-thawing can be of two different forms: internal destruction [1] and surface cracking [2]. Therefore, timely detection of cracks in concrete products and tracking the dynamics of their development is an important task to prevent destruction of concrete structures. Many new methods of concrete products control developed to solve this problem. One of these methods is based on the use of the generation of electromagnetic fields by heterogeneous non-metallic materials during mechanical excitation.

By now gained a large amount of experimental results on the study of the generation of alternating electric fields by mechanical excitation of heterogeneous non-metallic materials. Much of the research is devoted to studying the mechanisms of appearance of alternating electromagnetic fields in the quasi-static mechanical excitation. Most authors attribute the appearance of alternating electromagnetic fields with the processes of cracking. Labors [3–5] are devoted to research of electromagnetic emission in the process of formation cracks in different materials (such as rocks, grout, composite materials, ice and others). Along with the research of occurrence of electromagnetic fields by the quasi-static mechanical excitation over a number of years the research of generating the alternating electromagnetic fields by the pulse mechanical excitation of heterogeneous non-metallic materials is conducted [6, 7].

On the basis of the use of this phenomenon the non-destructive testing of structural and mechanical properties of concrete is being developed.

To identify the informative parameters of the electric response carrying information about the cracking evolution process in the concrete the information about the basic laws of mechanoelectric transformations and their physical basis is needed.

2. Physical bases of the method

The method based on mechanoelectrical transformations can be used to solve the problem of defect detection. The principle of this method is that the object under research is subjected to the elastic shock excitation which leads to propagation of the acoustic waves in the sample. The electric response is caused by the deformation and shift of the elastic wave due to the pulsed mechanical excitation of the signal sources in material. The previous research has established that piezoelectric inclusions play the crucial role in the mechanoelectrical transformation in concrete [8]. Piezoelectric quartz is contained in sand and gravel which were used to manufacture the samples. Under the mechanical stresses caused by acoustic waves, charge formation occurs on the faces of the piezoelectric quartz owing to direct piezoelectric effect. As a result, an external electric field occurs and it is detected by the signal receiver placed near the test object.

Piezoelectric source–dipole charge is proportional to the mechanical stress value induced by acoustic longitudinal at the location point of the piezoelectric source and the magnitude of the piezoelectric quartz module. The number of sources is determined by the number of grains of sand and coarse aggregate that have piezoelectric quartz in their structures. The aggregate grains are evenly distributed over the whole volume of the concrete sample. Piezoelectric axes of quartz inclusions have different direction relatively to the receiving point and the direction of the longitudinal waves formed in the sample during its shock excitation. The total electric field at the location point of receiving electrode is a result of vector addition of the fields from each source.

Current starts to flow through the input impedance of the measuring circuit as a result of the free charge carriers occurrence on the receiving electrode surface which were induced by an electric field (figure1).

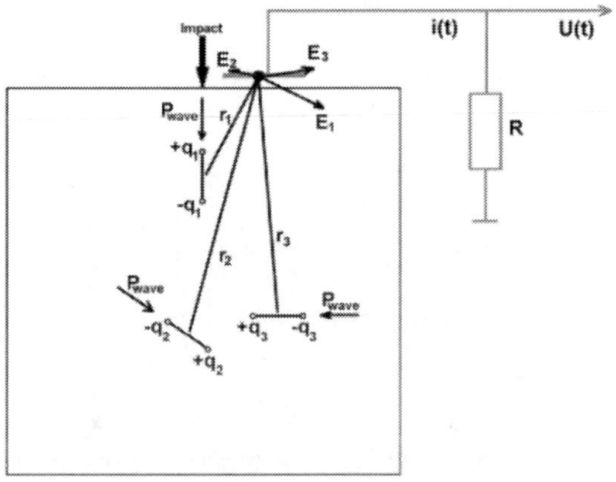

Figure 1. Measuring circuit.

An electric receiver is a metal plate located close to the sample surface within the range of the electric field generated by internal sources and grounded through the input resistance (*R*).

The charge (*Q*), induced on the surface of a conductor (electrical receiver) placed into electric field is:

$$Q = DS_d = \varepsilon_0 \varepsilon E S_d \quad (1),$$

where S_d – the electric receiver area; *D* – electric displacement vector.

Then the measured voltage from a single source is determined by:

$$U(t) = \varepsilon_0 \varepsilon \frac{\partial E(t)}{\partial t} R S_d \quad (2),$$

where R – input impedance.

Consequently, the value of the recorded electrical signal is determined by the change of the electric field strength in the area of the receiver, receiver area size and the value of the input resistance.

Using classical electrodynamics and mechanics relations the model of mechanoelectrical transformations in heterogeneous materials containing piezoelectric inclusions was developed [9]. Within the model the electric field rate change at acoustic excitation of piezoelectric sources was calculated:

$$E'(t) = \frac{dSMl_P}{4\pi\varepsilon_0\varepsilon L}\frac{V_y(t)}{r^3(t)}\left(\frac{3h^2}{r^2} - 1\right) \tag{3}$$

Inserting (4) into (3) we obtain that the value of the measured voltage from a single source is given by:

$$U(t) = \frac{dSMl_P}{4\pi L}\frac{V_y(t)}{r^3(t)}\left(\frac{3h^2}{r^2} - 1\right)\cdot RS_d \tag{4}$$

where S_d is the area of the measuring electrode; M is the elastic modulus; L is the model size in the excitation direction; $V_y(t)$ is the displacement rate in the excitation direction; h is the depth of the piezoelectric source position; r is the distance from the source to the receiving electrode; l is the thickness of the piezoelectric quartz crystals; d is the piezoelectric modulus of quartz; and S is the sample cross-sectional area; and R is the input impedance.

Therefore, the parameters of the electrical response and elastic waves are interrelated.

The presence of large amount of piezoelectric sources with different directions of the electric axes can reliably reflect the transformation structure of the acoustic waves parameters within the interaction with different configurations defects.

3. Technique of research

The experimental research was carried out at a laboratory hardware–software complex that performs pulse mechanical excitation of samples, and records electrical response signals. Figure 2 shows a photo of the measurement system used.

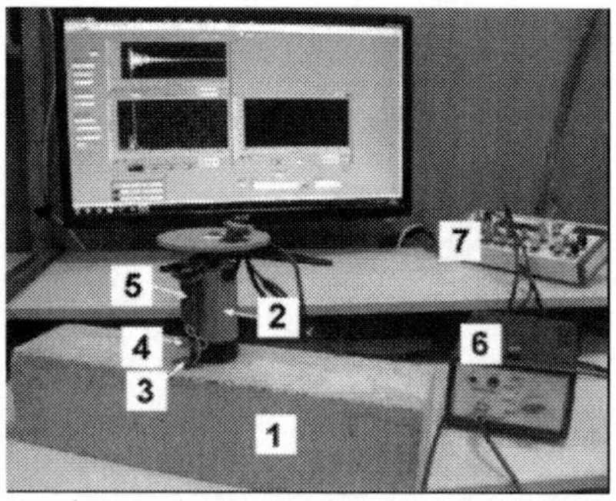

Figure 2. The measuring system photo: 1 – sample; 2 – electromechanical impact device; 3 – measuring electrode; 4 – compensation electrode; 5 – differential amplifier; 6 – power supply; 7 – NI board terminal block.

The pulsed mechanical excitation of samples (1) is performed with electromechanical impact device (2) with a normalized impact force. The electromagnet was used as the impact device. The differential electrical sensor was used for the measurement of electric response. Electrical measuring sensor (3) is placed at the distance of 2 mm from the sample surface, thus conditions for a non-contact measurement are created. The compensating receiver (4) is placed at the height of 30 mm from the measuring receiver. Then the signals come to the input of differential amplifier (5), are subtracted and amplified. The signals from the differential electrical sensor were registered using multifunction input/output board “NI PCI 6253” installed into the PC. The signal recording is performed with a special program in LabVIEW. The measurement technique of the electric response to mechanical excitation is described in more detail in [10].

Along with electrical signal measurement, we measured the acoustic signal using a standard Olympus V1011 piezoelectric transducer.

4. Experimental research and discussion

For this research the products made from heavy concrete sized of 100×100×300mm were used.

The samples manufacturing process complies with GOST 7473-2010. The mass ratio of cement / sand / coarse aggregate was 1:2:4, with a maximum aggregate size of 20 mm. The water-cement ratio was 0.6.

Studies of the influence of cyclic freeze-thaw on the parameters of electric and acoustic response were conducted. To accelerate of structure degradation at cyclic freeze-thawing, each sample was penetrated by water. Then cyclic freeze-thawing was performed. Freezing was carried out in a climatic chamber at –40 °C, thawing was carried out at 20 °C (± 5) °C and humidity of 95%.

After every fourth cycle of freeze-thaw the electric and acoustic response from concrete sample were measured. Twelve measurements of both responses at each stage of cyclic freeze-thawing were conducted. Responses were measured from four different sides of the sample. The sampling frequency of measurement was 1 MHz.

Along with electrical and acoustic signals measurement the samples surfaces were photographed for visual tracking of the surface cracking processes.

Figure 3 shows a photograph of the end surface of the concrete sample after 32 freeze-thaw cycles.

Figure 3. Photo of the sample surface.

As seen in figure 3 during the process of cyclic freezing and thawing on the sample surface formed a net of cracks.

Cracks have different directions and complicated configuration.

The carried out studies showed that during the cyclic freeze-thawing of concrete samples the temporal characteristics of the electric response change. Figure 4 shows typical electric responses of a concrete sample during cyclic freeze-thawing process.

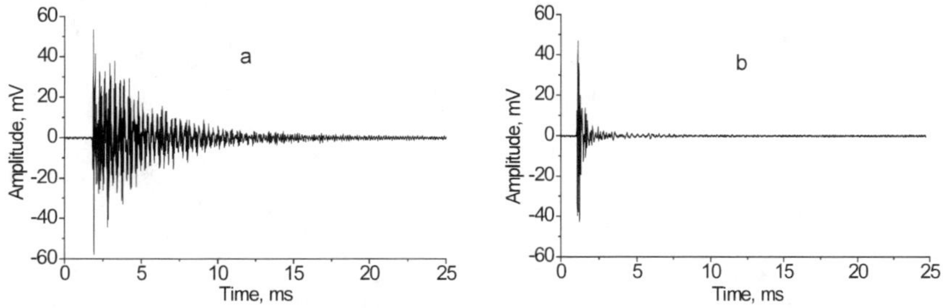

Figure 4. Changes in the electric responses of a concrete sample during cyclic freeze-thawing: a – the response of the sample in the initial state; b – the response of the same sample after 32 freeze-thawing cycles.

Cracks in the concrete sample start growing during the cyclic freeze-thawing. Scattering of elastic waves across cracks reflects their attenuation in time.

To study the behavior of the attenuation coefficient of the electric response energy from the concrete samples depending on the number of freeze-thawing cycles, the time-frequency analysis is used as described in [11, 12]. This technique allows tracking the signal energy attenuation as the time function in any selected range of frequencies. The experimental data were processed using the program routine developed in a LabView software.

Figure 5 shows the dependence of the electric response energy attenuation coefficient from the number of cycles.

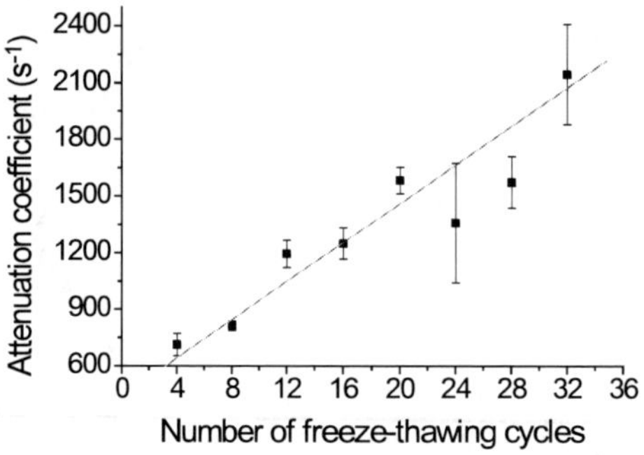

Figure 5. Dependence of the attenuation coefficient on the number of freeze-thawing cycles.

The dependencies of the attenuation coefficient (β) on the number of freeze-thaw cycles (n) are well approximated by a linear dependence of the form: $\beta = 437 + 51n$ with a correlation coefficient of 0.96.

In a similar way, the acoustic signals were processed from the samples. Figure 6 shows the dependence of the attenuation coefficient of the acoustic signal energy on the number of freeze-thawing cycles.

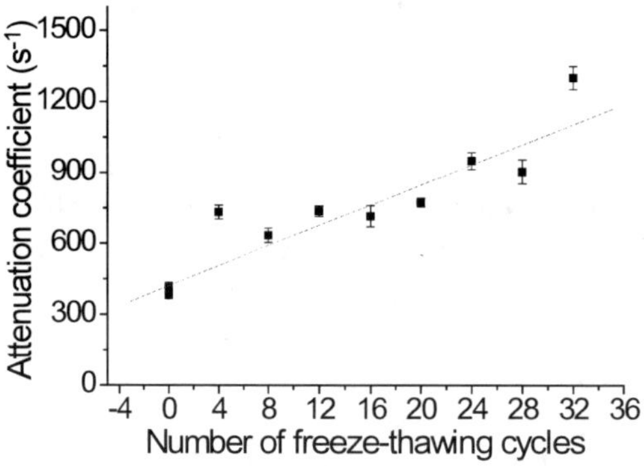

Figure 6. Dependence of the attenuation coefficient of the acoustic signal energy on the number of cycles.

As can be seen from a comparison of figures 5 and 6 dependencies have different tilt angle. That is, the increase in the attenuation coefficient of energy of electric response during the process of cyclic freeze-thaw is more intense compared to the acoustic response.

5. Conclusions

The purpose of research presented in this paper was the studying the laws of transformation of parameters of electric and acoustic response during the process of formation cracks caused by freeze-thaw cycling.

It is found that with increasing number of freeze-thaw cycles is increased attenuation coefficient of energy of both electrical and acoustic response. Studies have shown that with increasing the number of freeze-thaw cycles is observed more intensive increase of the attenuation coefficient of energy of electric response as compared with the acoustic response.

Electric response reliably reflects the interaction of longitudinal waves with cracks having different orientation due to the large number of piezo sources that have different directions of electric axes. In contrast, piezoelectric receiver detects the waves propagated in a perpendicular direction relatively to the receiver plane. As a result, the acoustic signal is not sensitive to cracks, which are perpendicular to the sensor plane.

The increase of attenuation coefficient of energy of electric and acoustic response can serve as diagnostic features for evaluation of violations in concrete during the process of cyclic freeze-thaw.

The work is done under the State task "Science".

References

[1] Setzer M J (2001) Micro-ice-lens formation in porous solid *J Colloid Interface Sci.* **243**(1) 193–201

[2] Sun Z and Scherer G W (2010) Effect of air voids on salt scaling and internal freezing *Cem. Concr. Res.* **40**(2) 260–270

[3] Kyriazopoulos A, Anastasiadis C, Triantis D, Brown C J (2011) Non-destructive evaluation of cement-based materials from pressure-stimulated electrical emission – Preliminary results *Construction and Building Materials* **25**(4) 1980–1990

[4] FifoltI D A, Petrenko V F, Schuls E M (1993) Preliminary study of electromagnetic emissions from cracks in ice *Philosopgical Magazine B.* **67** 289–299

[5] Koktavy P (2009) Experimental study of electromagnetic emission signals generated by crack generation in composite materials *Meas. Sci. Technol.* **20**(1) 015704

[6] Surzhikov A P, and Fursa T V (2008) Mechanoelectrical transformations during elastic impact excitation of composite dielectric materials *Technical Physics.* **53**(4) 462–465

[7] Fursa T V, Savelev A V, Osipov K Yu (2003) Interrelation between electromagnetic response parameters and impact excitation characteristics in insulators *Technical Physics Journal* **48**(1) 1419–1423

[8] Fursa T V, Dann D D (2011) Mechanoelectrical transformations in heterogeneous materials with piezoelectric inclusions *Technical Physics* **56**(8) 1112–1117

[9] Fursa T V, Lyukshin B A and Utsyn G E (2013) Relation between the electric response and the characteristics of elastic waves under shock excitation of heterogeneous dielectric materials with piezoelectric inclusions *Technical Physics* **58**(2) 263–266

[10] Fursa T V, Osipov K Yu, Lyukshin B A and Utsyn G E 2014 The development of a method for crack-depth estimation in concrete by the electric response parameters to pulse mechanical excitation *Meas. Sci. Technol.* **25**(5) 055605

[11] Quiviger A, Payan C, Chaix J-F, Garnier V and Salin J (2012) Effect of the presence and size of a real macro-crack on diffuse ultrasound in concrete *NDT&E Int.* **45**(1) 128–132

[12] Becker J, Jacobs L J and Qu J (2003) Characterization of cement-based materials using diffuse ultrasound *J. Eng. Mech.* **129**(12) 1478–1484

Method of Noncontact Calibration of the Robotic Ultrasonic Tomograph

V N Borikov[1], O V Galtseva[2] and G A Filippov[3]

[1]Professor, NDT Institute, Tomsk Polytechnic University, Tomsk
[2]Associate Professor, NDT Institute, Tomsk Polytechnic University, Tomsk
[3]Postgraduate student, NDT Institute, Tomsk Polytechnic University, Tomsk
E-mail: piano@tpu.ru

Abstract. The method of calibration of robotic ultrasonic tomograph with the construction of the trajectory of movement of the robot-manipulator on the object of control by using 3D-scanner is described. This method can significantly accelerate the process of calibration of tomograph and prevent possible displacement of the object during calibration. The algorithm of transition from use of a contact method of calibration of the tomograph to noncontact calibration is offered. Experimental data of application of this algorithm show a positive result: the time of research of object considerably decreases. Results of researches prove the practical relevance of the presented work and high efficiency of application of robotic ultrasonic tomography for nondestructive testing of objects of different forms.

1. Introduction

It is difficult to imagine the modern testing without ultrasonic control methods. Ultrasonic inspection does not destroy and does not damage the studied sample; this is the main advantage of this method [1 – 3]. It is also necessary to mention the ability of control of the products from variety of materials, both metal and nonmetals. Besides, it is possible to note the high speed of research at the low cost and the low danger to the person (in comparison with X-ray defectoscopy) and the high mobility of the ultrasonic defectoscope.

A few years ago the ultrasonic detectors were primarily hand-held devices; the difficulties of obtaining of bulk pictures of defects are encountered during using them, the interpretation of the data was virtually impossible. Such devices were succeeded by the ultrasonic tomographs which can reconstruct a three-dimensional picture of defects. Disadvantages of such devices are considerable constraints due to the shape of the surface of the studied object.

The rapid development of additive technology places the new demands to ultrasonic defectoscopes, since the shapes of objects, obtained with the help of three-dimensional printing technology is usually very complex and nonlinear. One solution of this problem is development of device which capable of monitoring the surface of parts of complex shape. This system is designed on a base of Tomsk Open Laboratory for Material Inspection (NDT Institute, Tomsk Polytechnic University).

2. Experimental

Robotic ultrasonic tomograph (figure 1) consists of a six-axle manipulator, immersion tank and block of electronics.

The precise definition of the relative position of the controlled object and the manipulator is one of the main problems of such devices. Today this problem is solved by a contact method.

The operator of installation should touch to the angles or distinctions of the controlled object by the special gauge needle. This method is labor-consuming; it has the low accuracy of definition of provision of object and it also is interfaced to direct contact of needle with controlled sample; it isn't always admissible.

The following algorithm is used for construction of trajectory of the movement of robot - manipulator on object of control. A special tool with sharp tip is mounted on a robot-manipulator. This tool is brought to pre-selected points of the object to determine the coordinates of these points (including the length of the tool). These coordinates are entered to the software and compared with the chosen object points (as a rule, angles). This process is quite laborious occupation; the slightest mistake leads to the need to start measuring again.

It is proposed to solve the problem through the use of *3D* - scanner; it determines the coordinates of all points of the object at the scanning process. Development in the field of *3D* - laser tomography is described in detail in [4]. The method of using the system of tomographic analysis with *3D* - representation of the data is presented in [5 – 6].

Special attention should be paid to the fact that the scanner has the system of coordinates (*SC*) for determination of coordinates of points, which differs from the coordinate system of the robot.

Let *Oabc* – the coordinate system of the robot; *Ox'y'z'* – the coordinate system of the scanner; *Oxyz* – the abutment coordinate system, it is located at the same point as the coordinate system *Ox'y'z'*, but it is oriented as the coordinate system of the robot (figure 1).

Figure 1. Complex of ultrasonic inspection.

Also we assume that a, b, c – the coordinates of point in the coordinate system of the robot, x', y', z' – the coordinates of point in the coordinate system of the scanner, x, y, z – the coordinates of point in the abutment coordinate system.

Then it is possible to write down:

$$\begin{aligned} a &= x + n \\ b &= y + k \\ c &= z + l \end{aligned} \quad (1),$$

where n, k, l – shifts of *SC* of the scanner relatively *SC* of robot.

Thus for systems of coordinates of *Ox'y'z'* and *Oxyz* the equation will be fair:

$$\begin{vmatrix} x \\ y \\ z \end{vmatrix} = C \begin{vmatrix} x' \\ y' \\ z' \end{vmatrix},$$

where C – the matrix of transformation of coordinates; it has the following form:

$$C = \begin{vmatrix} c_{11} & c_{12} & c_{13} \\ c_{21} & c_{22} & c_{23} \\ c_{31} & c_{32} & c_{33} \end{vmatrix}.$$

Then

$$x = c_{11}x' + c_{12}y' + c_{13}z';$$
$$y = c_{21}x' + c_{22}y' + c_{23}z';$$
$$z = c_{31}x' + c_{32}y' + c_{33}z'.$$

In accordance with this the system (1) is converted to

$$a = c_{11}x' + c_{12}y' + c_{13}z' + n;$$
$$b = c_{21}x' + c_{22}y' + c_{23}z' + k;$$
$$c = c_{31}x' + c_{32}y' + c_{33}z' + l.$$

It is necessary to scan the calibration tool tip for determination of coordinates of one point in *SC* of the robot and *SC* of the scanner (thus, to receive its coordinates in *SC* of the scanner); the coordinates in *SC* of the robot can be seen on the control panel.

It should be noted that the 12 unknowns (9 – guides of *cos* and 3 – displacement on respective coordinate) are present in the existing system of three equations. 12 equations required to solve for obtaining of all 12 unknowns. The grouping of these equations is made as follows after establishing of the equation for 4 points:

$$a_1 = c_{11}x_1' + c_{12}y_1' + c_{13}z_1' + n$$
$$a_2 = c_{11}x_2' + c_{12}y_2' + c_{13}z_2' + n$$
$$a_3 = c_{11}x_3' + c_{12}y_3' + c_{13}z_3' + n$$
$$a_4 = c_{11}x_4' + c_{12}y_4' + c_{13}z_4' + n$$

$$b_1 = c_{21}x_1' + c_{22}y_1' + c_{23}z_1' + k$$
$$b_2 = c_{21}x_2' + c_{22}y_2' + c_{23}z_2' + k$$
$$b_3 = c_{21}x_3' + c_{22}y_3' + c_{23}z_3' + k$$
$$b_4 = c_{21}x_4' + c_{22}y_4' + c_{23}z_4' + k$$

$$c_1 = c_{31}x_1' + c_{32}y_1' + c_{33}z_1' + l$$
$$c_2 = c_{31}x_2' + c_{32}y_2' + c_{33}z_2' + l$$
$$c_3 = c_{31}x_3' + c_{32}y_3' + c_{33}z_3' + l$$
$$c_4 = c_{31}x_4' + c_{32}y_4' + c_{33}z_4' + l$$

Journal of Physics: Conference Series **671** (2016) 012014 doi:10.1088/1742-6596/671/1/012014

A solution of the obtained equations allows obtaining complete information about the mutual position of the scanner relative to the robot - manipulator.

The easiest way to solve this system of equations is Cramer's method, whereby the solution takes the form:

$$C_{11} = \frac{D_1}{D};\ C_{12} = \frac{D_2}{D};\ C_{13} = \frac{D_3}{D};\ n = \frac{D_n}{D},$$

where

$$D = \begin{vmatrix} x_1^/ & y_1^/ & z_1^/ & 1 \\ x_2^/ & y_2^/ & z_2^/ & 1 \\ x_3^/ & y_3^/ & z_3^/ & 1 \\ x_4^/ & y_4^/ & z_4^/ & 1 \end{vmatrix};\ D_1 = \begin{vmatrix} a_1 & y_1^/ & z_1^/ & 1 \\ a_2 & y_2^/ & z_2^/ & 1 \\ a_3 & y_3^/ & z_3^/ & 1 \\ a_4 & y_4^/ & z_4^/ & 1 \end{vmatrix};\ D_2 = \begin{vmatrix} x_1^/ & a_1 & z_1^/ & 1 \\ x_2^/ & a_2 & z_2^/ & 1 \\ x_3^/ & a_3 & z_3^/ & 1 \\ x_4^/ & a_4 & z_4^/ & 1 \end{vmatrix};$$

$$D_3 = \begin{vmatrix} x_1^/ & y_1^/ & a_1 & 1 \\ x_2^/ & y_2^/ & a_2 & 1 \\ x_3^/ & y_3^/ & a_3 & 1 \\ x_4^/ & y_4^/ & a_4 & 1 \end{vmatrix};\ D_4 = \begin{vmatrix} x_1^/ & y_1^/ & z_1^/ & a_1 \\ x_2^/ & y_2^/ & z_2^/ & a_2 \\ x_3^/ & y_3^/ & z_3^/ & a_3 \\ x_4^/ & y_4^/ & z_4^/ & a_4 \end{vmatrix}.$$

Thus, the cloud of points of studied object in the coordinate system of ultrasound tomograph can be obtained by noncontact method, using the resulting matrix of the guides.

The proposed method allows making the transition from using the contact method of calibration of the tomogragh to noncontact method, which greatly reduces the time of research.

3. Results and considerations

Figure 2 shows the position of the studied object.

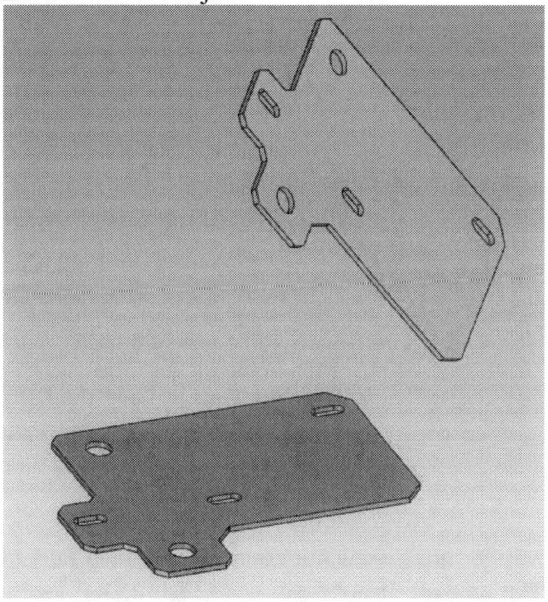

Figure 2. The simulation of working of algorithm.

The software MathCAD was chosen to validate the selected method. The proposed algorithm was fully implemented by the program MathCAD.

The software T-Flex CAD is additionally used. With the help of this software the object of study was modeled and the initial characteristics of this object were defined: translation and rotation relative to the basic coordinate system. Further on the object the four control points were selected, parameters of mutual translation and rotation are calculated by using these points. Map of reference points is shown in figure 3.

The result of running of the algorithm became the reliable matrix, which are the guides for this system of coordinates.

The same experiment was done on real sample by using three-dimensional scanner. The sample is placed in the tub, three-dimensional model of sample is known in advance. The process of reconstruction of three-dimensional cloud of points of the object is carried out by using a three-dimensional scanner; the scanner is fixed at known distance from the robot manipulator.

The reconstruction results are shown in figure 4.

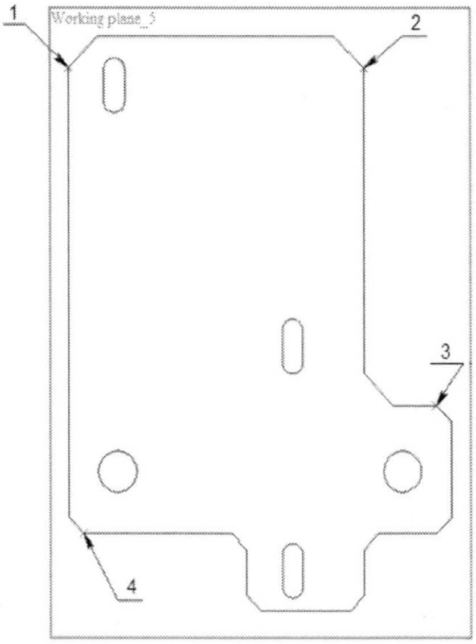

Figure 3. The map of reference points: 1…4 – control points.

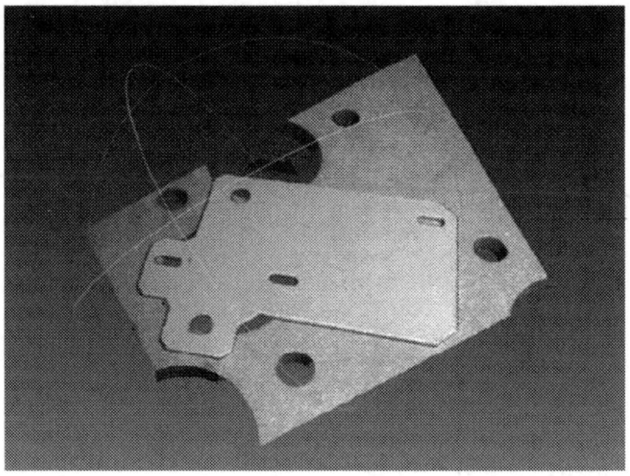

Figure 4. The cloud of points of experimental sample.

The result of the algorithm in this case is a guide matrix, which characterizes the relative position of the object of study in the coordinate system of the ultrasonic tomograph.

Use of this method of measurement shows a significant acceleration of process of calibrating of tomograph and the possibility of excluding of displacement of object at the calibration process.

4. Summary

The presented method allows making the transition from use of a contact method of calibration of the tomograph to noncontact calibration; it significantly reduces the time of the study of object.

The proposed measurement method can significantly accelerate the process of calibration of tomograph and prevent a possible displacement of the object during calibration.

References

[1] Berke M (2000) Nondestructive Material Testing with Ultrasonics *The e-Journal of Nondestructive Testing* **5**(09) http://www.ndt.net/article/v05n09/berke/berke1.htm

[2] Klyuev V V (1996) Development of nondestructive testing and technical diagnostics is a basis of safety *Proceeding of the 14th World Conference on Non-Destructive Testing* 109–112

[3] Kumar S and Mahto D (2013) Recent trends in industrial and other engineering applications of non destructive testing: a review *International Journal of Scientific&Engineering Research* **4** (9) 183–195

[4] Blais F (2004) Review of 20 years of range sensor development *Journal of Electronic Imaging* **13**(1) 231–243

[5] Atcheson B *et al* (2008) Time-resolved 3D capture of non-stationary gas flows *ACM Transactions on Graphics* **27**(5) Article number 132

[6] Robb R A, Hoffman E A *et al* (1983) High speed three-dimensional x-ray-computer tomography: The Dynamic Spatial Reconstructor *Proceedings of the IEEE* **71**(3) 308–319

Journal of Physics: Conference Series **671** (2016) 012015 doi:10.1088/1742-6596/671/1/012015

Rejection of Welded Connections According to EN ISO 10675: a New Approach to Increasing Objectivity

N P Aleshin[1], D I Galkin[2], A S Sorokin[3] and A V Khodzhaeva[4]

[1] Head of department of technologies of Welding and Diagnostics, Bauman Moscow State Technical University, Moscow
[2] Assistant professor of department of technologies of Welding and Diagnostics, Bauman Moscow State Technical University, Moscow
[3] Assistant professor of Department of Plekhanov Russian University of Economics, Moscow
[4] Research Officer of laboratory complex technology NDT, Scientific Centre of Russian Federation JSC «PA «CNIITMASH», Moscow
E-mail: aleshin@bmstu.ru, digalkin@sertink.ru, assorokin@mesi.ru, avkhodzhaeva@cniitmash.com

Abstract. The currently applied criteria of identification of imperfections based on results of Radiographic Testing have no unambiguous description in regulatory documents. Therefore, there is a lack of coherence between opinions of the experts who are interpreting images on radiographic films .Authors suggest dividing images of imperfections into classes, each of which includes imperfections with similar projections on radiographs. This allows to simplify the identification procedure and to increase repeatability of results of the interpretation which is carried out by different experts independently.

1. Introduction

The purpose of Radiographic Testing (RT) is to obtain images on a radiograph [1]. After comparing characteristics of the shadow image of a imperfection with the admissible values established by regulatory documents, the experts decide whether the welded connection complies or not with requirements provided by regulatory documents.

Characteristics of the image of imperfections on radiographs, determined by results of the analysis, are: imperfection type, length and width of the imperfection, summary length of imperfections within L (L- any 100 mm testing length, in millimetres), etc.

As a rule, there are area and shape distortions in images of imperfections on radiographs. Considering that, as well the absence of practical recommendations about the analysis of discontinuity images, it is impossible in most cases to define unambiguously the type of defect which created a specific projection on a radiograph [2, 3].

For example, projections of such imperfections as porosity and gas pores (2011 according to EN ISO 6520-1) and slag inclusions (301) are extremely difficult to be distinguished. There are strong similarities between projections of lack of fusion (401), cracks (100), elongated slag inclusions (301), shrinkage groove (5013), lack of penetration (402) [4].

Absence of accurate and clear criteria of sorting imperfections leads to essential differences in results of interpretation [5].

Study [5] contains statistical analysis of results of interpretation of 10 radiographs of welded connections. The interpretation was independently carried out by 9 experts, all of them RT-certified for 2nd or 3rd skill level and having a practical work experience from 3 to 30 years. The purpose of the study was to assess the influence of a subjective factor on results of interpretation.

The assessment of the degree of coherence between opinions of experts on identification of the revealed imperfections was carried out gradually:

- Taking into account the results of the interpretation carried out by the experts, the authors determined all statistically significant events (imperfections identified by more than 50% of experts).
- The authors then counted the coincident (and not coincident) decisions made by the experts on identification of the same imperfection projection. For each couple of experts and each discontinuity type, a contingency table (binary matrix) was composed as follows (see Table 1):

Table 1. Example of the contingency table (with real experimental values) made for Experts 1 and 2 and for single slag inclusion

Experts		2 no	 yes	total
1	no	1 (a)	3 (b)	4
	yes	2 (c)	2 (d)	4
total		3	5	8

- The contingency coefficient (four-point –ϕ-correlation), a quantitative measure of coincidence in sorting a discontinuity projection within each couple of experts (e.g. categorizing it as porosity and gas pores, including the extended ones, slag inclusions, including the extended ones, lack of penetration, lack of fusion, cracks, root concavity, shrinkage groove), was determined for each contingency table according to the formula (1) (see Table 1):

$$s_{ij} = \frac{a \cdot d - b \cdot c}{\sqrt{(a+b)\cdot(a+c)\cdot(b+d)\cdot(c+d)}} \quad (1)$$

- The obtained contingency coefficients were grouped within intervals characterizing "force" of coherence between opinions of experts: depending on value of contingency coefficient, the level of correlation between decisions of experts has been set as follows: [-1; 0.3) – weak coherence, [0.3; 1] –strong (acceptable) coherence [6].

2. Results and discussion

The static experiment carried out according to the described technique shows a weak coherence between opinions of experts: when identifying a slag inclusion (301), strong coherence is observed only in 27.3% of cases, when identifying a lack of penetration (402) – in 8.3% of cases. The acceptable coherence of experts' opinions when sorting projections of imperfections by types of considered defects was established only in 35.3% of cases.

It should be noted that rationing of the admissible sizes of defects makes sense only for a specific controlled object (part of object) and for established operative conditions. However, traditional RT only allows to define presence of discontinuity projections on radiographs [7, 8]. As shown above, radiographs do not allow to identify at the acceptable accuracy level even the discontinuity type, let alone such parameters as discontinuity curvature on all its surface, discontinuity depth, its orientation within the controlled object. However, those parameters are necessary for strength calculations [9].

Taking into account this fact, as well as the ambiguity of the existing criteria, acceptance levels based on RT results should be considered as a way of maintenance of technological discipline in the conditions of a specific industrial site and process. It allows to reject standard approaches to the RT-assisted quality evaluation based on defect type definition and to develop new criteria, whose application will increase repeatability of radiograph interpretation results. The developed criteria should comply with following requirements:

- Rules of identification of discontinuity images have to be unambiguous.
- The result of application of the developed criteria has to maximally coincide with results of the evaluation of quality (rejection) which is carried out according to the operating criteria.
- Application of the developed criteria has to provide significantly strong coherence of opinions.

To address the above problems, authors suggest dividing images of imperfections into classes, each of which includes defects of various types, but with shape-similar projections.

For this purpose, it is expedient to use the criteria used in Ultrasonic Testing (UT): if the maximum sizes ratio of a projection is more than 5, it is considered as extended, 4 to 3: elongated (longish), 2 to 1: roundish. This classification takes into consideration influence of various types of defects on reliability of welded connections.

Implementing the above approach allows to increase quantity of cases of strong coherence (coincidence of opinions). It is confirmed with the results of experiment [10] involving 6 experts who were carrying out interpretation both according to proposed criteria and to those contained in the operative Russian regulatory documents (RD-25.160.10-KTN-016-15 was used as the example). This fact makes expedient attempts of broader application of the offered criteria of identification, e.g. for increasing objectivity of EN ISO 10675-1 requirements. Previously, a number of problems should be noted in this document:

- The provided norms of quality evaluation (rejection) for linear porosity and clustered porosity based on area of the defective site are less rigid than those for the single layer porosity and gas pores, which does not match with strength calculations and with numerous experiments carried out in order to establish influence of these imperfections on reliability of a welded connection [11]. So, to match the Admissibility Threshold 2, the area of the defective site with single layer porosity and gas pores should not exceed 1.5%, with clustered porosity: 8%, with linear porosity: 4%.
- The standard contains no definition of clustered porosity and linear porosity. Besides, various norms of quality evaluation (rejection) for cases of multilayered and single layer distribution of imperfections are provided in the document. This cannot be unambiguously established by results of application of traditional RT.
- The standard provides various norms of quality evaluation (rejection) for lack of fusion and shrinkage groove (imperfections which often is determined only by results of RT) in spite of the fact that their projections on radiograph are similar.
- The norm of quality evaluation (rejection) used in the standard for linear porosity (sum of projected areas of indications related to $L \cdot w_p$, in % (A) – is not optimum. For example, at A=8% (maximum admissible value for single-layer clustered porosity at the Admissibility Threshold 3) the above norm considers as admissible the existence in welded connection of 17 pores with a diameter of 3 mm, situated in sequence at a 3 mm distance. We consider that the total length of porous clusters at L=100 mm would be a more appropriate parameter for rejection.

3. Conclusions

In order to eliminate the above problems in ISO 10675-1 and according to the accepted approach to the identification of discontinuity projections, we suggest to accept the quality evaluation criteria provided in Table 2.

Table 2. Norms of quality evaluation (rejection) for various image categories (as exemplified by Admissibility Threshold 3 according to ISO 10675-1)

№	category of defect images	defect types corresponding to a category (ISO 6520-1)	Offered criteria of assessment (rejection)
1	S (single)	2011	$A \leq 1.5\%$ at L=100 mm
2	C (cluster)	2013	$d \leq 0.3$ s, but not exceeding 3,0 mm
3	El(elongated)	2015, 2016, 301	$h \leq 0.3$ s, but not exceeding 3,0 mm $\Sigma l \leq s$, but not exceeding 50 mm within L=100 mm
4	Ex (extended)	402, 5013, 401, 2014	not admissible
5	Cr (crack)	100	not admissible

d:diameter of a roundish discontinuity projection
l, h:length and width of a discontinuity projection
s:weld thickness
L: basic length.

It is obvious that the quality evaluation (rejection) carried out according to the offered technique is more transparent and allows to disambiguate the ISO 10675-1 to a certain extent. Besides, the researches [10] conducted earlier allow to count on a bigger repeatability of results of control, which can be checked if necessary according to the technique described in the article. However, the practical use of the offered identification criteria is only possible in case there will be established, for categories of images, a set of norms who, being used, will produce results identical to those of rejection according to the operating criteria. As we see, the above mentioned problems existing in ISO 10675-1 will not allow to do so. For this reason, expediency of introduction of the offered criteria can be only comprehensively studied in the course of ISO 10675-1 revision.

References

[1] Halmshaw R 1995 *Industrial Radiography Second Edition* (Mortsel, Belgium)
[2] Fucsok F, Scharmach M, Müller C (2000) Human Factors: The NDE reliability of routine radiographic film evaluation *Proceedings of 15th World Conference on Non Destructive Testing* http://www.ndt.net/article/wcndt00/papers/idn740/idn740.htm
[3] Fucsok F, Muller C, Scharmach M (2002) Reliability of routine radiographic film evaluation – an extended roc study of the human factor *Proceedings of 8th European Conference on Non Destructive Testing*
[4] EN ISO 6520-1 2007 Welding and allied processes– Classification of geometric imperfections in metallic materials–Part 1: Fusion welding
[5] Aleshin N P, Galkin D I and Sorokin A S (2015) Statistical evaluation of the results of decoding radiographs of welds *Welding and diagnostics* **1** 11–14
[6] Beshelev S D, Gurvich F G 1980 *Mathematical and statistical methods of expert estimates.* (Moscow: Finance and Statistics)
[7] Murakami K (1998) Image Processing for Non-destructive Testing *Welding International* **4**(2) 144–149

[8] Kehoe A, Parker G A (1990) Image Processing for industrial radiographic inspection: image enhancement *British Journal of NDT* **32**(4) 183–190
[9] Das S, Mukherjee D, Shah B K (2010) Advanced digital radiography systems and image processing operations in radiography inspection *Inspectioneering Journal* **16** 42–46
[10] Aleshin N P, Galkin D I (2015) Improving the quality assessment criteria based on the results of radiographic testing of welded joints of pipelines *Welding and diagnostics* **4** 37–40
[11] Deev R F, Paskevich I R 1984 *Defects in welds* (Kiev: Nauk.dumka)

Using the Hybrid Nanoscope for Non-destructive Control of Nanostructural Materials

V Gelever[1], E Usachev[2] and A Manushkin[1]

[1]Researcher, Moscow State Technical University for Radioengineering, Electronics and Automation Russia, Moscow, Russia
[2]Chief Executive Officer, Moscow State Technical University for Radioengineering, Electronics and Automation Russia, Moscow, Russia
E-mail:gelgan@yandex.ru

Abstract. In Moscow State University of Information Technology, Radio Engineering and Electronics there was developed a hybrid nanoscope (HN), which is intended for the study and control of nanostructure materials using a variety of microscopes and spectroscopy. When operating as X-ray microscope, it allows to obtain information on micron and nanometer level about internal structure of different composites and hybrid materials without their destruction.

1.Introduction

For comprehensive study of structure and composition of nanostructural materials, the different types of electronic, probe and optical microscopes are used. Electronic and probe microscopes allow mainly getting the image of the surface of objects, but for a considerable part of nanomaterials, many of their properties are related to the underlying structure. To get the information about an underlying structure, edge cuts and fractures that are obtained with some certain methodologies are often studied. In addition, applying the layer etch of surface of an object by ionic bunch, it is possible to get information about the internal structure of objects. But these destroying and expensive methods do not give complete and operative information about an internal structure. There exist some problems, both as with preparation of objects and also with interpretation of the results obtained. Studying the internal structure of materials and getting three-dimensional images is possible by means of the x-rayed radiation. Such radiation does not practically interacts with objects and in many cases does not require special sample preparation. Investigation can be conducted on air, in a liquid phase, and in a vacuum. Because of it's moderate resolution (20–30нм on synchrotrons[1], 50нм on the demountable x-rayed tubes [2–5]), high cost, overall largenesses and high running expenses, today's nanofocal x-ray microscopy is rarely used.

2. The design of hybrid nanoscope

For complex control of nanostructured materials, a hybrid nanoscope (HN) has been developed, whose modular design allows in optimal way (with high resolution) combine the basic types of microscopes and spectroscopic detectors (figure 1). The basic microscope of the hybrid is a scanning electron microscope (SEM) in the desktop version. The main element of HN is an electron-probe module (EPM), which consists of a column with an electron gun, vacuum system components, x-ray electronic detectors, as well as a set of objects tables (figure 2). Column is the main functional element that provides a focused electron beam. In this instrument, column is the basic structural element of EPM. The column is a system of magnetic lenses with deflecting systems inside them. To ensure maximum density of electronic beam in used range of energy and size of the probe, an optimal focusing lens system was applied [6]. Selection of elastically scattered and secondary electrons was conducted through the objective (the last) lens (OL), which with the help of the detectors before OL focuses the electron beam on the object or target.

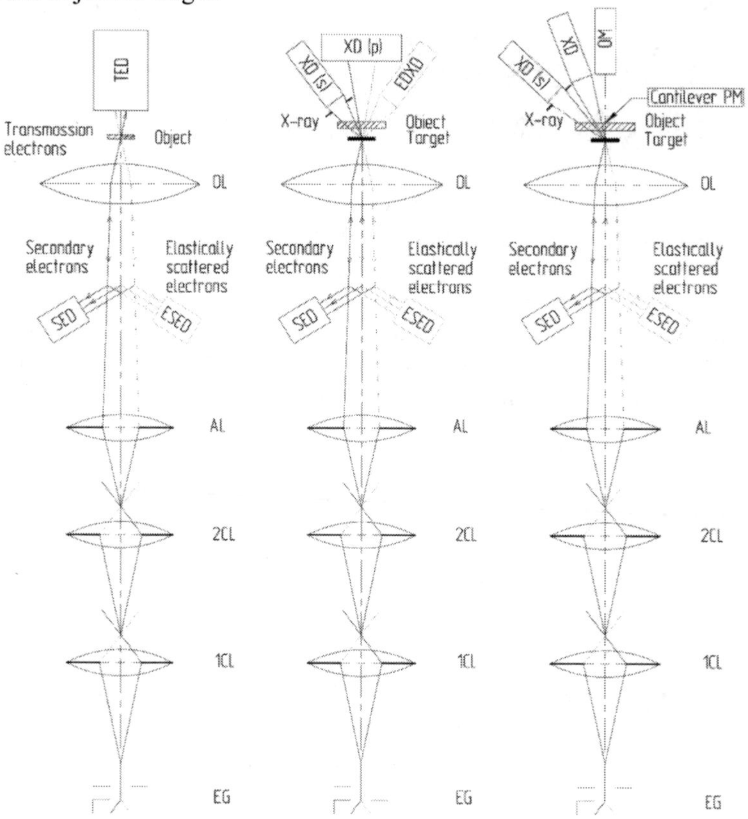

Figure 1. Schematic diagram of the operation HN in various modes:
TED – transmission electron detector; SED – secondary electron detector; ESED – elastically scattered electron detector; EDXD – energy despersive x-ray detector; XD(p) – x-ray detector for projection mode; XD(s) – x-ray detector for scanning mode; OL – objective lens; AL – aperture lens; CL – condensor lens; EG – electron gun.

Figure 2. Electron-probe module.

The nanoscope utilize tungsten thermionic cathode operating under accelerating potential 1-40kV and has following basic modes and parameters:

Table of the main modes and parameters

	Mode name	Environment	Limit permission (пм)
1.	Raster in secondary electrons	vacuum	2–3
2.	Raster in passed electrons	vacuum	1–2
3.	Raster in elastically scattered electrons	vacuum, air	10
4.	Projective in X-ray	vacuum, air	20–30
5.	Scanning in X-ray	vacuum, air	50

3. Modes of transmission X-ray microscope

The design of the scanning microscope has been optimized for operation in air in the mode of transmission x-ray microscopy with nano resolution . Since the X-ray regime is given a special attention, as the basic one an option is selected, when the electron gun is at the bottom, whereas OL and objects are at the top. Then, the objects may simply be put down on the target substrate. Thus, SEM is working in the transmission X-ray microscope (TXM) mode, while setting under the electron beam the thin metal targets on a vacuum tight substrate, which transmits X-rays on the air to the object and the X-ray detectors. For accurate and rapid focusing the electron beam on the target, a detector of secondary electrons from the target is used [7], since the nanoscale electron beams give low X-ray intensity and thus it is practically impossible to provide accurate and timely focusing of X-ray beam. The X-ray microscope can work in projection mode, when the electron beam is at a point on the target, and the X-rays passing through the object is registered with the coordinate-sensitive detector (figure 3a). Besides, , while scanning beam on the target and the use of X-ray detectors with variable aperture entrance for X-rays, one can to obtain in raster (inversion) mode the X-ray image of the object (figure 3b), which resolution is determined by the input aperture of the detector and the size of the focal spot. Using a number of detectors allows to obtain multiple images from different angles, adding them to increase the resolution and get a three-dimensional image. It is possible to use a hybrid system of detectors, in which the coordinate-sensitive detector is located on the axis for the projector mode , and other detectors are located aside, to give raster images at different angles. Moreover, among them can be energy-dispersive detectors.

 doi:10.1088/1742-6596/671/1/012016

To enhance X-ray spatial resolution, a near focus mode is used, when using substrates of micron and submicron thickness, a distance between the object and an electron beam (d_{o-f}) on a target is minimized. In this case, X-ray flux density is considerably increases on the object and the detector compared to common for X-ray microscopy values for d_{o-f} of several hundreds microns. This compensates for X-ray intensity reduction at nanoscale focal beams. Modern technologies allow to obtain thin micron and submicron vacuum- tight membranes of Be, Si, Si_3N_4, C etc. At the film with thickness of the order of 0.1 mkm, electrons with energy of 10-30 keV passe trough the film on an air, and backscattered electrons passed backward are registered with special detector. This allows control without a special treatment a surface of materials on an air, like as for atmospheric SEM JASM-6200 (Japan).

Using ultra-thin wafers for the targets allows you to vary within wide limits the distance the object - the focus (d_{ob-f}) and, if necessary, implement a phase contrast by selection d_{ob-f} in accordance with the focal spot. In addition, if the thickness of the object is greater than the thickness of the substrate, the area of the object at different distances from the focal spot will appear at various zoom levels. This allows you to add pictures of objects made with the calibrated displacements of the focal spot, and receive cross-sectional images of the object, i.e., implement a mode of digital laminography.

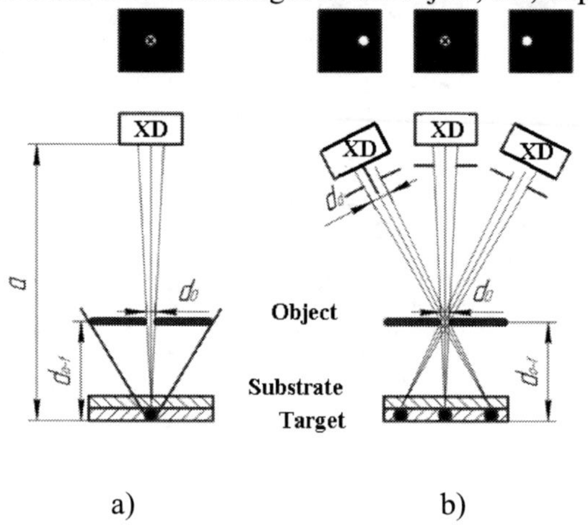

Figure 3. Projection and raster (with three detectors) modes of transmission x-ray microscope: d_0 – diameter of hole in an object; d_{o-f}– distance object - focus; a – target- detector distance; XD- X-ray detector; d_d –figur hole diameter at the input to the detector.

4. Conclusion

Developing of HN on own initiative is being conducted for several years. The main attention was paid to the design of EPM, that provides optimal formation of nanoscale electron beam in the accelerating voltage range of 1–40 kV. The work was made a small pilot party of devices has been made (figure 4), on which there were tested some design options of EPM . On the first of these, there were obtained some good results in X-rays (0.2 – 0.3mkm at 15kV) and electrons (less than 10 nm at 15 kV). Images zinc particles of different sizes in the organic film with the thickness of 270 mkm obtained in X-ray and a secondary electron in SEM are shown in figure 5. When comparing these images, the advantage in the X-ray image is apparent. The study with SEM of fracture of the film or layer-etching of the surface with the ion beam does not give a complete and timely information about the internal structure of the film as was do nein an X-ray mode. HN in X-ray modes may be efficiently used for obtaining information about the internal structure of the various embodiments of composite and hybrid materials in the micron and nanometer level. The main technical parameters of HN give a possibility to combine the features and functions of most imported different types of microscopes (figure 6) with parameters higher than that of each of them, and at the cost of the desktop microscopes. This is an import-substituting device of high level, that has yet no equals.

Figure 4. Appearance of hybrid nanoscopes (side view).

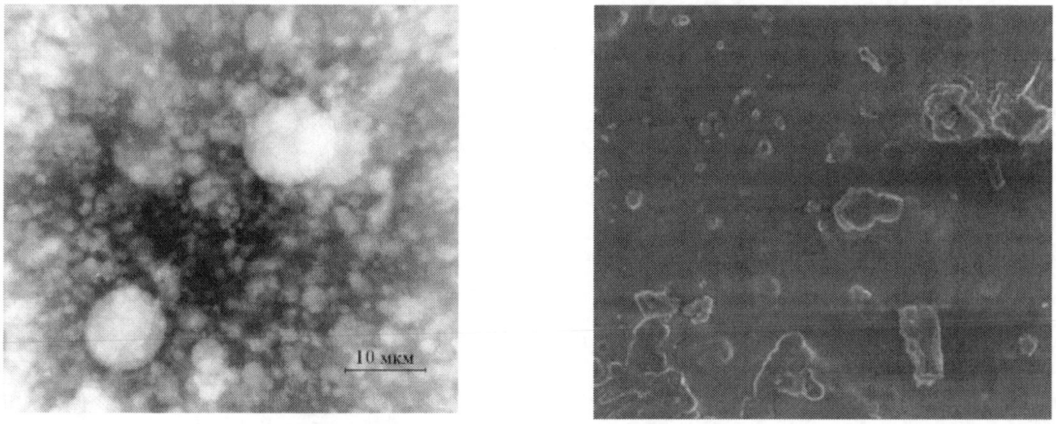

a) b)

Figure 5. Object – organic film of 270mkm thickness with Zn particles: a) TXM (X-ray); b) SEM(electrons).

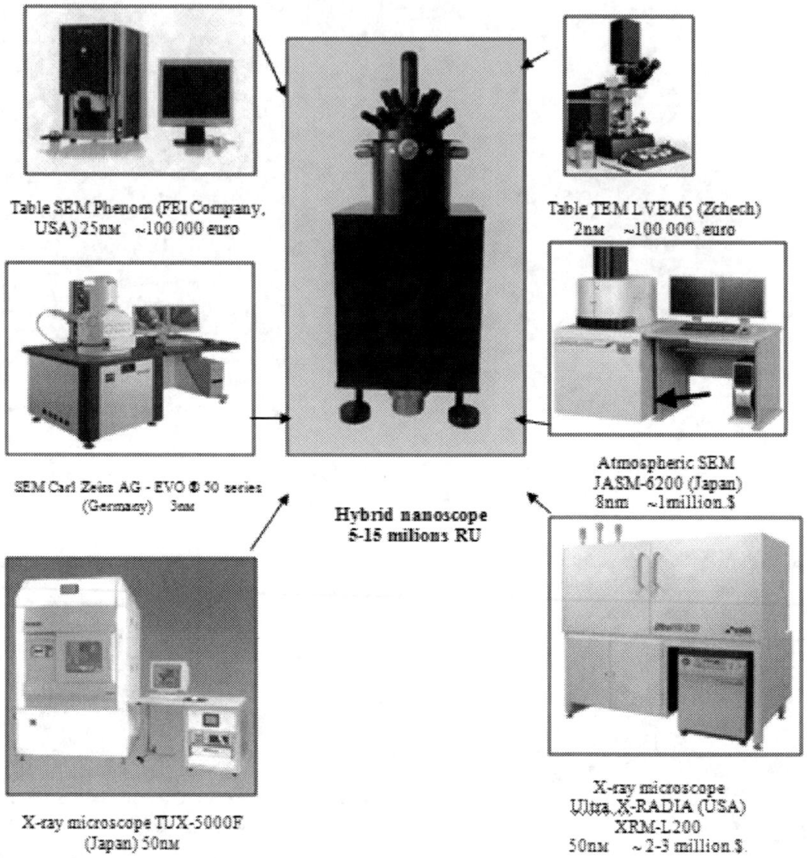

Figure 6. Hybrid nanoscope and devices, features and capabilities of which combines GN.

References

[1] Andrewsa J C, Brennana S *el al* (2008) A high resolution, hard X-ray bio-imaging facility at SSRL *Synchrotron Radiation News* **21**(3) 1–20 doi:10.1080/08940880802123043

[2] Saito Y, Hiromi Kai, Kohei Shirota and Keiji Yada (2008) High resolution X-ray inspection microscope equipped with a field emission gun and its application http://xrm2005.spring8.or.jp/program2/posterAbstract/101.Yasushi_Saito.pdf

[3] Kanji Y 2002 X-ray microinspection apparatus Patent JP 2004138460(A)-2004-05-13

[4] Lau S H *el al* (2007) Non destructive characterization of advanced package and die level defects with novel 3D X-ray micro and nanotomography systems *SEMI's Semiconductor Manufacturing Magazine* 1–10

[5] Minami K, Saito Y, Kai H, Shirota K and Yada K (2009) High resolution projection X-ray microscope equipped with fluorescent X-ray analyzer and its applications *Journal of Physics: Conference Series* **186** Article number 012010 doi: 10.1088/1742-6596/186/1/012010

[6] Gelever V D (2000) The lens system for optimal formation of submicron electron probes *Izvestiya Akademii Nauk. Ser. Fizicheskaya* **64**(8) 1601–1607

[7] Gelever V D Patent RU № 2 452 052/ Russian/ 27.12

Application of Finite Elements Method for Improvement of Acoustic Emission Testing

S Gerasimov[1], T Sych[2] and V Kuleshov[3]

[1] Professor of Department of Structural Mechanics, Siberian Transport University, Novosibirsk, Russia
[2] PhD Student of Department of Structural Mechanics, Siberian Transport University, Novosibirsk, Russia
[3] Professor of Institute of Non-Destructive Testing, Tomsk Polytechnic University, Tomsk, Russia
E-mail: 912267@gmail.com

Abstract. The paper deals with the acoustic emission sensor modeling by means of FEM system COSMOS/M. The following types of acoustic waves in the acoustic emission sensors are investigated: the longitudinal wave and transversal wave. As a material is used piezoelectric ceramics. The computed displacements are compared with physical model under consideration. The results of numerical and physical simulations of the processes of acoustic wave propagation in solebar of the freight-car truck are presented. The fields of dynamic displacements and stresses were calculated for improvement of acoustic emission testing method.

1. Introduction

The acoustic emission (AE) technology is widely used in field testing of high-pressure vessels, pipelines and castings. The AE-sensor constitutes the first part in an AE measurement chain and as of this is of particular importance [1]. A subsequent measurement system can only process signals which the AE-sensor picked up. An AE-sensor converts the surface movement caused by an elastic wave into an electrical signal which can be processed by the measurement equipment. The piezoelectric element of the AE-sensor should pick up faintest surface movements (i.e. have high sensitivity) and convert this movement most efficiently to an electrical voltage. In this paper we made an analysis of the behavior of the piezoelectric plate under the influence of the two kinds of waves - longitudinal and transversal. In the first case analyzes the response of the model against the front of the acoustic wave applied normal to the site of the piezoelectric plate that simulates the arrival of the longitudinal wave. In the second case - the response of the same model, but against the front tangential movement that simulates the arrival of transverse waves. Such an analysis is relevant, since mechanical response of the piezoelectric plate to the impact of different types of waves has not been studied.

2. Acoustic emission sensor simulation

The investigated model presented in figure 1. For Model 1 were selected physical properties of ceramics. Basic physical and mechanical characteristics for the model are as follows: elastic modulus $E = 2.2\cdot10^5$ MPa, the Poisson's ratio $\nu = 0.22$, density $\rho = 2.3\cdot10^3$ kg/m^3, wavelength $\lambda_{min} = 10$ mm, longitudinal wave velocity $C_l = 9780$ m/s.

Boundary conditions are presented in figure 1. At the edge of the lower edge of a ban on the movement of the axis *z*. Also, two nodes of edge A lying on the same diameter (nodes 21 and 221), ban the displacement along the x axis and rotations about the axes *x, y, z*. In all nodes of edge A Model 1 applied axis displacement U_z = 100 micrometers in the form of steps duration $1\cdot10^{-5}$ s, modeling, thus coming to the piezoceramic plates longitudinal acoustic wave (normal to the surface of the plates). The calculation is made in the FEM system COSMOS/M, module nonlinear dynamic analysis. The calculation parameters are fully compliant with the spatial and temporal sampling, detailed in [2]. These calculated results are shown in figure 2, where the fixed scale –300 ... 300 micrometers.

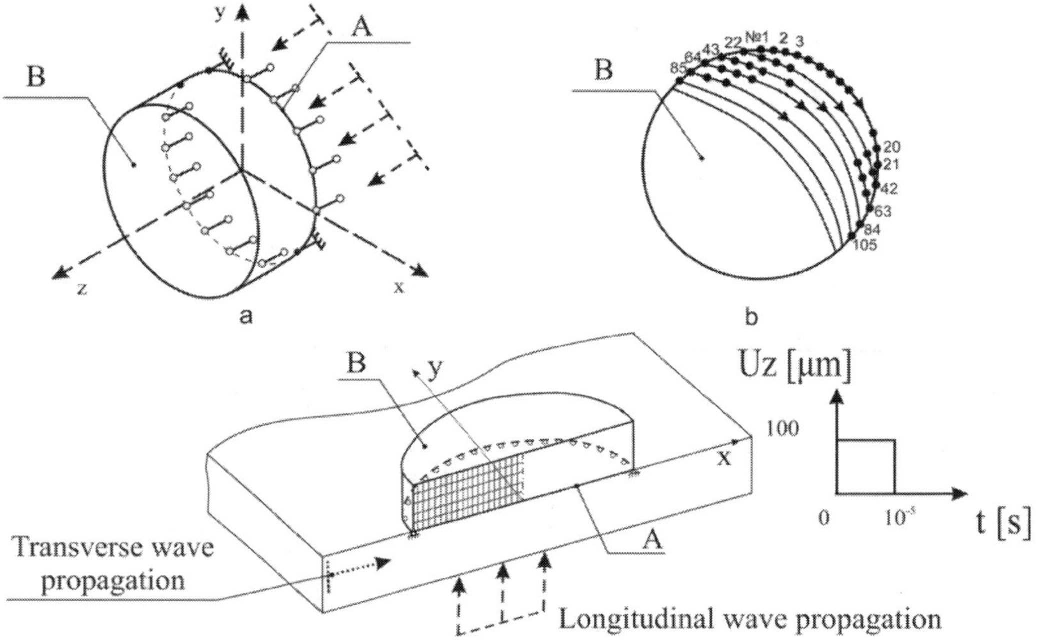

Figure 1. The FEM model of the AE sensor.

Figure 2a shows the distribution of displacements U_z of wave front extends from the edge A and reached the center of the model at time $2.5\cdot10^{-7}$ s. According to theoretical calculations, the arrival time of the waves on the edge B was $5\cdot10^{-7}$ s. Next figure 2b shows that the theoretical calculations and numerical simulation results are the same. Figure 2c observed maximum positive displacement at edge B. In the following times at $1\cdot10^{-6}$ s and $1.25\cdot10^{-6}$ s the process of reflection of the wave from the free edge B. At time $1.5\cdot10^{-6}$ s there are zero field displacements – the wave is reflected and passes the zero position. At time $1.75\cdot10^{-6}$ s we can observe negative maximum displacement of edge B. On figure 2h compression wave front moves to edge A, at time $2.25\cdot10^{-6}$ s reflection occurs and at time $2.5\cdot10^{-6}$ s the model once again has "zero" displacement. At time $2.75\cdot10^{-6}$ s there are observed positive movements of edge B and the process is repeated.

Thus, the arrival of the longitudinal acoustic wave on piezoelectric plate leads to demonstrated in figures 2a–p modes oscillations, the main strain is a "tension-compression". In the framework of this problem it is evident that vibrations of the plate in this fashion will result in maximum electrical signal to the piezoelectric plates, since the total deformation of the upper face at a certain time in such a fashion is maximized.

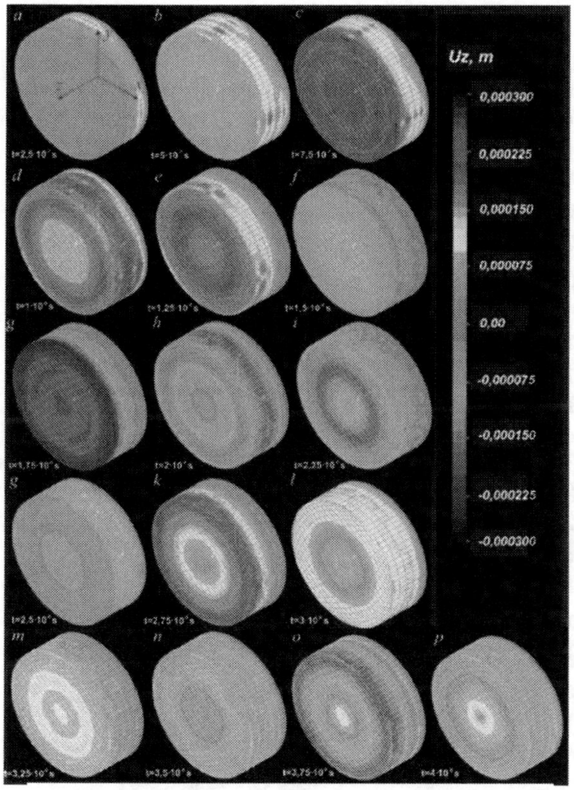

Figure 2. Displacements U_z for Model 1, with the arrival of longitudinal waves.

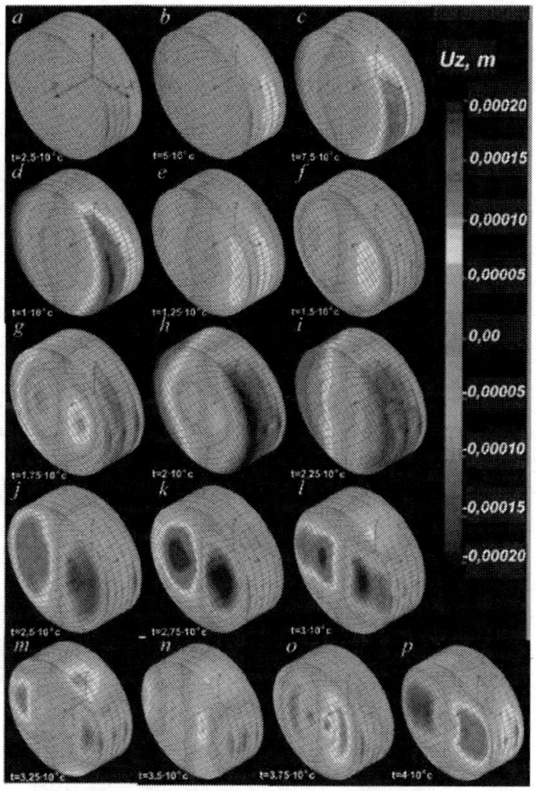

Figure 3. Displacements U_z for Model 1, with the arrival of a transverse wave.

The second experiment with the Model 1 consisted in changing the direction of perturbing displacement on the U_x. The rest of the boundary and initial conditions remain unchanged. Figure 3a-p analyzes the displacement U_z of model edge B.

On figure 3a at time $2.5 \cdot 10^{-7}$ s the wave comes from the edge A. On figure 3b the maximum wave extends to the middle of the model. On figure 3c the maximum wave comes to top face. On figure 3d the wave reaches the upper face and displacements U_z are maximized at two parts of the upper face. By the time $1.5 \cdot 10^{-6}$ s from the zone of maximum displacement shifted to the center, the diagram becomes more complex, with two-bend at time $2 \cdot 10^{-6}$ s. A maximum displacement U_z again observed at the edges, with the zone of negative (blue color) and positive (red color) movements are reversed when compared to figure 3d. Next, the positive and negative peaks of displacements are shifted again to the center of the model and the process repeats.

The results of mathematical modeling show that the oscillation mode determines the distribution of node displacements upper edge model. In real piezoelectric sensor the sputtered layer is the upper edge (the so-called electrode), where the electric potential is recorded. In the above examples the electric potential (total displacement U_z) is smaller at several orders of magnitude for the transverse waves than for longitudinal waves.

3. Experimental results

The non-destructive testing of thousands of solebars of the freight-car truck was done in 2012–2014 years. At the same time the AE sensor position during AE control significantly influences to the accuracy of the AE control. The AE sensors work more effective when their locations are in the zones of extreme displacements of the object surface. In this regard, relevant is the problem of determining such zones.

Journal of Physics: Conference Series **671** (2016) 012017 doi:10.1088/1742-6596/671/1/012017

The finite element model of a solebar of the freight-car truck were drawn in the software package SOLIDWORKS. The model accounts the most important and general structural elements and the nature of cross-sections. Acoustic emission signal was simulated by force with an amplitude of 1 kN and a duration of $1 \cdot 10^{-5}$ s [3,4].

The direction of applied force is shown in figure 4a by the arrow. The highest frequency of fastest generated acoustic waves mode consists of 100 kHz. In this case, the wavelength of the propagating acoustic waves is more than 6 cm. The selection of the applied force place was based on the frequent breaks statistics. Adaptive finite element mesh generating was provided in the field of geometric concentrators (holes, fillets, casting, etc.). The finite element size in these fields were mounted 5 mm, i.e. the density of the mesh in these fields was ~ 10-12 finite elements on a single characteristic size of the wave, which is in according with the requirements of [1]. Maximum step time discretization were chosen uniform and equal to 10^{-6} s, it's were provided 10 iterations per period of the pulse duration.

The model were consisted of 133 541 nodes, 73687 elements.

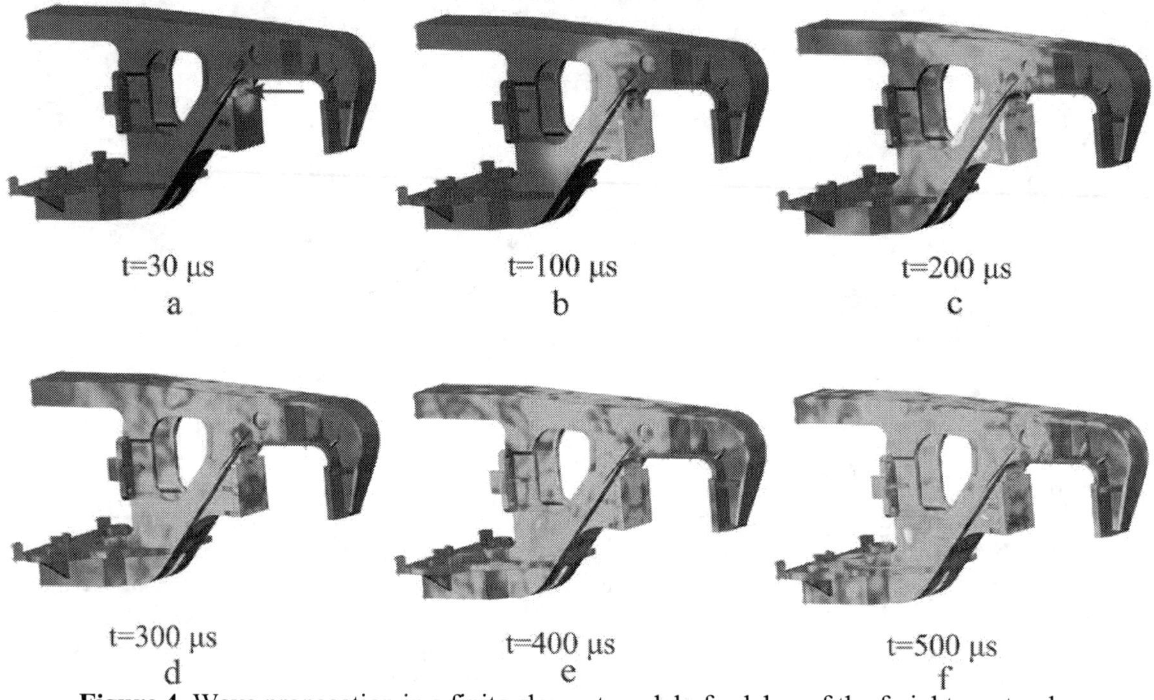

Figure 4. Wave propagation in a finite element model of solebar of the freight-car truck.

The theoretical speed of fastest generated acoustic waves mode was 5.9 mm/μs. Figure 4a – 4f shows the computing result of displacements of model surface. The node displacements are presented at time from 0 to 500 μs. According to preliminary estimates the computing acoustic wave propagates to 30 mm by 50 μs. Thus the results of numerical modeling of the acoustic wave propagation are conforming to theoretical estimates.

Figure 5 presents the regular places of AE sensors in the frontal plane *YZ* (a) and the computing results of the displacements U_x component of the solebar surface (b) at time $t=1 \cdot 10^{-3}$ s. The process is well established. The sign "+" in the figure 5b is indicated the installing places of AE sensors in accordance with existing technological instruction. The figure shows that AE sensors № 9(12), № 10(11), № 13(16) are installed on the solebar zones where the mean value of normal displacements U_x component tends to zero (green). While AE sensors are the most sensitive to the normal component of the displacements.

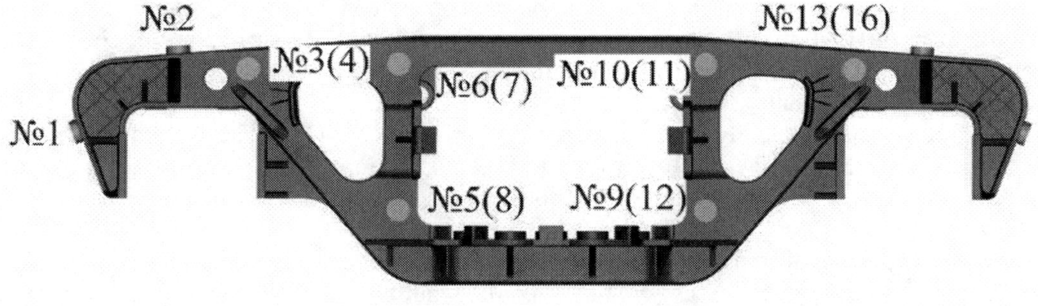

a

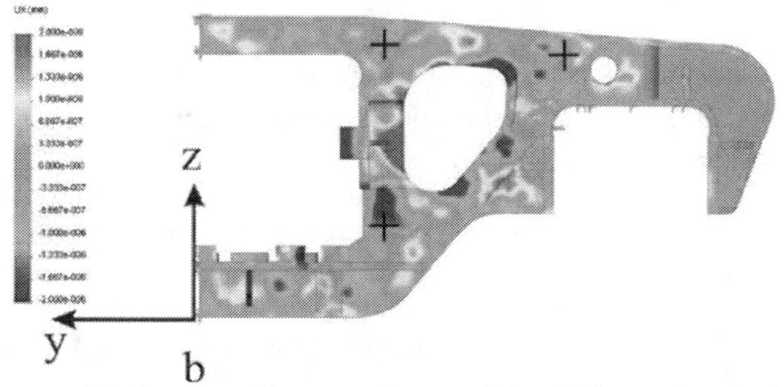

b

Figure 5. The installing regular places of AE sensors in accordance with existing technological instruction (a) and U_x displacements distribution at the time 0.001 s (b).

The analysis of the distributions of displacements U_x component in the finite element model allows to select nodes with high value of normal displacements (U_x component). The value of normal displacements in these nodes is significantly more than at the regular places of installation of AE sensors. The column 2 of table 1 consists the absolute value of the surface U_x displacement in the regular places of installation of AE sensors, the column 3 consists average displacements at the nodes in the zones in which it is recommended to transfer the centers sensors. Column 4 shows the ratio of values of the 2-nd and 3-rd columns.

Table 1. Computing results. Values of U_x displacements in regular and recommended places of solebar surface and its ratio

Sensors number	Normal displacements (U_x component) in the regular nodes, m, $\cdot 10^{-6}$ s	Normal displacements (U_x component) in the recommended nodes, m, $\cdot 10^{-6}$	The ratio
1	*2*	*3*	*4*
№9(12)	-1.11	-2.02	1.8
№10(11)	-0.69	-0.86	1.2
№13(16)	-0.22	-1.43	6.5

Using the finite element method the results of acoustic waves propagation in the object of complex shape were obtained. According computing results AE sensor No.9, 12 are recommended to transfer to z=41 mm; AE sensor No.10, 11 are recommended to transfer to z=26 mm and y=-26.5 mm; AE sensor No.13, 16 are recommended to transfer to y=49 mm, and z=2 mm.

The physical experiment was carried out for comparison of computing and nature results. The experiment includes the first case, the AE sensors were installed in the regular places, the second case, the AE sensors were installed in the recommended places.

The acoustic emission signals were generated by pencil-lead breaks. The experimentally, place of application of force was equivalently to the numerical experiment condition. The system and channels setup were made according to the existing technological instruction. Pencil-lead breaks are widely used as a reproducible source for test signals in acoustic emission applications (Hsu-Nielsen source) [5,6].

To analyze the influence of transferring the sensor installing places were taken 20 series of breaks for the first and second experiment cases. We analyzed the amplitude of the signals. The experimental results were shown the increasing the average level of experimental signals with comparing of the first and second experiment.

4. References

[1] Nasedkin A V, Shikhman V M and Zakharova S V (2011) Finite-element calculation of high-temperature acoustic-emission transducers *Russian Journal of Nondestructive Testing* **47**(7) 468–479

[2] Sych T, Gerasimov S and Kuleshov V (2012) Simulation of the propagation of acoustic waves by the finite element method *Russian journal of nondestructive testing* **48**(3) 147–152

[3] Hora P and Cerneva O (2010) Acoustic emission source modeling *Applied and Computational Mechanics* **4** 25–36

[4] Markus G and Richler S (2015) Finite element modeling of cracks as acoustic emission sources *Journal of nondestructive evaluation* **34** 4–17

[5] Hamstad M (2007) Acoustic emission source location in a thick steel plate by Lamb modes *Journal of acoustic emission* **25** 194–214

[6] Wotzka D, Boczar T and Fracz P (2011) Mathematical model and numerical analysis of AE wave generated by partial discharges *Optical and acoustical methods in science and technology* **120**(4) 767–771

Analysis of the Impact of Major Influencing Factors on the Waveform of the Surface Eddy Current Probe for Electroconductive Nonmagnetic Pipe Thickness

A E Goldshtein[1], V Y Belyankov[2]

[1] Professor, National research Tomsk Polytechnic University, Tomsk, Russia
[2] Graduate student, National research Tomsk Polytechnic University, Tomsk, Russia
E-mail: algol@tpu.ru

Abstract. The results of computer simulation of interaction between the magnetic field of the surface eddy current probe and a conductive pipe performed in COMSOL Multiphysics were used to determine the dependences of the probe on major influencing factors: pipe wall thickness, the gap between the probe and the pipe surface, material electroconductivity, the pipe wall curvature, areas with a smooth V-shaped change in the thickness and local thinning of a spherical shape, misalignment of the probe axis relative to the pipe surface and transverse displacement of the probe axis.

1. Introduction

Surface eddy current probes (SECP) are widely used to solve various problems of non-destructive testing: measuring the thickness of electroconductive objects and non-conductive coatings on metal surfaces, control of metal and alloy electroconductivity, inspection of products of different shapes, structuroscopy of parts of non-magnetic and ferromagnetic alloys.

The advantages of the surface eddy current probe are their versatility, the ability to control objects of plane, cylindrical and complex forms with one-sided access to the test object, locality that provides high resolution and precise determination of the defect zone when scanning the surface of the test object [1,2].

One of the essential inspection problems efficiently solved via SECP is measuring the wall thickness of pipes made of electroconductive nonmagnetic material and the thickness of the dielectric coatings on these pipes or the air gap between the probe and the pipe surface.

A practical example of SECP application is testing of the wall thickness of light-alloy drill pipes (LADPs) manufactured from duralumin D16T. The advantages of LADPs compared to steel pipes are low weight, low flow resistance and non-magnetic materials, the properties which are required for inclinometer survey.

The paper addresses the analysis of the interaction between the magnetic field of the SECP and the electroconductive pipe taking into account the main influencing factors to measure the wall thickness of the pipe and the gap between the SECP and the pipe surface.

2. Problem formulation

Figure 1 schematically shows the structure of the transformer SECP commonly used to inspect the wall thickness. The SECP comprises excitation coil w_1, measuring coil w_{21} and compensating coil w_{22}. With no test object present, their initial EMFs cancel each other out due to subtractive polarity of the measuring and compensation coils. If the electroconductive test object is located near the SECP, a waveform caused by the eddy current in the object arises at the probe output. In the general case, the amplitude and phase (complex components) of the induced EMF are determined by the amplitude and frequency of the excitation current, SECP design parameters, electromagnetic characteristics of the material and geometric parameters of the test object, and the relative position of the SECP and the test object.

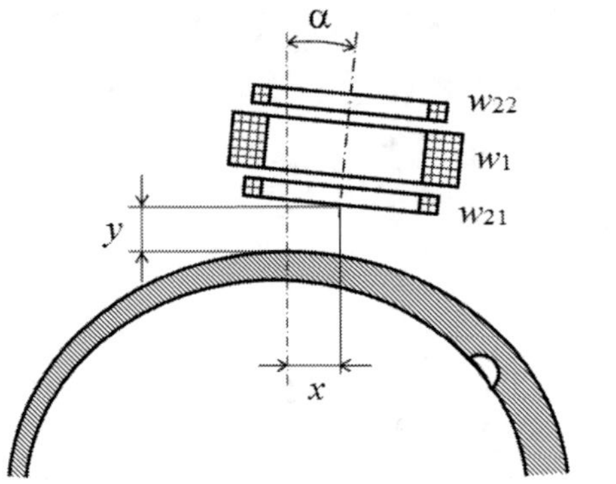

Figure 1. SECP interacting with the conductive pipe.

To calculate the results of the interaction between the SECP magnetic field and the electroconductive test object and to find the function of the influencing parameter conversion into the SECP waveform parameters, we can use the analytical models well proven to investigate different variants of interaction with a sufficiently high degree of compliance with theoretical and experimental results [1,3,4].

However, these analytical results have been obtained for a limited class of interactions typically involving axisymmetrical relative positioning of the probe and the test object of a regular geometric shape (plane, spherical, cylindrical, etc.). Therefore, the analytical results do not allow us to study the effect of different factors such as linear and angular displacement of the transducer symmetry axis with respect to the test object symmetry axis, irregular shape of the test object, local defects such as cracks or local thinning, the proximity of the test object edge on the SECP signal.

Numerical simulation methods are free of this drawback and the finite element method (FEM) is supposed to be the most appropriate for the considered class of problems. FEM is a numerical method for solving differential equations with partial derivatives and integral equations derived when solving the problems of applied physics. This method is widely used to solve the problems of solid mechanics, heat transfer, hydrodynamics and electrodynamics. FEM provides an example of versatility of the real problem solution, an object of any form can be treated, approximation of standard geometric shapes is not required, asymmetric problems taking into account the heterogeneity parameters of materials and environments can be solved [2].

FEM accuracy is mainly determined by the density of the mesh applied. Computational capabilities of modern computers allow processing of a large amount of data. In addition, a lower density mesh can be made in the areas with small values of the electromagnetic field gradient. Thus, FEM can provide high precision simulation with no significant restrictions on the complexity of the boundary conditions [5].

With regard to the problem of interaction between the SECP magnetic field and the conductive pipe, FEM allows the analysis of the effect of not only material conductivity σ, wall thickness T and the pipe outer diameter D, the gap between the SECP and the pipe surface y on the SECP output signal, but in contrast to the mentioned analytical models, it enables the analysis of the effect of linear x and angular α misalignment of the SECP and the pipe, non-uniform thickness and the presence of pipe wall local thinning (figure 1).

The SECP parameters set in the simulation were as follows (figure 1): the outer diameter of the excitation coil was 40 mm; the inner diameter of the excitation coil was 32 mm; the height of the excitation coil was 10 mm; the diameter of the middle turn of the measuring and compensation coils was 30 mm; the distance between the planes of the middle turns of the measuring and compensating coils spaced symmetrically relative to the excitation coil was 16 mm. The frequency of the excitation current was assumed to be 100 Hz.

The test object was a nonmagnetic pipe with a specific electric conductivity σ in the range of (10...25) MS/m, with a nominal outer diameter D = 147 mm and wall thickness T in the range of (5...15) mm. The distance between the SECP measuring coil and the pipe surface varied in the range of (3...15) mm.

3. Simulation of the interaction of the secp magnetic field and the test object using comsol multiphysics software

3D model was created using COMSOL Multiphysics, version 4.3a. When applying the mesh, the parameters of which directly affect the calculation accuracy, a higher density mesh was used for regions with a higher gradient of the change in the magnetic field (figure 2).

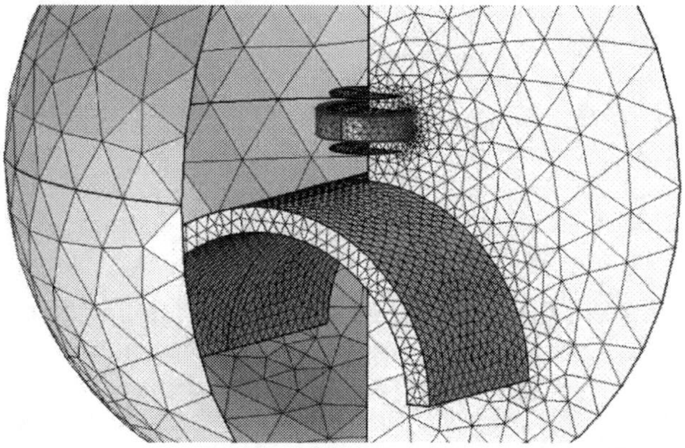

Figure 2. Mesh applied to the model in COMSOL.

To test the adequacy of the developed model, the results of computer and physical modeling of SECP interaction (Figure 1) with a duralumin pipe were compared. The outer diameter of the pipe was 147 mm and electroconductivity was σ = 16 MS/m. The results of computer simulation differ from those obtained in physical simulation by not more than 3% in the above range of the influencing parameters for pipe thickness T and air gap y. It indicates their high adequacy.

Figure 3 shows the hodographs of the relative induced SECP voltage for the change in the pipe wall thickness T, the gap y and the material conductivity σ obtained by numerical simulation for other values of the influencing parameters. These results qualitatively coincide with the results obtained using the analytical model [1]. The quantitative difference in the tested range of changes in the influencing parameters does not exceed 7%. These findings are generally consistent with the results obtained in [5].

Figure 4 shows the results obtained in numerical modeling of the effect of the test object surface curvature on the SECP signal. The dotted line shows the hodograph for the change in the gap y for a plane surface. The solid lines indicate the hodograph of the curvature radius R in the range (50...∞) mm and (-∞...-73) mm. Positive radius values correspond to the radius of the convex surface (outer pipe surface), and negative radius values indicate the concave surface (inner pipe surface). The analysis of the dependencies plotted in Figure 5 shows that the hodographs of the change in the test object surface curvature are close to a straight line intersecting the hodograph of the change in the gap at an angle of 5...10°.

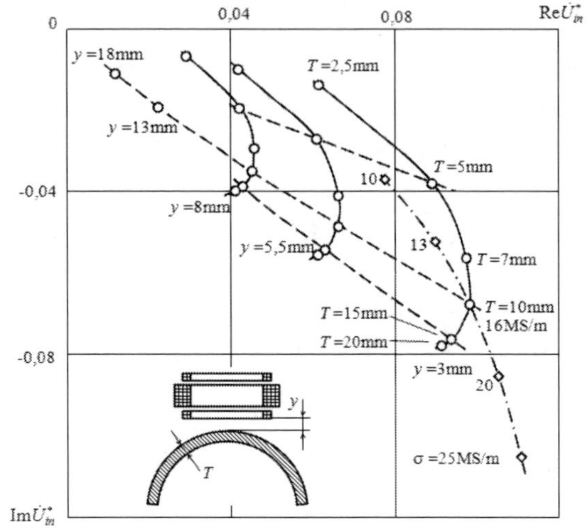

Figure 3. Hodographs of the relative induced SECP voltage for the change in the pipe wall thickness T, the gap y and the material conductivity σ.

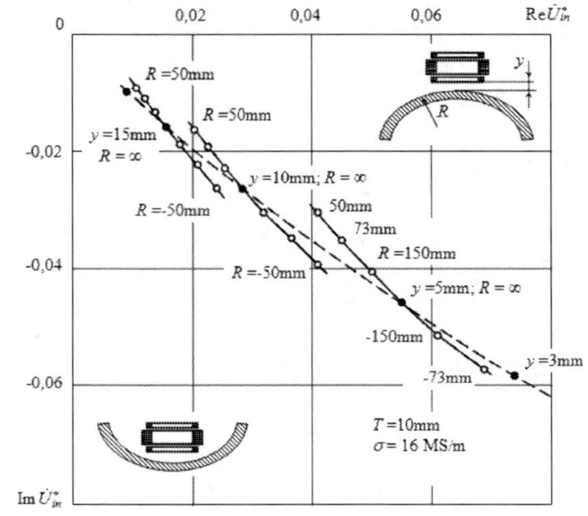

Figure 4. Hodographs of the relative induced SECP voltage for the change in the gap and the radius of the test object surface curvature.

Figure 5 shows the numerical results for the effect of the displacement x along the SECP longitudinal axis on the SECP waveform relative to the transverse pipe axis. Dashed lines show the hodographs of the change in x for three values of the pipe thickness T, and solid lines indicate the hodographs of the change in the gap y for the same thicknesses. The analysis of the modeling results shows that the hodographs for displacement x are close to a straight line almost coinciding with the hodographs for the change in the gap y. A significant change in the SECP waveform caused by displacement is found at x exceeding the half radius of the SECP coil turns.

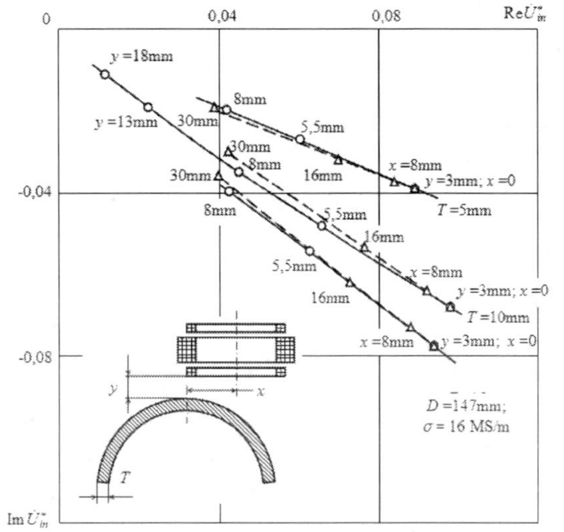

Figure 5. Hodographs of the relative induced SECP voltage for the longitudinal axis with respect to the pipe transverse axis.

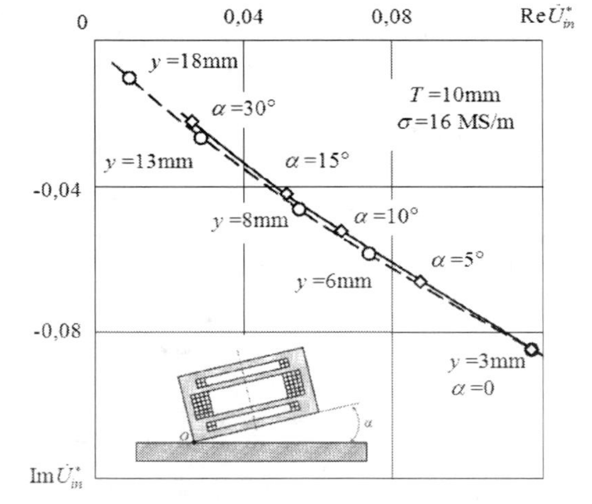

Figure 6. Hodographs of the SECP relative induced voltage for the change in the skew angle α.

Figure 6 shows the results obtained in numerical modeling of the effect of the angle α between the SECP butt end and the test object on the ECP signal. The test object was a conductive plate. It was assumed that the SECP is skewed relative to the common point O between the plate surface and the edge of the SECP body with a diameter of 44 mm and a height of 22 mm. The hodograph $\Delta\dot{U}^*_{in}$ for the change in the skew angle α (dashed line) is almost equal to that for the change in the gap y. For the considered SECP sizes, it can be assumed that the change in the induced voltage amplitude caused by the skewness of 5° is similar to the change in the gap y from 3mm to 5 mm.

In pipe operation, LADP in particular, the pipe wall wears unevenly due to mechanical and corrosive attack. In this regard, we studied the effect of V-shaped and local changes in the wall thickness on the SECP waveform.

The hodograph of the induced SECP voltage with respect to the change in the thickness of the V-shaped object in the normal cross section virtually coincide with the hodograph for the change in the thickness of the plane object. Hence, the signal of the SECP located above the object under this thickness variation and the value of the wall thickness T_0 along the SECP longitudinal axis corresponds to the SECP signal located above the plane object with a wall thickness of T_0.

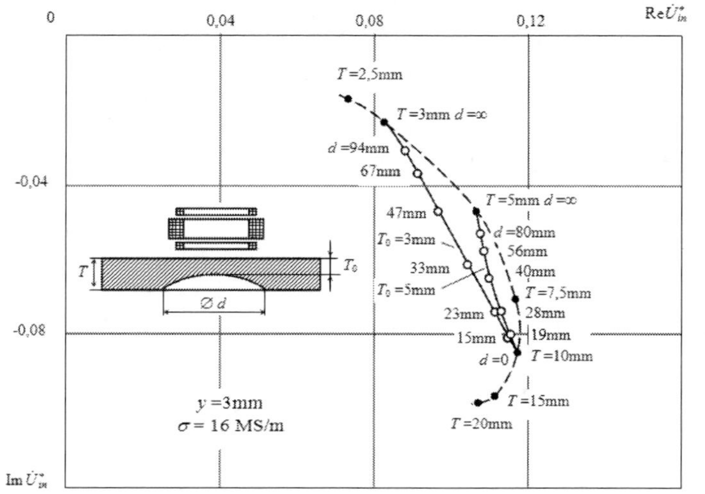

Figure 7. Hodographs of the SECP relative voltage for the change in the diameter of the spherically-shaped local thinning.

Figure 7 shows the results obtained in numerical modeling of the effect of local thinning on the SECP signal. In this case, similar to the previous one, the test object was a conductive plate taken for greater clarity. The local thinning was assumed to be of a spherical shape located on the inner surface of the plate. The depth dimensions were set by the two parameters: the minimum depth from of the outer surface T_0 and the circle diameter d on the inner surface. The analysis of the results shows that the hodographs of the induced voltage for the change in the thinning diameter d in the range from 0 to ∞ (solid lines) are close to straight lines connecting the points in the hodograph for the change in the plate thickness (dotted line) corresponding to thicknesses T and T_0. A sharp change in the SECP signal due to local thinning can be observed at d values commensurable with D_0 values.

4. Conclusion

Methods of numerical modeling, the finite element method in particular, are effective techniques to simulate the interaction of the SECP with an electroconductive test object in case the test object and the SECP coil are of a complex shape or the arrangement of the probe relative to the object is asymmetrical. A high degree of adequacy of the results obtained in computer simulation and physical simulation of this interaction in COMSOL Multiphysics is exemplified by numerical modeling of the SECP – electroconductive pipe interaction with respect to the problem of pipe wall thickness measuring. The performed analysis considers the effect of the major influencing factors on the SECP signal: pipe wall thickness, the gap between the probe and the pipe surface, material conductivity, pipe wall curvature, the presence of the areas with smooth V-shaped change in the wall thickness and local thinning of a spherical shape, skewness of the probe axis relative to the pipe surface, lateral misalignment of the probe axis.

References

[1] Klyuev V V 2003 *Non-destructive testing* **T2** (Moscow, Mechanical engineering)
[2] Shubochkin A E 2014 *The development and present state of the eddy current method of NDT* (Moscow, Spektr)
[3] Xuefei M, Yinzhao L (2013) Analytical solutions to eddy current field excited by a probe coil near a conductive pipe *NDT&E International* **54** 69–74 doi:10.1016/j.ndteint.2012.11.010
[4] Sandovskii V (2013) Measurements of the thickness and electrical conductivity of nonmagnetic plates by an eddy-current method *Measurement Techniques* **55**(10) 1201–1208
[5] Yating Y, Pingan D, Lichuan X (2008) Coil impedance calculation of an eddy current sensor by the finite element method *Russian Journal of Nondestructive Testing* **44** (4) 296–302
[6] Bennoud S, Zergoug M (2014) Numerical simulations of eddy current testing for plated aluminum parts *Australian Journal of Basic and Applied Sciences* **8** 47–50

A Long Pulse CuBr Vapor Oscillator for Laser Monitor Applications

F A Gubarev[1,2], M S Klenovskii[1,3]

[1] National Research Tomsk Polytechnic University, Tomsk, Russia
[2] V.E. Zuev Institute of Atmospheric Optics, Russian Academy of Sciences, Siberian Branch, Tomsk, Russia
[3] Institute of Electron Physics, National Academy of Sciences of Ukraine, Ukraine
E-mail: gubarevfa@tpu.ru

Abstract. The paper discusses the perspectives of application of a long pulse CuBr vapor oscillator for laser monitors. The influence of excitation conditions of the copper bromide vapor laser on the lasing pulse duration is discussed. It is revealed that increased pulse duration of lasing up to 200 ns and more is observed in unstable mode of the discharge. Laser monitors based on the long lasing pulse CuBr vapor oscillators can be demanded for distant objects monitoring.

1. Introduction

A number of non-destructive testing tasks need visualization of processes shielded by intensive background lighting. Such processes occur in interaction areas of a matter and powerful flow of energy (laser material processing, film deposition in a discharge, irradiation of biological tissues by a concentrated flow of energy, surface modification by exposure of electron beams), self-propagating high-temperature syntheses, discharge plasmas etc.

One of the powerful tools for observing objects and processes shielded by intensive background lighting is a laser monitor based on high-frequency monoatomic copper or copper bromide vapor lasers [1–7]. The temperature of objects under observation by means of such laser monitor can reach up to 40000 K and more [7]. The copper vapor lasers and brightness amplifiers based on them radiate at two different wavelengths in the visible range of the spectrum, namely 510.6 nm (green) and 578.2 nm (yellow).

The operating principle of a laser monitor is similar to the operating principle of a laser projection microscope [8] and is as follows. The object under observation is illuminated by an amplified spontaneous emission of a brightness amplifier. After that the reflected from the object radiation is amplified and projected onto a screen or the sensor of a digital CCD-camera. This type of a laser monitor scheme is widely used and called monostatic [3].

There are several embodiments of the monostatic laser monitor scheme that are shown in figure 1, *a* – short-focus lens is located close to the brightness amplifier (typical scheme), *b* – short-focus lens is located at a distance of several meters from the brightness amplifier, and *c* – long focal-length lens is located close to the brightness amplifier.

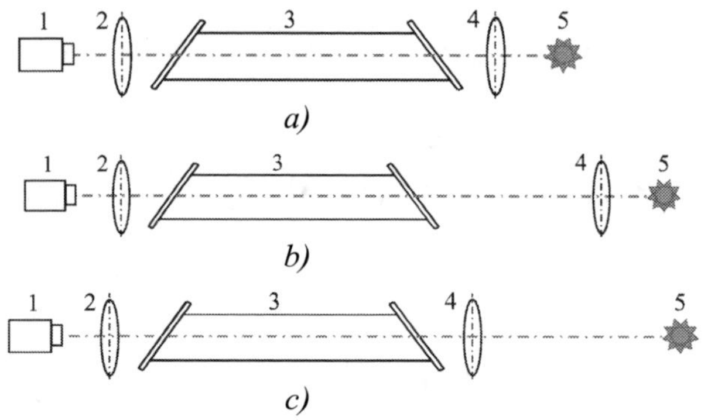

Figure 1. Laser monitor schemes: 1 – CCD-camera; 2, 4 – optical lenses; 3 – a brightness amplifier; 5 – an object under observation.

Visualizing distant objects (situated in a distance of a few meters or tens of meters) by means of a laser monitor is of special interest [9, 10]. The figure 1b and *c* shows the modifications of the typical scheme of a laser monitor for visualizing distant objects depending on the relative position of a brightness amplifier, optical lens, and an object under observation. The scheme in figure 1b has obvious limitations caused by the need of placing the lens in close proximity to the object under observation. Such a scheme is not acceptable for visualizing processes passing at high temperature or accompanied by the combustion product splashing. Moreover, when the lens is moved away from the brightness amplifier the area of view is reduced [11]. Thus, the most suitable variant of a laser monitor monostatic scheme (based on a single brightness amplifier) for observing distant objects is the one illustrated in figure 1c.

In a case of visualizing distant objects by means of a laser monitor the pulse duration of lasing can be of fundamental importance. Thus, in a case of monostatic scheme of laser monitor that is more economical in comparison with a bistatic scheme (a scheme with two brightness amplifiers [3]) the duration of existence of laser level population inversion limits the longest possible distance to an object under observation. The lasing pulse duration of copper vapor laser (copper bromide vapor laser) is influenced by such factors as peculiarities of active medium, composition and component ratio of working mixture, geometry of a gas-discharge tube (GDT), excitation conditions, in particular the discharge pumping pulse waveform. Typical lasing pulse duration of copper vapor laser is 20–40 ns [12–15]. In a case of CuBr laser the typical pulse duration of lasing can reach 80 ns [16].

The maximum possible pulse duration of lasing on self-terminating transitions is estimated accurate enough in [17]. According to the work estimations the maximum possible pulse duration of copper vapor laser in the self-terminating mode is limited to 175 and 206 ns for the transitions of 510.6 and 578.2 nm, respectively. Increasing the pulse duration of lasing more than estimated values is possible by means of controlled current restriction of the discharge by a series connection of an additional tacitron in the discharge circuit [18–20]. The record radiation pulse duration was achieved to be 230 ns for copper vapor laser [19] and 320 ns for CuBr laser.

Using systems of controlled current restriction of the discharge is not acceptable in low-cost applications. The aim of the current research is to investigate an influence of excitation conditions of CuBr laser on the pulse duration without using an additional system of controlled current restriction of the discharge. It should be noted that this research is not aimed at increasing the energy parameters of the laser but only at the possibility of lasing with radiation pulse duration of the order of 200 ns. In laser monitor applications the duration of amplification is more crucial than the power of radiation. Therefore, it is quite acceptable compromise to increase the duration of existence of population inversion to the detriment of power and energy of radiation.

2. Experimental technique

In the present research it was employed CuBr lasers with external heating the active region of the GDT which was also used in [21, 22]. The active part of GDT (90 cm in length and 5 cm in diameter) was inserted into an external heater to provide the capability of stabilizing heating temperature of the GDT. The heating temperature of the GDT was measured and controlled by a thermocouple and a microcontroller. Pressure of buffer gas (neon) in the GDT was 30 Torr.

Pumping active medium of GDT was carried out by means of a scheme of pulsed charge of a storage capacitor and its direct discharge by a thyratron TGI1-1000/25 [22]. In the experiments the PRF of discharge excitation was set by means of digital generator GW Instec SFG-72120 and varied in the range of 1–10 kHz.

As sensors of voltage on the GDT electrodes, current through the GDT, and laser pulse, were used P6015A Tektronix voltage probe, 8450 Pearson Current Monitor, and FC-22 coaxial photoelectric cell, respectively. The signals from the corresponding sensors were routed to inputs of WJ-324 LeCroy digital oscilloscope. Pulse duration of lasing was taken at the pulse base (at the level of 5% of the pulse amplitude).

3. Results and discussion

One of the most influencing parameters of operation is PRF of excitation. Figure 2 shows a result of influence of the PRF of discharge excitation on the pulse duration of lasing.

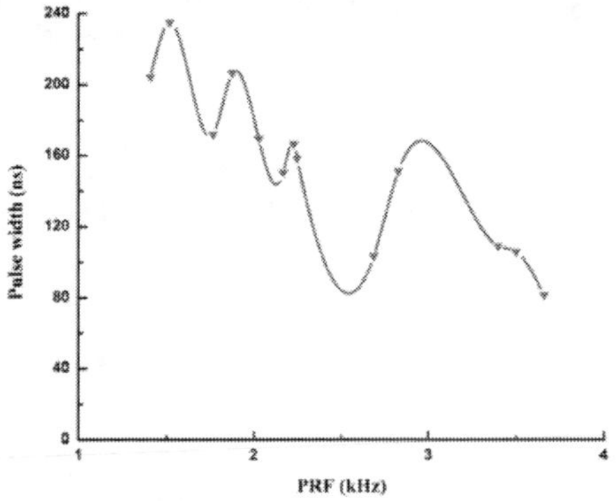

Figure 2. Influence of the PRF of excitation on the pulse duration of lasing. The heating temperature of the reservoir with CuBr powder is 390°C, voltage on the rectifier is 195 V, capacitance of storage capacitor is 4.95 nF.

As one can see (figure 2) at decreasing the excitation PRF the pulse duration of lasing is increased. Pulse duration of the CuBr laser achieved up to 200 ns and even more (figure 2). However, at the same time decreasing the excitation PRF resulted in a deterioration of the discharge shape. In this case the discharge contracts and moves (it is observed how the discharge current changes the area of flowing) inside the GDT. It should be mentioned that according to the work [23] the longer pulse duration of lasing was observed at higher value of the storage capacitor. Considerable reducing excitation PRF leads to disruption of the discharge. Higher storage capacitor value allowed working at lower PRF of discharge excitation and as a result to observe broader pulse of lasing.

According to our assumptions increasing the pulse duration of lasing supposedly may be caused by a spatial-temporal inhomogeneity of the discharge, namely by shifting the discharge within the cross section of the GDT to the colder areas where the conditions for realizing population inversion is more favorable. Thus, the population inversion occurs in different parts of the GDT active volume and in the different instant of time. But at the same time moving the discharge inside the GDT causes deterioration of laser beam quality.

In the work it was also carried out an estimation of laser beam profile quality by a simplified technique similar to [24]. Figure 3 shows photographs of laser beam profiles recorded by FastCam

HiSpec1 CCD camera. The exposition of CCD camera was 2 μs. Thus CCD camera recorded images formed by single pulses of lasing. The shown photographs correspond to different pulse durations of lasing. In the case when the pulse duration of lasing was 170 ns (figure 3*a*) the CuBr laser oscillation was characterized by a deformation of laser beam profile. Figure 3 allows comparing deformations of laser beam profile at excitation conditions corresponding different pulse duration of lasing. To use the active element of a laser as a brightness amplifier in a laser monitor the laser beam profile should be characterized as uniform as possible distribution. The longest obtained pulse of lasing with at the same time satisfactory radial distribution of laser beam profile was 100 ns. Laser pulse duration of 100 ns is sufficient for visualizing objects situated in a distance of 15 m from the brightness amplifier by a laser monitor (a case of the monostatic scheme).

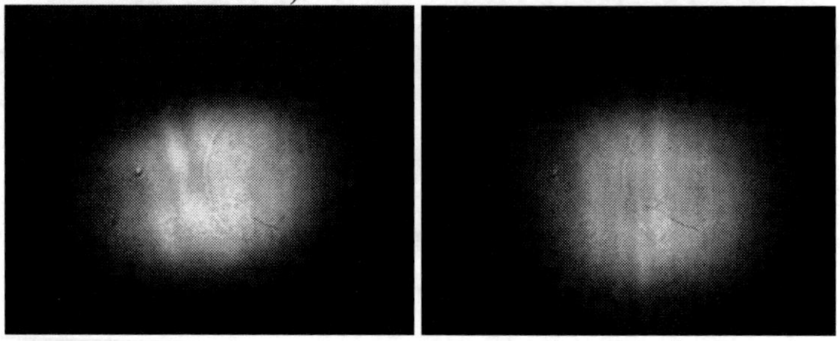

a *b*

Figure 3. The photographs of laser beam profiles corresponding to: *a*) CuBr laser pulse duration of 170 ns, PRF of excitation of 2 kHz; *b*) CuBr laser pulse duration of 100 ns, PRF of excitation of 2.6 kHz. The heating temperature of the reservoir with CuBr powder is 390°C, voltage on the rectifier is 195 V, capacitance of storage capacitor is 4.95 nF.

Figure 4 demonstrates dependence the pulse duration of lasing on the PRF of excitation at various heating temperatures of the reservoir with CuBr powder, i.e. at different copper bromide vapor content. The shown dependences just seemingly demonstrate the fact that increasing the pulse duration of lasing occurs in the case when the discharge transitions to the unstable mode. As one can see (figure 4) at increasing the concentration of copper bromide vapor the fluctuation of the pulse duration observes at higher values of the PRF. This fact correlates well with the corresponding shift of the critical frequency value at which the discharge transitions to the unstable mode.

In metal halide vapor lasers a pumping pulse provides not only the excitation of working atoms but also dissociation of metal halide molecules and heating the active medium. Therefore, this type of lasers is characterized by the presence of a lower lasing-action threshold of PRF which depends on the discharge excitation conditions. As one can see (figures 2 and 4) the lower lasing-action threshold for the current conditions is correspond to PRF of 1.5-2 kHz (see also [20]). The longest pulse durations of lasing are observed just at the low PRF (figures 2 and 4).

The waveforms analysis also showed that the lasing occurs with considerable delay after the thyratron triggering up to 200 ns (see also [23]). This delay is caused by the fact that decreasing PRF of excitation and energy deposited to the discharge lead to increasing pre-pulse resistance of the discharge and as a result the breakdown of the discharge becomes more difficult.

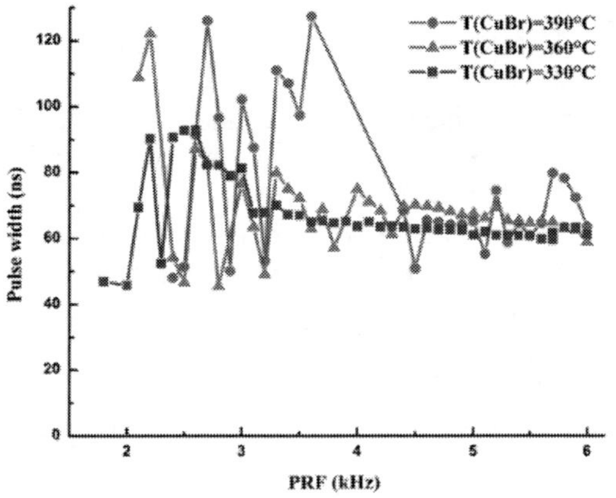

Figure 4. Influence of the PRF of excitation at various heating temperature of the reservoir with CuBr powder on the pulse duration of lasing. Voltage on the rectifier is 200 V, capacitance of storage capacitor is 1.65 nF.

4. Conclusions

Results of study the influence of excitation conditions of the copper bromide vapor laser on the pulse duration of lasing are presented. Increased pulse duration of lasing is revealed can be observed when the discharge transitions to an unstable mode. Such modes are achieved at low pulse repetition frequencies.

It is demonstrated that CuBr laser oscillation with pulse duration of lasing up to 235 ns at the laser pulse base can be obtained without using an additional system of controlled current limiting of the discharge. But at the same time the laser beam quality deteriorates appreciably as well. To use the active element of a laser as a brightness amplifier in a laser monitor the laser beam profile should be characterized as uniform as possible distribution. According to the results the longest obtained pulse of lasing with at the same time satisfactory distribution of laser beam profile was 100 ns. Laser pulse duration of 100 ns is sufficient for visualizing objects situated in a distance of 15 m from a brightness amplifier by a laser monitor (a case of the monostatic scheme).

Acknowledgments

The authors gratefully acknowledge Prof. G.S. Evtushenko of Tomsk Polytechnic University (Russia) for his interest to this work and helpful discussions.

The author, Miron Klenovskii, is expressing his gratitude to the Tomsk Polytechnic University for the opportunity to conduct scientific research in the TPU postdoctoral program.

References

[1] Batenin V M, Klimovsky I I and Selezneva L A (1988) Study of the surface of electrodes of a carbon arc during arcing *Soviet Phys. Dokl.* **33** 949–952

[2] Abramov D V, Galkin A F, Zharenova S V, Klimovsky I I, Prokoshev V G and Shamanskaya E A (2008) Vizualisation of laser radiation interaction with surface of glass and pyrocarbon by means of laser monitor *Bul. Tomsk Polytechnic Univ.* **312** 97–101

[3] Kuznetsov A P, Buzhinskij R O, Gubskii K L, Savjolov A S, Sarantsev S A and Terekhin A N (2010) Visualization of plasma-induced processes by a projection system with a Cu-laser-based brightness amplifier *Plasma Phys. Reports* **36** 428–437

[4] Trigub M V, Evtushenko G S, Gubarev F A and Torgaev S N (2011) Laser monitor with frame-by-frame imaging *Testing. Diagnostics* (**special issue**) 140–143

[5] Buzhinsky R O, Savransky V V, Zemskov K I, Isaev A A and Buzhinsky O I (2010)

Observation of objects under intense plasma background illumination *Plasma Phys. Rep.* **36** 1269–1271

[6] Evtushenko G S, Trigub M V, Gubarev F A, Torgaev S N and Shiyanov D V (2013) Laser monitor model based on CuBr-laser active volume for non-destructive testing and diagnostics *Testing. Diagnostics* **13** 42–45

[7] Evtushenko G S, Trigub M V, Gubarev F A, Evtushenko T G, Torgaev S N and Shiyanov D V (2014) Laser monitor for non-destructive testing of materials and processes shielded by intensive background lighting *Rev. Sci. Instrum.* **85** 033111-1–5

[8] Petrash G G 1991 *Optical Systems with Brightness Amplifiers* (Moscow: Nauka)

[9] Buzhinskij O I, Vasiliev N N, Moshkunov A I, Slivitskaya L A and Slivitsky A A (2002) Copper vapor laser application for surface monitoring of divertor and first wall in ITER *Fusion Engineering and Design* **60** 141–155

[10] Buzhinskij O I, Otroshchenko V G, Slivitsky A A and Slivitskaya I A (2003) Videoscope on the basis of copper vapor laser for spatially-temporal diagnostics of tokamak discharge chamber internal components *Plasma Devices and Operations* **11** 155–160

[11] Fedorov K V, Trigub M V and Evtushenko G S (2015) Laser monitor for remote object visualization *Int. Siberian Conf. on Control and Communications (SIBCON), IEEE* in press

[12] Batenin V M, Bokhan P A, Buchanov V V, Evtushenko G S, Kazaryan M A, Karpukhin V T, Klimovsky I I and Malikov M M 2011 *Lasers on Self-Terminating Transitions of Metal Atoms* 2nd ed (Moscow: Fizmatlit)

[13] Soldatov A N and Solomonov V I 1985 *Self-Terminating Gas Discharge Metal Vapor Lasers* (Novosibirsk: Nauka)

[14] Little C E 1998 *Metal Vapor Lasers: Physics, Engineering and Applications* (Chichester: John Willey & Sons Ltd)

[15] Grigor'yants A G, Kazaryan M A and Lyabin N A 2005 *Copper Vapor Lasers: Construction, Characteristics, and Applications* (Moscow: Fizmatlit)

[16] Yudin N A, Sukhanov V B, Gubarev F A and Evtushenko G S (2008) On the nature of phantom currents in the active medium of self-contained metal atom transition lasers *Quantum Electron.* **38** 23–28

[17] Bokhan P A (2011) On pulse duration of self-terminating lasers *Quantum Electron.* **41** 110–114

[18] Soldatov A N (1993) Physics and technology of copper-vapor lasers with controlled parameters *Atmospheric and Oceanic Optics* **6** 650–658

[19] Evtushenko G S 1996 Applications of Pulsed Metal Vapour Lasers in Atmospheric Optics *Pulsed Metal Vapour Lasers (NATO ASI Series* vol 5*)* ed C E Little and N V Sabotinov (Dordrecht: Kluwer Academic Publishers) pp 445–452

[20] Fedorov K V, Fedorov V F, Gubarev F A and Evtushenko G S (2014) Long-pulse copper bromide laser *International Conference of Young Specialists on Micro/Nanotechnologies and Electron Devices, EDM* 333–336

[21] Gubarev F A, Evtushenko G S, Vuchkov N K, Sukhanov V B and Shiyanov D V (2012) Modeling technique of capacitive discharge pumping of metal vapor lasers for electrode capacitance optimization *Rev. Sci. Instrum.* **83** 055111–115

[22] Dimaki V A, Sukhanov V B, Troitskii V O and Filonov A G (2012) A stabilized copper bromide laser with computer-controlled operating modes and a mean lasing power of 20 W *Instrum. Exp. Techniq.* **55** 696–700

[23] Gubarev F A, Klenovskii M S 2015 About CuBr laser pulse duration in an unstable discharge *16th Int. Conf. of Young Specialists on Micro/Nanotechnologies and Electron Devices (EDM), IEEE* 292–296

[24] Gubarev F A, Klenovskii M S and Li Lin (2015) Radial distribution of a single-pass amplified radiation in the active elements of CuBr lasers *IOP Conf. Series: Mater. Sci. Eng.* **81** 012116

The Statement of Design and Application Questions for the Gyroscope with a Gas-dynamic Suspension of Ball Rotor in the Navigation Support Drilling System

A N Golikov[1] and A A Ignatovskaya[2]

[1] Chief of laboratory, NDT Institute, Tomsk Polytechnic University, Tomsk
[2] Postgraduate student, NDT Institute, Tomsk Polytechnic University, Tomsk
E-mail: nastena@tpu.ru

Abstract. The general issues of creation the inertial navigation system are considered. Analysis of the possible implementation of the gyroscope with a gas-dynamic suspension of ball rotor for using it as a sensor in the information and measuring complex is provided. The permissible layout construction is proposed. The software system variants for the mathematical modelling of the main gas-dynamic bearing characteristics are considered. Mathematical modelling is essential to extend and optimize the design parameters of the developed gyroscopic construction. Some results of modelling, theoretical evaluation and preliminary experimental studies are summarized.

1. Introduction

In the process of well construction measurement accuracy and efficiency of data submission largely determine the construction cost as a whole. Consequently, the creation of reliable measuring systems with an acceptable accuracy, which based on inertial sensors, is a very urgent task. Such systems must suffice the requirements for the drill body orientation in the process of the well development [1].

Currently, the borehole orientation systems which use the magnetical elements are known very well. Accuracy of such systems is highly depends on the presence of the magnetic anomalies near from the flux gate sensors. In contrast, the main advantages of the gyroscopic sensor elements that determine their application in the borehole orientation systems are immunity to ferromagnetic masses and magnetic anomalies, as well as positioning the borehole axis in relation to the geographic coordinate system. But, inertial sensors application in the borehole orientation systems shows their low reliability under the high level of external exposure factors because of their support and other mechanisms. The most stable, in such conditions, can be gas-dynamic bearings as a support of the sensor rotor.

Appliance of the gas-dynamic bearings is mainly designated by the features that characterize the natural gas lubrication [1]. Gases, essentially, have the less viscous than liquid. Environmental temperature has a little impact on them. Even less they affected by the environmental pressure. Such viscosity stability and its small value open up the great possibility for using such type of bearings in the devices which operate at the high speeds over a wide temperature range.

Gas bearings may also be used in the field of high radioactivity as organic lubricant in such conditions loses its operating properties. Moreover, gas-dynamic bearings, compared to other, have no limitation on the service life because of wear absence during operation. Also, these bearings exclude the number of negative phenomena caused of the dense vibrations spectrum as in the case of the ball bearings [1].

2. The statement of the main problem

The main document that sets out requirements for the geophysical equipment is the standard 26116-84. According to this document, during the design process of borehole navigation system their size is strictly limited as the diameter of the well is a standard value. The second decisive factor is the operating conditions, which are regulated by this normative document too [1].

The list of defined parameters which is determined by the system is ordinary: determination of the azimuth, zenith angle and the angle of tool rotation. Accuracy requirements for inertial navigation system elements, in particular for gyroscopes, have a value in the range of 0.25–1 angular degree. It is necessary for the storage accuracy or recreation in another way the reference inertial orientation [2].

The main goal of this research is to develop the sensor, which based on the gyroscope with a gas-dynamic suspension of ball rotor. And inside of this basic goal there are different tasks. At this moment, the most important is a modelling of the main bearing characteristics. It is essential to determine the optimal parameters for the estimation of the technical realization the position sensor or an angular velocity sensor, which can be used in the borehole orientation system.

Accuracy and reliability of the gyroscopic device are requested to ensure due to sufficient reserves of bearing capacity, stiffness of gas-dynamic bearings and, therefore, ensure the gyroscopic device durability [3].

It is expected that the application of this type of sensor in the navigation support drilling system will improve measurement accuracy, reduce human and financial costs without quality loss in the implementation of promising drilling types.

The basic operation principle of borehole orientation system, where developed gyroscopic device can be used, is the following: location of the drilling tool is defined in relation to the geographically oriented coordinate system. For its creation two reference vectors are used: the gravity acceleration at the measuring point and the angular velocity of the Earth's rotation. In order to determine the projection of the gravity acceleration, navigation system must contain, at least, two accelerometers. Information about the angular velocity of the Earth's rotation comes from the gyroscopic device. In this case, the gyroscope is a two-part angular velocity sensor.

Further, by means of the certain algorithms, orientation parameters of the drilling tool are calculated relative to the reference coordinate system [4]. It is important that the sensor will be efficient and retain their accuracy characteristics at the stopping time of the borehole tool for measurement. If these requirements are achieved it would lead to a fundamentally new level in matters of metrological support drilling.

The factors, which have mentioned above, define the basic requirements for the design and operating conditions of inertial sensors, show that the computational simulation of the bearing characteristics coming to the fore because of several reasons. Firstly, the formulation of the experiment is extremely complicated by the need to provide special materials and technical base. Secondly, manufacture of support elements is difficult. The geometry of the bearing parts has the high demands.

On the other hand, preparation of the settlement programs is also very complex and multifaceted task, because of successful realization is necessary to have certain skills in programming and applied mathematics. There are a number of universal software tools that are used to solve some problems of gas dynamics, such as LS-DYNA, ABAQUS, ANSYS modules - CFX and Fluent, Flow Vision and others. Their usage is conditioned by the fact that they help to solve some problems of gas lubrication. When they are used in this field, there is a set of specific problems, which do not allow using them effectively. This is the basic complexity of the gas-dynamic bearing characteristics mathematical modelling.

The general gyroscope layout view and the construction are shown in figure 1 and figure 2.

The main parts of the gyroscope layout are two hemispherical bowls 7 and rotor 1, which is placed in the cavity of the bowls. Rotor is a standard ball bearing and has an axial bore, where movable elements of angle sensor 4 are pressed. Mating part of the angle sensor 3 is positioned in one of the bowls. The diameter of the bowls for $5 \div 10$ microns more than the actual diameter of the ball rotor. The rotor diameter is 28.587 mm. This difference provides the initial gap. Ball rotor is rotated by the three-phase asynchronous electric stator 2. Its power is 40 V at frequencies of 500 or 1000 Hz.

As the rotor rotates, the gas is drawn into the initial gap between the bowls and the rotor. With further gas influx into the bearing gap the overpressure is formed. Whereby, the rotor "floats" and when the rotor rotates at nominal speed and it provides the constant gas lubrication.

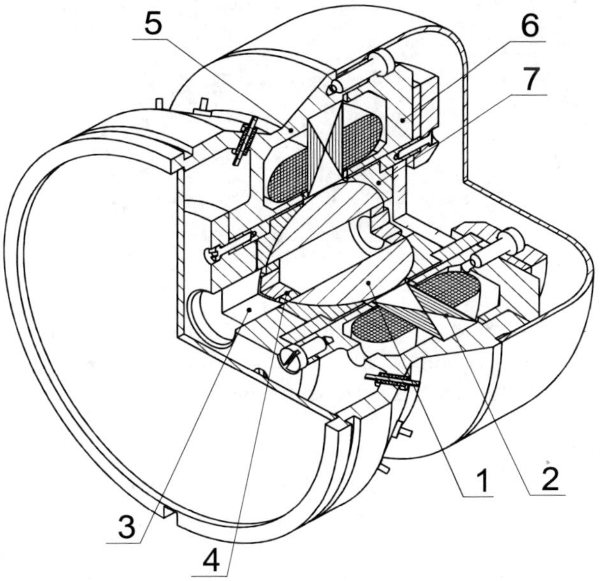

Figure 1. The construction of the gyroscope layout with a gas-dynamic suspension of ball rotor: 1 – ball rotor, 2 – stator of the electric driver, 3 – two-part angle sensor, 4 – mating part of the angle sensor, 5 – device case, 6 – device cover, 7 – hemispherical bowl.

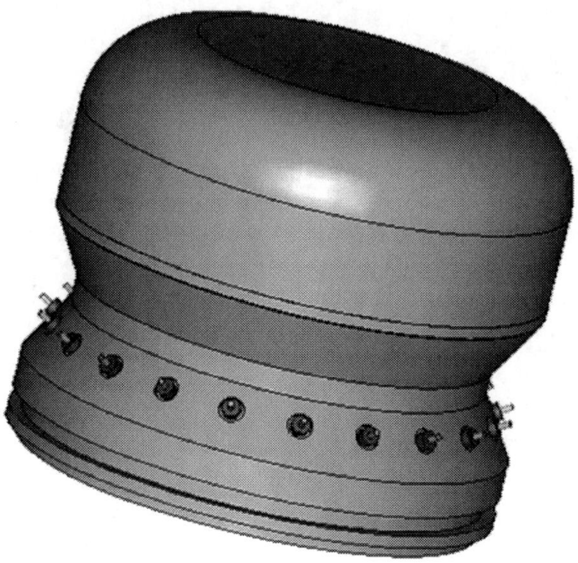

Figure 2. The general view of the gyroscope layout

In the general view do not shown the elements of controlling the ball rotor position.

Layout overall: diameter – 77 mm, length – 68 mm. The dimensions are not final.

3. Results and considerations

The dependence of the pressure distribution in the bearing gap was determined by the mathematical modelling in the software envelope ANSYS 15.0 - Fluent and estimations of the main gas-dynamic bearing characteristics. Also, the influence of the relative eccentricity and radial clearance to the minimum rotor speed at which the bearing starts to work in gas lubrication mode was established. Moreover, the estimation of the viscous friction moment value and stiffness of the suspension at different values of the radial clearance in the bearing were assessed too.

The range of the bearing gap values is from 2 to 20 microns by selecting the diameters of the ball rotors. It was found that the lifting force will be decreased because of clearance reduction and, as well as, pressure reduction in the bearing. The optimal gap values in the bearing up to 2 microns. In this case, the stable operation of the gas lubrication at a given load (at the moment, this load is a own weight which equal to 0.8 N) will be achieved. If the gap value is equal to 2 microns it will mean that we can get a pressure of 3102.8 Pa on a smooth surface of the bearing. This value has an equivalent of the bearing lifting force is 1.24 N at the rotating speed 16 000 rev/min. According to these estimations, the rotor floats at the speed of 300 rev/min. The results are not effective. If the gap is less than 2 microns, we can get the pressure, which is several times greater than the pressure at 2 microns. Such values are guaranteed to provide a margin of the bearing capacity, which is currently unavailable.

The maximum value of the viscous friction moment achieves the value 0.0023 Nm. The magnitude of this parameter makes it possible to evaluate the effect of air resistance to heat and sensitivity of the device. The primary method of the reducing the influence of viscous friction is the introduction of the gaseous medium [5]. It can greatly reduce the heat loss and increase the number of revolutions, thereby providing a stable value of the angular momentum.

As for the stiffness of the suspension, it largely shows the subsidence of the ball rotor [6]. It can lead to an undesirable eccentricity between the ball rotor and the hemispherical bowls. The values, which were achieved, are relatively low. It is also explains by the smooth surface of the bearing. Bearing shaping can be as a primary solution for increasing the rigidity of the bearing. Also, for achieving the maximum suspension stiffness, the clearance must be done as low as possible.

4. Conclusion

Mathematical modelling of the gas-dynamic characteristics for the ball gyroscope shows the convergence of the estimating and full-scale simulation results. It shows that this realization of the gyroscope with a smooth profile of the bearing is inefficiency in terms of bearing capacity.

Further research will be focused on a testing of physical and mathematical model of the gas-dynamic bearing characteristics, increasing of the bearing capacity and rigidity. It will also be connected with an optimization of the ball gyroscope construction in particular bearings profiling, selection of the gaseous medium, and issues related to a control of the ball position. It is necessary for a further application of the gyroscope as a sensor in the underground navigation.

References

[1] Kuzma A A, Plekhanov M S, Golikov A N (2012) Research design and some elements for gyroscope with gas-dynamic suspension of rotor ball *Testing. Diagnostics* **(13)** 167–172

[2] Binder Ya I, Gutner I E, Mezentsev A P , Molchanov A A (2003) Present-day information-measuring systems for subsurface navigation and orientation *Gyroscopy and Navigation* **1**(40) 110–123

[3] Belyanin L N, Golikov A N, Martemyanov V M, Samojlov S N (2002) Experience of the gyroscopic inclinometer creation *Geofit* **(2)** 327

[4] Dubinin A V, Smolyan KV (2012) Dependence of load carrying capacity of gas-dynamic support of dynamically tuned gyroscope upon gaps in gas-dynamic support *Support Engineering Journal: Science and Innovation* **3** 41–42

[5] Agrawal G L (1997) Foil air/gas bearing technology – an overview *ASME 1997 International Gas Turbine and Aeroengine Congress and Exhibition* 1–11

[6] Whipple R T P (1951) Theory of the spiral grooved thrust bearing with liquid or gas lubricant *Atomic Energy Research Establishment* 1–24

A Simple Approach of Presampled Modulation Transfer Function Measurement Tested on the Phoenix Nanotom Scanner

D Ivashkov[1], A Batranin[2] and T Mamyrbayev[3]

[1] Master student at Institute of Physics and Technology, Tomsk Polytechnic University , Tomsk, Russia
[2] Junior researcher at Institute of Nondestructive Testing, Tomsk Polytechnic University, Tomsk, Russia
[3] Research engineer at Institute of Nondestructive Testing, Tomsk Polytechnic University, Tomsk, Russia
E-mail: dvi2@tpu.ru

Abstract. In this paper presampled modulation transfer function of the 2D images obtained on the Phoenix Nanotom scanner was investigated with different measurement set-ups. Three parameters were chosen to investigate their influence on modulation transfer function: source-detector distance, tube current and binning mode. A simple method for modulation transfer function determination of digital imaging detectors from edge images was applied. The following results were achieved and briefly discussed: modulation transfer function improves with increase of the source-detector distance, slightly improves with increase of the current and remains constant for different binning modes. All measurements were carried out in University of Applied Sciences Upper Austria at Wels campus.

1.Introduction

In digital radiography as well as in tomography modulation transfer function (MTF) is one of the basic performance measures describing the signal transfer characteristics of the system as a function of spatial frequency [1, 2]. It can provide different information depending on the experiment results and measurement method applied. Therefore, it is used to investigate the capabilities of the new system, to find the weakest component of the scanner as a signal processing chain [3] or to investigate stability of the system and its behaviour while changing its parameters (as in this case), etc. It should be mentioned, that in this case the qualitative analysis was provided since the quantitative analysis demands rather complex positioning procedures [4]. See also the accuracy disclaimer section for additional information.

2.Materials and methods

In this study, presampled MTF was investigated which is a function of the signal transfer before the detector's sampling stage. In this approach the line profiles obtained from the image of slightly angulated edge are rearranged in a way to neglect the influence of the detector sample grid on the MTF. Therefore, the result function is independent on the signal patterns. Details of the image acquisition procedure as well as the calculation flow of presampled MTF can be found in the corresponding articles [4, 5]. Although some details were not fulfilled due to lack of suitable positioning instruments, as will be described further. The test object used was a utility cutter made of steel approximately 0.5 mm in width (see figure 1).

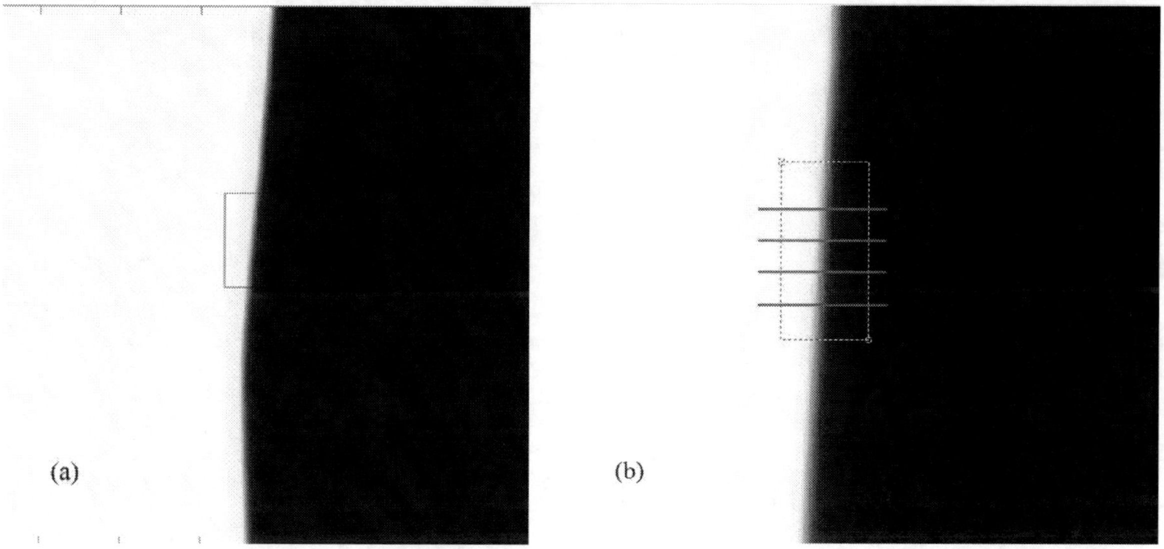

Figure 1. Example of the initial image with ROI for the MTF evaluation (a); linear profiles are obtained along the red lines (b).

During the investigation, a voxel size (VS) adjustment procedure was carried out using the ruby ball bar (see table 1). The error appeared to be relatively small and can be neglected for this particular task. Also we considered the error to be independent on the measurement parameters. It was decided then to use the voxel size stated by the acquisition software as the pixel size of the obtained 2D images.

Table 1. Parameters and results of the pixel size adjustment. SOD stands for source-object distance

Experiment	SDD (mm)	SOD (mm)	Stated VS (μm)	Measured VS (μm)	Nominal error (%)
1	500	50	5	5.075	1.5
2	200	20	5	5.04	0.8

3.Accuracy disclaimer

As it has been mentioned, it was not possible to provide positioning of the object needed for accurate quantitative analysis. For estimating of the inherent error the influence of two parameters on the result was estimated:

- Incline of the object in the detector plane (figure 2).
- Rotation of the object around the rotation axis of the drive (figure 3).

As can be seen from these pictures there is a significant influence of these parameters on the obtained MTF value, which implies a systematic error – apparently a bias to the worse MTF. This makes accurate quantitative analysis impossible. However since the object positioning was constant during the experiments (except for the SOD) it is possible to provide qualitative analysis.

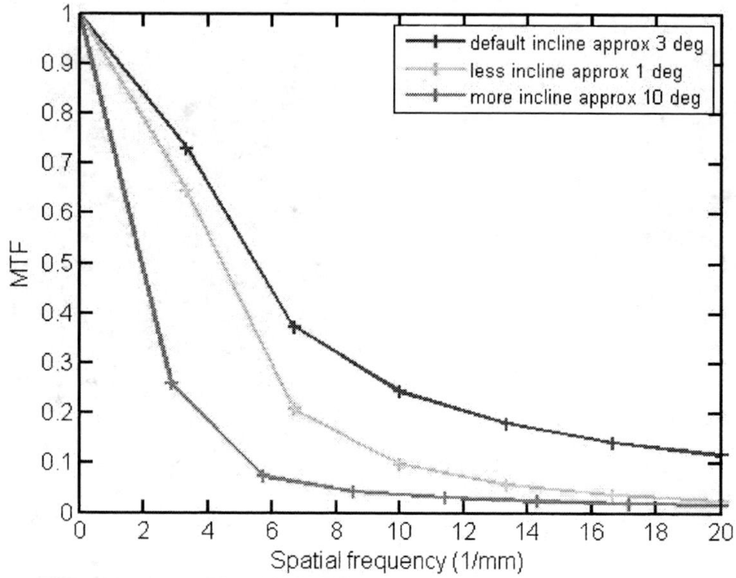

Figure 2. MTF curves for different incline of the test object in the detector plane: 3° (default) – blue, 1° – green, 10° – red.

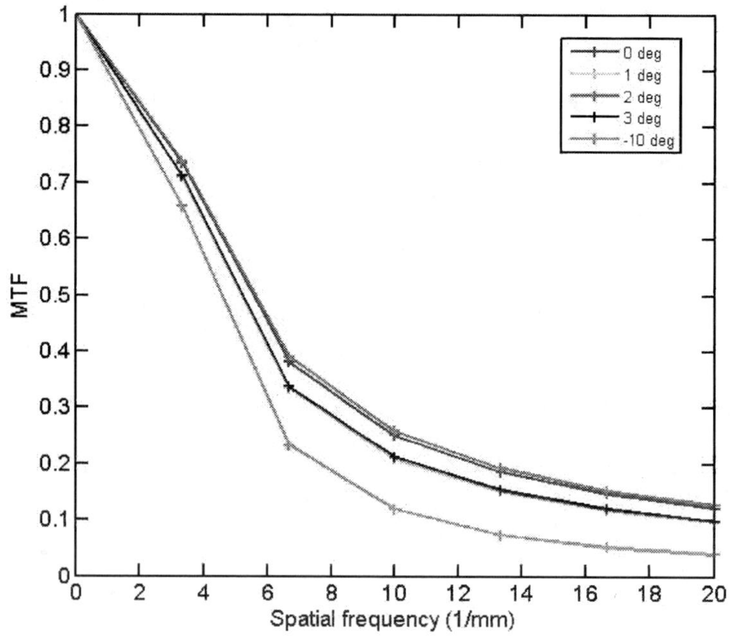

Figure 3. MTF curves for different rotation of the object around the rotation axis of the drive: 0° (default) – blue, 1° – green, 2° – red, 3° – black, -10° – purple. The highest values of MTF on 0 and 2 degrees are due to non-uniform shape of the cutter.

4.Experiments

All experiments within one group were provided in random order to avoid any potential disturbances due to unknown factors. Averaging of 50 projections was used to improve the signal-to-noise ratio.

4.1SDD impact study

In this case only the SDD was considered as a variable. Source-object distance (SOD) and integration time were changed for every experiment to make the geometric magnification and the mean grey value to be the same. In case of grey values that was made approximately. Any other parameters remained the same (see table 2). Results of measurement are shown on figure 4.

Table 2. Parameters for the SDD impact study

Target	U (kV)	I (uA)	Binning	V (um)		
Molybdenum	45	370	2x2	5		
Experiment	1	2	3	4	5	7
SDD (mm)	200	250	300	350	400	500
SOD (mm)	10	12,5	15	17,5	20	25

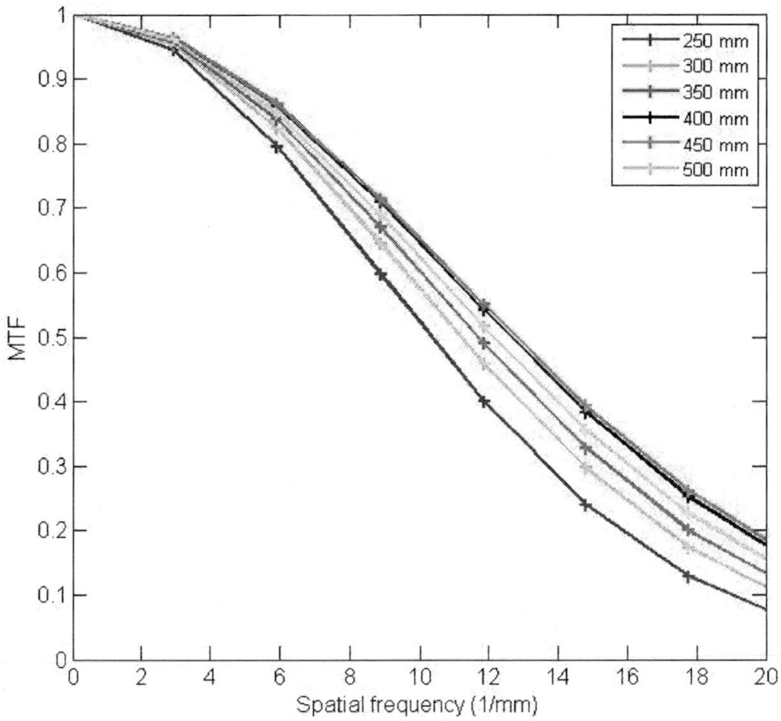

Figure 4. MTF curves for different SDD values: 250 mm – blue, 300 mm – green, 350 mm – red, 400 mm – black, 450 mm – purple, 500 mm – light blue. MTF improves along with increase of SDD.

4.2Study of the tube current impact

In this case only the current was considered as a variable. Integration time was changed for every experiment to make mean grey value approximately be the same. Any other parameters remained the same. Mode 1 was used since it has been stated by the manufacturer that the tube power doesn't influence the focal spot size. Therefore, no apparent decrease in resolution should be expected as the current changes. We used different current values (see table 3) and have found that MTF even improves slightly when current increases (see figure 5).

Table 3. Parameters for the current impact study

Target	U (kV)	SDD (mm)	SOD (mm)	binning	VS (um)
Molybdenum	80	300	12	1×1	2
Experiment	1	2	3	4	5
I (uA)	180	150	120	90	60

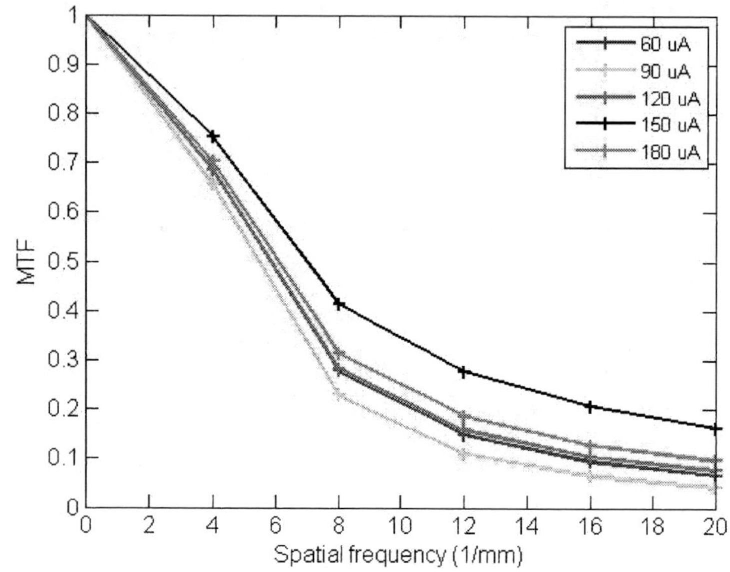

Figure 5. MTF curves for different current values: 60 uA – blue, 90 uA – green, 120 uA – red, 150 uA – black, 180 uA – purple. MTF slightly improves along with increase of the tube current.

4.3 Study of the binning mode impact

In this case only the binning was considered as a variable. SOD and integration time were changed for both experiments to make geometric magnification to be the same and the mean grey value to be the same approximately. Any other parameters remained the same (see table 4). As shown on figure 6, binning does not influence on MTF noticeably.

Table 4. Parameters for the binning mode impact study

Target	U (kV)	SDD (mm)	I (uA)	VS (um)
Molybdenum	80	300	150	2
Experiment	1	2		
Binning mode	1×1	2×2		
SOD (mm)	12	6		

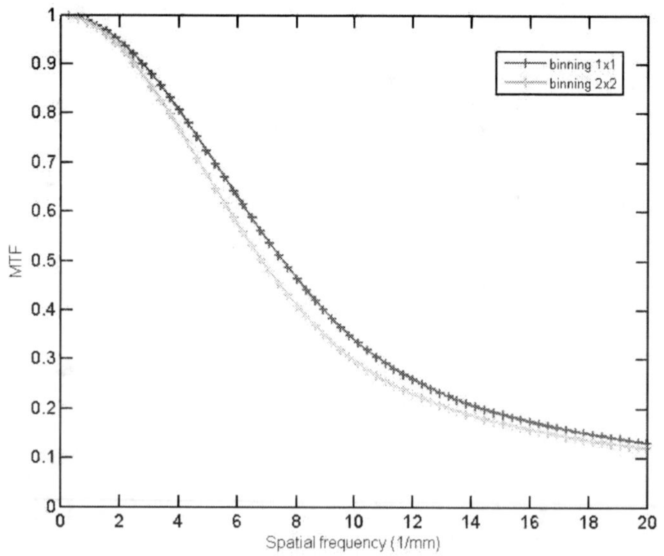

Figure 6. MTF curves for different binning modes: binning 1×1 – blue, binning 2×2 – green. MTF varies slightly with change of the binning mode.

5. Discussion

To conclude here are the main results of these experiments and possible explanations:

1) MTF increases with the SDD becoming greater. The effect is more pronounced for higher spatial frequencies. The possible reason is increase of scattering influence along with decrease of SDD.
2) Higher tube current slightly improves MTF. This apparently proves manufacturer's claims that mode 1 provides focal spot more or less independent of the tube power. Therefore the reason for the slight improve of the MTF with increase of the current should have something to do with non-linear response of the detector for different integration times.
3) MTF remains almost the same for different binning modes. The slight decrease in the resolution for the binning mode 2×2 is probably due to the higher scattering because of the less SDD value.

Acknowledgments

This work was financially supported by The Ministry of Education and Science of the Russian Federation in part of the science program.

References

[1] Williams T L 1999 The optical transfer function of imaging systems (Bristol, Institute of Physics Publ.)

[2] Batranin A, Ivashkov D and Stuchebrov S (2015) Performance evaluation of micro-ct scanners as visualization systems *Advanced Materials Research* **1084** 694–697

[3] Buzug T M 2008 *Computed tomography: from photon statistics to modern cone-beam CT* (Berlin, Springer Science & Business Media)

[4] Buhr E, Günther-Kohfahl S and Neitzel U (2003) Simple method for modulation transfer function determination of digital imaging detectors from edge images *Proceedings of SPIE - The International Society for Optical Engineering* **5030** 877–884

[5] Samei E, Flynn M J and Reimann D A (1998) A method for measuring the presampled MTF of digital radiographic systems using an edge test device *Med. Phys.* **25**(1) 102–113

Narrowband Registration of Ultra-Wideband Signal

A V Kamenev[1], R N Satarov[2], V P Yakubov[3], S E Shipilov[4]

[1]Bachelor student, the department of radiophysics, Tomsk State University, Tomsk, Russia
[2]Candidate of Physico-mathematical Sciences, Jr. Researcher the department of radiophysics, Tomsk State University, Tomsk, Russia
[3]Doctor of Physico-mathematical Sciences, professor the department of radiophysics, Tomsk State University, Tomsk, Russia
[4]Candidate of Physico-mathematical Sciences, associate professor the department of radiophysics, Tomsk State University, Tomsk, Russia
E-mail: s.shipilov@gmail.com

Abstract. The research of various parameters influence on the Fresnel reflector focusing properties in particular spatial sizes, the focal point position, etc. is conducted. The version of using a flat Fresnel reflector for receiving ultra-wideband signals with the use of Wiener filtering is suggested.

1. Problem of receiving UWB signals

Main distinctive feature of ultra-wideband (UWB) signals is a very wide band of occupied frequencies and they usually have pulse signals form. UWB signals become more widely used in practice: radiolocation, radiotomography, communication, etc. Preferential using of narrowband electronics is the deterrent factor in this development path. If the problem of generating UWB signals is almost solved, the problem of receiving such signals is still actual. Using stroboscopic receivers or high-speed ADCs are two alternative approaches in solving this problem but both of them are quite complicated. The solution of the problem with using narrowband systems and technologies of solving inverse problems is suggested in present work. Fresnel mirror is considered as an example of narrowband device. The ability of amplifying signals without high energy cost is its main advantage [1–3].

2. Flat Fresnel reflector modeling

Let a flat wave propagates in free space with a given Decart coordinate system. It incident at a certain angle to the axis z. $\mathbf{k}=(-k\sin\alpha, 0, -k\cos\alpha)$ – a wave vector, $k = 2\pi f/c$ – its module (figure 1).

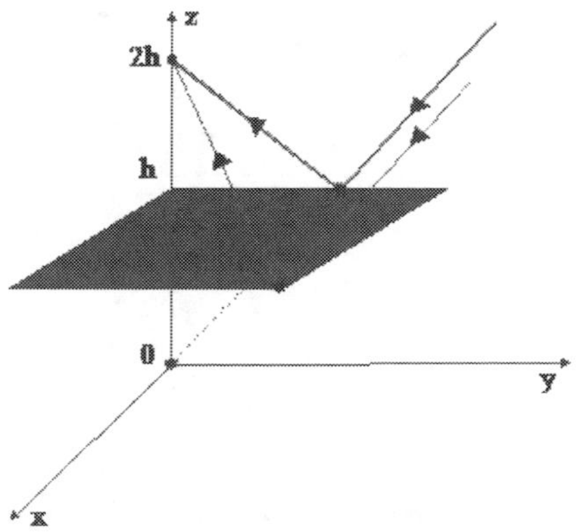

Figure 1. Objective's geometry.

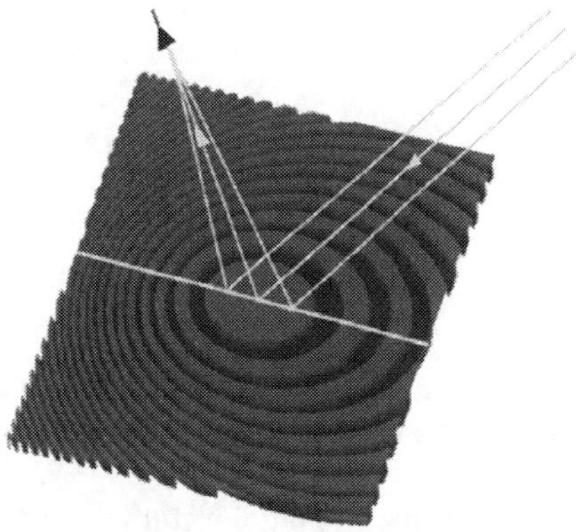

Figure 2. The profile of the reflector.

The origin of the coordinates is combined with a mirror imaged point of the expected focus $\mathbf{r}_0 = (0,0,0)$. The reflector is located on height h. A phase shift of the reflectors axial line of is: $\varphi_0 = \mathbf{kr}_0 = 0$. The phase shift of the reflected wave at an arbitrary point $\mathbf{r} = (x, y, h)$ is determined as

$$\varphi = kr + \mathbf{kr} = k(\sqrt{x^2 + y^2 + h^2} - x\sin\alpha - h\cos\alpha).$$

Formula [4–5] $\Delta\varphi = \varphi - \varphi_0 = m\pi,\quad m = 0, 1, 2,...$ determines the location of Fresnel zones. Even values m correspond to in-phase zones, the odd m to the antiphase ones. In general, the zones shape represents concentric ellipses. The effect of focus can be obtained by opening even zones or obscuring the odd ones. Besides changing the contribution of antiphase components to the opposite sign would lead to in-phase addition of all Fresnel zones [6]. This can be done by changing the phase of the odd or even zones (corrugation reflector surface), what is represented in figure 2. As a result it will provide significant field intensity increasing in the focal point.

If the incident angle is fixed $\alpha = 0°$ Fresnel zones will take the form of concentric circles. Calculations of the field intensity distribution in focal area are made using Huygens-Fresnel-Kirchhoff method [5]:

$$E(\mathbf{r}) = -2\iint_S E_0(\mathbf{r}_s)w(\mathbf{r}_s)\frac{d}{dz_s}G(\mathbf{r} - \mathbf{r}_s)\,\mathrm{dS},$$

$w(\mathbf{r}_s)$ is equal to 1 for in-phase Fresnel zones and -1 for antiphase onesравен 1 и -1,

$G(\mathbf{r} - \mathbf{r}_s)$ – Green's function:

$$G(\mathbf{r} - \mathbf{r}_s) = -\exp\{ik\}/4\pi|\mathbf{r} - \mathbf{r}_s|.$$

This formula represents convolution integral and calculates using direct and inverse Fourier transforms and Weyl spectral decomposition for normal derivative of the Green's function [5]:

$$\frac{d}{dz}G(\mathrm{r}) = \frac{1}{2\pi}\iint \exp(iz\sqrt{k^2 - u^2 - v^2})\exp\{i(ux + vy)\}\,dudv.$$

3. Field distribution in focal area

The investigated reflector was calculated for following parameters: estimated frequency – $f = 24$ GHz, incident angle – $\alpha = 45°$ and height of the focus – $h = 15$ cm. Consider the field intensity distribution in the focal plane for various cases.

The figure 3 shows the distribution of the reflected radiation intensity, depending on focusing height. It's obvious that focus area is rather localized in space.

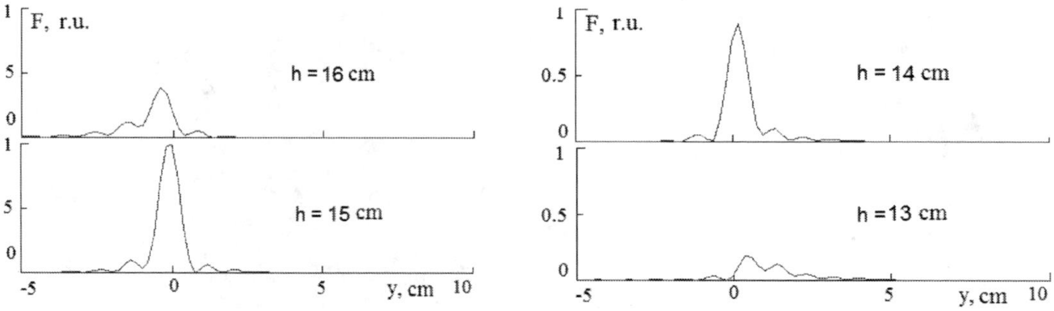

Figure 3. The distribution of field intensity, depending on focusing height.

The observation point displacement leads to a rapid decrease of the field intensity F. It shows that flat Fresnel reflector has a good spatial focusing ability. If the same device is used for other incident angles, the focal point will move in the corresponding direction (figure 4).

For particular case at angle $\alpha = 45°$ changing frequency f leads to shifting the focal point along the focus line for investigated reflector (figure 5).

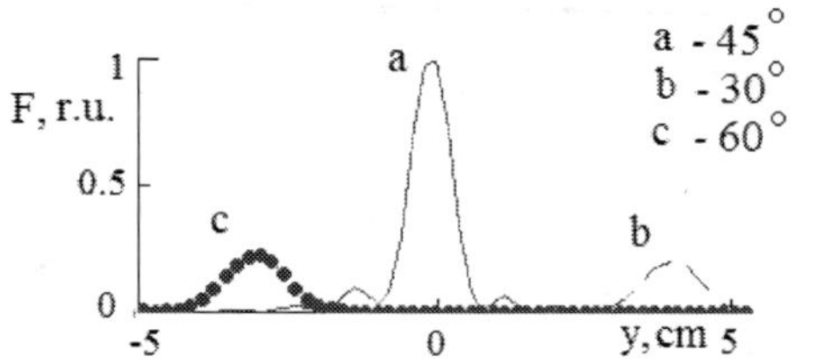

Figure 4. Field intensity distribution for different incident angles.

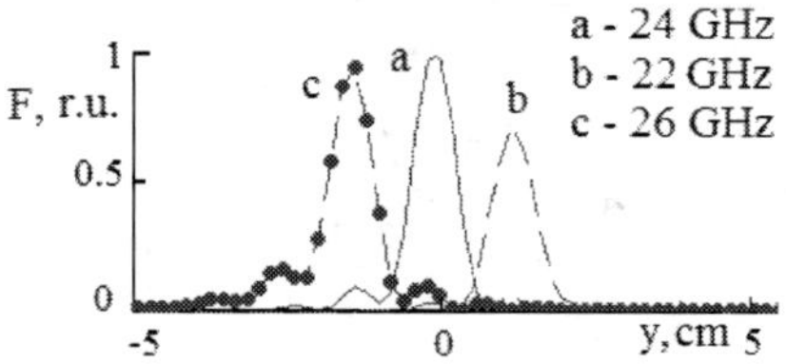

Figure 5. Field intensity distribution for different frequencies.

Such spatial behavior of the focal point is similar for parabolic reflector but there is a significant distinction. Position of the focal point defines only by the geometry of the reflective surface and there is no frequency dependence for parabolic reflector [7].

4. UWB signal and Fresnel focusing

Currently, UWB pulses are used extensively in many industries. But there are some problems with receiving such signals which are connected with their short duration. Consider the opportunity of using flat Fresnel reflectors for registration UWB pulses.

Consider the UWB signal incident normally to the reflectors surface. Its duration is $\tau = 0.2$ ns (figure 6a). The spectrum of this signal is quite wide and it occupies the band from 0 to 40 GHz (figure 7a). Registration of such a signal is a big problem.

Journal of Physics: Conference Series **671** (2016) 012022 doi:10.1088/1742-6596/671/1/012022

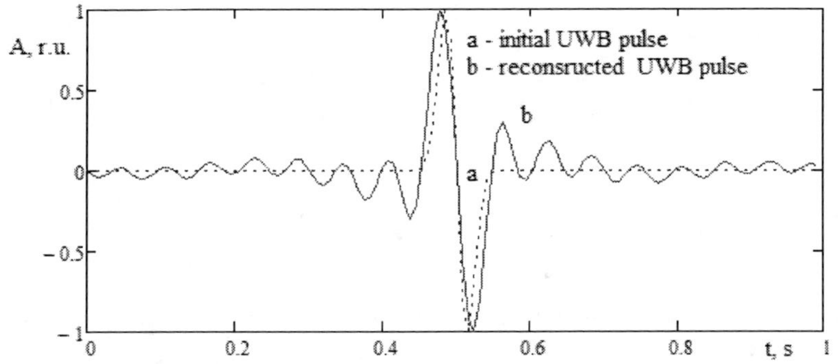

Figure 6. Timing UWB signal form.

Further investigations are connected with UWB signal interaction with narrowband device - Fresnel reflector. Figure 8 represents frequency response of Fresnel reflector for signals which were received in focal point. Operating frequency was changed $f_0 = 10$ GHz to combine it with UWB pulse spectrum maximum. It is obvious that Fresnel reflector is typical narrowband device but it has bandwidth near the combination frequencies $3f_0$ and $5f_0$. If the received signal at the focal point bandwidth is limited 0-20 GHz, the output UWB signal will be similar with "long" sinusoid whose frequency is $f_0 = 10$ GHz. Digitization of such signal can be carried out with the using a narrowband quadrature receiver. This signal carries information about original UWB signal despite of its narrowband. This will be used to restore the input UWB signal.

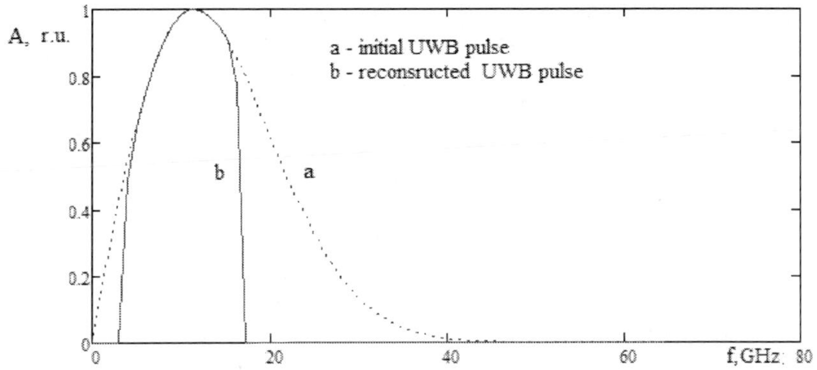

Figure 7. Signals spectrum.

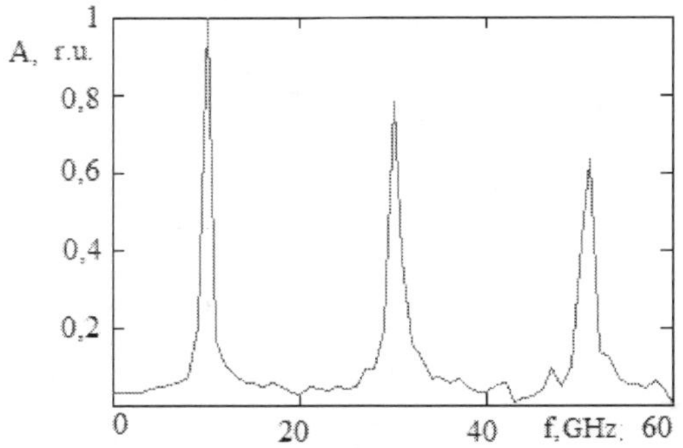

Figure 8. Frequency response of Fresnel reflector.

Output narrowband signal is a result of UWB signal spectrum filtration with known frequency response of Fresnel reflector. UWB signal spectrum reconstruction can be carried out by conducting the inverse operation with using regularized Wiener filtering (figure 6b). This operation corresponds to the solution of the inverse problem of UWB pulse reconstruction [8] or otherwise, compression of narrowband signals.

5. Conclusion

The proposed approach allows using Fresnel reflector for receiving UWB signals from ordinary narrowband devices.

Present work is done for increasing the competitiveness of Tomsk State University.

References

[1] Di Benedetto M G. and Giancola G 2004 *Understanding Ultra Wide Band Radio Fundamentals* (New York City, Pearson Education)

[2] Ghavami M, Lachlan Michael, Ryuji Kohno 2007 *Ultra Wideband Signals and Systems in Communication Engineering, 2nd Edition* (New York City, John Wiley&Sons)

[3] Aster C R, Borchers Brian, Thurber Clifford 2013 *Parameter Estimation and Inverse Problems, Second Edition* (United States, Academic Press)

[4] Goodman Joseph W 1996 *Introduction to Fourier Optics* (United States, McGRAW-HILL)

[5] Lokshin G R 2009 *Osnovi radiooptiki* (Dolgoprudniy: Intelekt)

[6] Webb G W (2011) Variable reference phase in diffractive antennas: review, applications, new results *IEEE Antennas and Propagation Magazine* **53**(2) 77–94

[7] Yakubov V P, Shipilov S E, Antipov V B, Ponamorev S V, Kamenev A V (2014) Ploskiy reflector Frenelya *Izvestiya vuzov. Fizika* **57**(9/2) 123–126

[8] Yakubov V P, Shipilov S E, Suhanov D Ya, Klokov A V 2014 *Radiovolnovaya tomografiya: dostijeniya I perspektivi* (Tomsk: Izd-vo NTL)

Special Aspects in Designing High – Frequency Betatron

A A Filimonov[1], S V Kasyanov[2], V A Kasyanov[3]

[1]Senior research worker, lab 41, INDT, Tomsk Polytechnic University, Russia, Tomsk
[2]Senior research worker, lab 41, INDT, Tomsk Polytechnic University, Russia, Tomsk
[3]Head of laboratory 41, INDT, Tomsk Polytechnic University, Russia, Tomsk
E-mail: kassianov2011@yandex.ru

Abstract. The article is devoted to designing the high – frequency betatron. In high – frequency betatron most important problem is overheating of the elements of the body radiator unit. In an article some directions of solving this problem are shown.

1. Introduction

Last years, it can be seen the increasing market requests on betarons, specialized for custom inspection systems[1,2]. Betatron's advantages are:
- relatively low cost,
- low maintenance costs,
- low requirements for servicing personnel,
- ability to work in dual – energy mode.
- low power consumption good for mobile inspection systems.

Most serious betatron's disadvantages are low output dose rate in comparison with linac and lager dimensions of an accelerator, that causes increasing the weight of radiation protection shielding.

One of the most effective method of increasing the dose rate of a betatron is increasing the pulse repetition rate. This method will not cause increasing the weight and dimensions of the betatron radiator.

In institute of introscopy we gathered some experience on designing betatrons with high repetition pulse rate (Table 1). One of the most successful models is betatron for inspection system with energy 3 Mev and pulse repetition rate 400 Hz. Also there were designed betatrons on energies 5, 6.5, 7.5 MeV with pulse repetition rate 300 Hz, and 9 Mev with pulse repetition rate 400 Hz. Most exotic variant – betatron on energy 5 Mev with pulse repetition rate 900 Hz was designed for one American customer, it was experimental model. This model served as a basement to check some technical decisions and didn't go in commercial series

Table 1. Betatrons with high pulse repetition rate, *E* – experimental models, *C* – commercial models.

E/freq	3	5	6.5	7.5	9
300			C	C	
400	C	E			C
900		E			

At a first sight, we can think that the question of designing the betatrons for custom inspection systems with pulse repetition rate 400 Hz was successfully solved, however it's worth to say that present models of high frequency betatrons have some unsolved problems, which reduces the areas of using them.

One of the most important problems is keeping normal temperature conditions.

As a rule, while designing betatrons for customs inspection systems it needs to minimize dimensions of the radiator, to reduce weight of a shielding and the weight of all system. According this direction, all electronic circuits are removed from the radiator to separate block. Removing electronic circuits from radiator also helps to decrease radiation loads on electronics components. Finally, radiator of the betatron for custom inspection system consists of magnetic guide, poles, central inserts block, magnetizing winding, and HV transformer with acceleration tube, situated between poles of a magnet.

According to our experience, normal work of a tube with HV injection unit, that connected through thyristors keys and uses double stage pulse forming circuit can be obtained for frequencies up to 400 - 500 Hz. Some tubes can work on frequency 600 Hz [3]. Higher frequency will cause the overheat of anode of the injector and output of the betatron will reduce to zero. The betatron on frequency 900 Hz that i mentioned before, could work on 900 Hz repetition rate not more than 10 seconds.

To go further in increasing the frequency of pulses we can apply the reduction of injection current using special forming circuits or using fully controlled injector [4]. The same time, reducing the injection current causes reduction of electron capture time and increasing the current of an injection. These questions for detailed experimental research.

Let's see the thermal picture of working betatron (figure 1).

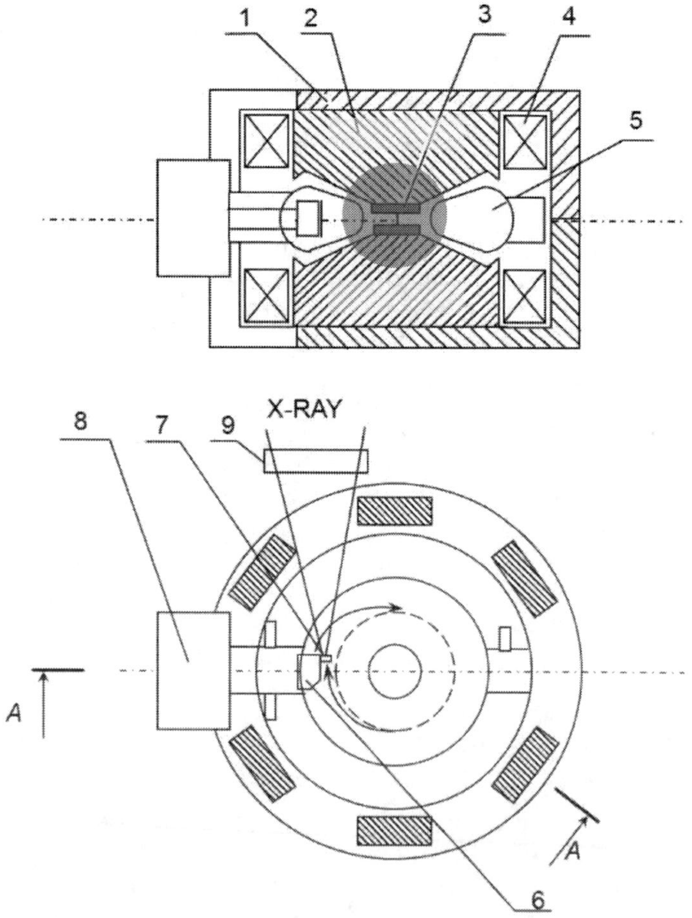

Figure 1. Construction of a betatron: 1 – magnetic guide; 2 – pole; 3 – central inserts block; 4 – magnetizing winding; 5 – tube; 6 – injector; 7 – target; 8 – HV transformer; 9 – dose rate monitor,

90 – 95 C in 20 minutes 400 Hz,

60– 65 C.

Most heat dissipation goes in electromagnet of betatron - in magnetizing winding and in the magnetic guide. The value of these energy losses gives more than half of all losses in units of betatron. That is why we should seriously think about choosing materials for electromagnet to improve temperature mode of a betatron. At first, improving the temperature mode of a betatron is reaching with using special transformer steel with lowest specific losses.

Usually, in inspection system the betatron radiator is housed in a massive shielding container. Most appropriate variant of cooling in a mobile inspection system is a axial or loop expulsion with air.

As usual, there is no problem with temperature mode of a magnetizing winding - you just need to choose right high frequency litzendrat wires and right fan for cooling it.

More problems we have in cooling magnetic guide, exceptionally central inserts block. Central inserts block consist of separate disks of magnetic guide material. These disks are separated with non - magnetic fiberglass material, which have high radiation resistance, high heat conductivity and high working temperature.

As it can be seen on figure 1, central inserts block can't be cooled by air flow, produced by cooling fan. In fact, cooling the central block goes through heat conduction from central inserts block to poles of the electromagnet. Cooling the poles goes through their side, and also through heat conduction from ends of a poles to the reverse magnetic guide. In betatrons with relatively low energy of accelerated electrons the gradient of temperature between reverse magnetic guide and the pole is not high and has a value about several degrees. While designing high frequency betatrons on energies 7 Mev and higher it's necessary to take additional actions to cool down the poles.

Good positive improvement is using additional heatsinks on a top and on a bottom of a radiator. Experiments with temperature mode of 9 MeV betatron shows, that using additional heatshinks allows to decrease the average temperature of a poles on 20 – 25 %. Also it gives effect for central insert block, however, its value is not so high, the temperature comes lower on 4–5 degrees.

Central inserts block is the most heat loaded element in betatron, and it's extremely undesirable to increase its temperature more than 100 C, higher values of temperature will cause the damage of it. In standard conditions with cooling fan the temperature of a central inserts block will reach 90 - 95 C after 20 minutes for betatron on 9 MeV with 400 Hz pulse repetition rate. And the poles will be heated to 60 - 65.

It's impossible to reduce temperature mode of central inserts block, using only air cooling, and we need more effective constructions for heat dissipation.

There was experiment on 9 MeV betatron to dissipate heat power from central inserts block with heatpipe.

Heatpipe is a vacuum closed device [5], that passing heat using locked evaporation condensate cycle with the contact with source of a heat from one side and heat receiver from other side (see figure 2). Heat energy is taken from source to the vapor area, and goes to boil the liquid, which comes to vapor. Heat comes in steam to other side of the tube, to the condenser area. Heat is taken by heat receiver (heatsink) and liquid condensates. The condensated liquid returns to vapor area through the capillary structure or just under the gravity force. So, instead of conduction heat transfer by electrons, that we have in solid metal caliduct, in a heatpipe the natural-convection heat transfer is used.

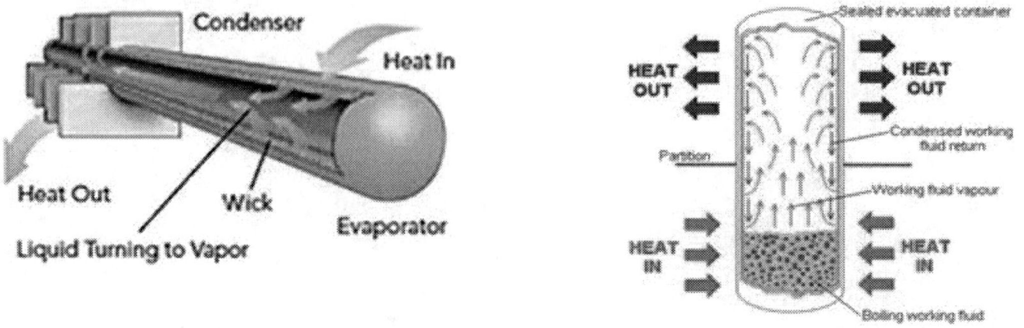

Figure 2. Construction and principles of work of a heatpipe.

Which advantages does the heatpipe give? At first, it's ability to move hundreds and thousands of watts of heat power. If we will evaporate several grams of liquid in second, it's possible to pass heat flow with value of several kilowatts.

Another interesting feature is ability to concentrate heat energy by using several heatpipes and configuring them for different. In a temperature range from 20 to 100°C wide range of heat transfer agents can be used - alcohol, halocarbon, ammonia, distillate water, ester.

In our case we used heatpipe from copper, 8 mm in diameter and 300 mm in length, filled with water for 8 – 10 percent of volume. There is a lowered pressure in pipe, allows to decrease temperature of boiling to 30–40°C. This heatpipe in using with heat source with temperature from 60 to 90°C can pass from 80 to 150 watts of heat.

The core of the central inserts block can be reached through the central bolt, which was made empty inside, and heatpipe was set inside it (figure 3).

There was a heatsink at the condenser area with additional fan. As a result, the temperature of the central inserts block reached 95 C after 60 minutes of work. After 15 minutes with radiation off it came to 60 C. In comparison, without using heatpipe temperature of the central inserts block reached 95 C after 20 minutes of work.

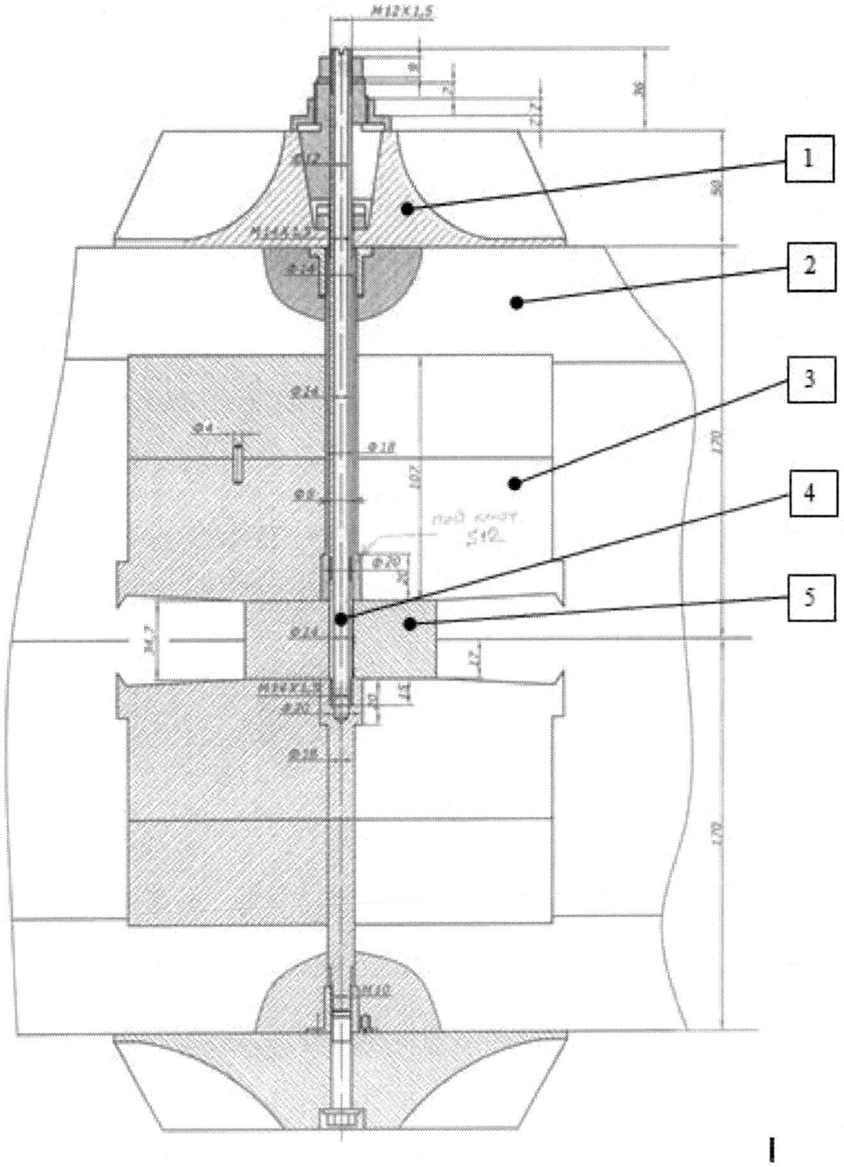

Figure 3. Heatpipe cooling system: 1 – heatsink, 2– reverse magnetic guide, 3 – pole, 4 – heatpipe, 5 – central inserts block.

Also there were experiments on cooling central inserts block with help of liquid cooling system. To take heat from the core of a central inserts block also central bolt was used. There is a special construction of a central bolt, where is the cooling liquid coming through it. Autonomic system of liquid cooling used pump, expansion tank, and a heatsink with fan.

For 9 Mev 400 Hz maximum temperature of the central block reaches 85 C in cycle 45 min on / 15 min off

The result is similar to using heatpipe. Heat pipe is more autonomic and reliable, but liquid cooling system is easier to repair and can be positioned in any angle, at the same time heatpipe should be positioned closer to vertical position, otherwise it will not work properly.

Autonomic liquid cooling system is used in defectoscopy betatron on 9 MeV. It helped to decrease working temperature in the central inserts block from 110 C to 85 C.

These experimental results will determine the ability to design betatrons, that can have stable work on a frequencies up to 400 Hz.

References

[1] http://portal.tpu.ru/departments/nii/nii_in/eng/Developments/SSB
[2] Stein M M, Kasyanov V A, Chakhlov V L, Macleod J, Marjoribanks P, Hubbard S (2004) Small-size betatrons for radiographic inspection *Palais des Congres. 16-th world conference on nondestructive testing* TS4.7.4
[3] Bel'tyaev Yu N, Pushin V S 1973 *Choice of the parameters of the injection system with a higher excitation frequency for a small betatron* (Proceedings of the scientific and research institute on nuclear physics, electronics and automatics of the Tomsk polytechnical institute, Atomizdat, Moscow) 15–17 [in Russian]
[4] Pushchin V S, Chakhlov V L (1975) Controllable injection for a high-frequency betatron *Ministerstvo Vysshego i Srednego Spetsial'nogo Obrazovaniya SSSR* 58–59
[5] http://www.thermacore.com/thermal-basics/heat-pipe-technology.aspx

Journal of Physics: Conference Series **671** (2016) 012024 doi:10.1088/1742-6596/671/1/012024

Betatron Application in Mobile and Relocatable Inspection Systems for Freight Transport Control

S V Chakhlov[1], S V Kasyanov[2], V A Kasyanov[3], S P Osipov[4], M M Stein[5], A M Stein[6], Sun Xiaoming[7]

[1]Head of laboratory 40, NDT Institute, Tomsk Polytechnic University, Russia, Tomsk
[2]Senior research worker, lab 41, NDT Institute, Tomsk Polytechnic University, Russia, Tomsk
[3]Head of laboratory 41, NDT Institute, Tomsk Polytechnic University, Russia, Tomsk
[4]Senior research worker, lab 40, NDT Institute, Tomsk Polytechnic University, Russia, Tomsk
[5]Head of laboratory 43, NDT Institute, Tomsk Polytechnic University, Russia, Tomsk
[6]Engineer, lab 43, NDT Institute, Tomsk Polytechnic University, Russia, Tomsk
[7]Director of PowerScan Ltd, Beijing, China.
E-mail: copperman@rambler.ru

Abstract. Accelerators with energy level up to 4 MeV having high level of penetration ability by steel equivalent are the popular to control oversize cargo transported by road, by railway and by river. Betatron's usage as cyclic induction accelerator has some advantages in comparison with linear accelerators and other sources. Tomsk Polytechnic University has developed many types of betatrons, most of them are being produced by separate affiliated company " Foton ". Article is shown the results of application of the betatrons in inspection custom systems.

Inspection systems using X-ray units and accelerators of ionizing radiation is one of the most rapidly developing directions in digital radiography during the last decades. Accelerators with energy level up to 4 MeV having higher level of penetration ability by steel equivalent are the most popular to control oversize cargo transported by road, by railway and by river.

Betatron's usage as cyclic induction accelerator has some advantages in comparison with linear accelerators and other sources. Betatron has a small focal spot which improves the resolution of the system. Narrow-beam betatron radiation is easy to collimate so a spurious background of scattering radiation is much lower than in the systems based on other sources. The energy of accelerated electrons could be regulated in relative units from about 0.3 to 1. Moreover, the betatron control system allows programming the radiation spectrum which could be used to determine the density of the scanned object. Tomsk Polytechnic University has developed many types of betatrons, most of them are being produced by separate affiliated company “Foton”. Technical specifications of betatrons for using in customs inspection systems are shown in Table 1.

Table 1. Technical specifications of betatrons

Parameters	MIB-2.5	MIB-5	MIB-7.5	MIB-9
Peak energy of bremsstrahlung radiation, MeV	1 – 2.5	2 – 5	2–7.5	2–9
Peak dose rate at 1m from the target, cGy/min	2	4	7.5	20
Pulse repetition rate, Hz	400	400	300	400
Power consumption, kVA	2	3	3	5
Radiator weight, kg	57	80	115	150
Total weight of units, kg	120	180	220	260
Size of focal spot, mm	0.25×2	0.25×2	0.25×2	0.25×2
Maximal controlled thickness (of steel), mm	150	200	270	330
Dimensions of the radiator, mm	Ø320×220	Ø360×230	Ø380×315	Ø450×340
Angle of beam in a vertical plane, deg	60	52	52	60

All betatrons, designed for inspection systems, have separate radiator with cylindrical – shaped form. This form minimizes radiation protective shielding. External views of radiators are shown on figure 1 and figure 2.

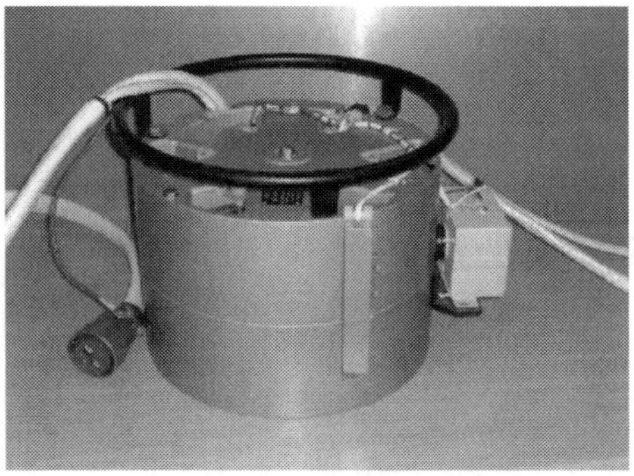

Figure 1. Betatron MIB–5.

Figure 2. Betatron MIB–9.

Customs inspection system, based on two accelerators with energy of 7.5 MeV, started to work in Bangunan Sultan Iskandor in BuHit Chaqar, November, 2012. This complex is supposed to control 180 trucks per hour or 2000 – per day.

This system has high scanning speed – up to 15 km/h, that was achieved by using ionizing radiation detectors of high-performance as well as the latest algorithms of data processing. This system also allows getting three-dimensional shadow images to increase the efficiency of freight inspection. External views of an inspection system are shown on figure 3.

Figure 3. Custom inspection system in Bangunan Sultan Iskandor in BuHit Chaqar, based on two betatrons with energy of 7.5 MeV.

One of the most important advantages of this system is low dose rate on the object of control. It is about 50 microroentgen per one scanning. This parameter is close to normal radiation background. This can be possible because of low dose rate of betatron and high sensitivity of the detector array. Comparison with different customs inspection systems can be seen in Table 2.

In the future, the use of betatron with dual-energy mode will make possible to recognize materials contained in the test object. This technology could improve the information content of the entire complex and would enable to identify the explosives and drugs with higher effectiveness.

Table 2. Comparison with different customs inspection systems.

Parameters	Rapiscan Eagle	Silhouette Scan Mobile CaB 2000	HCV-Mobile	DRS CargoScan 5000M	System in BuHit Chaqar
developed by	Rapiscan security Products (USA)	Smith Heimann (Germany)	Smith Heimann (Germany)	Adani, TPU (Russia, Belorussia)	TPU, Power Scan, (Russia, China)
maximum thickness of penetration in steel, mm	300	110	180 (270)	300	250
resolution, mm	3	3	3	2.5	5
sensitivity, %	2	No data	No data	3	2
scanning speed, m/s	0.3	1	0.2	0.4	3
x-ray source	6 MeV linac	3 MeV betatron	3(4) Mev linac	5 MeV betatron	7.5 MeV betatron
dose per one scanning, mR	3	0.15	1(2)	No data	Less than 0.05
type	Relocatable	Mobile	Mobile	Mobile	Relocatable

Developing and improving the accelerator's dual-energy mode to increase efficiency in identification and detection materials prohibited for transportation, is carried out on the experimental platform – the ICC, Tomsk Polytechnic University. Algorithms and operating modes tested on the platform are applied and mastered in the operating complex. On a pictures below some results of using betatron in dual-energy mode it can be seen (figure 4).

Journal of Physics: Conference Series **671** (2016) 012024 doi:10.1088/1742-6596/671/1/012024

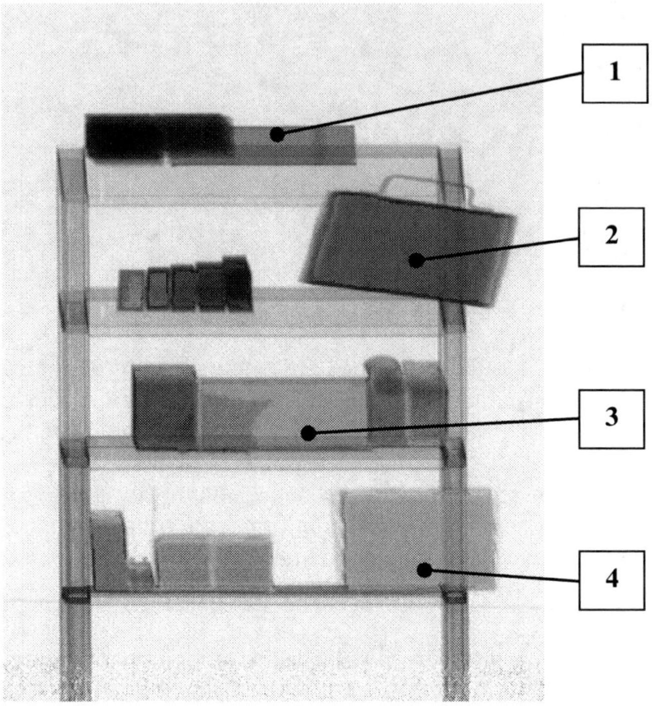

Figure 4. Scanning in dual-energy mode: 1 – Lead; 2 – Steel; 3 – Aluminum; 4 – Plexiglas.

Aslo, MIB – 9 betatron was used in mobile inspection system ASD – 65201, which had successfully passed probational tests, working in Adler, Sochi, during Olympic Winter Games 2014.

Figure 5. Mobile inspection system in work during Olympic Winter Games 2014 in Sochi, Russia.

In summary, it is worth to say, that betatron is a good choice to be considered as a x-ray source in customs inspection systems, and this is an optimal decision for using in mobile and in relocatable inspection systems.

References

[1] http://www.eideticscorp.com/THScan/summary.htm
[2] http://www.smith-heimann.de
[3] http://www.rapiscansystems.com/
[4] http://www.adani.by/prod_cargoscan.php
[5] http://www.vividusa.com/products/cargoscreen.htm

[6] Kasyanov V A, Chakhlov V L, Stein M M, Macleod J, Marjoribanks P, Habbard S (2004) Small-size betatrons for radiographic inspection *Palais des Congres. 16-th world conference on nondestructive testing* TS4.7.4
[7] Pincu R, Kleinberger-Riedrich O (2012) A review Digital Radiography in the service of security *Proceedings of 18th World Conference of Nondestructive Testing* http://www.ndt.net/article/wcndt2012/papers/110_wcndtfinal00110.pdf
[8] Heisemann B J, Lippert J, Stierstorfer K (2003) Density and atomic number measurements with spectral X-ray attenuation method *Journal of Applied Physics* **94**(3) 2073–2079
[9] Orogodnikov S, Petrunin V (2002) Processing of interlaced images in 4-10 MeV dual energy customs system for material recognition *Physical Review Special Topics-Accelerators and Beams* **5**(10) 67–77 doi: 10.1103/PhysRevSTAB.5.104701
[10] Volkov V G, Zrelov Yu D, Kasyanov V A, Filimonov A A, Chakhlov V L, Shtein M M, Marjoribanks P (2010) Small-Size Betatron for the Energy 2.5 MeV for Field inspection. Report № 1.4.35 *10th European Conference on Non-Destructive Testing: reports* 14598665

Refinement of Techniques Metallographic Analysis of Highly Dispersed Structures

A Khammatov[1], D Belkin[2] and N Barbina[3]

[1]Engineer, Institute of Non-Destructive Testing, Tomsk Polytechnic University, Tomsk, Russia
[2]Director, Institute of Non-Destructive Testing, Tomsk Polytechnic University, Tomsk, Russia
[3]Quality engineer, Institute of Non-Destructive Testing, Tomsk Polytechnic University, Tomsk, Russia
E-mail: hammatov@tpu.ru

Abstract. Flaws are regularly made while developing standards and technical specifications. They can come out as minor misprints, as an insufficient description of a technique. In spite the fact that the flaws are well known, it does not come to the stage of introducing changes to standards. In this paper shows that in the normative documents is necessary to clarify the requirements for metallurgical microscopes, which are used for analysis of finely-dispersed.

1. Introduction

The methods of metallography analysis based on comparing with standard scales still remain up to date, despite intensive development of digital image processing and a broad range of computer programs for metallographic analysis.

The major advantages of comparative method are its operational efficiency (several seconds are required for analysis) and cost efficiency. Besides, during analysis process it is necessary to estimate the form of structural components, their distribution and combinations that are difficult to describe quantitatively. Thus comparative methods have become widely spread in practical metallography among production laboratories in spite the fact that these methods have considerable disadvantages. The major method disadvantages are the following: low precision at quantitative estimations and an operator's subjective apprehension.

The total microscope magnification is considered to be one of the basic observation conditions in GOST but without additional requirements to its resolving power or lens aperture [1–9]. That is why it is magnification which is taken into account by a specialist while choosing the equipment. Sometimes it is difficult to make a case for additional equipment requirements due to the absence of these requirements in the corresponding GOST.

The classification of the most disperse steel structures (perlite, sorbite and martensite) observed in an optical microscope is defined according to GOST 8233-56 [1]. Perlite with dispersibility up to 0.6 μM is referred to grades 1, 2, 3, 4 (sorbitic pearlite, latent lamellar pearlite, thin-lamellar pearlite, fine lamellar pearlite). The most dispersed structure is characterized by the fact that specialists are unable to distinguish separate particles (flakes, needles, grains) with the 0.2 μM size by 1000^{x} magnification. It is necessary to get 0.2 μm magnification (eyes resolving ability is 0.2 mm with the distance of 200-250 mm) in order to achieve useful magnification. It is possible to achieve such kind of resolution in lightning with violet colour-filter by using lenses with aperture not less than 1.25. Useful 1000^x magnification in more comfortable colour range is achieved by increasing aperture up to 1.4–1.5 values.

At the same time one of the trends in the development of microscope equipment has become the increase of ergonomics and the efficiency increase of setting changes and magnifications. Therefore, the use of dry (not immersion) lenses allows to save time that is wasted on a sample and a lens cleaning after analysis and during the change to weaker lenses. However, the resolving power of such kind of lenses is less than resolving power of immersion lenses. The use of lenses with increased working distance allows to increase the efficiency of change resolution without the threat of damaging a monitored object and the frontal lens. However, the working distance increase is often achieved by the number aperture decrease.

On the other hand, high magnification ratio can be achieved by combination of a 40^x lens with 0.55-0.65 aperture and a 25^x ocular. The resolving power of such kind of systems does not exceed 0.45 μM.

Thus metallographic microscopes with maximum magnification of 1000x are being produced in increasing frequency. This magnification is achieved by the use of lenses with N.A.=0.55÷0.95 aperture that corresponds to resolving power of d=0.7÷0.35 μm relatively.

GOST 8233-56 (Union State Standard) [1] is one of the foundational documents that classifies the structure of steels. It practically has not been changed since its creation.

The aim of the work is to highlight the necessity for changes that have to be introduced into GOST 8233. In particularly, the necessity for detailed requirements to metallographic microscopes and to the replacement of photo-standards.

2. Experiment

The use of lenses with lower resolving power can be justified while observing relatively large structural constituents with the size of 0.5 μM and more: (the determination of non-metallic constituents, lamellar structure, the ratio of large structural constituents and etc.). However, the evaluation structure results highly depend on microscope resolving power while estimating finely-dispersed structures especially when 1000^x magnification is required. If using dry lenses so the structure of lamellar pearlite with dispersibility less than grades 3 and 4 can be correlated to grade 1 according to GOST 8233-56 (figure 1a).

a b

Figure 1. Photo-standards of lamellar pearlite GOST 8233: a – grade 1, b – grade 2.

Let us give an example. Figure 1 shows photo-standards of lamellar pearlite with grades 1, 2 according to GOST 8233-56. Further figure 2 represents photographs of one and the same part of bearing steel as that was annealed on lamellar pearlite. These photographs were taken with microscope

Olympus BHMJ using the following lenses: Plan 40 LWD with 0.55 aperture, Plan 100 with 0.9 and 1.25 aperture. Photography was carried out with Lab 200C US camera, the lightning was organized with the use of JS-18 (yellow) colour-filter.

The metallographic templates preparation was carried out according to a standard technique that includes grinding, polishing and further etching with a reagent (2% solution of nitric acid) in spirit. Grinding was conducted on an abrasive wheel with 450 μm grit and on paper with diamond paste. At the same time, the paper paste with grit of 60/40, 14/10 and 7/5 was being changed sequentially. Polishing was conducted on artificial chamois with diamond paste of 0.5/0 grit.

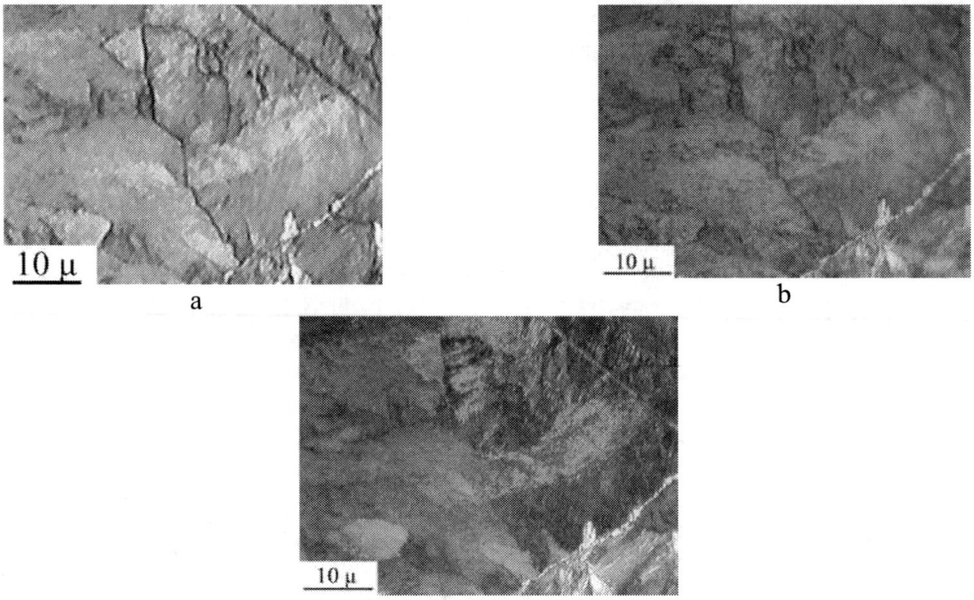

Figure 2. Microphotographs of bearing steel, taken with the use of lenses a – 40^x (N.A.0.55), b – 100^x (N.A.0.9), c – 100^x (N.A.1.25).

The grains (pearlite colony) which can be graded as 1 according to GOST 8233-56 are observed in the first two cases (figures 2a and 2b) as pearlite lamellar structure can not be distinguished. The grains take the form of distinct lamellar structure. It means that the distance between cementite plates in pearlite is no more than 0.2 μm. A such kind of structure can be graded as 2 according to GOST 8233-56.

Spoilage metal products can be let pass at the controlling stage because of this mistake. According to technical specifications under number 14-15-254-91 steel rope wire structure must have not less than 30 % of lamellar pearlite and graded as 1 according to GOST 8233-56.

In order to prevent mistakes while choosing equipment it is necessary to introduce to GOST 8233 the following points: the section about equipment and the correspondence table of magnification ration, the aperture of a lens and resolving ability.

Table 1.the correspondence between lenses aperture and resolving power in magnification ratio

Microscope magnification ratio	**Numerical aperture value of a lens not less than**	**Resolving power* with lightning through blue colour-filter (450 nm)**
100^x	0.15	2.0 μm
200^x	0.30	1.0 μm
500^x	0.65	0.4 μm
1000^x	1.25	0.2 μm

* The resolving power is a minimal distance in which two points can be distinguished separately

Besides the obvious flaw in GOST 8233-56 is a relatively low quality of photographs in working samples and the absence of size scales in these samples. As can be seen in figure 1 image sharpness in photo-standards leaves much to be desired especially in digital versions. According to the article authors‘ experience in the usage of this GOST on different factories it can be added that photographs in different paper copies of an original edition have different quality and that can be related with out of date photoprinting technology. It is sometimes hard to distinguish principle difference between the first three grades of lamellar pearlite.

The refinement of photo-standards quality can be achieved by means of digital scanning of original photographs (negatives) or by means of microphotograph changes into new ones.

3. Conclusion

It was proved the necessity of evaluation technique refinement of highly dispersed structures. In particularly, it is necessary to introduce a suggestion to Federal Agency on Technical Regulating and Metrology about the change in GOST 8233-56. The change should include requirements to microscope resolving power and to lens aperture with 1000x magnification or it should include a correspondence table of minimal aperture values with different magnification ratios.

It was shown the necessity of photo-standards revision and the necessity of microstructures scales replacement that can be done partly or completely.

References

[1] GOST (State Standard) 8233-56 *Steel. Microstructure standards* (Moscow: Standards publishing house)

[2] GOST (State Standard) 1778-70 *Steel Metallographic techniques for the determination of nonmetallic inclusions.* (Moscow: Standartinform)

[3] GOST (State Standard) 3443-87 *Iron casting with various graphite forms. Techniques of structure determination* (Moscow: Standartinform)

[4] GOST (State Standard) 5639-82 *Steels and alloys. Techniques of grain size determination* (Moscow: Standards publishing house)

[5] GOST (State Standard) 5640-68 *Stee.. Metallographic technique for the evaluation of sheets and strands microstructure* (Moscow: Standards publishing house)

[6] GOST (State Standard) 9391-80 *Sintered hard alloys. Techniques of hardness and microstructure determination* (Moscow: Standards publishing house)

[7] GOST (State Standard) 21073.1-75 *Nonferrous metals. Grain size determination by means of microstructure scale comparison* (Moscow: Standards publishing house)

[8] GOST R ISO 4967-2009 *Steel. The determination of nonmetallic inclusions content. Metallographic technique with the use of standard scales* (Moscow: Standartinform)

[9] STO (Standard organization) 1723082.27.100.005-2008 *Basic elements of steamers, turbines and pipe leadings of thermo power plants. The control of metal conditions. Standards and specifications»* (Moscow: NP "INVEL")

Journal of Physics: Conference Series **671** (2016) 012026 doi:10.1088/1742-6596/671/1/012026

Technique of Automated Control Over Cardiopulmonary Resuscitation Procedures

A Sh Bureev[1], E Yu Kiseleva[2], M S Kutsov[3] and D S Zhdanov[4]

[1]Head of laboratory "Instrumentation", Tomsk State University, Tomsk, Russia
[2]Engineer, Tomsk State University, Tomsk, Russia
[3]PhD student, Tomsk State University, Tomsk, Russia
[4]Jr. researcher, Tomsk State University, Tomsk, Russia
E-mail: eka.kiselyova@gmail.com

Abstract. The article describes a technique of automated control over cardiopulmonary resuscitation procedures on the basis of acoustic data. The research findings have allowed determining the primary important characteristics of acoustic signals (sounds of blood circulation in the carotid artery and respiratory sounds) and proposing a method to control the performance of resuscitation procedures. This method can be implemented as a part of specialized hardware systems.

1. Introduction

Sudden cardiac death is the leading cause of death in the world. Thus, in the US, the number of sudden cardiac death is 1 case per 1000 people per year, while in Russia - up to 2 cases per 1000 people. Emergency assistance for people with cardiac failure - sudden cardiac death - remains an urgent problem not only in emergency medicine, but also in therapy and cardiology in particular [1, 2, 3].

According to the statistical data, up to 90% of the patients who faced clinical death could survive, if they received adequate medical care within the first 5 minutes after cardiac arrest. Unfortunately, even at large medical institutions, physicians often cannot perform the cardiopulmonary resuscitation (CPR) procedure in compliance with recommendations.

According to the data of All-Russia Non-Government Organization Russian Red Cross, the share of people who underwent special pre-doctor (first) aid courses is extremely low. For example, only 392 people were trained in first aid principles in the territory of Tomsk Oblast in 2013 (with the population of 1,070,128 people). In this situation, it is essential to develop a device that will be able to recommend the order of procedures for resuscitators and automatically monitor the adequacy of performed resuscitation measures by controlling and coordinating the CPR procedure.

2. Method

The performance of resuscitation measures can be controlled by means of analyzing acoustic blood circulation sounds registered at the carotid bifurcations and airflow sounds in the resuscitated patient's trachea.

The force of noises occurring at the carotid bifurcations is proportional to the volume of blood passing these bifurcations. Similarly, the volume of air is proportional to the intensity of respiratory sounds [4].

It is possible to experimentally establish a functional relation between the dynamics of blood circulation in the carotid arteries and blood circulation sounds at the carotid bifurcations and between the intensity of airflow in the upper respiratory tracts and respiratory sounds in the trachea. This functional relation has the following properties:

- non-negative and step-by-step character;
- direct proportionality;
- normability.

The step-by-step character of the functional relation is conditional to a clear border between the presence and the absence of blood circulation and respiration sounds.

Direct proportionality is conditional to the following fact: the more intensive blood circulation and respiration are, the louder the sounds are.

The normability property consists in the following: during the CPR measures, observed blood circulation and respiration sounds will have quite distinct values; therefore, the loudness scale and calibration will depend on the current set of measured values and sensitivity of devices in each measurement [5].

The human resuscitation cycle can be represented in the form of a chronological sequence of CPR measures (see the figure 1):

T_0 – moment of time, when the resuscitation personnel attaches an auxiliary device on a patient's body;

T_i – time of actual initiation of the cardiopulmonary resuscitation. The auxiliary device is turned on at this moment of time;

T_{CPR} – timing cycle of the CPR measures consisting of sequential intervals T_{CCCM} and T_{ALV} and two intervals Tch;

T_{CCCM} – time interval of closed-chest cardiac massage (CCCM);

T_{ALV} – time interval of artificial lung ventilation (ALV);

T_{CH} – time interval of checking for the presence of primary vital functions;

T_{PC_CCCM} – time interval of performance control for CCCM [6].

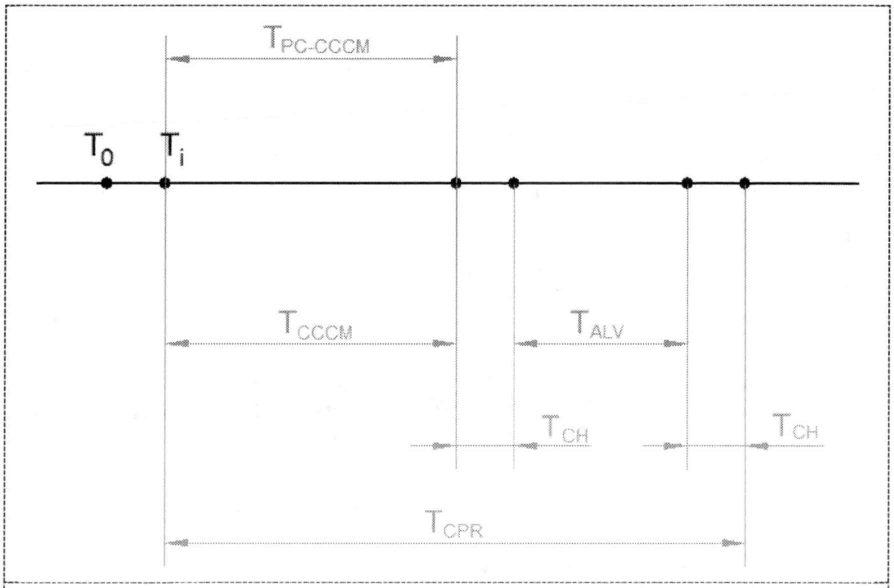

Figure 1. Chronological sequence of the CPR measures.

In order to assess the dynamics of changes in the blood flow velocity at the carotid bifurcations, it is possible to measure changes in blood circulation sounds registered by means of acoustic sensors. In this case, it is necessary to analyze only the sounds caused by blood flow at the bifurcations of the common carotid arteries. Laminar blood flow is transformed into turbulent flow at this place, which generates sounds.

The following is necessary for this purpose:

- registering a basic level of blood circulation sounds at the bifurcations of the common carotid arteries over a time interval (T_0, T_{CH}) prior to the initiation of CPR;
- registering the dynamics of changes in blood circulation sounds within T_{CCCM}, when the resuscitation personnel performs closed-chest cardiac massage (CCCM), and simultaneously assessing the efficacy and correctness of CCCM by evaluating the speed of chest compressions;
- performing a check for the presence of independent blood circulation and airflow sounds (T_{CH}).

In further CPR measures are necessary, points 2-3 are repeated cyclically and consistently with the chronological sequence of the CPR procedure.

The dynamics of changes in the airflow speed in the patient's respiratory tracts can also be assessed by means of registering and analyzing acoustic information received from a sensor located at the level of the resuscitated patient's trachea. The presence of independent airflow in the upper respiratory tracts can be detected by analyzing respiratory sounds generated in the process of artificial lung ventilation, when air passes through the resuscitated patient's throat or independent breathing is restored.

The following is necessary for this purpose:

- determining a basic level of airflow sounds in the upper respiratory tracts within a time interval (T_0, T_{CH}), when resuscitation measures have not been initiated yet;
- acquiring the dynamics of changes in airflow sounds within T_{ALV}, when the resuscitation personnel performs artificial lung ventilation (ALV), and simultaneously assessing the efficacy and correctness of ALV by analyzing the flow of air passing the resuscitated patient's trachea;
- performing a check for the presence of independent blood circulation and airflow sounds (Tch).

In further CPR measures are necessary, points 2-3 are repeated cyclically and consistently with the chronological sequence of the CPR procedure.

Besides controlling the CRP procedure, this technique also allows detecting positive (emergence of independent blood and air flow and increase in these parameters) or negative (disappearance of independent blood and air flow or reduction in the intensity of blood circulation and airflow) dynamics of the patient's health during the resuscitation procedure [7].

3. Summary

The analytical research has been focused on the detection of the intensity of respiratory sounds in the patient's trachea and registration and analysis of the intensity of blood circulation sounds at the bifurcations of the common carotid arteries. The research has revealed that the use of hardware auscultation systems is the simplest and most effective method from the viewpoint of technical and software implementation. This technique can be used to develop a device for automated control over the CPR procedure. It will allow assessing not only the correctness of CRP but also the efficacy of this procedure thanks to the control over parameters of the patient's health.

Aknowledgments

The work has been performed under Agreement #14.578.21.0078 (unique agreement identifier #RFMEFI57814X0078) on the subject Device for Control over Cardiopulmonary Resuscitation Procedures between National Research Tomsk State University and the Ministry of Education and Science of the Russian Federation.

References

[1] Bokerija L A, Bokerija O L, Kislicina O N (2010) Randomized clinical research aimed at prevention of sudden cardiac death: principles and results *Annaly aritmologii* **2** 5–13 (in Russian)

[2] Kobylkin I S (2012) Review of quality control devices for performing of chest compressions during CPR *Medicinskij alfavit. Neotlozhnaja medicina* **4** 58–62 (in Russian)

[3] Field J M, Hazinski M F, Sayre M R, Chameides L, Schexnayder S M, Hemphill R, Samson R A, Kattwinkel J, Berg R A, Bhanji F, Cave D M, Jauch E C, Kudenchuk P J, Neumar R W, Peberdy M A, Perlman J M, Sinz E, Travers A H, Berg M D, Billi J E, Eigel B, Hickey R W, Kleinman M E, Link M S, Morrison L J, O'Connor R E, Shuster M, Callaway C W, Cucchiara B, Ferguson J D, Rea T D, Vanden Hoek T L (2010) Part 1: executive summary: 2010 American Heart Association Guidelines for Cardiopulmonary Resuscitation and Emergency Cardiovascular Care *Circulation* **122** (3) S640–S656

[4] Bonow R O, Mann D L, Zipes D P, Libby R 2005 *Braunwald E.. eds. Braunwald's Heart Disease* (Philadelphia Pa: Elsevier Saunders)

[5] Cooper J A, Cooper J D, Cooper J M (2006) Cardiopulmonary Resuscitation: History. Current Practice, and Future Direction *Circulation* **114**(25) 2839–2849

[6] Safar R 1989 *History of cardiopulmonary cerebral resuscitation*, in; Kaye W.. Bircher N.. eds. Cardiopulmonary Resuscitation (New York, NY: Churchill Livingstone) pp 1–53

[7] Zoll P M, Linenthal A J, Gibson W *el al.* (1956)Termination of ventricular fibrillation in man by externally applied electric countershock *The New England journal of medicine* 727–732

Investigations and Non-destructive Testing in New Building Design

V Klimenov[1,2], A Ovchinnikov[3], S Osipov[4], A Shtein[5], A Ustinov[3] and A Danilson[3]

[1]Professor, NDT Institute, Tomsk Polytechnic University, Russia
[2]Professor, Tomsk State University of Architecture and Buildings, Russia
[3]Research Assiatnt, Tomsk State University of Architecture and Buildings, Russia
[4]Leading Reseacher, NDT Institute, Tomsk Polytechnic University, Russia
[5]Engineer, NDT Institute, Tomsk Polytechnic University, Russia
E-mail: nauka@tsuab.ru

Abstract. Mechanical rebar couplers are preferable in the advanced building construction and structural design of antiseismic elements. The paper presents destructive inspection techniques used to investigate stress fields (tensile and compressive) and deformation curves for mechanical rebar splicing. The properties of mechanical rebar splicing are investigated by the non-destructive testing digital radiography. The behavior of real connections (column-to-column, beam-to-column) is studied under static and dynamic loads. Investigation results allow the elaboration of recommendations on their application in the universal prefabricated antiseismic structural system developed at Tomsk State University of Architecture and Building, Tomsk, Russia.

1. Introduction

In building design and construction, the longitudinal connection of reinforcing bars is often needed since their length is restricted by the transportation conditions and should not exceed 12 m. In Russia, welded and lap splices are widely used. Welded splice has a number of disadvantages connected with electricity costs, attraction of qualified specialists, and labour intensity. Another type of rebar connection is lap splice that also has disadvantages lying in rebar overrun due to overlaps and installation of additional shear reinforcement in rebar connections [1–4]. In Russia, mechanical rebar splicing is used in in-situ concrete and prefabricated reinforced concrete. However, the increased requirements for the operational safety and earthquake resistance of building constructions as well as lowering costs of building and construction works pose the problem of connecting construction members with prefabricated reinforced concrete [5]. A series of experiments was carried out to solve this problem including the detection of the stress and strain state of rebar joints under compression and tension, measured connections between the reinforced concrete members of the universal prefabricated antiseismic structural system under static and dynamic loads.

The main quality parameter of mechanical rebar splicing is its strength uniformity. The strength uniformity condition is satisfied when the fracture load induces the fracture by rebar clear opening rather than by a coupler. Since the real rebar connections are exposed to different loads, the laboratory investigations included the tensile and compressive tests for rebar splicing and static and dynamic load tests for rebar fragments.

Unlike the welded splice which requires a complete weld inspection, the mechanical rebar splice allows it at a stage of the elaboration of technology or by selected node connections. At the same time,

Journal of Physics: Conference Series **671** (2016) 012027 doi:10.1088/1742-6596/671/1/012027

in building constructions of the high responsibility level, the rebar splice inspection can be carried out only by non-destructive testing (NDT) digital radiography [6, 7].

NDT digital radiography of the mechanical rebar splice allows not only to estimate the strength properties of these connections but also to forecast their functional performance including, for example, seismic loads.

2. Materials and metho

The schematic view of reinforcing bars spliced with a crimping coupler system designed to withstand tensile and compressive loads, is presented in figure 1. Hot-finished weldless tubes of the type C20E2C steel are used in the capacity of mechanical couplers for rebar splicing. The outer and inner diameters of the tubes are 51 and 32 mm, respectively. These tubes are manufactured in accordance with the Construction Standards and Regulations of the Russian Federation. This type of rebar connection is designed for the repeated load applications. The load tests of this rebar connection can result in the failure of both the rebar section and the coupler. Rebar was spliced on the rebar coupler crimping machine PA-80 with the operating pressure of 800 atm. Figure 2 demonstrates mechanical rebar splice performed by the coupler after its crimping.

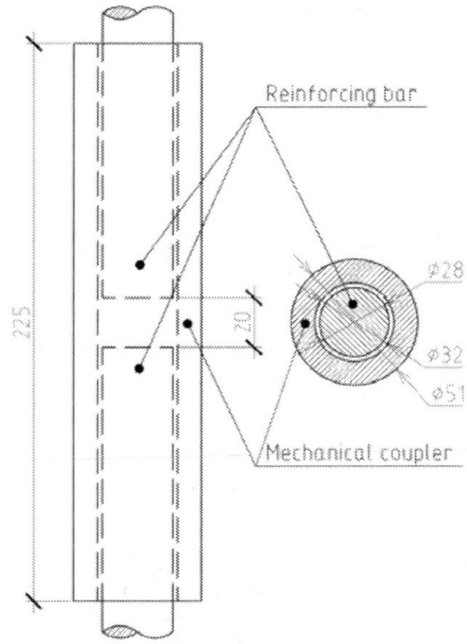

Figure 1. Schematic view of reinforcing bars spliced with a crimping coupler system designed to withstand tensile and compressive loads.

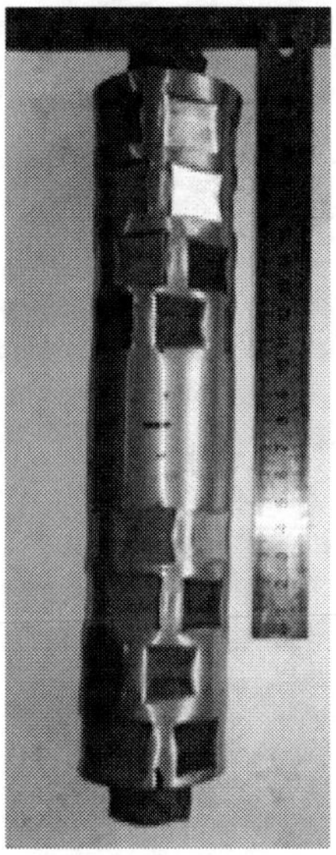

Figure 2. Photograph of mechanical rebar splice.

Vic-3D digital image correlation system allows detecting deformation fields on the specimen surface. Figure 3 contains the photographs of compressive and tensile load tests on the Universal Testing Machine 4500. Such parameters as the surface geometry, displacements, relative deformations (ε_{xx} and ε_{yy} by X and Y axes respectively and the main relative displacement ε_1), displacement and deformation ratios, and surface curvature were determined.

Figure 3. Photographs of load tests: *a* – tensile load test on Tensile Testing Machine MR-500; *b* – compressive load test on the Universal Testing Machine 4500.

The strength characteristics of mechanical rebar splicing were tested both in tension and compression. Tensile tests were carried out on 500 kN Tensile Testing Machine MR-500 (Russia) as shown in figure 3a. The maximum achieved loading rate was 1 kN/s.

Compressive tests were carried out on 4500 kN Universal Testing Machine 4500 (USA) as shown in figure 3b. The maximum achieved loading rate was 1 N/mm^2 *per* second.

3. Tensile test results

The tensile deformation curve of rebar splice coupler is presented in figure 4. The analysis of this curve shows that elastic deformations of rebar coupler are developed within the load range from 0 to 235 kN. At 185 kN tensile force, the stress experienced in the section of reinforcing bar is 300 MPa that corresponds to 0.6 $\sigma_{0.2}$. At this stress value, the plastic deformation in the coupler equals to zero. Figure 5 shows the tensile deformation of rebar coupler.

Tensile tests of the mechanical rebar splicing show that its fracture occurs at the yield strength of rebar with the formation of a neck that indicates to the rebar transition to plastic deformation. The value of deformability of the rebar coupler equaled to $2.48 \cdot 10^{-9}$ m/N, is detected at the conventional yield strength of 500 MPa achieved by the reinforcing bar.

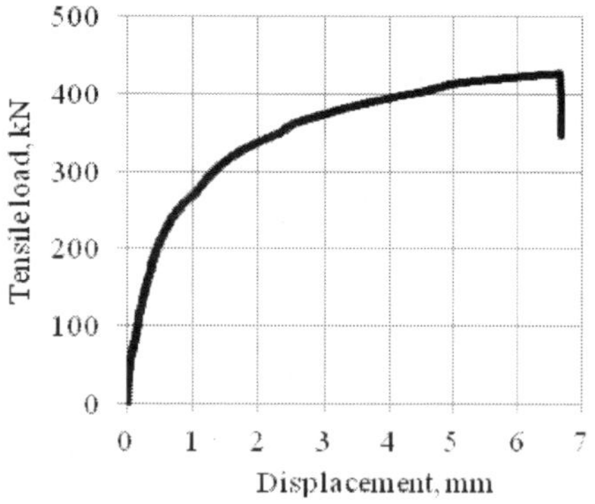

Figure 4. Deformation curve of rebar coupler.

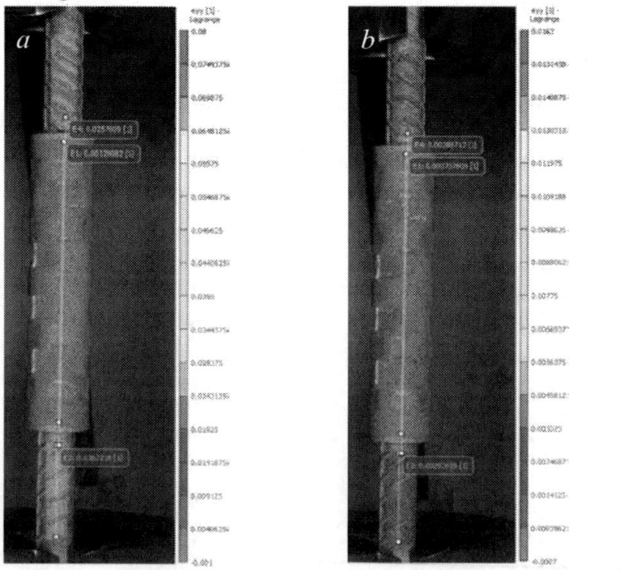

Figure 5. Surface deformation of rebar coupler: *a* – elastic deformation; *b* – inelastic deformation.

4. Compressive test results

The compressive deformation curve of the rebar coupler is given in figure 6. The analysis of this curve shows that elastic deformations of the rebar coupler are developed within the load range from 0 to 320 kN. At 185 kN compressive force, the stress experienced in the section of reinforcing bar is 300 MPa that corresponds to 0.6 $\sigma_{0.2}$. At this stress value, the plastic deformation in the coupler equals to zero.

The compressive deformation of the rebar coupler surface is shown in figure 7. The maximum plastic deformations occur in the coupler between the internal crimps.

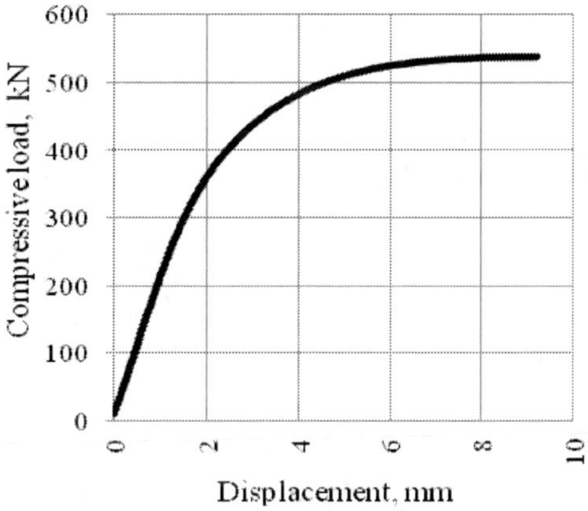

Figure 6. Deformation curve of rebar coupler connection with 20 mm gap.

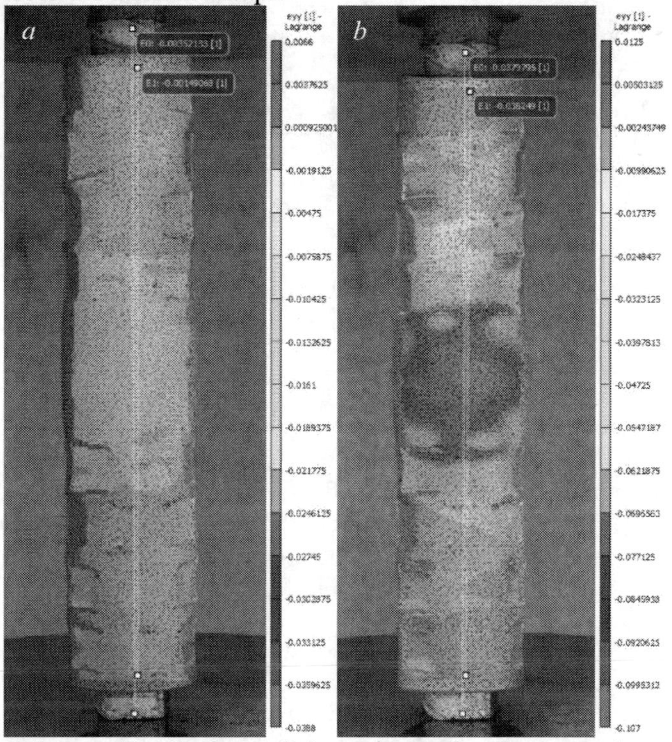

Figure 7. Surface deformation of rebar coupler: *a* – elastic deformation; *b* – inelastic deformation.

5. Non-destructive testing of reinforcing bar connection

The development of the reinforcing bar connecting technology using the analyzed technique is impossible without the NDT methods [6]. Due to the properties of test object, the only method that can be used for its investigation is the radiation method [7, 9]. The recent development of digital radiography [8] characterized by the relative simplicity and expressiveness, allows imaging by the Remote RadEye200 detector module with 1024×1000 pixel image area with 96 μm pixel size. Radioisotopes, X-ray generators and betatrons can serve as the sources of photon radiation. A possibility of the flaw detection analysis is assessed by radiographic images generated for the two test objects, namely rebar fragments. These images are obtained by the digital radiographic system with the maximum photon energy of 160 keV delivered by the X-ray source. In order to reduce the low-energy photon generation, the photon flux is filtered by 2 and 4 mm thick copper plates. Since the maximum photon energy used in this experiment, is far from the optimum, the resulting radiographic image is obtained by averaging 16 images to enhance the signal-to-noise ratio. Rebar locates at the maximum distance from the detector array, therefore, the geometric enhancement is insignificant. The optimum pixel size comes to 88–90 μm.

Figure 8 presents the shadow radiographic images of reinforcing bars connected by the mechanical couplers having the outer diameters of 47 and 51 mm respectively. The analysis of these images allows making a number of conclusions concerning the parameter assessment of tested rebar connections.

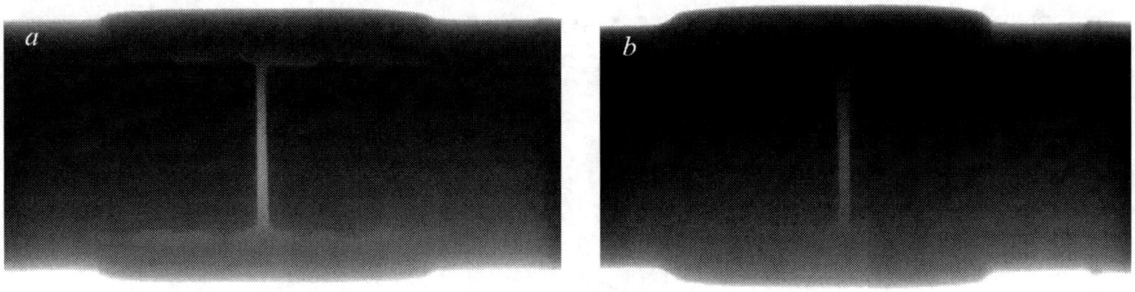

Figure 8. Radiographic images of reinforcing bar connected by the coupler with outer diameter: *a* – 47 mm; *b* – 51 mm.

1. A distance between the two rebar fragments connected by mechanical couplers can be measured with high accuracy (not less than 0,1 mm). In figure 8*a*, this distance varies between 0,704 and 1,6 mm, while in figure 8*b* it varies between 1,6 and 2,05 mm.
2. The face plane rebar deviations from each other can be measured with pixel accuracy, and, accounting for the boundary effects, with sub-pixel accuracy.
3. Light spots observed on the contact boundaries between rebar fragments and mechanical couplers indicate to the low-quality connection. Light spots in figure 8*a* are quite visible, while in figure 8*b*, the light stripe within the contact boundary is faintly visible only on the one side of the image.
4. The additional geometric parameters of the rebar splice can be estimated, such, for example, as the finite length of mechanical coupler and its section.

It is worth noting that the flow detection analysis of the reinforcing bar connection should be supported by the appropriate technical standard documentation. Otherwise, this documentation should be elaborated and contain radiographic images of mechanical rebar splicing compared with their mechanical test results.

6. Experimental

Static load tests of node connections in the universal prefabricated antiseismic structural system developed at Tomsk State University of Architecture and Building (TSUAB), Tomsk, Russia, were carried out in accordance with the Construction Standards and Regulations of the Russian Federation and the test program developed at the Department of Reinforced Concrete and Masonry Structures, TSUAB.

The load test of the beam-to-column connection is conducted stage-by-stage, in equal amounts, at 10% of the fracture load to determine such parameters of node connections as strength, stiffness, and crack resistance. A 10 min interval is selected for the stress relaxation in node connections, while for the detremination of the abovementioned parameters a 30 min interval is selected.

Figure 9. General view of beam-to-column connection: *a* – mechanical coupler; *b* – static load test.

Strength, stiffness, and crack resistance tests [10] of the beam-to-column connection assembled according to the universal prefabricated antiseismic structural system show that the rebar mechanical splice provided by the crimping coupler system is quite reliable for the node connection.

7. Conclusions

1. The suggested mechanical rebar splicing demonstrated its reliability providing the failure outside of the joint along the rebar cross-section rather than by the coupler.
2. The X-ray investigation allowed checking the assembly quality of the suggested mechanical rebar splicing using non-destructive testing that considerably simplifies the procedure of the quality control.
3. Mechanical couplers used in building constructions provide the reliable rebar connections allowing to design and build antiseismic structures in hazardous regions exposed to both static and dynamic loads.

Acknowledgements

The authors acknowledge the financial support from the Ministry of Education and Science of the Russian Federation (Agreement N 02.G25/31/0022) and the Russian Foundation for Basic Research (Project N 13-08-98027). Thanks also go to Margaret Vorobyeva, translator TSUAB. This work was carried out within the Governmental Decree N 218.

References

[1] Madatyan S A and D'yachkov V V (2007) *Beton i zhelezobeton* **4** 16–20
[2] Matkov N G, Balyuchik E A, Gubin V I, Rudenko A E and Uljutinov A V (2003) *Beton i zhelezobeton.* **4** 6–10
[3] D'yachkov V V 2009 *Properties and use of mechanical couplers for reinforcing bars* (Moscow: Stroitel'stvo)
[4] D'yachkov V V (2006) *Beton i zhelezobeton.* **1** 11–14
[5] Guidance documents of operating organization N 0657-2006 Regulations for rebar mechanical connections in building and nuclear plant constructions
[6] Einav I, Artem'ev B, Azizova E and Azizova A 2012 *Non-destructive testing in construction Teaching aid* (Moscow: Spektr)
[7] SNiP 17625-83 1983 *Reinforced concrete structures and products. Radiation method of detection of concrete layer thickness and rebar size and arrangement*
[8] Alkhimov Yu V, Shtein A M , Kas'yanov S V, Babikov S A, Batranin A V and Osipov S P (2013) *Testing. Diagnostics* **13** 31–42
[9] Shtein A M, Cheprasov A I, Klimenov V A, Knyazev S V, Chakhlov S V and Belkin D S (2013) *Russian Physics Journal* **56**(1-2) 267–270
[10] Hadi M N S and Tran T M (2015) *Materials and Structures* 19

A Comprehensive Study of Underground Animals Habitat

A V Klokov[1], A S Zapasnoy[1], A S Mironchev[2], V P Yakubov[3], S S Shipilova[4]

[1]PhD, associate professor, department of radiophysics, Tomsk State University, Russia
[2]Graduate student, department of radiophysics, Tomsk State University, Russia
[3]Doctor of Science, professor, department of radiophysics, Tomsk State University, Russia
[4]Senior Lecturer, Institute of International Education and Language Communication, Tomsk Polytechnic University, Russia
E-mail: zas_rff@sibmail.com

Abstract: This paper describes a method of studying the natural habitats of underground animals by the example of zokor. The purpose of the research is to find habitation of animals using unmanned aircraft and investigate networks of tunnels and burrows with ground penetrating radar "OKO-2". Geolocation data were processed by techniques developed by the authors.

1. Introduction

Zokor is a mole rat rodent which lives underground and borrows tunnels with their powerful front claws. This is a medium-sized rodent with body length of 6 – 11 inches and a tail up to 6.5 inches. Most part of their lives zokors spend underground digging complex multi-level tunnels of 160 – 330 ft length [1]. Zokors are the active object of soil biota, one of the components in biological diversity of land animals. The active observation of zokors' behaviour is performed by trapping and opening their burrows. But, zokors are easily injured when being trapped. Moreover, the natural balance is destroyed by opening the burrows. In that case, the alternative non-destructive technique for underground investigation is supposed to be more efficient.

However, there are some difficulties related to the fact that the animals habitat is quite vast, and the animals don't live in packs or groups, they live individually [2]. The evident signs of the fact that a certain place is occupied by zokors are a large amount of ground piles cast out by the animal in the process of tunneling. Thus, starting the study of these animals a large territory should be covered to find the places with molehills along. So, saving the search time is a quite actual demand for the scientists because the detection of every borrow as well as counting the whole number of zokors are not always successful.

Tomsk State university conducted the summer ground penetrating radar (GPR) research of agricultural fields in Kozhevnikovski district, Tomsk region to detect the zokor habitats.

2. Experiments

To define the experimental field, the unmanned aircraft SUPERCAM-250 produced by Finko Ltd, Russia, was used. The aircraft got off the ground at 820 ft over the area, where the zokor habitats were supposed to exist. The area covered by air photographing was about 1.2×1.2 miles. The total time of flight was about 30 min, the identifying the areas inhabited by zokors according to the image analysis took 20 min. So, the 50-min experiment provided the research of the 1.2×1.2miles area, the detection of zokor habitats as well as a preliminary number of animal units. Also, the most suitable area for the GPR investigation was selected. Then, the aircraft was launched again to go on taking images, one of which is presented in figure 1.

Two pieces of the field with evident traces of animal activity – zokors' holes – were sounded. The size of the ground pieces were 4×4 m and 7.5×4 m, respectively. As the ground penetrating equipment, the radar subsystem OKO-2 with bipolar radiopulses was used. The average pulse frequency in emission band was 1700 MHz. The rated depth of sound penetrating was about 1 m, and the depth resolution was about 3 cm [3]. That resolution should allow us to detect the zokor's tunnels and probably the animals as well.

Figure 1. The GPR investigation.

To obtain the subsurface image, the GPR position on the surface should be precisely recorded. The relative position along the sounding line was provided by GPR tools. A rotation counter attached to a measuring wheel used to monitor linear motion. The crosswise motion to be regulated by ourselves so each piece of ground was marked out before the GPR sounding.

According to the antenna directivity pattern and the radar pulse duration, the longwise and crosswise scanning pitches were defined as 0.022 m and 0.2 m, respectively. The ground marking was performed with nylon cord: firstly, a rectangular of the predefined size was constructed, then, the nylon cord was strained at the defined pitch between the reference points longwise (figure 1). There are a number of possible errors in that approach. One of them is referred to the zero leveling as the GPR was not moved close to the ground. The second one is the need in data adjustment along the GPR trajectory from cross directions. To minimize the errors, a steel tape ruler was placed transversely to the GPR trajectory (figure 1) at the beginning of the trace. As the tape ruler was tightly placed to the ground, it was considered as the zero level and the reference point for the longwise GPR traces as well.

3. The GPR data proceeding

There were 21 parallel longwise traces marked out on each piece of ground. The reading of the scattered field was done at 350 points on each trace. To increase the signal-to-noise ratio (SNR), the reading was done 9 times at each point. That corresponded to the three-times increasing the SNR. The row GPR data are presented in figure 2. The figure 2 shows that there are shifts in the zero level signal from the tape ruler (marked with the oval sign) along the GPR trajectory. This shift occurred due to the different initial GPR positions at the cross traces.

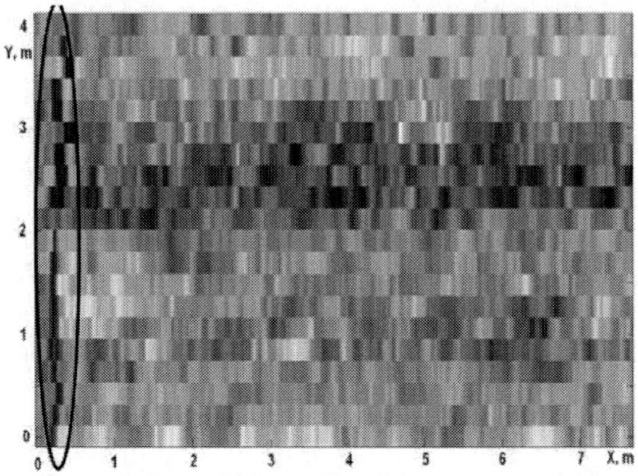

Figure 2. Row GPR profile of the test ground (the tape ruler position marked with the oval).

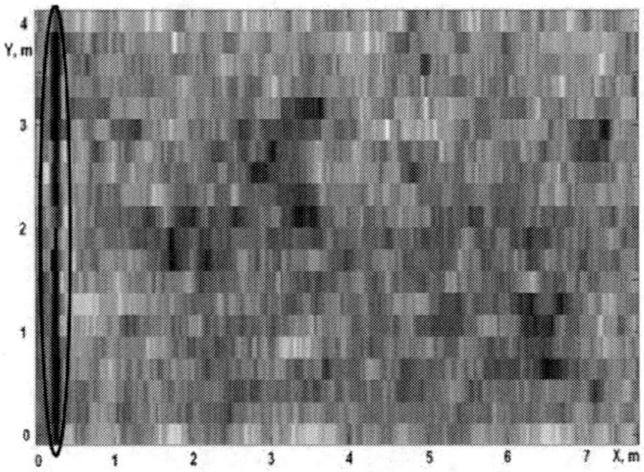

Figure 3. Row GPR profile of the test ground after the data adjusting (the tape ruler position marked with the oval).

The further data interpreting was possible only when the traces were adjusted against the tape ruler position. To perform the necessary adjustment, an area around the tape ruler was selected, then, the correlation rates were calculated for traces relatively each other. The rates were transformed to the trace shift references. The result of the adjustment is presented in figure 3. There, the signal from the tape ruler is straight across the traces and without essential shifts.

The next processing stage was to perform the focusing. So, the exponential attenuation of radiation when penetrating into the soil should be taken into account. In the test soil, there is a certain quantity of water which causes the signal attenuation. The radiation attenuation will be averaged over the background environment.

The operation sequence is as follows: at every GPR position, time records of scattered signals were transformed into the analytical signal amplitudes.

The analytical signal for the pulse signal was defined as a complex signal $z(t) = x(t) + iy(t)$,where the real part was identified with the measured pulse $x(t)$, and the imaginary part resulted from the application of the Hilbert transform [4]:

$$y(t) = \frac{1}{\pi} Vp \int_{-\infty}^{\infty} \frac{x(\tau)d\tau}{t-\tau}.$$

The waveform of this signal was evaluated as the complex signal module at each time point. Numerically, this transform was performed using the Fast Fourier Transform (FFT) and the negative frequency muting in the signal band [5]. Then, all the analytical signals obtained were averaged (figure 4). Supposing the exponential signal attenuation, the average ratio should be approximated by the inclined line using the semi-log scale. In figure 4, the corresponding inclined line was drawn using the least square method [6]. The figure 4 shows that deeper than 0.7 m (2.3 ft) the exponential shape was interrupted, and the signal seemed to increase. However, the increase is due to the accumulated intensity of measurement noises. This means that the data at the depth of 0.7 m and deeper are not reliable. So, the attenuation adjustment with renormalization of measured signals to the average exponential decrease provided the alignment of data from all layers in the range from 0 to 0.7 m.

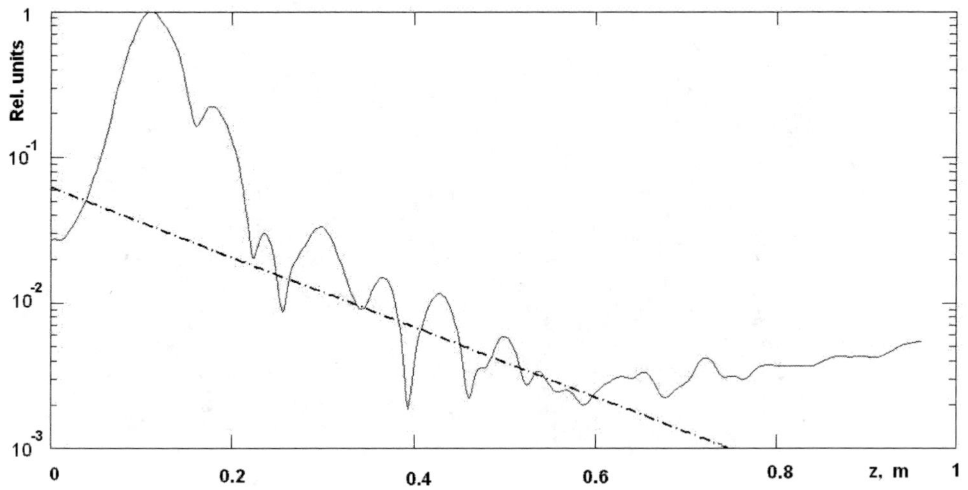

Figure 4. The exponential attenuation of signal.

The last stage in GPR data processing includes the layer-by-layer focusing. The most widely used methods of focusing are the method of diffraction summations and the Stolt migration [7]. Both of them have their advantages and disadvantages as well. The Stolt migration method is faster, as it operates in frequency domain using FFT algorithm. While the method of diffraction summations operates in time domain, but it makes the focusing easier to understand [8].

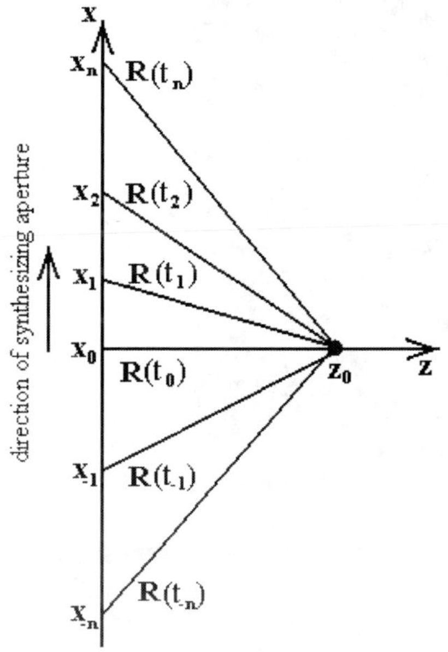

Figure 5. Different travels of reflected signal from the target in different points of sounding system.

As a result of diffraction of each point reflector located in a homogeneous medium is displayed in the form of a hyperbola [7]. In this case the true position of the point scatterer corresponds to the apex of hyperbola. Migration moves reflections to their true positions and mitigates the effect of diffraction, thus increasing spatial resolution and yielding image.

Each reflected pulse travels a distance $R(t_k)$, as shown in figure 5, and has a time delay τ_k. We can calculate the distance and the time delay by using the Pythagorean theorem:

$$R(t_k) = \sqrt{z_0^2 + (x_0 - x_k)^2}$$

Expressed from this equation z_0 obtain:

$$z_0^2 = R(t_k)^2 - (x_0 - x_k)^2$$

This is a hyperbola equation and z_0:

$$z_0 = \sqrt{(\tau_k c)^2 - (x_0 - x_k)^2}\ .$$

We can obtain in tree-dimension case:

$$z_0 = \sqrt{(\tau_k c)^2 - (x_0 - x_k)^2 - (y_0 - y_k)^2}\ .$$

The method of diffraction summation is based on the summation of amplitudes along the hyperbolic paths and has the following sequence. 1) Sequent scanning of the test volume points. 2) Calculation of the diffraction hyperbolic curve for each point. 3) Summation of measured signal data along the hyperbolic curve [9].

The results of GPR measurements are presented in figures 6 and 7 for the ground pieces 7.5 x 4 m and 4×4 m, respectively.

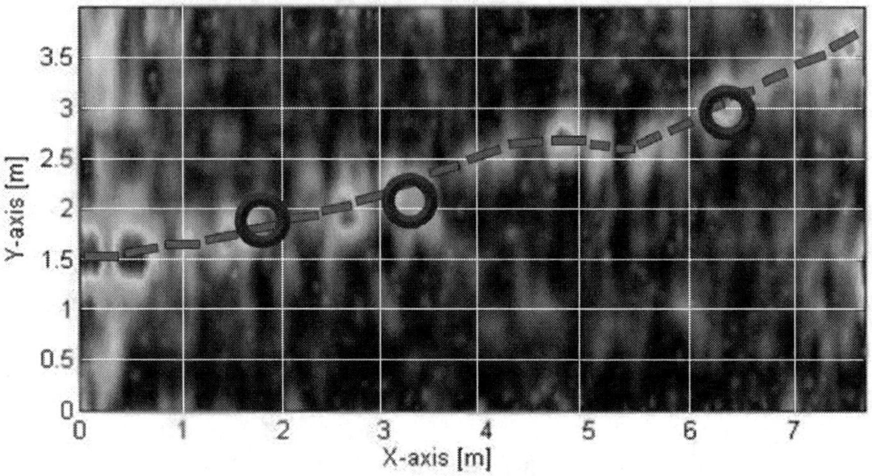

Figure 6. The GPR crosscut of a horizontal zokor's tunnel at the depth of 12 ft.

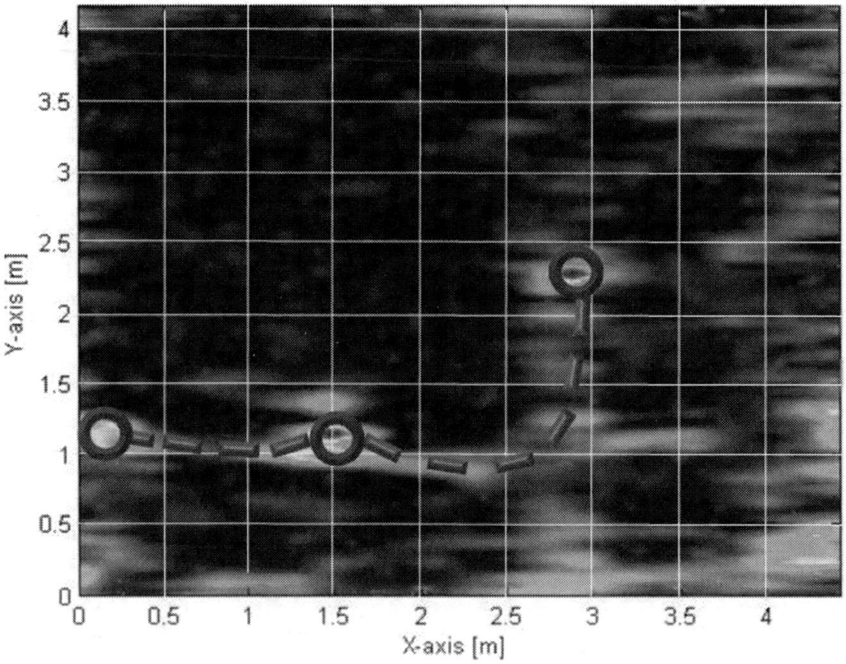

Figure 7. The GPR crosscut of a horizontal zokor's tunnel at the depth of 12 ft.

The horizontal tunnel is displayed as a dash line. The burrow location observable on the surface is marked with circles.

4. Conclusion

The conducted research demonstrated the opportunity of the GPR application for studying the habitat of underground animals. This is considered to be important and promising direction in geolocation as it prevents the damage caused trapping and opening animal's burrows and helps protect and preserve the animal's natural habitat as well.

This research was supported by grant № 8.1.12.2015 from “The Tomsk State University Academic D.I. Mendeleev Fund Program”.

Journal of Physics: Conference Series **671** (2016) 012028 doi:10.1088/1742-6596/671/1/012028

References

[1] Laptev I P 1958 *Mlekopitayushchiye tayozhnoy zony Sibiri (Mammals of Western Siberia Taiga)*– (Tomsk: TSU Publishing)

[2] Musser G G and Carleton M D 2005 *Superfamily Muroidea in Mammal Species of the World a Taxonomic and Geographic Reference* (Baltimore, Johns Hopkins University Press) pp. 894–1531

[3] Instruktsiya po ekspluatatsiyi radiotechnitcheskogo pribora podpoverkhnostnogo zondirovaniya OKO-2 (Manual for the ground penetrating radar OKO-2) 2011 (Moscow, «Logitcheskiye systemy»)

[4] Hilbert D 1953 *Grundzüge einer allgemeinen Theorie der linearen Integralgleichungen* (Chelsea Publ. Comp., New York)

[5] Bracewell R N 2000 *The Fourier transform and its applications 3rd ed* (Boston: McGraw Hill)

[6] Bretscher Otto (1997) *Linear algebra with applications* (4th ed.)(Prentice Hall)

[7] Yilmaz O, Doherty S M, Yilmaz O 2001 *Seismic data analysis: processing, inversion, and interpretation of seismic data* (Tulsa, OK: Society of Exploration Geophysicists)

[8] Klokov A, Sato M (2010) Comparison between the method of diffraction summations and Stolt migration *Proc. of the 123rd SEGJ Conference* 168–171

[9] Yakubov V P, Shipilov S E, Sukhanov D Ya, Klokov A V 2014 *Radiovolnovaya tomographiya: dostizheniya i perspectivy (Radio-wave tomography: achievements and prospects)* (Tomsk, NTL)

Journal of Physics: Conference Series **671** (2016) 012029 doi:10.1088/1742-6596/671/1/012029

Calibration Methods of Force Control Diagnostic System of A Rolling Stock on the Run

A Kolomeets[1], S Bekher[2]

[1]Junior Researcher, Siberian Transport University (STU), Novosibirsk, Russia
[2]Assistant Professor, Siberian Transport University (STU), Novosibirsk, Russia
E-mail: andreykolomeec@yandex.ru

Abstract. In article the technique of calibration of system of dynamic control of a rolling stock is described. The scheme of carrying out and the device for calibration of dynamic control of a rolling stock is provided. Dependence of dynamic force on temporary parameters of blow is defined.

1. Introduction

The control systems of a wheel attack (hereinafter referred to the System) based on the elastic contraction of a rail measure the magnitude of force with a limited accuracy [1]. At the present moment there are no certified calibration methods for such systems in our country. This paper presents calibration methods of the control systems of the attack wheel to rail [2] in the field conditions (hereinafter referred to the Method). The Method determines a confidence error for force measurement as ±5% with a confident coefficient of P = 0.95.

The System calibration is performed by the direct comparison method. The System calibration can be primary and periodic with the same range of activities for both. While calibrating the following operations are conducted: a visual inspection, a testing, a metrological examination, a measurement of actual value of dynamic force measured [3,4].

While testing a rolling stock runs through the System control section and measurement data comes from all System channels [5]. Calibration is performed on each sleeper and in the intercrib work across the control section (figure 1a). The ranges of forces applied are 25–50 кN, 50–75 кN, 75–100 кN, 100–125 кN, 125–150 кN.

2. Experiment

To produce a dynamic force an impact device is used (figure 1b), it consists of a hammer and a centering tube. The hammer is a set of plates (figure 1b; position 6 and 7) and striker pins (figure 1b; position 5) put on a stud M10 and fixed with a screw (figure 1b; position 1).

Plates and striker pins are made of steel 45 or steel with a higher yield point and provide a hammer weight variation in the range of 5–15 kg at a pitch of 2 kg. The diameter of plates and striker pins is (98±1) mm. It is measured with a caliper ШЦ-150 or other measurement tools with the same accuracy; maximum permissible error is ±0.5.

The lower plate in the hammer is made of duralumin D15 or other alloy with similar density (±20%) to meet $l > 2 \cdot d$; it should prevent hammer sticking while falling. A hammer weight is determined with a permissible error of ±1 % by weighing on the platform scales ВСП-15.2-4ТК or scales with the same accuracy; received data is stated in the record sheet. It is acceptable to weigh components by themselves and then to determine a hammer weight by totaling provided that a combined error does not exceed a maximum permissible error.

A centering tube allows an ordered motion of a hammer. It is made of a polypropylene with a coefficient of sliding friction on steel not more than 0.15 or other material with a lower friction ratio.

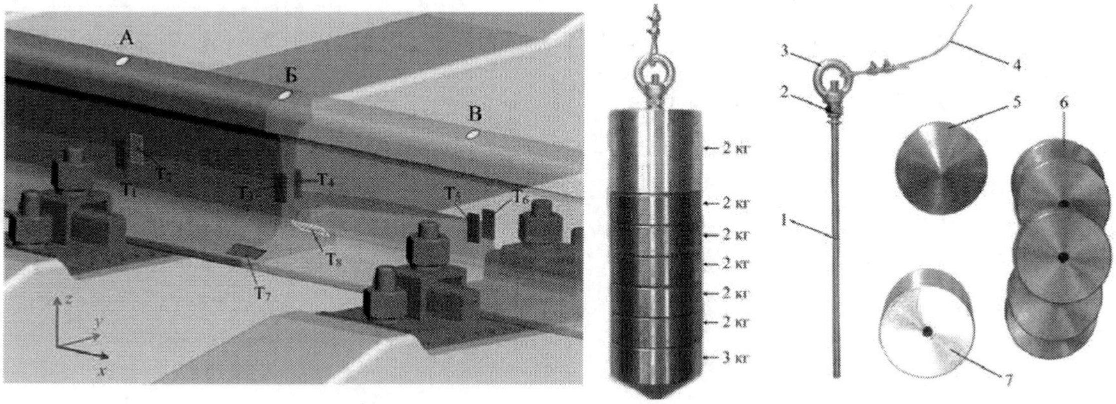

a) b)

Figure 1. Scheme of a control section (a) and an impact device (b) to produce a dynamic force.

The tube ID is (100±1) mm, tube length is min 2 m. There are four guide lines in 90° at the end of the tube to position devices on the rail. The tube is fit on the rail upright (out-of-upright is max 20 mm/m); uptightness is controlled with a block level (its length is 300 mm, readability is 0.1 mm/m) as per GOST 9392–75 or other devices with similar accuracy. Three through-holes of (10+0.5) mm in diameter are made in the tube to fix the weight at the height of (500±2) mm, (1000±2) mm, (1500±2)mm. Through-hole height is controlled with a 2.5 m measuring reel as per GOST 7502–98 or other devices with similar accuracy with max 1% permissible error.

Before and during testing a maximum height of a hammer above a top of rail, fall time and time of hammer-and-rail contact are measured and stated in the reports. Fall time and time of hammer-and-rail contact are measured with a digital-storage oscilloscope and/or an acceleration indicator.

To time the blow with a circuitry an electrical circuit is made up, it consists of a) a digital-storage oscilloscope or other similar electrical signal measuring devices with time permissible error of ±0,01 sec or with similar accuracy; b) a hammer; c) a rail in the control section; d) a source of constant voltage with 1–5 V; e) fixed resistors with nominal electrical resistance $R_1 = R_2 = R_3 = (200 \pm 10)$ Ohm (figure 2a).

An electrical circuit on the 1st channel of the oscilloscope captures the fall time on the top of a rail and the hammer-and-rail contact time equal to the blow time. An electrical circuit on the 2nd channel records the time of a hammer start. An electrical circuit on the 3d channel registers the fall time on the $2 \cdot \Delta L$ height. An electrical circuit on the 4th channel of the oscilloscope captures the fall time on the ΔL height.

When the hammer is falling, a digital-storage oscilloscope records signals shown in figure 3. The time of a hammer run to a blow t_1, t_2, t_3, t_4 and a blow time τ_1 are determined / measured by received signals. Measurement data is stated in the due form reports.

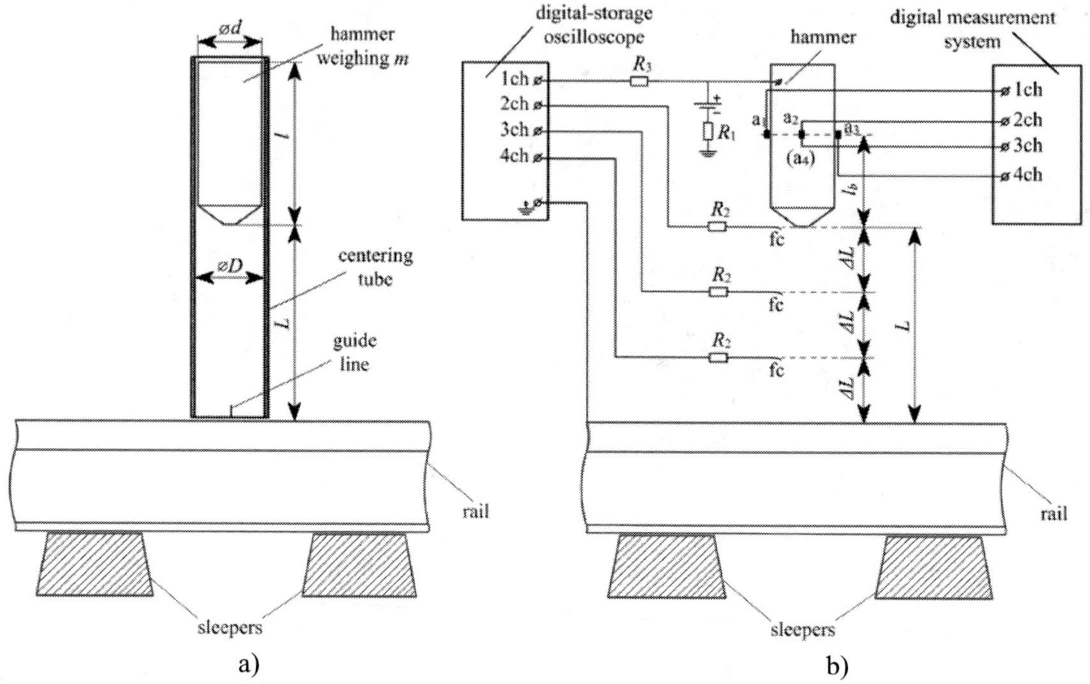

Figure 2. Scheme of a calibration device (a), scheme of calibrating with measurement systems (b): R_1, R_2, R_3 – resistance, a_1, a_2, a_3, a_4 – acceleration indicators, l_b – distance between a striker pin and a center of hammer weights, *fc* – flexing electrical contact.

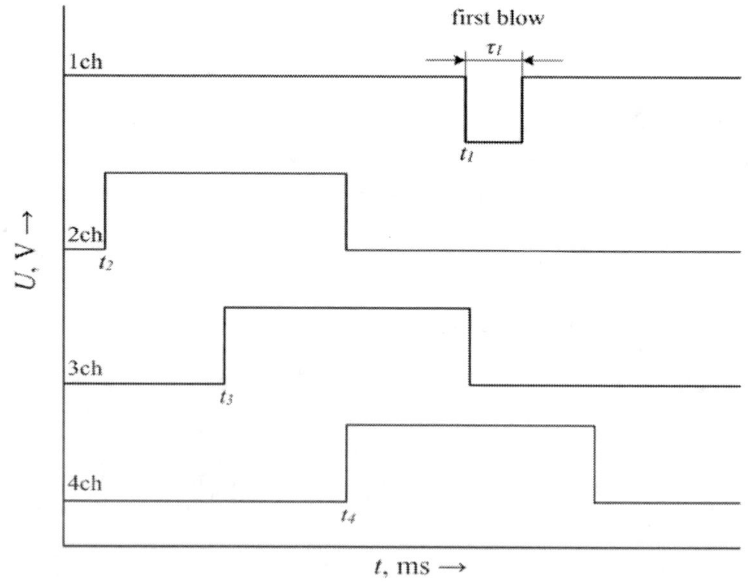

Figure 3. Signals recorded on Channels 1,2,3,4 at hammer falling.

A hammer weight center is calculated with CAD facilities in order to time the blow with an acceleration indicator. Maximum permissible error for finding a hammer weight center is ±10 mm. Four acceleration indicators (with range of -2000g – +2000g m/s^2, resolution limit max 10g) are fixed

Journal of Physics: Conference Series **671** (2016) 012029 doi:10.1088/1742-6596/671/1/012029

at the center-of-weight-level height on perimeter every (90±5)° to measure vertical acceleration (figure 4). Signals received from acceleration indicators are registered and averaged. Duration τ2 of a hammer blow on the rail is determined at the 5% level of peak acceleration. Measurement data is stated in the due form reports. Average time is calculated by formula:

$$\tau = \frac{\tau_1 - \tau_2}{2}$$

Uncertainty of blow time is evaluated and stated in the report. Average force of a blow is calculated by formula and then stated in the report:

$$F_m = \frac{m \cdot \sqrt{2 \cdot g \cdot L}}{\tau},$$

where g = (9,81±0,01) m/s^2 - reference value.

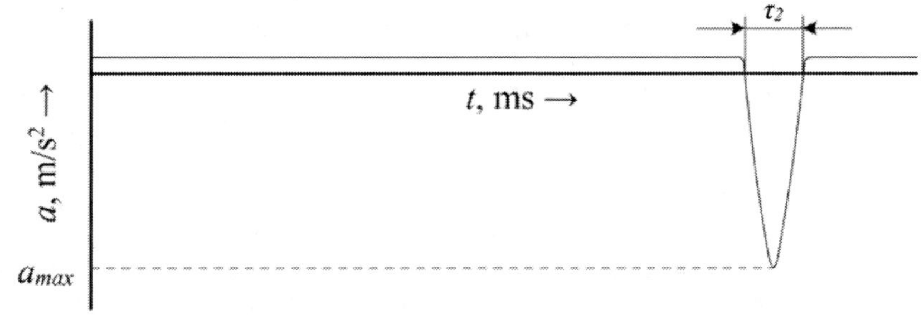

Figure 4. Vertical acceleration – impact time relation.

3. Conclusion

The force determination error is calculated by difference between hammer acceleration and nominal magnitude of gravity depending on fall time then it is registered in the report. The range of permissible confidence error of force generation is calculated with confident coefficient of P = 0.95. An error of average dynamic blow force generation does not exceed 5%.

While dynamic force measuring, safety instructions are to be followed. At least two specialists conduct the operations. If measurements are performed in the field conditions, specialist wear orange reflective jackets and there must be a signalman on the railway track. Calibration of the control systems of the attack wheel to rail is performed at temperature from -25 to +35 °C, atmospheric pressure of 84–107 kPa (630–800 mmHg), relative humidity of 30–80%.

4. References

[1] Stepanova L N, Bekher S A and Kochetkov A S (2009) Measurement error analysis of moving wheel vertical forces to rail *Testing. Diagnostics* **12** 54–60

[2] Otto L (2008) Monitoring railway wheel defects with rail installed measuring device *Publications of the Finnish Rail Administration* **12** 90

[3] Sekuła K, Kołakowski P (2013) Identification of Dynamic Loads Generated by Trains in motion using Piezoelectric Sensors *Proceedings of ISMA2010 including US201* 1099–1118

[4] Sharpe P (2010) Measuring of the road-bed rigidity *Railway Gazette International* **9** 190–194

[5] Stepanova L N, Kabanov S I, Bekher S A and Kolomeets A O (2013) Fast strain-gauge system for freight car wheel roll surfaces defects monitoring in movement *Sensors and Systems* **10** 38–44

Definition of Time Induction of Self-Ignition of the Substance on the Prognostic Extrapolation Depending on the Basis of Indicators Fire and Explosion Hazard

A Sechin[1], O Kyrmakova[2], S Osipenko[3]

[1]Professor, Institute of Non-Destructive Testing, Tomsk Polytechnic University, Tomsk, Russia
[2]Postgraduate student, Institute of Non-Destructive Testing, Tomsk Polytechnic University, Tomsk, Russia
[3]Postgraduate student, Institute of Non-Destructive Testing, Tomsk Polytechnic University, Tomsk, Russia
E-mail: olia_917@mail.ru

Abstract. In this article the research directed on development of a technique of definition of time of induction of the self-ignition of substances and materials which is an indicator of the beginning of development of an emergency is conducted. The experiment consisting in supervision over process of self-ignition of coal and oil deposits was the basis for research. On the basis of experimental data the curve expressing analytic - expected dependence of size of temperature of ignition on induction time was constructed. Proceeding from graphical representation of process, functional dependence of time of induction on a temperature indicator was received: $y = 16920 \cdot x^{0.537}$. By means of known indicators of such substances as bitumen oil oxidized (the combustible solid substance received by oxidation of residual product of oil refining) and tar oil (the combustible solid substance which is residual product of oil refining) and the received algorithm, verification of reliability of the received dependence and a technique of definition of time of induction of spontaneous ignition of deposits of oil in general was carried out. The practical importance of the conducted research is that having data on time of induction of process of self-ignition, by means of preventive measures becomes possible to avoid and prevent accidents in oil and oil processing branches, at the same time loss of property and loss of human life.

1. Introduction

The phenomenon of spontaneous ignition of substances and materials can be defined as direct emergence of burning of substance in lack of a source of ignition. It causes interest of scientists since the end of the 19th eyelid, and to this day. The greatest attention is paid now to questions of calculation of time of induction of process of spontaneous ignition which has a practical interest: prevention of emergence of industrial accidents. Coal mining and processing of coal, and also oil and oil processing - the most presented influence of this process. They are united by one property of the recycled materials: spontaneous ignition under certain conditions. As shows the analysis of self-ignition of coal volumes and deposits, process this repeated factor and one of major factors – time of induction of process of spontaneous ignition. The analysis of the reasons of accidents which occurred in parks of a reservoir for storage of oil products and oil products [1, 2], allowed to show that about 40% of accidents happen during cleaning and repair of tanks, one of which reasons - direct ignition of oil pools as relevance of the reasons of a case in point and requirement of more attentive studying. Supervision of some materials, from the growing old provisions of time, I showed that their chemical activity doesn't decrease, and they remain in the same degree of potential danger as were [3, 4].

It is known that indicators of fire-and-explosion hazard of substances and materials decide on the purpose of receiving basic data for development of systems on ensuring fire safety and explosion safety [5]. But developers quite often come up against a situation when the lack of physical and chemical data on substance doesn't allow to carry out by calculation methods of the complex analysis of fire-and-explosion hazard of this or that substance turned in technological process.

The known approaches to definition of conditions of thermal self-ignition of substances and materials offered by a number of the leading scientific centers of Russia are also not very successful.

Based on thermodynamic processes which happen in substance at its heating, the assumption was made that time of induction of these processes can be described by some one dependence – the speed of accumulation of heat in system. Temporary characteristics by which researchers the indicators of fire-and-explosion hazard of indicators of substances and materials are guided, were analysed from thermodynamics positions. The result of the analysis showed that it is possible to be guided by such indicators as group of combustibility, spontaneous ignition temperature, decay temperature, temperature of flash and temperature of the beginning of thermal destruction or sublimation of substance.

The problem which was solved in this work, consisted in definition on known temperature indicators of fire-and-explosion hazard of methodical approach in definition of time of induction of process of the self-ignition of substances and materials characterizing the beginning of development of some emergency process.

2. Calculated part

Process of self-ignition of coal [6,7] and oil deposits [8] was the basis for algorithm of research. These substances have a different set of indicators of fire-and-explosion hazard (Table 1 [5]).

Table 1. The list of the indicators of fire-and-explosion hazard presented to the analysis

Name of substance	Indicator of fire-and-explosion hazard of substance				
	Group of combusti bility	Self-ignition temperature	Decay temperature	Flash point	Onset temperature of thermal decomposition or subliming
Coal of brand CC of the Kuzbass pool	+	+	+	–	+
Oil deposits	+	+	+	+	+

From table 1 it is visible that for coal of brand CC there is no possibility of use of an indicator – flash Temperature. Circuitry of the instrument on which experiments with samples of coal of brand CC of the Kuzbass pool were made, by determination of temperature of self-ignition is presented in figure 1.

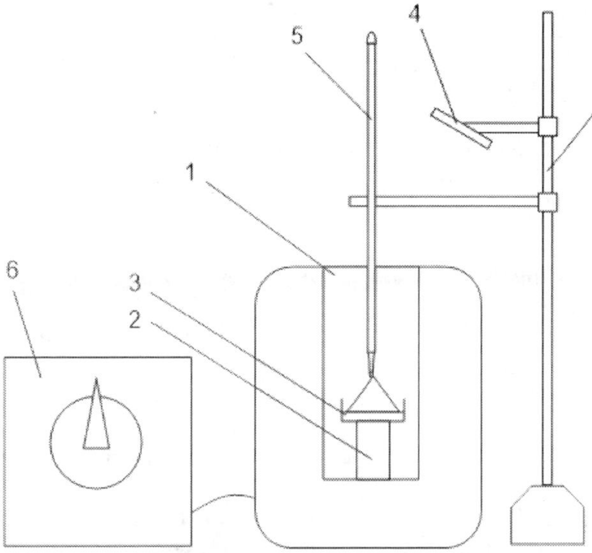

Figure 1. Circuitry of the instrument for determination of parameters of self-ignition of substances and materials: 1 – the electric furnace (V=1000ml); 2 – quartz support; 3 – a porcelain support for substance; 4 – viewing mirror; 5 – mercury thermometer; 6 – laboratory autotransformer; 7 – support.

Researches were conducted according to "A method of experimental determination of temperature of self-ignition of strong substances and materials", presented to industry standard 12.1.044-89.

The received results of the conducted researches are presented in Table 2.

Journal of Physics: Conference Series **671** (2016) 012030 doi:10.1088/1742-6596/671/1/012030

Having carried out the analysis of definition the fire-and-explosion hazard characteristics for the considered combustible strong substance – coal of the Kuzbass pool of brand CC, according to characteristics, known in literature, and also by results of the conducted researches (Table 2), the curve expressing analytic-expected dependence of size of temperature of ignition on time of induction (figure 2) was constructed.

Table 2. Sample temperature growth rate when carrying out experiment

Indicator temperature, °C		Induction time, *s*	Temperature growth rate, °C/s
Flammability class	900	35	26
Self ignition	535	60	9
Decay temperature	350	1200	0.3
Decay temperature	224	7920	0.03

The method of expected extrapolation – operation of extrapolation of a trend with use of the Microsoft Excel program was applied to processing of the received results.

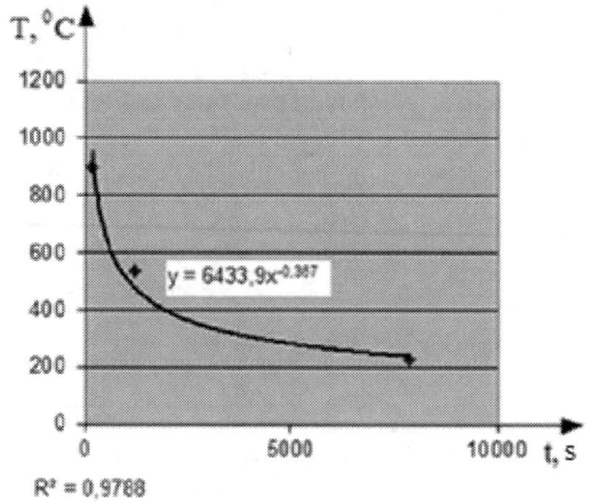

Indicator temperature, °C		Induction time, s
Flammability class	900	35
Self ignition	535	1200
Decay temperature	224	7920

Figure 2. Expected and extrapolation dependence of size of temperature of ignition on temperature growth rate at present time T/s, according to characteristics of fire-and-explosion hazard of coal of brand CC.

The size of reliability of approximation (*R2*) on the available initial temperature indicators makes 0.9788 that is considered admissible.

The function describing the received expected and extrapolation curve has an appearance of exponential function:

$$y = k \cdot x^n \tag{1}$$

where *k, n* – the constants determined by expected and extrapolation dependence on the basis of experimental data.

For coal of the Kuzbass pool of brand CC the following constants were defined: k=6433.9, n=-0.3674.

Thus, expression 1 is the expected and extrapolation dependence constructed on experimental data and allowing to define induction time before emergence of burning at a certain temperature. We will write down expression 1 in a form, convenient for production workers

$$T = 6433.9 \cdot \tau^{0.3674} \tag{2}$$

where T – substance temperature at the available speed of heating, °C; τ – a high-speed indicator of time of induction, *s*.

This expression 2 can represent great practical value for branches of coal mining, consumption and processing.

Journal of Physics: Conference Series **671** (2016) 012030 doi:10.1088/1742-6596/671/1/012030

3. Results

The received algorithm of research was applied to studying of process of self-ignition of the oil deposit taken from an oil separator. Possibility of this step locates that the passing mechanism on a spontaneous ignition curve for coal and a deposit of oil is identical, that these substances in the course of the transformation pass physical and chemical phases similar each other.

Thus, for the studied oil deposit which part are, including oxides of metals, the complex of researches on the scheme presented earlier and construction Expected and extrapolation dependence according to characteristics of fire-and-explosion hazard (Table 3, figure 3) was carried out [5].

Table 3. Characteristics of fire-and-explosion hazard of deposits of oil

Indicator temperature, °C		Induction time, s	Temperature growth rate, °C/s
Flammability class	900	52	31.7
Self ignition	550	63	9
Flashover	150	4500	0.03
Decay temperature	100	10800	0.01

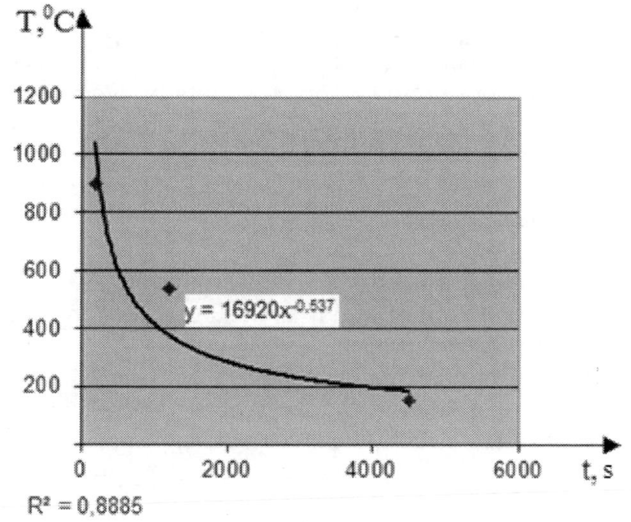

Figure 3. An expected and extrapolation curve according to characteristics of fire-and-explosion hazard of deposits of oil.

The received expected and extrapolation curve has sedate dependence:

$$y = 16920 \cdot x^{-0.537}$$

Having algorithm of calculation of time of induction for such substances as bitumen oil oxidized (the combustible strong substance received by oxidation of residual product of oil refining) and tar oil (the combustible strong substance which is residual product of oil refining), carried out an inspection of reliability of the received dependence and a technique of definition of time of induction of spontaneous ignition of deposits of oil in general [1, 3].

For the bitumen oxidized a formula for calculation of conditions of oil self-ignition:

$$\lg t = 2.2912 - 0.122 \cdot \lg \tau \ [4],$$

where t – temperature equal 204^{0}C; τ – time of induction of substance.

Value of the second parameter is calculated:

$$lg\tau = (lgt - 2.2912)/(-0.122) = -0.151$$

$$\tau = 10^{-0.151} = 2542 \text{ s.}$$

For tar oil:

$$lg\, t = 2.3106 – 0.122 \cdot lg\, \tau [4],$$

where t – temperature equal 204^{0}C; τ – time of induction of substance.

Value of time of induction is calculated:

$$\lg\tau = (\lg t - 2.3106)/(-0.122) = -0.0079$$

$$\tau = 100.0079 = 3665 \text{ s.}$$

Then put the used temperature indicator into the sedate dependence of an expected and extrapolation curve of characteristics of fire-and-explosion hazard of oil received by us.

$$y = 16920 \cdot x^{-0.537},$$

where y – a temperature indicator, x – induction time.

Respectively, time of induction made 3742 pages. Being guided by R^2=0.8885 – the size of reliability of approximation, it was received that

$$x = \tau \cdot R^2 = 3325s.$$

Comparison of the received results with settlement data for bitumen and tar, showed that temporary indicators of induction form range $2542<\tau(s)<3665$.

It is obvious that the size received by the developed technique τ = 3325s, taking into account the amendment, is admissible.

4. Summary

Thus, are received reliable and applicable in practice of oil and oil processing branches sedate dependence of time of induction on a temperature indicator: y=16920$x^{-0.537}$ and technique of its receiving.

The practical importance of the received results is in possibility of understanding, representation and the forecast of probable emergencies and acceptances of the appropriate measures directed on prevention of their emergence that eliminates or significantly reduces the damage caused by an emergency situation, which is expressed as in violation of technological process at the enterprise, financial expenses, and in death of people and in an adverse effect on territory ecology.

Thus, as a result of the conducted research:

It is offered, on known temperature indicators of fire-and-explosion hazard, the methodical approach in definition of time of induction of process of self-ignition of substances and materials characterizing the beginning of development of some emergency process.

The algorithm of definition of time of induction of emergence of explosive concentration in processing equipment is received, the result is presented in the form of a power function y=16920x-0,537, and allowing to predict development of certain events, the processes happening when conducting works in oil and oil processing branches that gives the grounds for acceptance of any precautionary actions.

References:

[1] Suchkov V P, Bezrodnyi I F, Vyaznikovtsev A V (1992) The fires of tanks with oil and oil products (Moscow, Chiiteneftehim)

[2] Chang J I , Lin C C (2006) A study of storage tank accidents *Journal of Loss Prevention in the Process Industries* **19** 51–59

[3] Nazarenko O B, Amelkovich Y A, Sechin A I (2014) Characterization of aluminum nanopowders after long-term storage *Applied Surface Science* **321** 475–480

[4] Nazarenko O B, Sechin A I, Amelkovich Yu A, Ilyin A P (2014) Prospects of using nanopowders as flame retardant additives *Advance Materials Research* **872** 123–127

[5] Baratov A N, Korolchenko A Ya, Kravchuk G N 1990 Fire-and-explosion hazard of substances and materials and means of their suppression: reference media (Moscow, Chemistry)

[6] Kessels W, Wuttke M, Wessling S, Xuan L (2005) Coal Fires Between Self-Ignition and Fire Fighting: Numerical Modeling and Basic Geophysical Measurements *International Conference of Spontaneous Coal Seam Fires: Mitigating a Grobal Disaster* **4** 467–484

[7] Medek J and Weishauptová Z (2003) Mechano-activation as Initiation of Self-Ignition of Coal *Energy & Fuels* **17** 159-163

[8] Emir Ceric 2012 *Crude oil, processing and production* (Sarajevo, IBC)

Journal of Physics: Conference Series **671** (2016) 012031 doi:10.1088/1742-6596/671/1/012031

Low-Frequency Acoustic Signals Propagation in Buried Pipelines

A L Ovchinnikov[1], B M Lapshin[2]

[1] Senior Researcher, OOO Fonon Research and Production Company, Tomsk, Russia
[2] Head of Laboratory, Institute of Non-destructive Testing, National Research Tomsk Polytechnic University, Tomsk, Russia
E-mail: ovchinnikov_al@inbox.ru

Abstract. The article deals with the issues concerning acoustic signals propagation in the large-diameter oil pipelines caused by mechanical action on the pipe body. Various mechanisms of signals attenuation are discussed. It is shown that the calculation of the attenuation caused only by internal energy loss, i.e, the presence of viscosity, thermal conductivity and liquid pipeline wall friction lead to low results. The results of experimental studies, carried out using the existing pipeline with a diameter of 1200 mm. are shown. It is experimentally proved that the main mechanism of signal attenuation is the energy emission into the environment. The numerical values of attenuation coefficients that are 0,14- 0.18 dB/m for the pipeline of 1200 mm in diameter, in the frequency range from 50 Hz to 500 Hz, are determined.

1.Introduction

Recently, basic research in the field of acoustic leak detection were aimed at the solution of the problems related to the study of the characteristics of signal propagation, mainly, to the study of the field mode structure, as well as the tasks, designed to improve the localization curacy. So, theoretical study of waveguide modes propagation in buried pipelines with liquid was conducted, phase velocities, spectra and modes energy distribution in the pipeline were calculated in [1–3]. The theoretical results, obtained in [3], concerning AE signals propagation in plastic pipes are experimentally proved in [4]. For effective AE signals extraction from noise an algorithm of using Wavelet transformation is proposed in the paper [5]. In [6–8] various ways of increasing the accuracy of leakage localization using cross-correlation method are considered.

The aim of this work is to study low-frequency acoustic signals propagation in the pipes with liquid and an estimation of signal attenuation. Also, the studies could be useful in development of long-range detection equipment for inline objects (pigs), which are moving with the flow of the pumped liquid.

Already published papers concerning the study of sound attenuation in the pipes deal with consideration of the attenuation in the frequency range from 10 to 100 kHz. From [9] it is known that for the majority of liquids in a wide range of frequencies there is a quadratic dependence of the sound attenuation coefficient on frequency. However, according to [10] attenuation coefficients, measured using a real pipeline in the frequency range of 10–50 kHz, are accurately approximated by a linear dependence. The values, obtained in [10], significantly exceed theoretically calculated attenuation coefficients, caused by the presence of viscosity, thermal conductivity and liquid pipe wall friction.

Paper [1] shows that using the numerical solution of the equations of hydrodynamics and dynamic elasticity theory for a three-layer waveguide: pumped liquid - a pipeline wall – soil, under certain conditions, in addition to signal attenuation due to internal losses, acoustic signals could be attenuated by energy emission into external environment. Nowadays, there are no accurate analytical and numerical solutions to estimate the signal attenuation caused by emission during its propagation in complex three-layer waveguide. However, there is a number of works devoted to the cylindrical shell sound emission in simpler conditions [11–13]. For example, [13] shows theoretically and experimentally that the cylindrical shell emitted power increases with increasing shell diameter, whereas emission coefficient in the low frequency region increases with increasing frequency. In authors' point of view, these regularities will be implemented during signals propagation in complex layered medium, but quantitative assessment of the ratios, obtained in these papers, will give incorrect results.

2.Experimental installation

The scheme of carrying out research on excitation and propagation of acoustic signals is shown in figure 1. The main element of this scheme is the section of the trunk pipeline with a diameter of 1220 mm and a wall thickness of 15.2 mm. This is the maximum diameter of the pipeline, used for oil transportation. Pipeline construction is made in accordance with Russian formal building regulations. The pipe is filled with West Siberian oil, which was moving at a velocity of 0.85 m/s.

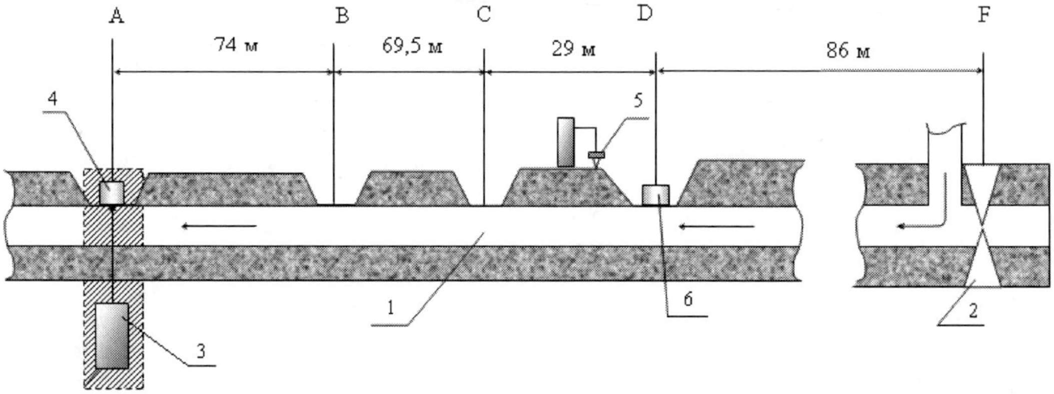

Figure 1. The scheme of research of low-frequency acoustic signals propagation using operating pipeline: 1 – pipeline; 2 – pipe valve; 3 – an electric signal generator; 4 – emitter; 5, 6 – receiving transducers; A, B, C, D – pits; F – pipe valves.

For direct access to the pipe, the areas without soil cover (pits) were kept. The insulation coating is reinforced film of polymeric tapes (2-layered polymeric insulating tape with 0.6 mm thickness of each and the protective wrap with a thickness of 0.6 mm). Backfilling material is watered sand. The ground water level is up to the middle of the pipe. One side of the pipeline section had a pipe valve and pipe rotating is 90^0, the other side had the closest valve at the distance of 4.0 km away. The pressure in the pipe during the testing period was 1.7 MPa.

Acoustic vibrations were excited using the emitter of the electromagnetic type, developed by the authors. Signals recording was carried out using two transducers of different types: electrodynamic type transducer (CV-10C brand), industrially manufactured, and the piezoelectric transducer, developed by the authors. Industrially produced receiving transducer CV-10C has the following characteristics: natural frequency is 10 Hz; the second resonance frequency is 70Hz; conversion factor is 18 $V{\cdot}m^{-1}sec$. The ohmic resistance of the inductance coil is 285 Ohms. Piezoelectric transducer, developed by the authors, has the frequency range from 5 to 2000 Hz and sensitivity of 100–150 $mkV{\cdot}Pa^{-1}$.

Receiving transducers were installed on the pipe wall directly through the insulation film layer or on the soil near the pipeline.

The signal from piezotranducers output goes to the preamplifier. Noise level of the amplifier, given to the input, in the frequency band 1 kHz, did not exceed 0.3 mV. Voltage amplification gain was 10 (20dB). The signal from the preamplifier output went to a recording device.

3. Experimental results and their discussion

During the research receiving transducer was installed on the pipe wall (figure 1) in D pit, and an acoustic signal emitter with the generator were alternately installed in pits C, B and A. In addition, single impulses were emitted into the pipe with a frequency range from a few hertz to a few kilohertz.

Several ways to receive the signal at a fixed distance from the emitter were tested: from clean pipeline wall, without the acoustic contact lubricant; from pipeline wall with contact lubricant; through 2-layered film insulation without contact lubricant; through 2-layered film insulation with contact lubricant. In all these cases, we did not observe any significant change in the signal amplitude or a phase. Therefore, later, receiving transducer was installed on the pipe through two-layered film insulation without lubricant.

Oscillogram and calculated with their use frequency spectrums, obtained in the tests under the contact signal input at the probing signal emitter location in the pits C, B and A, are shown in figures 2, 3.

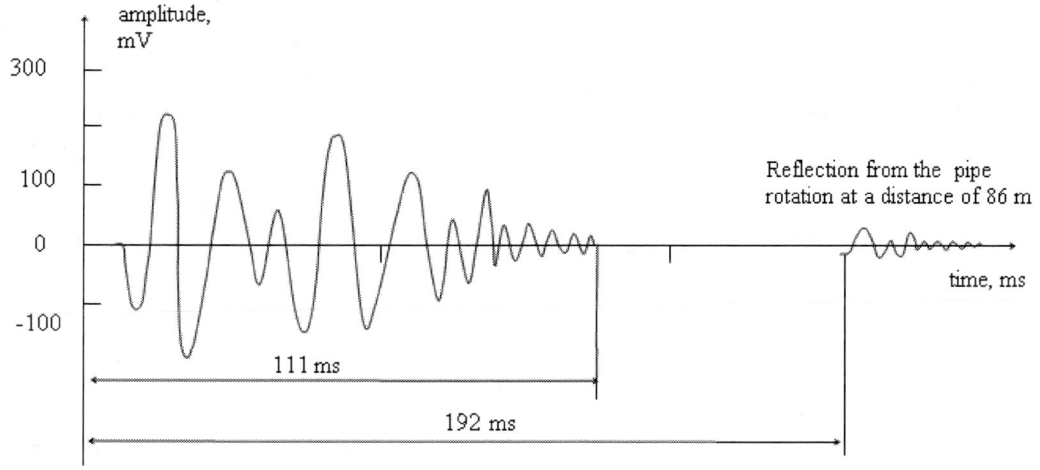

Figure 2. Oscillogram of the signal when the emitter is far from the receiving transducer in 29 m.

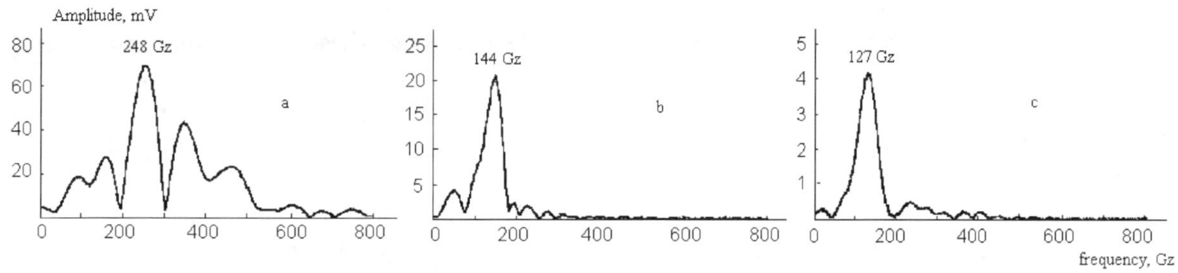

Figure 3. Frequency spectrum of sounding signal at different distances between the emitter and the receiving transducer (a–29 m, b–98.5 m, c–72.5 m).

When the emitter is located in B and C pits, there are the cases of signal reflection from the valve F (figure 2). It helps to estimate the velocity of signal propagation

$$V = \frac{2S}{\Delta t} = \frac{86 \times 2}{(192 - 27.4) \times 10^{-3}} = 1045 \text{, m/sec}$$

Where S is the distance from receiving transducer to the valve, m.

The value of the obtained sound speed matches with the speed of the pressure wave, obtained using well-known Korteweg equation:

$$c = \sqrt{\frac{K}{\rho}} \left(\sqrt{1 + \frac{Kd}{E\delta}} \right)^{-1} = \sqrt{\frac{1.35 \cdot 10^9}{810}} \left(\sqrt{1 + \frac{1.35 \cdot 10^9 \cdot 1.2}{2 \cdot 10^{11} \cdot 0.0152}} \right)^{-1} = 1042.8 \text{, m/sec}$$

Where K– liquid (oil) cubic elasticity modulus – $1.35 \cdot 10^9$ Pa; E – pipe (steel) material elasticity modulus – $2.0 \cdot 10^{11}$ Pa; ρ – liquid (oil) density–810 kg/m^3; d, δ – inner diameter and 1.2 m. and 0.0152 m. pipes wall thickness, respectively.

Fitting calculated and experimentally measured velocities proves that the acoustic signals in the pipe are spreading along the liquid. The Signals with other propagation velocities were not observed.

Figure 3 shows that the oil pipe, while transmitting low frequency acoustic oscillations, behaves itself as a low-frequency filter. This could be seen if comparing the frequency characteristics of the signals. When the emitter was located in C pits (figure 3a 29 m. distance) 248 Hz ± ΔF components prevailed in the frequency spectrum, but at emitter position in A and B pits (the distance is 172.5m and 98.5m) these components are already absent. This is in accordance with published theoretical and experimental results concerning stronger attenuation of high-frequency signals.

For determination of the acoustic power introduced into the pipe, it is necessary to determine the acoustic power, detected with the transducer at the distances of 29, 98.5 and 172.5 m. from the emitter, and, then, to approximate obtained values at the point of pipe signal input. In this case, the wave in the pipe is considered to be flat, and pipeline wall to be thin. Taking that into account, the pressure, detected with the transducer on the outer pipe wall, will be taken equal to the pressure of the liquid inside it. Instantaneous intensity of acoustic wave (instantaneous flow density of power) is determined by the known ratio [9]:

$$I(t) = (\rho c)^{-1} p(t)^2, \tag{1}$$

Instantaneous intensity is connected with acoustic power by equation:

$$P_x = \frac{1}{T} \int_0^T \int_S I(t) dt ds = \frac{1}{\rho c T} \int_0^T \int_S p(t)^2 dt ds \tag{2}$$

While integrating (2) over time and area, believing that a wave is flat, for the acoustic power at x distance from the emitter we get

$$P_x = \frac{\pi R^2}{\rho c T} \int_0^T p(t)^2 dt = \frac{\pi R^2}{\rho c T} (\gamma K_u)^{-2} \int_0^T U(t)^2 dt \tag{3}$$

where ρ–density of liquid, c–sound velocity in liquid; R – pipeline radius; T – duration of pressure impulse; γ – transducer sensitivity (in our case, 100–150 mkV/Pa); K_u– preamplifier voltage amplification gain; $U(t)$ – receiving transducer voltage dependence on time.

Substituting in equation (3) the following parameters R = 0.6 m; c = 1045m/s; ρ = 810kg/m3; K_u = 10; γ = 10^{-4} V/Pa; and integrating numerically voltage dependences shown in figure 3, as a the result, we get the following values of acoustic power at the distances of 29, 98.5 and 172.5 meters from the emitter:

P_{29}= 77.75 mW; $P_{98,5}$= 5.91 mW; $P_{172,5}$= 0.26 mW;

Approximating the data with a cubic spline interpolation, we get the following values of acoustic power at the pipe signal input: P_0=140.46 mW

Integral attenuation coefficients by signal power are found using the equation

$$\alpha = \left| \frac{10 \lg \left(P_{s1} / P_{s2} \right)}{s1 - s2} \right|, \text{ dB/m} \quad (4)$$

where *s1*, *s2* – distances from the emitter to input point. Calculated results based on the equation (4) are shown in the Table 1.

Table 1. Experimentally obtained attenuation coefficients of low-frequency pipeline signals

s1–s2	69.5	74	143.5	172.5	98.5	29
α, dB/m	0.16	0.18	0.17	0.16	0.14	0.089

Table 1 shows that the attenuation coefficients, obtained using experimental data (the first three points) have similar values, the difference is 0.01 dB / m (~ 6%). Approximately, the same values are obtained using the calculation based on the approximated value of the acoustic power at the pipe signal input point (the last point has the largest diversity, which is caused by approximation error).

This suggests that the resulting P_0 value corresponds approximately to the real power introduced into the pipe. In this case, factor of electrical power conversion into acoustic power is $\eta=(140.46\ 10^{-3})/860=1.6\ 10^{-4}$

The calculations show that at mechanical strike a small part of the energy is transformed into acoustic signal energy, which can be detected on the external pipeline wall. Probably, the main striking energy is transformed into waves energy, that are reflected from the outer and inner pipeline walls and do not penetrate the pipeline inside.

At the strike from the pipeline internal side, it could be expected that the main striking energy will be concentrated inside the pipeline, but because of the large acoustic impedances difference of the pipe wall and pumped liquid, only a small part of the energy will be transformed into waves, detected from the outside. More accurate value of the conversion factor can be obtained by solving the three-dimensional equations of cylindrical shell moving and the hydrodynamic equations.

To check the signal attenuation caused by energy emission into the soil, receiving transducer was installed on the pipe through the soil layer, and, then, was removed from the pipe axis along the soil surface. Measurements were carried out at the emitter position at B point (100m). The disappearance of the signal was observed only at receiving transducer location at the distance of more than 10m from the pipeline axis.

While approaching to the pipe, the signal increases to a maximum above the pipeline axis. In this case, signal delay dramatically increases compared with that, occurred at the receiving transducer location on the pipe wall. When installing the sounding emitter in C pit (29m), active area around the pipe increases up to 12–13m. In this case, only one signal was caught by recording device. This shows that the signal, propagating through the pipeline, emits an acoustic energy into the environment (into the soil), which, in authors' opinion, is a reason of high signal attenuation.

References

[1] Lapshin B M, Ovchinnikov A L (1998) Examination of the propagation of elastic waves in pipes with a liquid in acoustic emission leak testing *Russian Journal of Nondestructive Testing* **34** (7) 494–500

[2] Long R, Cawley P, Lowe M (2003) Acoustic wave propagation in buried iron water pipes *Proceedings of the Royal Society A: Mathematical, Physical and Engineering Sciences* **459**(2039) 2749–2770

[3] Muggleton J M, Brennan M J, Pinnington R J (2002) Wavenumber prediction of waves in buried pipes for water leak detection *Journal of Sound and Vibration* **249**(5) 939–954

[4] Muggleton J M, Brennan M J, Linford P W (2004) Axisymmetric wave propagation in fluid-filled pipes: Wavenumber measurements in in vacuo and buried pipes *Journal of Sound and Vibration* **270**(1) 171–190

[5] Ahadi M, Bakhtiar M S (2010) Leak detection in water-filled plastic pipes through the application of tuned wavelet transforms to Acoustic Emission signals *Applied Acoustics* **71** (7) 634–639

[6] Gao Y, Brennan M J, Joseph P F (2009) On the effects of reflections on time delay estimation for leak detection in buried plastic water pipes *Journal of Sound and Vibration* **325**(3) 649–663
[7] Lapshin B M, Ovchinnikov A L, Chekalin A S (2011) Features of the search for leaks in pipelines of heat networks using the acoustic-emission method *Russian Journal of Nondestructive Testing* **47**(10) 710–717
[8] Ozevin D, Harding J (2012) Novel leak localization in pressurized pipeline networks using acoustic emission and geometric connectivity *International Journal of Pressure Vessels and Piping* **92**(4) 63–69
[9] Golyamina I P 1979 *Ultrasound: small encyclopedia* (Moscow, Sov. Encycl.)
[10] Nikolaeva E D 1991 *Development of acoustic emission method of continuous control of underwater pipelines tightness: PhD. thes.* (Tomsk, NDT SRI, TPU)
[11] Vovk I V (1982) Radiation of sound by coaxial cylindrical shells *Akusticheskij Zhurnal* **28**(4) 470–473
[12] Kremleva O A, Shkundin S Z (1998) A method for calculating acoustic fields in a finite cylindrical channel with a flow *Acoustic Journal* **44**(1) 84–88
[13] Retling E V (2012) Concerning acoustic radiation of a cylindrical shell *VolgGASU Internet-series. Ser .: Polythematic* **1**(20) 1–6

Study of the Properties of Plessey's Electrocardiographic Capacitive Electrodes for Portable Systems

A A Uvarov[1], I A Lezhnina[2], K V Overchuk[3], A S Starchak[4], Sh D Akhmedov[5] and I A Larioshina[6]

[1] Assistant, NDT Institute, Tomsk Polytechnic University, Tomsk
[2] Associate Professor, NDT Institute, Tomsk Polytechnic University, Tomsk
[3] Postgraduate student, NDT Institute, Tomsk Polytechnic University, Tomsk
[4] Postgraduate student, NDT Institute, Tomsk Polytechnic University, Tomsk
[5] Professor, Federal State Budgetary Scientific Institution «Research Institute for Cardiology», Tomsk
[6] Assistant, Tomsk State University, Tomsk
E-mail: inna84-08@mail.ru

Abstract. Cardiac diseases are still most widely spread in all regions of the world. And more and more devices are invented to satisfy increasing requirements of the patients. One of the perspective technologies in cardiac diagnostics is capacitive sensing ECG electrodes.
This article describes a study of the properties of electrocardiographic capacitive electrodes PS25255 from Plessey Semiconductors for portable systems as well as some undocumented parameters of these sensors. We developed special cardiograph using Plessey's electrodes and applied to the number of patients with ischemic heart disease. We paid our attention mostly to the correct transition of the ST segment as it has critical impact on the diagnostics of ischemic heart disease.

1.Introduction

The concept of capacitive electrodes for biopotentials sensing has known for a long time and at present many patents are registered. But the lively interest for them appeared only last few years thanks to modern miniature analog components, which made possible placing the necessary electronics in the electrode, thereby improving its performance.

Traditionally, Ag / AgCl electrodes with wet conductive gels are used for biopotential records. The standard Ag / AgCl electrode has been well described and studied for many decades [1–3]. Most of its properties are well studied [4], and there is sufficient number of empirical data describing its characteristics, such as low-frequency noise and drift [2]. By using these electrodes with appropriate preparation an excellent signal may be taken.

2.Theory

The main principles underlying the helium electrodes are also well known. Despite decades of research in the field of alternative technologies of biopotentials sensing [5–8] for ECG, standard wet Ag / AgCl electrode is still broadly used for clinical and research applications. Every year billions of clinical electrocardiograms are recorded with disposable glue electrodes, while dry electrodes are limited to a niche of not medical (scientific) application, for example for fitness monitoring and toys.

Thanks to recent developments, applied to the capacitive electrodes of the Plessey Semiconductors, it becomes possible to use dry electrodes for clinical tasks.

The connection between a capacitive electrodes and skin may be described as in the case with other types of electrode, the layered conductive and capacitive structure with combinations of a series of parallel RC elements. Figure 1 shows examples of these structures, with different values of conductivity and capacitance. For each of these types of electrodes, usually one of the sections of RC dominated and electrical connection can be provided in one piece with the conductivity gc and parallel capacitance Cc, or simplified relations method $Yc\ (j\omega) = gc + j\omega Cc$.

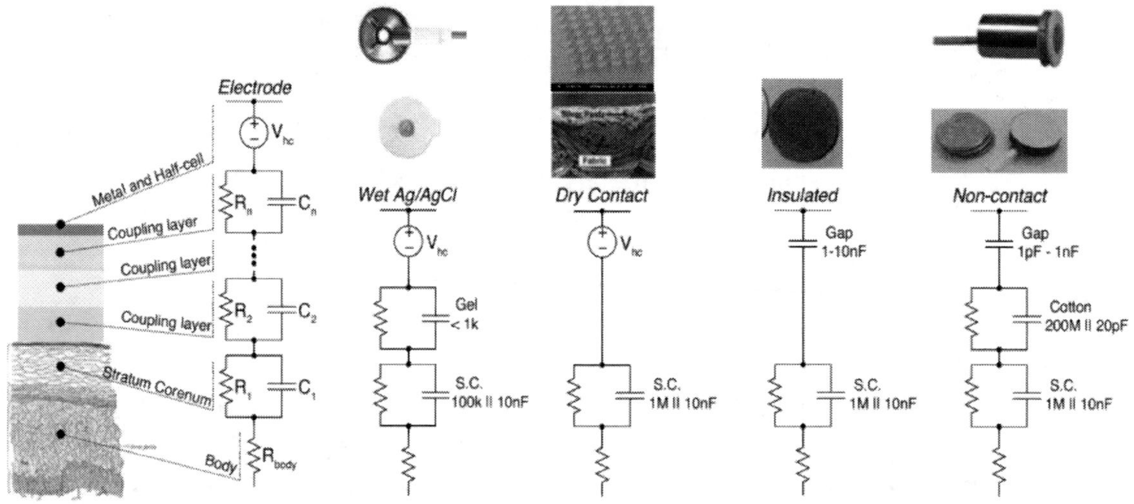

Figure 1. The electrical communication skin-electrode interface for different topologies of electrodes, including wet contact gel-based Ag/AgCl, with dry contact MEMS and the metal plate, a thin film of an insulated metal plate, the metal plate and wireless contact through clothes such as cotton. The insets shows examples of practical electrodes for each category [9].

All modern capacitor electrodes have the general the structure presented in figure 2. The structure of the electrode includes an amplifier with a high input resistance of about GOhms or even TOhms. The amplifier has active shield to protect input from interference from other sources, and includes a capacitive relation between the source and output of the amplifier.

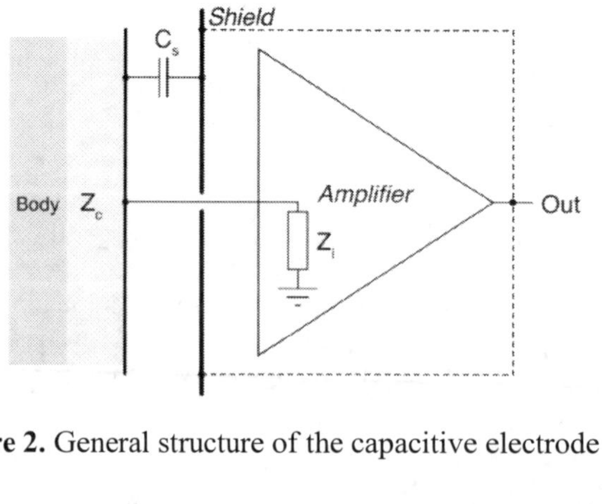

Figure 2. General structure of the capacitive electrode [9].

3. Experiment

For study of properties and suitability of capacitive electrodes from Plessey in clinical practice we designed a testing device for ECG registration form the patient's chest. Registration is made at the same time with capacitive electrodes and standard self-adhesive AgCl electrodes for the further comparison and quality control. Research carried out in the Cardiological Clinic of Tomsk, Russia on patients with ischemic heart disease. Figures 3 and 4 shows examples of registered signals.

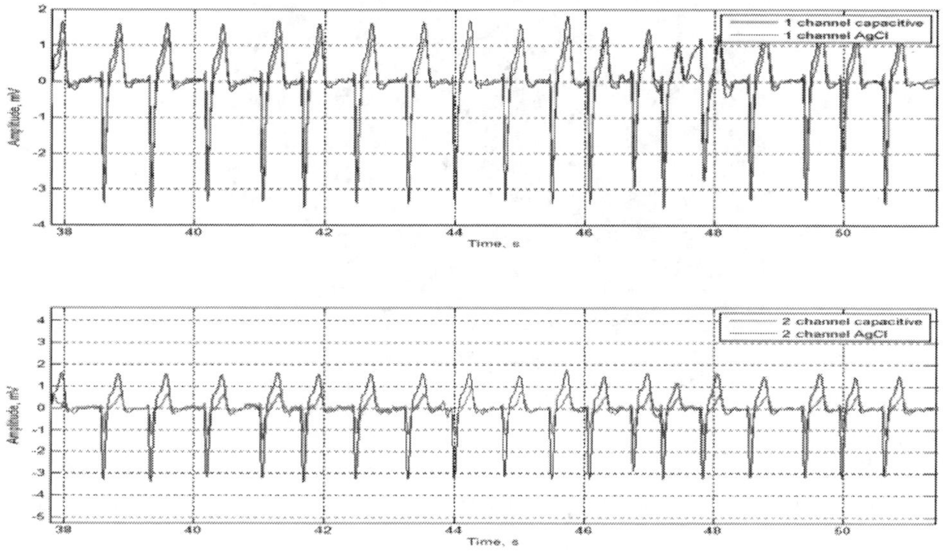

Figure 3. Example of a record from the patient in a fixed condition.

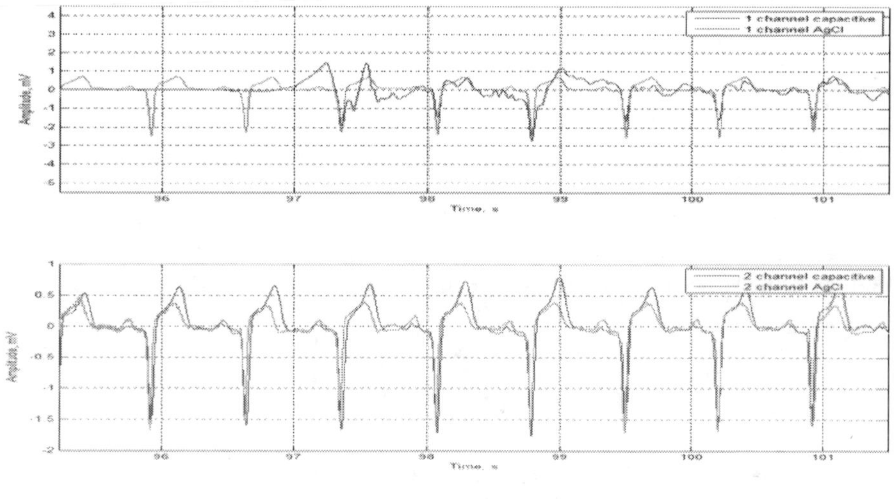

Figure 4. Example of a record from the patient during body rotation.

The experiment shows that capacitive electrodes sufficiently conduct the ECG signal with a comparable quality referring to the standard electrodes. However, the relative motion of the electrodes and the body, and friction of the electrodes with body surface leads to artifacts in the received signals (figure 4), which are a major barrier against adoption of dry electrodes in the mobile clinical practice.

During the research, at the capacitive electrodes were seen the transient processes about 1 second. This phenomenon was observed when the capacitive electrode is disconnected from the patient's skin for a moment and only after the transition process electrode remains to the normal condition. To confirm the presence of transient processes we carried out the experiment by using the square wave generator and an oscilloscope. Figure 5 presents the transient process of capacitive electrodes. It is worth noting, this characteristic was not presented in the standard technical documentation.

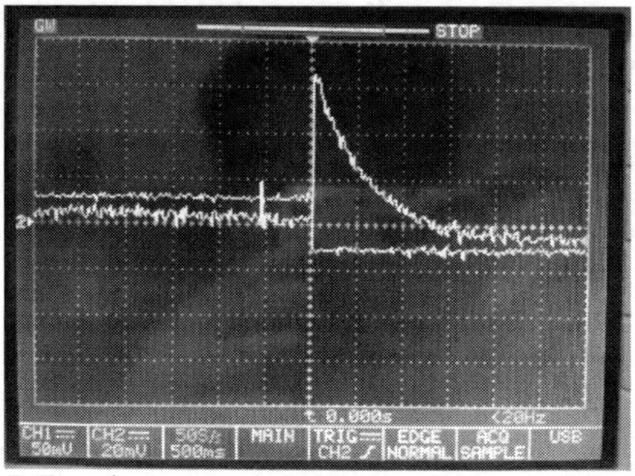

Figure 5. Transient process electrodes of the company Plessey.

Also, experiments were conducted to test the bandwidth. As a result, it was found that even outside of the documented bandwidth and up to 100 kHz electrodes' gain and phase shift is not changed.

4. Conclusion

Considering all of the above we can conclude that capacitive electrodes may be used in clinical practice, but they are needed to be improved to stabilize the signal from artifacts caused by patient motion. The reduction of patient's motion influence can be achieved by compensation of the relative movements of the electrode using an external mechanical structure, or by measuring the characteristics of the sensing circuit and following adaptive filtering.

REFERENCES

[1] Searle A and Kirkup L (2000) A direct comparison of wet, dry and insulating bioelectric recording electrodes *Physiological Measurement* **21**(2) 271–283

[2] Huigen E, Peper A and Grimbergen C A (2002) Investigation into the origin of the noise of surface electrodes *Medical and biological engineering and computing* **40**(3) 332–338

[3] Spach M S, Barr R C, Havstad J W and Long E C (1966) Skin-electrode impedance and its effect on recording cardiac potentials *Circulation* **34**(4) 649–656

[4] Baba A and Burke M J (2008) Measurement of the electrical properties of ungelled ecg electrodes *International Journal of Biology and Biomedical Engineering* **2**(3) 89–97

[5] Bourland J D, Geddes L A, Sewell G, Baker R and Kruer J (1978) Active cables for use with dry electrodes for electrocardiography *Journal of Electrocardiology* **11**(1) 71 – 74

[6] Lopez Jr A and Richardson P C (1969) Capacitive electrocardiographic and bioelectric electrodes *IEEE Transactions on Biomedical Engineering* **16**(1) 99

[7] Taheri B A, Knight R T, Smith R L (1994) A dry electrode for EEG recording *Electroencephalography and Clinical Neurophysiology* **90**(5) 376 – 383

[8] Matsuo T, Iinuma K, Esashi M (1973) A barium-titanate-ceramics capacitive-type EEG electrode *IEEE Transactions on Biomedical Engineering* **20**(4) 299 –300

[9] Chi Y M, Jung T-P, Cauwenberghs G (2010) Dry-contact and noncontact biopotential electrode: methodological review *IEEE Reviews in Biomedical Engineering* **3** 106–119

The Method of Manufacturing Nonmetallic Test-Blocks on Different Sensitivity Classes

N P Kalinichenko[1], A N Kalinichenko[2], I S Lobanova[3], A A Zaitseva[4], E L Loboda[5]

[1] Associate Professor, Institute of Nondestructive testing National research Tomsk Polytechnic University, Tomsk, Russia
[2] Associate Professor, Institute of Nondestructive testing National research Tomsk Polytechnic University, Tomsk, Russia
[3] Senior Lector, Institute of Nondestructive testing National research Tomsk Polytechnic University, Tomsk, Russia
[4] Master Degree Student, Institute of Nondestructive testing National research Tomsk Polytechnic University, Tomsk, Russia
[5]Head of Department of Physical and computing mechanics, National research Tomsk State University, Tomsk, Russia
E-mail: konarevai007@tpu.ru

Abstract. Nowadays in our modern world there is a vital question of quality control of details made from nonmetallic materials due to their wide spreading. Nondestructive penetrant testing is effective, and in some cases it is the only possible method of accidents prevention at high-risk sites. A brief review of check sample necessary for quality evaluation of penetrant materials is considered. There was offered a way of making agents for quality of penetrant materials testing according to different liquid penetrant testing sensibility classes.

1. Introduction

Nondestructive testing and technical diagnostic are an industry around the world, an essential part of manufacturing and operation of technical appliances: thousands of specialists provide daily screening test of nonconforming components at manufacturing and timely detection of dangerous cracks on technical equipment in operation, first of all cracks which are dangerous for life, health and environment (safety).

The most effective method for nondestructive testing of large areas is capillary, especially that with complex geometry and in case of mass production. Penetrant testing is good because it allows detecting a fault at the initiatory stage of manufacturing and at all stages of product life cycle. Liquid penetrant inspection technology is comparatively simple and does not need difficult and expensive equipment.

Nowadays on Russian market there is a great diversity of penetrant materials for different application conditions, with different behavior on single-type defects and quality-price ratio.

Material quality is assessed according to flaw-detection efficiency of different types and dimensions.

Properties of penetrant, cleaner and developer are tested together with complex testing of penetrant materials at conditions close to developing conditions on real details.

Sensibility, working efficiency and quality comparison of penetrant material sets are assessed with the help of check samples with fine cracks and test-panels.
Sample diversity used in different countries is almost similar. Samples are defective details with natural cracks or plates, disks and other products with intentional cracks. Samples organized as plates with brittle metallic, galvanic or chemical cover or with the surface hardened with chemical and thermal processing – nitrogen hardening are widely used. Cracks are formed with the help of samples deformation by bending or stretching, or punch pressing in with cylindrical or spherical form of contact surface. Obtained cracks characteristics such as opening width, length and depth, which as a rule is equal to the thickness of nitration case, are measured and put down into certificate [1–7].

Unsatisfactory features of such samples are cracks in nitration case which appear in uncertain areas with indefinite distribution density on a unit of surface of check sample, their number and distance between them is not defined. Also manufacturing possibility of a sample containing defects with definite regularity of opening width and depth from imposed load is excluded. Instability of charge in a sample causes variable depth and opening width of cracks along their length. Sample bending during its charge leads to curves (as it has permanent deformation) and that in its turn leads to free change of opening width of cracks along their length. Besides, quality testing of cavity defects cleaning from polymerizing remaining residues, frequently used penetrant materials, is difficult in check samples described above; there is no possibility of physical processes visualization occurring at liquid penetrant testing technology. As capillary methods allow diagnosing products made from any materials including glass, ceramics, plastic materials, it would be necessary to have check samples free from discontinuity described above.

2. Versions of the nonmetallic test-blocks for penetrant testing

Nowadays there is information about nonmetallic check samples with given characteristics of defects [4]. Such samples are simple and cheap in manufacturing, defects are received with given characteristics, there is no corrosion of material at sample cleaning, and penetrant materials residues remained in defect cavity are easily replaced. One more circumstance allows to do sample cleaning qualitatively which in its turn prolongs their operational lifetime.

There is one more problem at penetrant materials testing connected with the necessity of using two check samples (check and reference) with cracks of similar characteristics and almost similar dimensions on each sensibility class. But now there are check samples with one crack on the market. Therefore, according to GOST 18442-80 [8] it is necessary to have minimum six check samples on three most important sensibility classes.

The way of nonmetallic check samples manufacturing offered by the authors allows to implement defects on three capillary testing sensibility classes on one sample.

The following components are necessary for manufacturing: plate with sprayed metal coating (thickness of a layer about 1 μm), silver leaf (brass leaf) and foil, thickness 7 – 20 μm. The main technological stages are as follows.

Slots of similar depth are cut perpendicular to the basement on two opposite sides of rectangular form. Slots are necessary for placement foil in them, preliminary covered with a thin coating of epoxy adhesive, as shown in [9, 10].

Flat foils are set into slots of a form and a form is filled with epoxy adhesive for the lower part of foil strings to be put into adhesive for 1 – 2 mm, and the upper parts are left without being filled with adhesive.

After full consolidation of epoxy adhesive an obtained block is taken out of a form. After that mechanical processing (grinding, polishing) of the surface is done. As a processed sample is almost transparent, there is a possibility of width testing of remained lower parts of foil strings, and finally the depth of a defect will be tested.

After obtaining the required characteristics (necessary roughness of surface, the width of remaining lower parts of foil strings), etching of foil with ferric chloride solution is done.

Use of back surface allows obtaining high quality working surface as more bubbles formed at epoxy adhesive manufacturing are remained on sample facial surface. Also ultrasonic bath is used to decrease a number of bubbles formed during tar being mixed with hardener.

Obtained cracks are tested for width opening.

Check sample made in accordance with the given technology allows assessing the ability of flaw detective materials to defects detection, and slotting realization in the middle of a panel – to compare the sets with one another, as showing in figure 1.

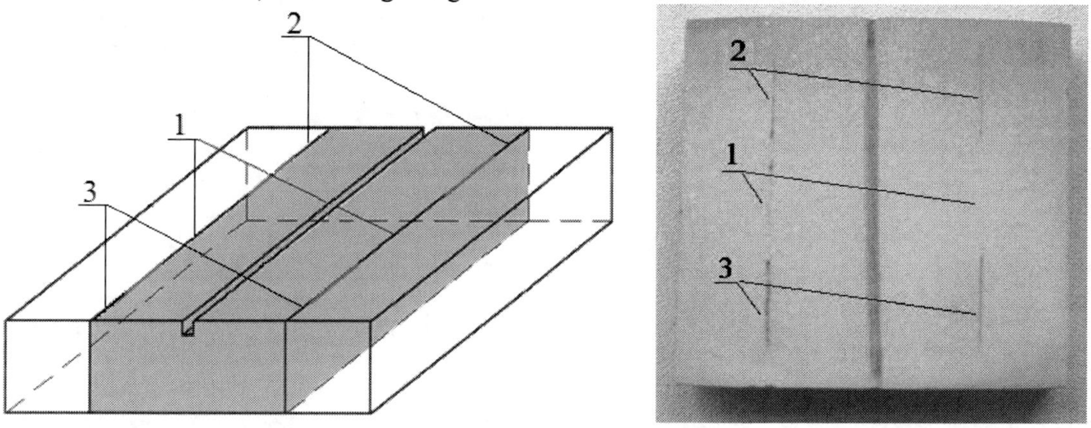

a) b)

Figure 1 Nonmetal test panel with cracks (1, 2, 3) according to I, II, III sensitivity classes:
a) Schematic, b) Image of test-panel after comparing two penetrant material sets.

3. Summary

Thus, variants of sample realization for penetrant nondestructive testing from nonmetals that are under consideration allow:

- placing some defects (cracks) with given characteristics (width opening, length, depth) on one sample. For example, one crack with opening on I sensitivity class, i.e. less than 1 μm, another – with opening on II class, i.e. from 1 to 10 μm, the third – with opening on III class, i.e. from 10 to 1000 μm and so on;
- decreasing the necessary number of samples on different sensitivity classes needed for users;
- realizing and using a test – a panel as a universally applicable check sample as the defects are done with standard characteristics (on different sensitivity classes);
- increasing resource of using every sample as: in this case there is no filling of cracks cavity with oxidizing products and samples of oxygen concentration of metal; samples are done from almost transparent material (epoxy adhesive) and it is easy to follow the quality of samples washing (or its absence) after each use from residues of applied penetrant materials which could not be done with samples from metals.

References

[1] William J. M (2004) Patent US 6729175 B2
[2] ISO 3452-3:2013 (2013) Nondestructive testing Penetrant testing Part 3: Reference test blocks
[3] James R. Alburger (1975) Patent US 3791198 A
[4] Frank J. Vicki, Setsuo Shimizu (1986) Surface penetrant inspection test piece having varying thickness plating US 4610157 A
[5] William J. Martin (2001) Patent US 6311538 B1
[6] Paul Dunnwald (2011) Patent Non-destructive liquid penetrant inspection process integrity verification test panel WO 2011097146 A1
[7] ISO 3452-1:2013 (2013) Non-destructive testing Penetrant testing Part 1: General principles

[8] GOST 18442-80, Nondestructive testing, Capillary methods. General Requirements (Moscow, Publish House of Standards)

[9] Kalinichenko N P, Kalinichenko A N, Lobanova I S and Borisov S S (2013) Methods for the manufacture of nonmetallic reference specimens for liquid penetrant inspection *Russian Journal of Nondestructive testing* **49**(11) 668–672 doi 10.1134/S1061830913110041

[10] Kalinichenko N P, Kalinichenko A N, Lobanova I S, Popova A Yu and Borisov S S (2014) Manufacturing technology and an investigation of samples for testing instruments for the capillary nondestructive testing of nonmetals *Measurement Techniques* **57**(5) 484–488 doi 10.1007/s11018-014-0485-1

About the Electron Charge Accelerated in the Small-size Betatron MIB-4

E L Malikov[1], A P Shestak[1], M M Rychkov[2], K V Sukharnikov[3]

[1]Researcher, NDT Institute, Tomsk Polytechnic University, Russia
[2]Head of laboratory, NDT Institute, Tomsk Polytechnic University, Russia
[3]Post-graduate student, NDT Institute, Tomsk Polytechnic University, Russia
E-mail: malikov@tpu.ru

Abstract. It is assumed that the electron charge accelerated in small-size betatons is of the same order as that in the classical betatron. However, the parameters of the interpolar space of small-size betatons significantly differ from the parameters of the classical betatron. We can expect that the value of the accelerated electron charge will be different. The paper presents the results of the measurements of the electron charge accelerated in a small-size betatron MIB-4. It is shown that the electron charge accelerated is this betatron is larger than that in the classical betatron.

1. Introduction

Depending on the requirements to radiation parameters and applications, betatrons can be divided into three types – stationary, high-current, and small-size betatrons. There are also special types of betatrons with a constant magnetic field, iron-free betatrons, cylindrical betatrons and modified betatrons, however, they are not widespread. The so-called "classical" betatrons with energies up to 18–35 MeV designed for industrial radiographic non-destructive testing are referred to stationary betatrons. High-current betatrons produce energies up to 15–50 MeV with the radiation intensity 2–3 orders of magnitude higher than that of the "classical" betatrons of the same energy. A significant increase in the radiation intensity is due to special technical solutions used in high-current betatons, which allocates them to a separate type of accelerators. These systems are used for imaging fast and explosive processes in enclosed volumes and for activation analysis of materials. Small-size betatrons produce energies up to 10 MeV [1]. Small-size betatrons are characterized by relatively low weight and high mobility. At the moment, this type of betatrons is most common and most demanded on the Russian and world markets. Small-size betatrons are widely used in industrial radiographic non-destructive testing, industrial tomography [2], medicine [3] and inspection systems [4] which have become particularly widespread in recent years.

Along with other parameters, the practical value of a betatron is assessed by the bremsstrahlung intensity. The betatron bremsstrahlung intensity is proportional to the quantity of the accelerated electron charge. Each type of betatrons provides the maximum accelerated charge, and its value is determined by the dimensions of the interpolar space and injection voltage. In the classical betatron, the accelerated electron charge is $5 \cdot 10^9$ electrons or 0.8 nC. In the high-current betatron, the accelerated electron charge is 10^3 times greater and amounts to $5.56 \cdot 10^{12}$ electrons or 890 nC. The data on the electron charge accelerated in small-size betatons is not provided in the literature. The electron charge accelerated in small-size betatons is assumed to be of the same order as that in the classical betatron. However, small-size betatrons are essentially different from classical betatrons in terms of the interpolar space parameters. Hence, the value of the accelerated electron charge is expected to be different.

This paper presents the measurement results obtained for the electron charge accelerated in the small-size betatron MIB-4.

2. Theoretical part

At the beginning of acceleration, the maximum equilibrium charge which can be maintained by the magnetic field of the betatron after the end of the injection process is determined by the relation

$$Q = \frac{E_i}{2eR_0}\left[\left(\frac{E_i}{E_0}\right)^2 - 1\right]S_0$$

where e –is electron charge; R_0 is equilibrium orbit radius; E_0 is the rest energy of the electron; E_i is the total energy of injected electrons; $E_i = E_0 + e \cdot U_i$, U_i is injection voltage; S_0 is the cross-sectional area of the equilibrium beam, i.e. the electron beam for which the Coulomb repulsion forces are balanced by the focusing forces of the magnetic field.

Voltage injection for MIB-4 and most small-size betatrons is 40 kV. The radius of the equilibrium orbit R_0 and the cross-sectional area of the equilibrium beam S_0 can be calculated by constructing a potential function of the magnetic field. The potential function is the most explicit and complete characteristic of the focusing properties of the betatron magnetic field. The expression of the potential function V through the vector potential A is written in the form

$$V = \frac{e}{2mc^2} \cdot A^2 = \frac{e}{2mc^2}\left[\frac{r \cdot \bar{H}(r)}{2}\right]$$

where $\bar{H}$ is the mean value of the magnetic field in a circle of radius r.

The radius of the equilibrium orbit can be determined by the minimum of the potential function. Since the potential function has a minimum in the equilibrium radius R_0, the graphs of the potential function are referred to as "potential wells". An essential characteristic of this potential well is the area of its cross section by the radial plane, i.e. the area covered by the equipotential line. The equipotential line line equation takes the form

$$z = \frac{1}{k} arcch \frac{A}{A_0[aJ_1(kr) + bN_1(kr)]}$$

where $J_1(kr)$ is a Bessel function of the first kind; $N_1(kr)$ is the Neumann function of the first kind; A_0 is the vector potential at the equilibrium radius.

The equipotential at the injection level sets the boundaries of the equilibrium charge zone, i.e. the area of the equilibrium beam cross-section.

After measuring the parameters of the magnetic field, the maximum equilibrium charge was calculated which is able to *maintain* the magnetic field of the betatron MIB-4.

The actually accelerated electron charge was measured by the method which implies measuring the electron current directly from a tungsten target with the phase shift from the equilibrium orbit corresponding to energy of 1 MeV. At electron energies significantly higher than those specified, the current target will be different from the true one for several reasons: firstly, the electron path in the material will be larger than the thickness of the target, therefore, a portion of the accelerated electrons will pass through the target and will not be recorded with the measuring instrument; secondly, under electron bombardment of the target, the target material releases secondary electrons. If some of these electrons escape from the target, the measuring instrument will record the decreased current from the target. In this case, we used a tungsten target with a thickness of 0.6 mm. For the energy of 1 MeV, the average linear electron path in the target material is 0.4 mm, which fully satisfies the measurement requirements.

In order to determine the true number of electrons collided with the target, the electrons reflected from the target material are to be considered. The reflection of electrons with energies varying from several tens of keV to several MeV for target materials in a wide range of Z has been studied in detail and the reflection coefficients has been provided in a number of studies [5–7].

The reflection coefficient of electrons is defined as following:

$$\beta = \frac{(n_{reg} - n_{init})}{n_{init}} = \frac{n_{reg}}{n_{init}} - 1$$

where $n_{init.}$ is the number of electrons recorded by the detector with no reflecting sample (in our case, it is the target) and $n_{reg} = n_{init} + n_{refl}$ is the number of particles recorded by the same detector at the same point with regard to the reflected electrons, that is in case of the presence of the reflecting sample (target); n_{refl} is the number of reflected electrons.

The reflection coefficient is β≈0.8 for tungsten and the energy of accelerated electrons of 1 MeV.

The correction for reflection in the form of $(1\text{-}\beta)^{-1}$ should be introduced in the expression to determine the charge dumped onto the target. Then, assuming that the current pulse from the target is sinusoidal, the charge of the dumped beam can be calculated with the formula

$$Q_{M} = \frac{I_{m}t}{1{,}57(1-\beta)} = \frac{U_{m}t}{1{,}57R(1-\beta)}$$

where I_m is peak value of the current pulse from the target; t is duration of the current pulse; β = 0.8 is reflection coefficient of the electrons from the tungsten target; U_m is peak value of the voltage drop across resistance R through which with the measured current of the electrons "dropped" onto the target flows.

3. Experimental part

The measuring disc shown in figure 1 was used to measure the magnetic field in the pole gap of the betatron electromagnet.

Figure 1. Measuring disc.

The conductors with the bundled ends in the form of twisted pairs are fixed in the grooves of the measuring disk. At one point of time, the emf values can be obtained from one measuring coil only. To measure the voltage in the coils, a digital oscilloscope WP 725Zi was used to record the data from two coils, namely the coil being measured at the given moment and thesupporting coil common to other coils. The objective of the approach is to eliminate errors due to the field amplitude fluctuations from one pulse to another. The recorded arrays in increment of 2 ns were transferred to PC for processing.

The graph for the potential function of the magnetic field of the betatron MIB-4 is presented in figure 2, and the graphs of the magnetic field equipotentials are shown in figure 3. The graph shows that the function minimum is observed at radius R=4.5 cm, and thus the radius of equilibrium orbit of the betatron MIB-4 is R_0=4.5 cm.

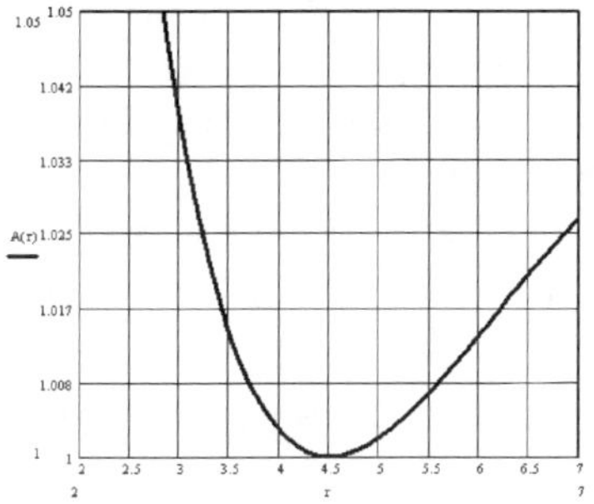

Figure 2. Graph of potential function of the magnetic field of betatron MIB-4.

The measurements of the magnetic field characteristics of betatron MIB-4 showed that the area of equilibrium beam is $S_0=3\cdot10^{-4}\text{m}^2$, and the maximum charge the field is capable of retaining at the time of injection is $Q=2.68\cdot10^{-9}$C that corresponds to the number of particles $N=1.68\cdot10^{10}$.

Journal of Physics: Conference Series **671** (2016) 012034 doi:10.1088/1742-6596/671/1/012034

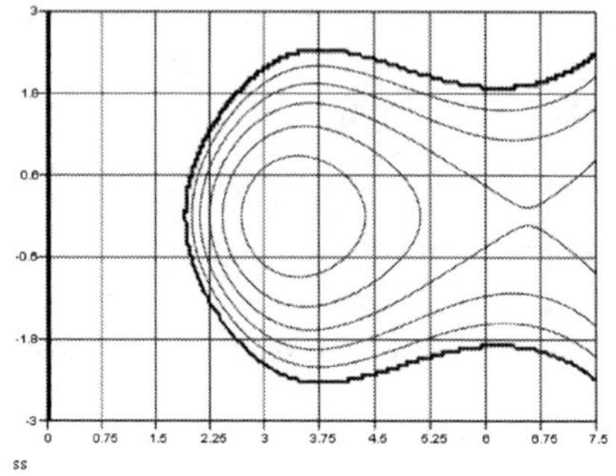

Figure 3. Equipotential graphs for the magnetic field of betatron MIB-4.

The measurements of the accelerated electron charge were carried out directly from the tungsten target of the small-size betatron MIB-4. The small-size betatron MIB-4 is a commercially available accelerator. The kinetic energy of accelerated electrons in the small-size betatron MIB-4 is 3.5 MeV, the total energy is 4 MeV, respectively. The betatron is equipped with soldered acceleration chamber RBK3-3. The chamber is designed to generate bremsstrahlung and it is equipped with the tungsten target of 0.6 mm thick attached to the anode injector. After the electrons reach the given energy, pulse is generated in the bias coil to extend the equilibrium orbit and to "dump" the accelerated electrons onto the tungsten target to generate X-ray bremsstrahlung. The measurements were carried out at the phase shift corresponding to the energy of accelerated electrons equal to 1 MeV in order for all accelerated electrons to be "decelerated" by the target material and recorded by measuring apparatus.

Oscillogram of the current pulse from the target of small-size betatron MIB-4 under electron "bombardment" with the energy of 1 MeV is shown in figure 4.

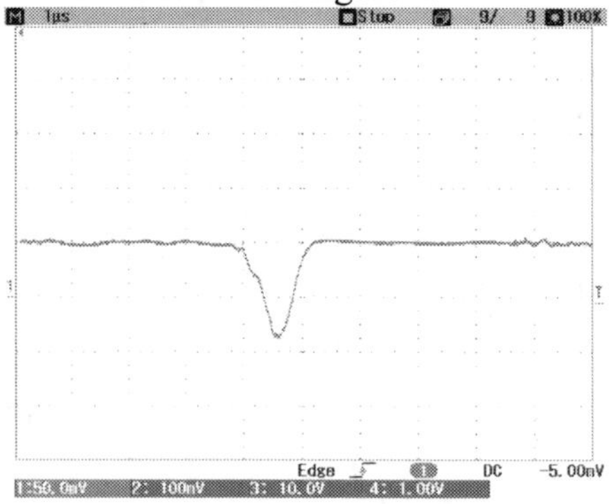

Figure 4. Oscillogram of the current pulse from the target.

The voltage pulse from the target of the small-size betatron MIB-4 generated by the electron current across the resistance of 51 Ω, with the bias phase of accelerated electrons corresponding to 1 MeV, has the amplitude equal to 0.0875V with the duration of 1.2 microseconds. In this case, the electron charge is:

$$Q_{M} = \frac{U_{m} t}{1{,}57R(1-\beta)} = 1.6 \cdot 10^{-9}\ \mathrm{C}$$

That corresponds to the number of accelerated electrons N=1.0×10^{10}.

4. Conclusion

The principle of geometric similarity was applied to develop a classical betatron, that is when a basic well-established accelerator model is taken as a basis, and accelerators with different energy are developed by changing geometric dimensions. The principle of geometric similarity proved to be inapplicable to develop small-size betatrons. If the classical betatron B-18 with energy of 18 MeV is taken as a basis to develop a geometrically reduced replica betatron with the energy of 4 MeV, the cross-section of acceleration chamber will be unacceptably small causing significant decrease in the accelerated electron charge. The reduced dimensions of the acceleration chamber will affect its electric strength and will not allow application of sufficiently high voltage injection which will also cause decrease in the accelerated electron charge. Small-size betatrons were developed after the development of high-current betatrons. To develop small-size and high-current betatrons, the optimum dimensions of the interpolar space for the classical betatrons had to be withdrawn volume for the focusing power action. Small-size betatrons are geometrically reduced replica accelerators of high-current betatrons in terms of the parameters of the interpolar space.

The effective electron capture into acceleration allows bringing the accelerated electron charge to the possible maximum for the given volume of the focusing power action and voltage injection. Special schemes called contractors have been developed for this purpose. At the initial stage of acceleration the radius of electron rotation differs from that of the equilibrium orbit, and some time is required for the positions of radii to coincide. The contractor coils are installed above and below the equilibrium orbit. The magnetic flux generated by the current pulses flowing through the contractor coils makes it possible to affect the rate of change in the electron rotation radius at the moment of injection, thereby affecting the efficiency of electron capture into acceleration. The contractors have shown high efficiency. They are used in all newly manufactured small-size betatrons.

Thus, the increased dimensions of the interpolar space to volume of the focusing power action and the contractor application, thereby increasing the capture efficiency, made it possible to bring the accelerated electron charge in small-size betatrons up to 10^{10} particles. The number of accelerated electrons in small-size betatrons is significantly greater than that achieved in the majority of best examples of classical betatrons ($5\cdot 10^9$ electrons), but less than in high-current betatrons (10^{12} electrons). However, the accelerated charge in the small-size betatron MIB-4 is significantly greater than the charge which the betatron magnetic field is able to retain, namely Q=2.68$\cdot 10^{-9}$C that corresponds to the number of particles N=1.68$\cdot 10^{10}$.

References

[1] Access: http://www.jme.co.uk/JME-Products-Betatron.aspx
[2] Vaynberg E I, Kasyanov V A, Chakhlov V L, Stein M M (2004) *Experience of using small-size betatron Mib-5 in the structure of industrial computed tomograph for testing of heavy-wall products* Palais des Congres. 16-th world conference on nondestructive testing TS2.26.2
[3] Lochman D J (1959) Therapy in cancer of the head and neck with the betatron electron beam *The American Journal of Surgery* **98** (6) 847–850
[4] Bjorkholm P, Cargo J J 2004 *Examination using the X-ray Radiation of High Energy* (Cargo Security Intelligence)
[5] Seliger H H (1952) The backscattering of positrons and electrons *Physical Review* **88**(2) 408–412
[6] Dressel R W (1966) *Retrofugal electron flux from massive targets irradiated with a monoenergetic primary beam* Physical Review **144** (1) 332–343
[7] Ebert P J, Lauron A F, Lent F M (1969) *Transmission and Backscattering of 4.0- to 12.0-MeV Electrons* Physical Review **183**(2) 422–430

Problems of Objects Identification in Three-dimensional X-ray Tomography

T Mamyrbayev[1], A Batranin[2] and D Ivashkov[3]

[1] Research engineer at Institute of Nondestructive Testing, Tomsk Polytechnic University, Tomsk, Russia
[2] Junior researcher at Institute of Nondestructive Testing, Tomsk Polytechnic University, Tomsk, Russia
[3] Master student at Institute of Physics and Technology, Tomsk Polytechnic University, Tomsk, Russia
E-mail: talgat.mamyrbayev@yandex.ru

Abstract. X-ray computed tomography is a unique non-destructive technique which is suitable for dimensional measurements and quality testing. X-ray tomography is applied widely not only in the traditional medical area, but in natural science and engineering fields. The result of tomography is a volumetric map of attenuation coefficients which has, in fact, a limited practical sense. Usually, segmentation that converts the map into binary objects by means of threshold value is applied. In this paper we discuss the some key features of segmentation applied to a test object which is gravel composed of unconsolidated rock fragments. Gravel appeared to be a challenging object for segmentation and image recognition. One of effective approach for image processing of tomographic data, among many possible ones, is presented in this paper.

1. Introduction

X-ray XCT is a non-destructive technique for imaging the internal structure of solid objects. XCT is useful for a wide range of materials, such as rocks, bones, ceramic, metals and soft tissues.

The interaction of X-rays with matter is a base of XCT technique. X-rays are attenuated due to Beer's law for complex materials after they pass through an object and this attenuated radiation is registered by the detector. In XCT, a set of shadow projections of the object are obtained from different angles. The set of projections is then reconstructed and a volumetric map of attenuation coefficients is obtained. For tomographic reconstruction different mathematical algorithms and software based on them are used. The purpose of dedicated software is not only reconstruction, but also reduction of artifacts and basic imaging operations, that are required for further processing such as segmentation and quantification [1, 2]. A map of attenuation coefficients has a limited practical sense. Usually, reconstructed volume is processed by segmentation. This operation converts the map into binary objects by means of threshold value chosen manually or by some algorithm. On this first step, one can encounter unavoidable difficulties which are inherited from nonlinearity and complexity of X-ray and matter interaction. After segmentation, the most common procedure is quantitative analysis that deals with separate binary objects produced by segmentation.

2. Materials and methods

The high energy computed tomography scanner was assembled at Institute of Nondestructive Testing, at Tomsk Polytechnic University. The research group developed the present setup reported recently their first successful results in field of XCT scanners assembling and evaluating [3, 4].

The main components of the system are a scintillator based X-ray detector, an X-ray tube and a rotary stage. The setup uses the MXR-451HP X-ray tube (produced by Comet, Switzerland) as an X-ray source. The voltage of the tube is in 20–450 kV range and the focal spot has two modes and can be 0.4 or 1 mm in size. The X-ray detector is XRD 1622 flat panel (produced by Perkin Elmer, USA) with 2048×2048 pixels array based on TFT technology. The pitch of pixels is 200 μm. Scintillator is made of CsI crystals. Rotary stage provides smooth rotation by step down to 0.01 degree and linear translation of the specimen between the detector and the tube to adjust required magnification and, thus, voxel size.

We used gravel mixture as a test object to demonstrate segmentation and image processing as well as some possibilities of quantitative analysis. Scanning parameters were: accelerating voltage of 100 kV, anode current of 3 mA, angular step of 0.5 degree, frame exposure of 1 s and voxel size of 101 μm. It required approximately 45 minutes to acquire the data set of shadow projections with total number of 720. In addition, we used copper filter 0.2 mm in thickness to reduce beam hardening artifacts.

Usually, XCT consist of the three main steps: data acquisition, reconstruction and segmentation [5]. The results obtained on each step are shown on figure 1.

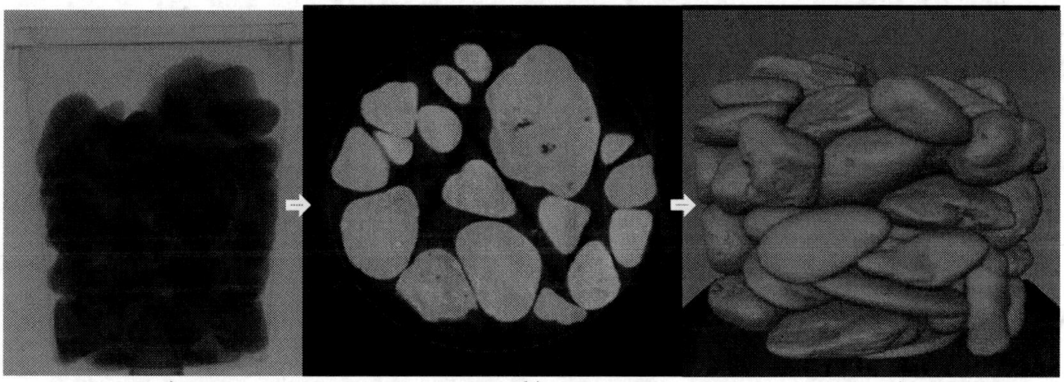

a) b) c)

Figure 1. The three main step of XCT: acquisition results in shadow projections (a); reconstruction results in tomographic slice (b); segmentation results in binary objects, seen separately within the volume of interest (c).

The internal structure of the investigated rock fragments was reconstructed by means of the Nrecon software for cone-beam scanning geometry developed by Bruker microCT [6]. To enhance the image quality of slices a few correction options were used, namely, beam hardening correction and ring artifact reduction [7]. In result, a set of 8-bit grayscale slices was obtained.

3. Results

Segmentation is an operation applied to a grayscale image to convert it into a binary image by using some segmentation algorithm and some threshold level. The gravel mixture appeared to be a challenging object for segmentation procedure. As seen on figure 2, there are 15 rock fragments on the slice. However, a simple segmentation by Otsu algorithm with threshold value is equal 71 reveals only 11 fragments, because the gap between some of them is not pronounced enough. We can change the threshold value, for example, to 117. In this case, we are getting a proper number of fragments, but distorted picture of objects within the slice.

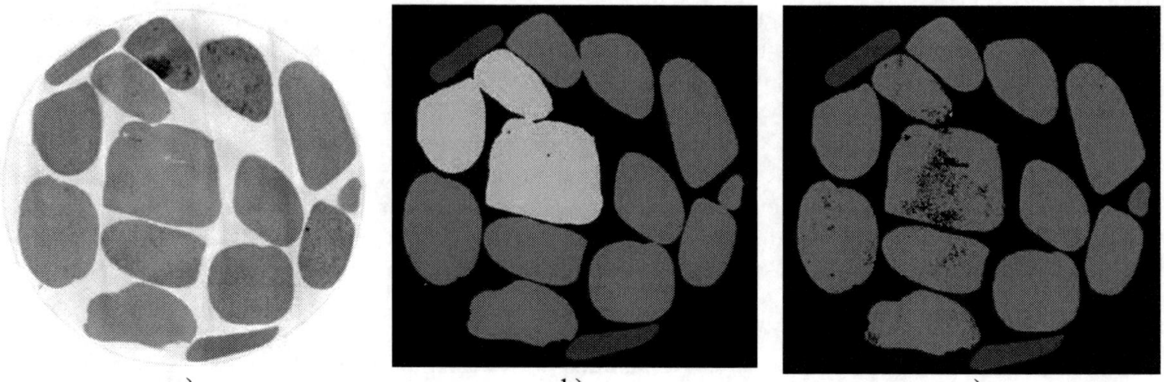

a) b) c)

Figure 2. The slice in grayscale shows 15 fragments (a); segmented image with threshold on 71 shows 11 fragments (b); segmented image with threshold on 117 shows 15 distorted fragments (c).

The reason for this is illustrated on figure 3. Any tomographic image suffers from partial volume effect [6], which means a border between an object and the outer media is a slope, not a sharp edge. In some regions gaps between objects are pronounced enough to separate them, in others they are not. In addition, internal structure of rock fragments is inhomogeneous. Resuming, one-step simple segmentation is not enough to identify rock fragments in the mixture. It requires much more complex approach.

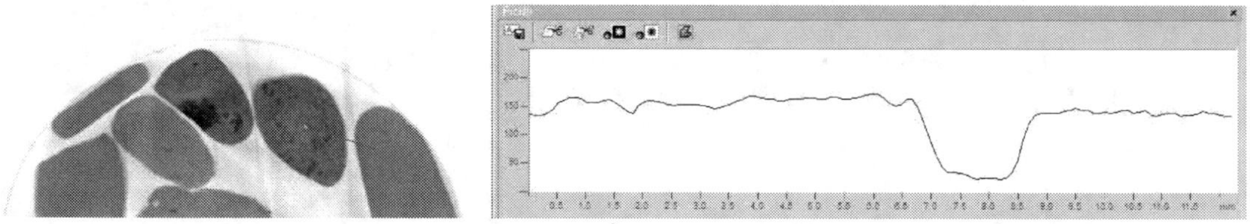

a)

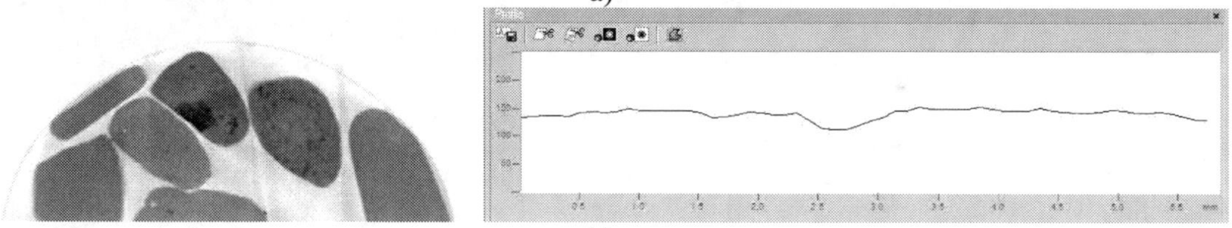

b)

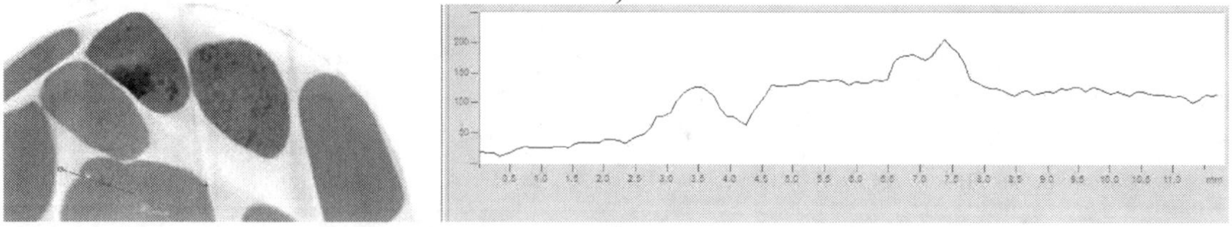

c)

Figure 3. The gap between fragments is pronounced (a); the gap is hardly detected (b); the density, i.e. gray value, within a fragment is inhomogeneous.

The image segmentation and visualization was performed by the VSG Avizo Fire 9 software [8]. We used several morphological operations for image processing, namely, despeckle, thresholding, dilation, separation, and additional operation: labeling, label analysis and volume rendering. Despeckle tool was used to remove "bad" (or "died") pixels and the noise from the image. The size of removed pixels was 300 μm. Global threshold was used to produce binary images. Thresholded images were dilated which uses structuring element for probing and expanding the shapes contained in the input images. Finally, dilated images were separated by watershed segmentation. The combination of these morphological operations made it possible to further quantitative image analysis. The complete scheme of image processing is shown on figure 4.

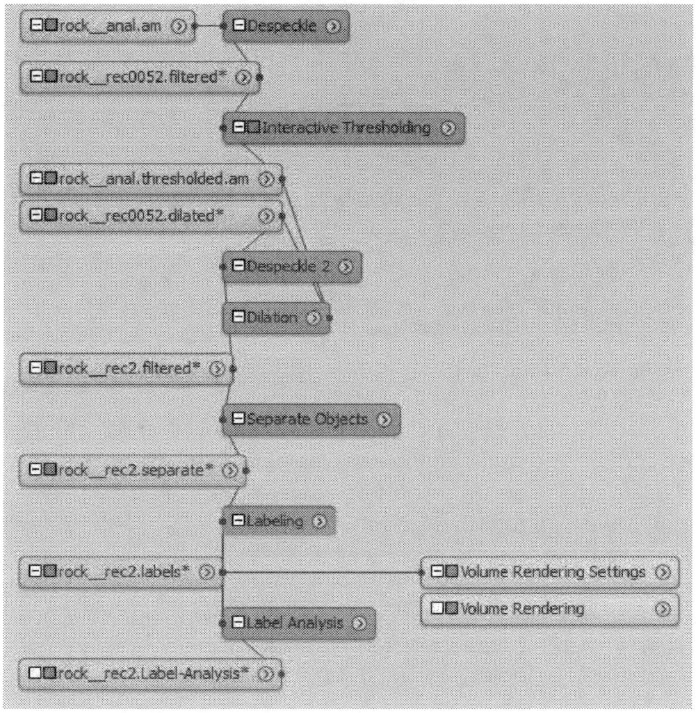

Figure 4. The scheme of image processing applied to gravel mixture from despeckle step to quantitative analysis.

Figure 5 shows the difference between filtered image (the second step on the scheme on figure 4) and segmented image in XY plane (the next to last step on the scheme).

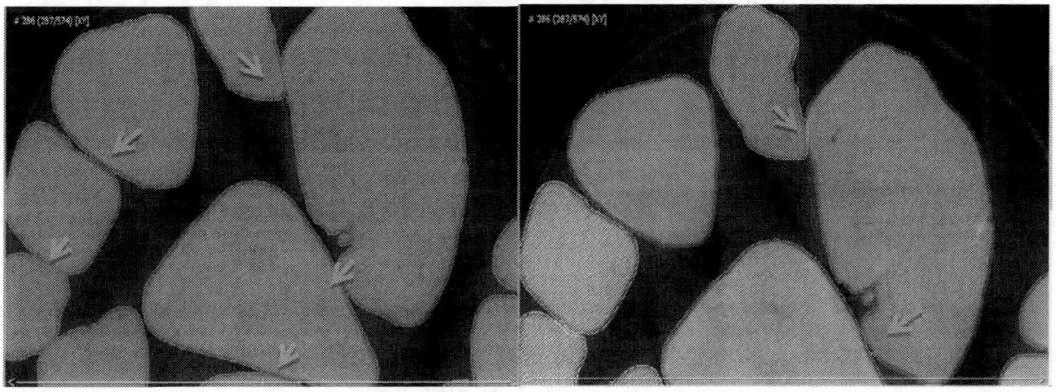

Figure 5. Image processing: a filtered image (left) and the same image after segmentation and labeling (right).

As seen on figure 5, on the filtered image borders between rock fragments are absent; consequently, the number of fragments is far from the true value. On the contrary, segmented image shows a realistic result. Segmented image was labeled to visualize the shape of each fragment.

Labeling algorithm was implemented to the segmented binary images. Then the rock fragments were separated according to their volume and visualized using different colors as shown on figure 6.

Figure 6. Different rock fragments labeled by different colors.

After visualization of a 3D object from 2D slice images, the quantitative analysis and volume distribution was performed. For this purpose number of rock fragments was calculated from the 3D data set, and then they were organized into 10 groups according to their volumes (see figure 7). The number of rock fragments counted automatically was equal to their number in reality.

Besides volume distribution, one can estimate the volume of objects as a percentage to the volume of interest, which can be defined, for example as the space within the container. In addition, spatial orientation of rock fragments is a possible option. Resuming, relevant segmentation opens as many ways to quantitative analysis as one would need.

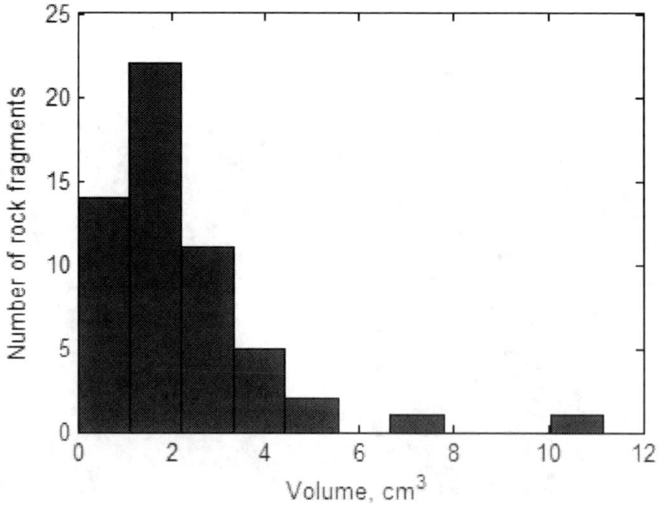

Figure 7. Rock fragments' volume distribution.

4. Conclusion

XCT is a powerful technique for investigation the internal structure of objects. In this paper, we presented the image processing and image analysis with using gravel as a test object. The main steps of XCT are demonstrated. Data was acquired by means of high energy computed tomography set-up. The reconstruction slices from shadow projections was enhanced by applying beam hardening and ring artifact corrections. Reconstructed data set was segmented to obtain binary objects. The number of rock fragments was correctly counted and their volume distribution was calculated. The described method can be applied for investigation of porous and non-uniform materials to perform quantitative analysis.

Acknowledgements

This work was financially supported by The Ministry of Education and Science of the Russian Federation in part of the science program.

References

[1] Kaczmarczyk J *et al (*2010) The interpretation of X-ray computed microtomography images of rocks as an application of volume image processing and analysis *18th International Conference in Central Europe on Computer Graphics, Visualization and Computer Vision, WSCG 2010 - In Co-operation with EUROGRAPHICS, Full Papers Proceedings* 23–30

[2] Elke Van de Casteele *et al (*2007) X-ray microtomography as a non-destructive tool for stone characterization in a conservation study *4th international conference on NDT* (Crete, Greece) URL: http://www.ndt.net/article/hsndt2007/files/ElkeVandeCasteele.pdf

[3] Batranin A V, Chakhlov S V, Grinev D V, Kapranov B I and Klimenov V A (2013) Design of the X-Ray micro-CT scanner TOLMI-150-10 and its perspective application in non-destructive evaluation *Applied Mechanics and Materials* **379** 3–10

[4] Batranin A, Ivashkov D and Stuchebrov S (2015) Performance evaluation of micro-CT scanners as visualization systems *Advanced Materials Research* **1084** 694–697

[5] Baker D R (2012) An introduction to the application of X-ray microtomography to the three-dimensional study of igneous rocks *Lithos* **148** 262–276

[6] Bruker microCT home page, available on: http://bruker-microct.com/home.htm

[7] Ketcham R A, Carlson W D (2001) Acquisition optimization and interpretation of X-ray computed tomographic imagery: applications to the geosciences *Computers & Geosciences* **27**(4) 381–400

[8] Westenberger P *et al (*2012) Fiber Orientation Visualization with AvizoFire *Conference on Industrial Computed Tomography (ICT)* (Wels, Austria) http://www.ndt.net/article/ctc2012/papers/107.pdf

Numerical Simulation of the Microtron Electron Beam Absorption by the Modified ABS-Plastic

S G Stuchebrov[1], I A Miloichikova[2] A L Melnikov[3] and M A Pereverzeva[3]

[1] Senior Researcher, Tomsk Polytechnic University, Tomsk, Russia
[2] Engineer, Tomsk Polytechnic University, Tomsk, Russia
[3] Bachelor student, Tomsk Polytechnic University, Tomsk, Russia
E-mail: stuchebrov@tpu.ru

Abstract. Each specific task of the electron beam application imposes requirements for the beam profile and shape. One of the methods allows achieving high accuracy and low cost of the filters production is the 3D print method. The required properties of the electron beam interaction with the material can be achieved by using the modified plastic filaments. In this paper, the results of the model creation of the electron beams interaction with the ABS-plastic doped with different concentrations are presented. The depth dose distributions of the electron beam in the modified ABS-plastic are sown. The electron beam profiles and the electron beam distribution in the modified ABS-plastic are illustrated.

1. Introduction

Nowadays, the electron beams in common with photons and neutron have a wide range of application in the medical science and industry [1–8]. Each specific task of the electrons application imposes requirements for the beam profile and shape. In this case it is necessary to create the new method of the electron beam modulation. Certainly, the advanced directions of the development are such as that allow increasing the accuracy and rapidity of the filters production and reducing their cost.

Using 3D print methods allows achieving low cost, high accuracy and speed of the electron beams filter production [9–10]. One of the most common materials for 3D-printing is the ABS-plastic [11]. Adding a mixture of the heavier elements in a plastic filament allows to influencing on the nature of the electron beam interaction with the material. The first stage of the research in this direction is the theoretical estimation of the electron beams interaction with the various materials.

In this investigation the theoretical analysis of the electron beams interaction with the ABS-plastic doped with different concentrations of lead by means of numerical simulation TPU microtron extracted electron beam with energy equal to 6.1 MeV was carried out.

2. Materials and methods

2.1. Test materials

The test material density for the numerical simulation was selected according to the real anatomical structure density, for example, muscular tissue density ≈ 1.06 g/cm^3; skin density ≈ 1.1 g/cm^3; bone tissue 2.4 g/cm^3 [12].

In this research, the ABS-plastic doped with different concentrations of lead was used as test materials. The test materials density was calculated by:

$$\rho = \frac{\rho_{ABS} \cdot \rho_{Pb}}{k \cdot \rho_{ABS} + (100 - k) \cdot \rho_{Pb}}, \quad (1)$$

were ρ_{ABS} – the ABS-plastic density; ρ_{Pb} – the lead density; k – ratio of the lead mass concentration in the test material.

The dependence of the test material density from the lead mass concentration is presented on the figure 1.

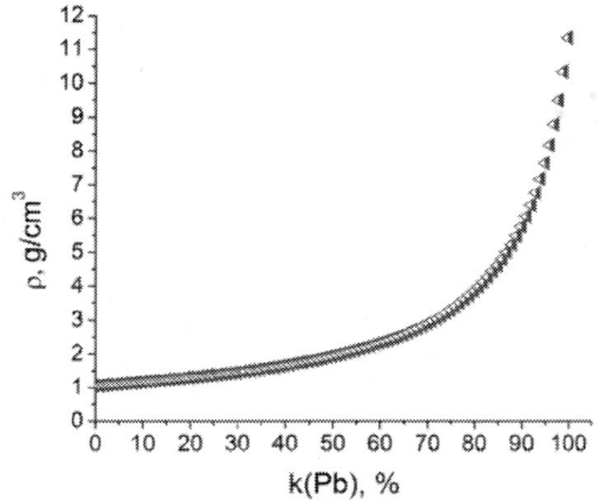

Figure 1. The dependence between the test material density and the lead mass concentration.

2.2. Emitting source

The TPU microtron extracted electron beam, with the following characteristics was used as the emitting source: beam size at output ≈ 2.0 mm^2; beam divergence – 0.1 rad; electron energy – 6.1 MeV.

2.3. Simulation program

In this investigation the program "Computer laboratory (PCLab)" version 9.6 was used for the model creation of the TPU microtron extracted electron beam. In this software package the Monte Carlo method applies for the numerical simulation of the propagation process of electrons, protons, photons and positrons in matter with specified characteristics [13].

2.4. Experiment geometry

In the numerical simulation the irradiation source has the following parameters: particle type – electron; source type – normal plane disc (diameter – 2.0 mm); electron energy – 6.1 MeV. These parameters correspond to the actual TPU microtron beam. The source was placed in front of the beryllium output window (diameter – 40 mm; thickness – 50 μm). The beam shape analysis was carried out in the ABS-plastic doped with different concentrations of lead.

3. Results and discussions

In the first phase of the experiment the depth dose distribution of the electron beam in the ABS-plastic doped with different concentrations of lead was obtained (figure 2). In the figure 3 an example of the electron beam profile obtained by simulation are shown at the 5 cm distances from microtron output window.

The figures 2, 3 show that small change of the heavy elements concentration in the plastic filament significantly affects on the character electron beam interaction with the material.

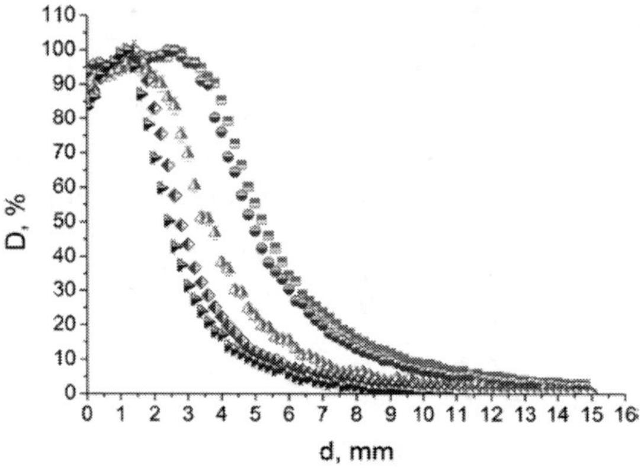

Figure 2. The depth dose distribution of the electron beam in the test materials.

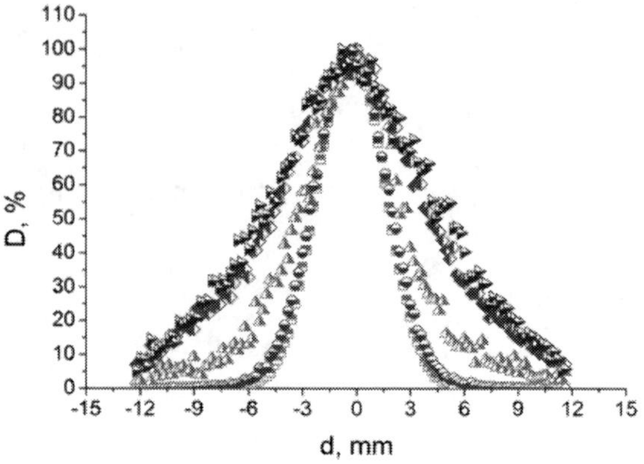

Figure 3. The electron beam profiles in the test materials at the distance from microtron output window equal to 5 cm.

The modified ABS-plastic density: ■ – 1.06 g/cm^3; ● – 1.1 g/cm^3; ▲ – 1.5 g/cm^3; ◆ – 2.0 g/cm^3; ► – 2.5 g/cm^3.

In the figures 4, 5 examples of the electron beam distribution in the horizontal plane obtained by simulation in the modified ABS-plastic are presented. The dose results were averaged and normalized to the maximum simulation dose in the layer.

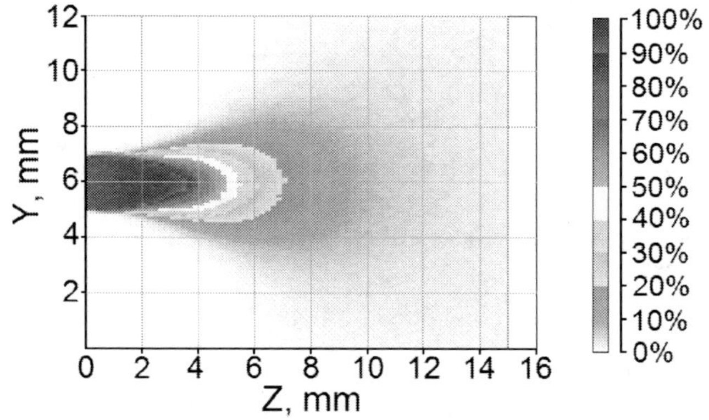

Figure 4. The electron beam distribution in the horizontal plane (the modified ABS-plastic density – 1.1 g/cm^3).

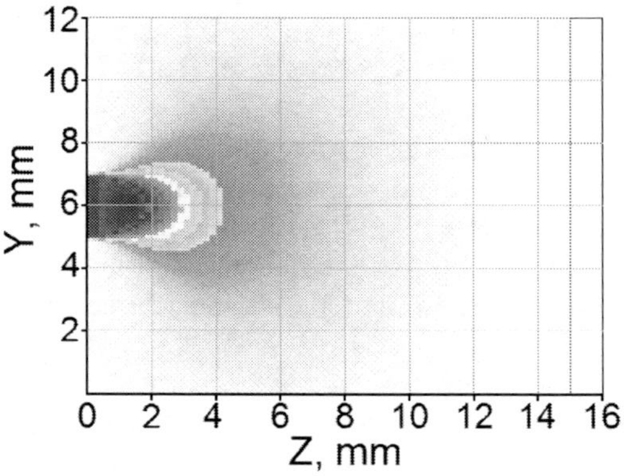

Figure 5. The electron beam distribution in the horizontal plane (the modified ABS-plastic density – 2.5 g/cm^3).

In the figures 6, 7 examples of the electron beam shape in the vertical plane at the distance from microtron output window equal to 5 cm in the modified ABS-plastic are shown. The dose results were averaged and normalized to the maximum simulation dose in the layer.

Presented examples of the calculated cross sections of the radiation dose distributions (figures 4–7) show that the radiation absorption is faster and the layers with the highest dose are closer to the surface on the filter with higher density. Additionally, the higher density of the material leads to the higher electron beam scattering, that is in agreement with the theory.

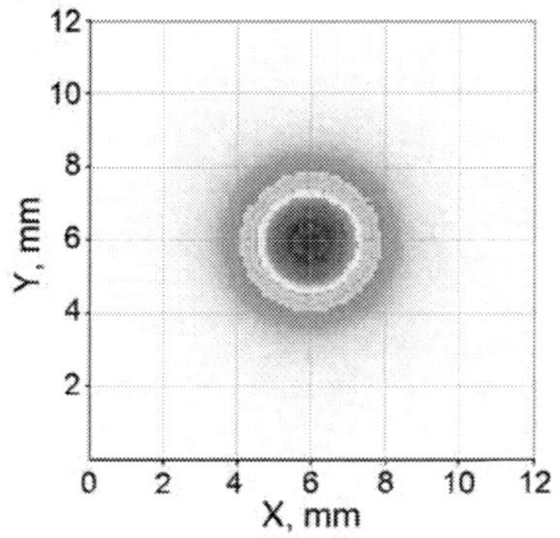

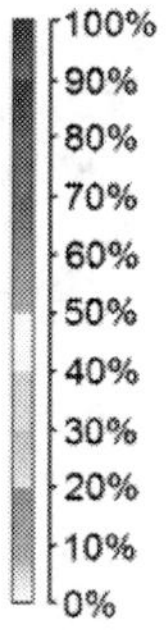

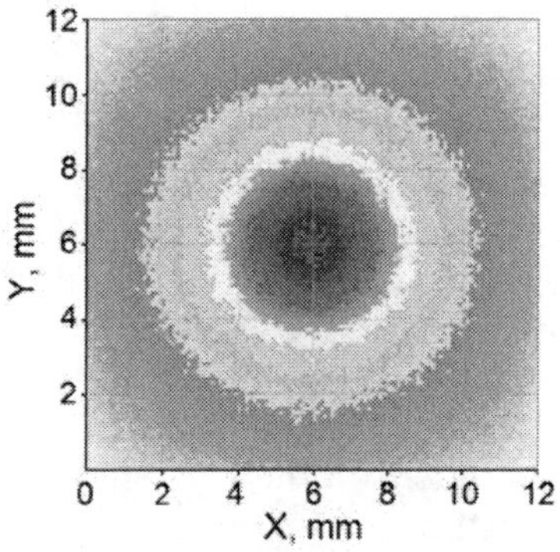

Figure 6. The electron beam shape in the vertical plane at the distance from microtron output window equal to 5 cm (the modified ABS-plastic density – 1.1 g/cm^3).

Figure 7. The electron beam shape in the vertical plane at the distance from microtron output window equal to 5 cm (the modified ABS-plastic density – 2.5 g/cm^3).

4. Summary

In this investigation the theoretical analysis of the electron beams interaction with the modified ABS-plastic by means of numerical simulation TPU microtron extracted electron beam was carried out. The results show the possibilityof the "Computer laboratory (PCLab)" using for the analysis of the real electron beam interaction with the modified materials.

It is shown that small change of the heavy elements concentration in the plastic filament significantly affects on the character electron beam interaction with the material. The next phase of this investigation it is necessary to produce the ABS-plastic doped with different concentrations of lead, then to obtained the real density and the Hounsfield's units of the modified ABS-plastic and then to estimate the radiation resistance of the materials. Than if it will be needed to correct the numerical model.

Acknowledgements

This work was financially supported by the Ministry of Education and Science of the Russian Federation in part of the science program (N 5.1485.2015).

References

[1] Burdovitsin V A, Klimov A S, Medovnik A V and Oks E M (2010) Electron beam treatment of non-conducting materials by a fore-pump-pressure plasma-cathode electron beam source *Plasma Sources Sci. Technol.* **19**(5) 055003

[2] Gribova O V, Musabaeva L I, Choinzonov E L and Mukhamedov M R (2012) The clinical course of thyroid cancer following its combined treatment with the use of fast-neutron therapy in the patients at the high risk of relapses *Vestnik otorinolaringologii* **5** 91–92 [in Russian]

[3] Choinzonov E L and Musabaeva L I 2006 *The intraoperative electron and gamma-ray teletherapy of the malignant neoplasms* (Tomsk: NTL)

[4] Information on http://www-pub.iaea.org/MTCD/publications/PDF/P1433_CD/datasets/papers/ap_ia-12.pdf

[5] Melnikov A A, Vasilyev S A, Musabaeva L I, Velikaya V V, Gribova O V, Startseva Zh A, Urazova L N, Lebedev I N and Choynzonov E L (2012) Cytogenetic effects of neutron therapy in patients with parotid gland tumors and relapse of breast cancer *Experimental oncology* **34**(4) 354–357

[6] Weisenburger A, An W, Engelko V, Heinzel A, Jianu A, Lang F, Mueller G and Zimmermann F (2009) Intense pulsed electron beams application of modified materials *Acta Physica Polonica A* **115**(6) 1053–1055

[7] Musabaeva L I, Lisin V A and Velikaia V V (2014) Radiation-induced reactions in breast cancer patients after breast-conserving surgery with intraoperative and external beam radiotherapy *Radiatsionnaia biologiia, radioecologiia* **54**(5) 474–478 [in Russian]
[8] Hogstrom K R and Almond P R (2006) Review of electron beam therapy physics *Physics in Medicine and Biology* **51**(13) R455–R489
[9] Christ S, Schnabel M, Vorndran E, Groll J, Gbureck U (2015) Fiber reinforcement during 3D printing *Materials Letters* **139** pp 165–168
[10] Information on http://www.ptcommunity.com/system/files/pdf/ptj3910704.pdf
[11] Information on http://www.appliancedesign.com/ext/resources/AM/Home/Files/PDFs/themoplastics.pdf
[12] Information on http://www.scrollseek.com/training/densitiesofdifferentbodymatter.html
[13] Bespalov V I 2015 *Computer laboratory (Version 9.6)* (TPU: Tomsk) 115 p [in Russian]

Hydrostatic Vibratory Drive of the Test Stand for Excitation of the Amplitude-Modulated Vibrations

A I Nizhegorodov[1,4] A N Gavrilin[2,5], B B Moyzes [3,6]

[1] Professor, Irkutsk national research technical university, Irkutsk, Russia
[2] Associate professor, Tomsk Polytechnic University, Tomsk, Russia
[3] Associate professor, Tomsk Polytechnic University, Tomsk, Russia
E-mail: [4]nastromo_irkutsk@mail.ru, [5]tom-gawral@list.ru, [6]mbb@tpu.ru

Abstract. The article reviews the problems arising during the development of the test stand hydrostatic vibratory drive, which synthesize controlled amplitude-modulated vibrations required testing of vibration strength and vibrostability of technological devices. The newly developed modification can adequately simulate the transport vibration and vibration of the operating power-supply units of technological machinery vibration by means of implementing of a continuous frequency spectrum of the vibration exposure in the desired frequency range.

1. Introduction

It was found that polyharmonic vibrations superposition [1, 2] is one of the effective methods for vibration strength and vibrostability testing of technological devices.

It provides the simulation of hard harmonic vibrational impacts which the most frequently occurred during operation. The method can be implemented by the set of the driving oscillator for sinusoidal signals [2] of the technological devices, synthesizing frequency-modulated and spectrally rich vibrations [3] or by controlled amplitude-modulated vibrations (AM-oscillations) [4].

The spectrum of the AM-oscillations can be represented as a wide line spectrum [4, 5], which is quite satisfactory oscillatory process that simulates the transport vibration and vibration of the operating power-supply units of technological machinery. AM-vibrations envelopes can be defined by different functions, prompting to modify their spectral composition [4–9].

The mechanism of impact on any technical devices or machinery (hydraulics and pneumatics, electronic and radio-relay devices, board computers and other mechanical and electrical devices) is due to the occurrence of resonance elements of these products and the corresponding dynamic load, leading to failure, failure of normal functioning (vibrostability) and even mechanical failure of structural components of such devices (vibration strength) [2, 10, 11].

Regardless of the dynamic structure of the vibratory system, the amplitude modulation of harmonic vibrations is made up of the deformation of the envelope amplitude (modulating function), and can be illustrated by a block diagram of an amplitude modulation channel (figure 1a) [4]. The spectrum of the AM-oscillations can be represented as a wide line spectrum [4], which is quite satisfactory oscillatory process that simulates the transport vibration and vibration of the operating power-supply units of technological machinery. AM-vibrations envelopes can be defined by different functions, prompting to modify their spectral composition [4].

The input signal (modulating function) specified in the operator form $x_0(p)$, after passing through a variable coefficient block $\sin(\omega t)$, will have the following form:

$$x(p) = 0,5\left[x_0(p + j\omega) + x_0(p - j\omega)\right], \tag{1}$$

where ω – carrier frequency of AM vibrations, $p = \frac{d}{dt}$ – differentiation operator.

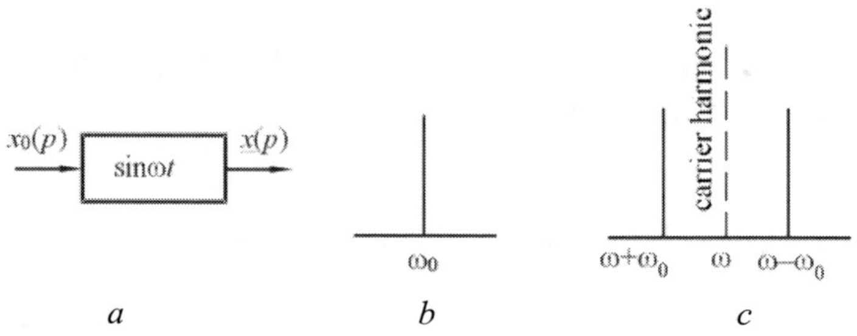

Figure 1. Block diagrams: *a* – amplitude modulation channel; *b* – input vibration spectrum; *c* – output vibration spectrum.

The output vibrations spectrum becomes evident when substituted $p=j\omega_0$:

$$x(j\omega) = 0,5\left[x_0(\omega_0 + \omega)j + x_0(\omega_0 - \omega j)\right]. \tag{2}$$

Therefore, the spectrum of the AM-oscillations is equal to half-sum of the spectra of the modulating function when shifted to the frequency rate of the variable coefficient sin (ω*t*), i.e. at the carrier vibrations frequency. If the modulating function comprises only one harmonic (figure 1b) $x_0(t) = sin\omega_0 t$, then the AM vibrations consist of two harmonics (figure 1c) – the cumulative and differential:

$$x(t) = 0.5\left[\sin(\omega + \omega_0)t - \sin(\omega - \omega_0)t\right]. \tag{3}$$

The spectrum of the modulating function when the amplitude varies with frequency ω_0 harmonically and the spectrum of the AM-oscillations contains only $\omega \pm \omega_0$ (figure 1c) frequency side. It is obvious that the carrier harmonic divides the modulating spectrum which set of harmonics varies infinitely. Therefore spectrally saturated oscillatory processes can be carried out by means of the amplitude modulation in the vibration test stands.

Consider the content of the AM frequency vibrations (figure 2).

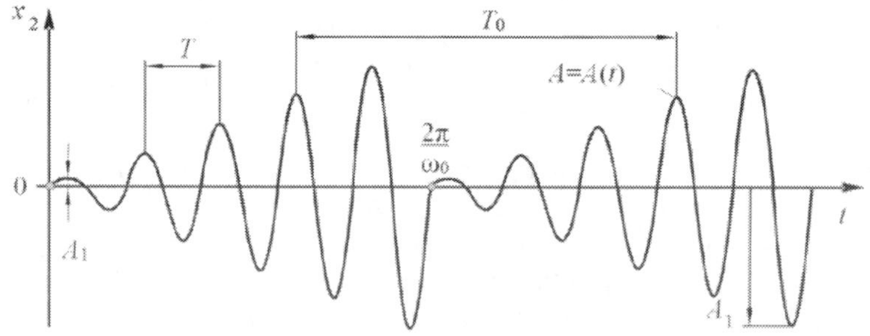

Figure 2. Amplitude modulation with a linear law of the amplitude change: *T* and T_0 – periods of the carrier and modulating functions, accordingly.

When changing the modulating function according to linear law:

$$x = kt \text{ at } 0 < t < \frac{2\pi}{\omega_0},$$

where k – the rate of increase of the modulating function, A – the amplitude of the first harmonic, the frequency spectrum of AM-vibrations will have the form shown in figure 3.

The spectrum shown in figure 3 is full of side harmonics, the level of harmonics vary slightly with decrease, this indicates the harmonics high energy [4].

The interval covering the spectrum in the range of from ω–$4\omega_0$ to $\omega + 4\omega_0$ at the carrier frequency, e.g., 50 Hz (314 rad / s) and a ratio ω_0/ω=5, is almost a decade (the ratio of the frequency of extreme harmonics differs by 10 times).

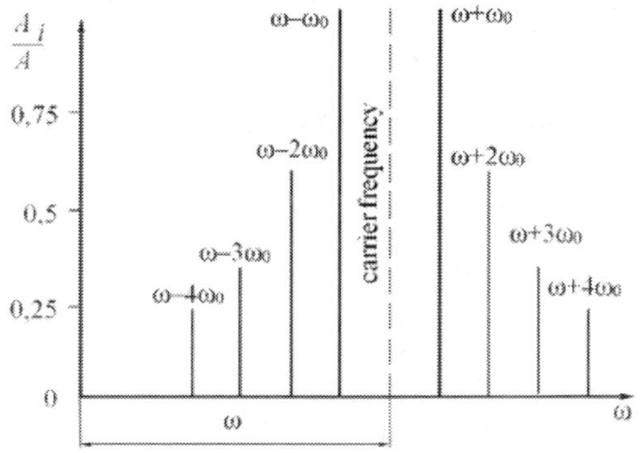

Figure 3. The range of the AM vibrations in the linear modulating function.

2. Hydrostatic vibratory drive of the test stand for excitation of the amplitude-modulated oscillations

To implement this method the hydrostatic vibratory drive of the test stand was developed. The hydraulic of the test stand (figure 4) incorporates a plunger generator *1* with a drive hydraulic motor *2* and the load hydraulic cylinder *3*.

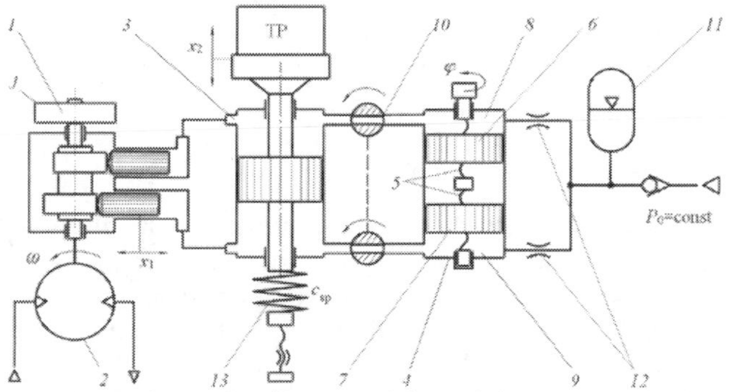

Figure 4. Hydrostatic vibratory drive of the test stand scheme.

The additional double-piston hydraulic cylinder design incorporates the follows: a crossed-axis helical gear *5* and the pistons *6* and *7* on the threaded multidirectional screw. When the screw rotates the pistons either come closer together increasing the volume of the chamber *8* and *9*, or move apart reducing the given volumes.

The load hydraulic cylinder *3* and *4* additional hydraulic cylinders are connected through the rotary valve *10*, which can be moved to one of two positions – «open» or «closed» by means of a rotary cylinder (not shown in figure 4). The hydraulic fluid leakage is made up by the boost pump (figure 4 not shown). The average initial pressure in the system P_0=const is maintained by the pneumatic pressure tank *11* and throttles *12*. There is a spring *13* with the adjusting screw to compensate the weight of the test product. When the directional valve *10* is closed, the generator *1* creates alternating fluid flow in the hydraulic lines which are connected with the hydraulic cylinder 3 with the angular frequency ω and oscillates the vibrating table x_2 with test product (TP) with a small amplitude of a below resonance mode A_1 ~ 0.15...0.3 mm (figures 2 and 5).

The rigidity of the vibratory system determined mainly by «hydraulic spring» (deformable liquid volume contained in the cavities 8 and 9 of the hydraulic cylinder 4), is calculated according to the formula

$$c_1 = c_{sp} + 2F_2^2 \frac{E}{W_0}, \tag{4}$$

where c_{sp} – spring rigidity*13*, F_2 – the effective area of the load hydraulic cylinder piston, E – modulus of fluid elasticity, W_0 – liquid volume in the hydraulic cylinder chambers and connecting hydraulic lines.

However natural (resonance) frequency of the vibrating system is determined by the expression:

$$\omega_{p1} = \sqrt{\frac{c_{sp} + \dfrac{2F_2^2 E}{W_0}}{m_{mp} + m_{tp}}}, \tag{5}$$

where m_{mp} – the mass of the table with moving parts of the hydraulic cylinder, m_{tp} – the mass of the test product, it is in the *a*-point of the frequency characteristic *1* (AFC 1, figure 5), and it is much more excitation frequency ω.

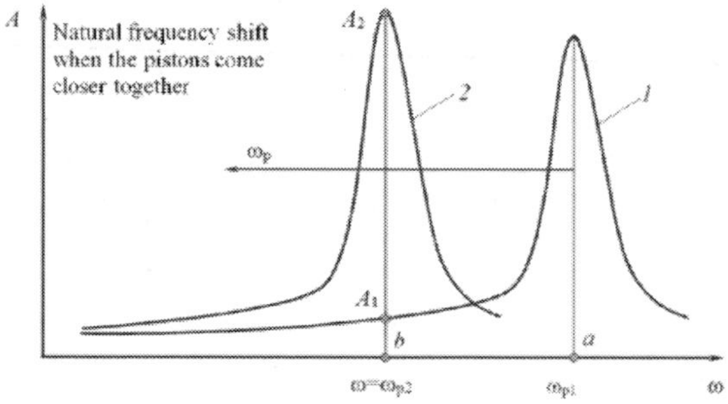

Figure 5. Changing of the frequency characteristic of the resistance system when the distributor is switching.

This below resonance mode is the initial point of the transient amplitude-modulating process corresponding to the switching of the distributor *10* in the «open» position.

At that moment, the rigidity (4) and the natural (resonance) frequency of the system (5) is instantly reduced by the adding extra liquid volume contained in the cavities 8 and 9 of the additional hydraulic cylinder 4 to the values evaluated by:

$$c_2 = c_{sp} + 2F_2^2 \frac{E}{W_\Sigma}, \tag{6}$$

$$\omega_{p2} = \sqrt{\frac{c_{sp} + \dfrac{2F_2^2 E}{W_\Sigma}}{m_{mp} + m_{tp}}}, \tag{7}$$

where W_Σ – total liquid volume in the chambers of load and additional hydraulic cylinders and connecting hydraulic lines.

Now the natural frequency of the oscillating system is shifted to the point *б* (AFC 2, figure 5) and becomes equal to the excitation frequency ω. The transient process begins (amplitude modulation) which is accompanied by an increase of the amplitude from A_1 to A_2 (figures 2 and 5), with the period of the modulating function, equal to T_0.

Let's consider the block-diagram of the oscillating system (figure 6).

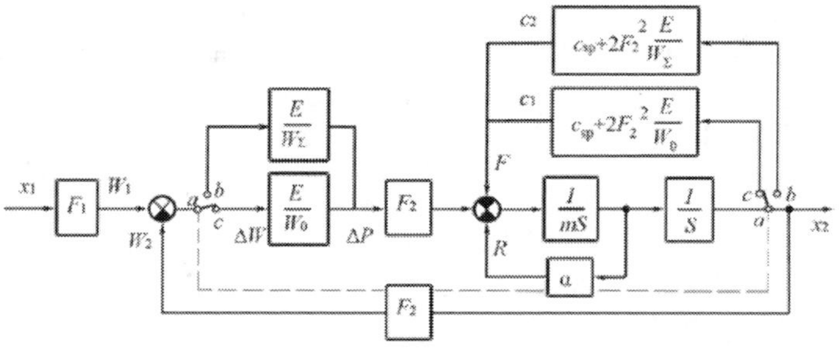

Figure 6. The block-diagram of the oscillating system.

There is linear system in each of the modes (below resonance and resonance), if we neglect the change of the modulus of elasticity and liquid volume ΔW. The Non-linear element is a valve *10* (figure 4) and it's imitator marked as *abc* (figure 6).

The switches connection *a-c* corresponds to below resonance mode, when the fluid volume determines the system rigidity c_1 equals W_0. Other switches position *a-b* corresponds to the resonant mode when the amount of liquid which determines the system rigidity is equal to c_2 W_Σ.

The important factor is the duration of the transient process t_{pt}. The transient process time which equals to the period T_0 will depend on the quality factor of the system.

Unlike the technological vibration machinery, for test stands, the quality factor estimated by dynamic factor k_d [12, 13] is very high – $k_d = 10\ldots12$ as the resistance to vibrations (the dissipative force *R*, figure 6 reduced to the rod of the load hydraulic cylinder) have only friction in the liquid in the movable joints of the load hydraulic cylinder and air friction.
Control for the transient process duration and the period of the modulating function can be carried out only by controlling the valve switching duration from closed position to open. The liquid throttling in the valve at the initial stage of its opening will lead to the increase of transient process time t_{pt}.

The amplitude modulation mode creates a line spectrum with a very high level of energy of the fundamental (carrier) and side harmonics but the resonant frequencies of the test product elements may occur in the range between neighboring harmonics. To implement the most effective test conditions, it is necessary to set the shutting motion (fluctuation) to the carrier frequency ω, simultaneously changing the resonant frequency of the oscillating system (7), while maintaining the equality $\omega = \omega_p$ by drawing the pistons *6* and *7* of the auxiliary cylinder *4* (figure 4) together and apart when changing a parameter at a specific algorithm. So it is possible to implement a wide continuous spectrum of a decade or more, scanning the test product in the required test mode frequency range. This process can be regarded as the identical to amplitude frequency modulation of oscillations [14].

Figure 7 shows the spectrum of an AM-oscillations with fluctuating carrier frequency for the case corresponding to the linear spectrum shown in figure 3 in the coordinates $S(\omega) - \omega$, where $S(\omega)$ – the process spectrum density. It also consists of a carrier and multiple side harmonics symmetrically arranged relative to the carrier frequency ω. The usual AM oscillation spectrum structure persists but now it is continuous.

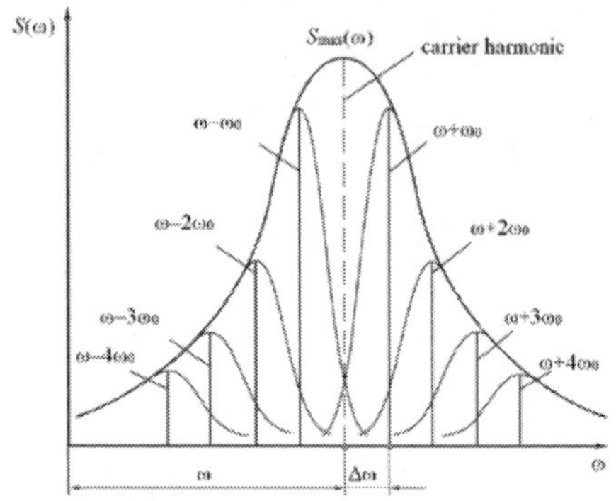

Figure 7. The spectrum of the AM-oscillations with the swinging (fluctuating) carrier frequency.

3. Conclusion

Further development of the concept of hydrostatic drive for the vibrational test stands requires creation of the dynamic excitation systems of random narrow-band vibration with the high level of harmonics spectrum density. This type of testing machines are capable to synthesize the random oscillations, simulating as much as possible the transport vibration and the vibration of the operating power technological machinery at high lifting capacity due to the energy and control possibilities of the volume hydraulic drive.

4. Acknowledgment

The research work was supported by Grant for Research of BP exploration operating company limited) and performed by research team of Department of Automation and Robotics in Mechanical Engineering of National Research Tomsk Polytechnic University and Department of Road Construction Machinery and Hydraulic Systems of Irkutsk National Research Technical University.

References

[1] Kljuev V V 2010 *Devices and systems for measuring vibration, noise and shock* (Moscow, Mashinostroeniye)
[2] Kljuev V V 1982 *Test engineering* (Moscow, Mashinostroeniye)
[3] Nizhegorodov A I (2014) Hydraulic drive testing stand with commutating device *Stroitel'nye i dorozhnye mashyny* **4** 31–34
[4] Shatalov A S 1977 *Automatic Control Theory* (Moscow, Vysshaya shkola)
[5] Goinaraghi F, Kuo B C 2010 *Automatic Control Systems* (Wiley Online Library)
[6] Groves R L, Pipho G A 1983 *Test stand for testing hydraulic devices* United States Patent 4368638
[7] Nizhegorodov A I, Gavrilin A N, Moyzes B B Hydraulic Power of Vibration Test Stand with Vibration Generator Based on Switching Device *Key Engineering Materials* Article in Press
[8] Gavrilin A N, Moyzes B B, Zharkevich O M Design and technological methods to reduce vibration activity of elements in technological systems *Journal of Vibroengineering* Article in Press
[9] Gavrilin A N, Chuprin A E, Moyzes B B, Halabuzar E A (2014) Land-based sources of seismic signals *Proceedings of 2014 International Conference on Mechanical Engineering, Automation and Control Systems, MEACS 2014* Article number 6986947
[10] Tjablikov Ju E, Karamyshkin V V 1979 *Features of interaction of the vibrating system of a given structure with an energy source: in Elastic and hydroelastic vibrations of machines and structures* (Moscow, Nauka)
[11] Grudinin V G (2011) Study of the effect of additional bonds in mechanical oscillating systems *Vestnik of Irkutsk National Research Technical University* **2** 34–40
[12] Shkalikov V S 1980 *Vibration and shock measurement* (Moscow, Standards publishing)
[13] Halit E 2000 *Acceleration, Vibration, and Shock Measurement* (Abingdon, Taylor & Francis Group, CRC Press LLC)
[14] Solodovnikov V V 1967 *Technique cybernetics. Automatic control theory* (Moscow, Mashinostroeniye)

Contribution of Compton Scattering in Problems Associated with Measuring the Surface Density of Radiation Protection Coatings

B I Kapranov[1] and M A Mudrov[2]

[1] Section Head, Laboratory No. 40, National Research Tomsk Polytechnic University, Tomsk, Russia
[2] Principal Engineer of Accreditation, Monitoring And Diagnostics Regional Centre, National Research Tomsk Polytechnic University, Tomsk, Russia
E-mail: mudrov@tpu.ru

Abstract. The paper analyses the ratio between the number of Compton-scattered quanta excited in elements of radiation protection coatings by isotope Am-241 and the number of fluorescence quanta excited in in the same elements, taking into account the fluorescence yield coefficient and the distance between the energy of primary radiation quanta (60 keV) and the energy of absorption K-edge. The paper shows that the number of scattered quanta exceeds the number of fluorescent quanta from K-level in all elements. The obtained values define the ratio of scattered and fluorescent quanta in any radiation protection coatings with a known chemical composition.

1. Problem of Controlling Protective Properties of Radiation Protection Coatings

Radiation protection coatings are widely used in the aviation and space engineering for protection against ionizing radiation. This issue is especially important for instrumentation, in particular for improving the operational reliability of electronics exposed to external X-ray radiation.

Instrument modules are protected with a special coating to effectively attenuate X-ray radiation in the energy range of up to 100 keV. The standard requirements for such coatings include mechanical strength, maximum attenuation efficiency and minimum weight. In order to fulfill these requirements, special coating compositions [1], production or application methods [2], and methods for measuring attenuating properties [3] are developed. Such coatings are most often made of RZK-6 (rare-earth concentrate) and RZO-4 (rare-earth oxide), which contain oxides of rare-earth elements [4, 5]. They are efficient due to the fact that absorption K- jumps occur in the energy range of up to 100 keV. The resulting attenuation coefficient is higher than the attenuation coefficient for these energies in lead, while the density of such materials is lower than lead density.

In order to control the quality of such coatings, the fluorescent method is most often used. This method is aimed at irradiating the coating-base combination with primary γ- or X-rays and recording the intensity of the fluorescent quantum flux from elements of the coating. In this case, the measured parameter is the surface density ρ expressed in g/cm^2. With the increase in the surface density, the intensity of the fluorescent quantum flux grows. The devices developed for such control are based on the one-sided access geometry [6–9]. The long-lived isotope Am-241 is most often used as a primary radiation source. The development of methods and equipment for control of such coatings is always associated with the problem of the ratio between intensities of fluorescence and Compton scattering radiation fluxes. The Compton radiation is generated both in elements of the coating and in the base covered with such a coating. The resulting secondary flux has a complex energy spectrum. The type of the spectrum depends on a number of factors, including geometrical thickness of the coating and the base; additional excitation of elements by radiation scattered from the base; subexcitation of K-lines of elements with a lower atomic number by K-lines of elements with a higher atomic number; scattering of fluorescent quanta by the base from the elements.

This paper studies the contribution of Compton scattering to the total secondary flux occurring in elements of the coating excited with γ-quanta of Am-241. To solve this problem, let us consider the geometry shown in figure1.

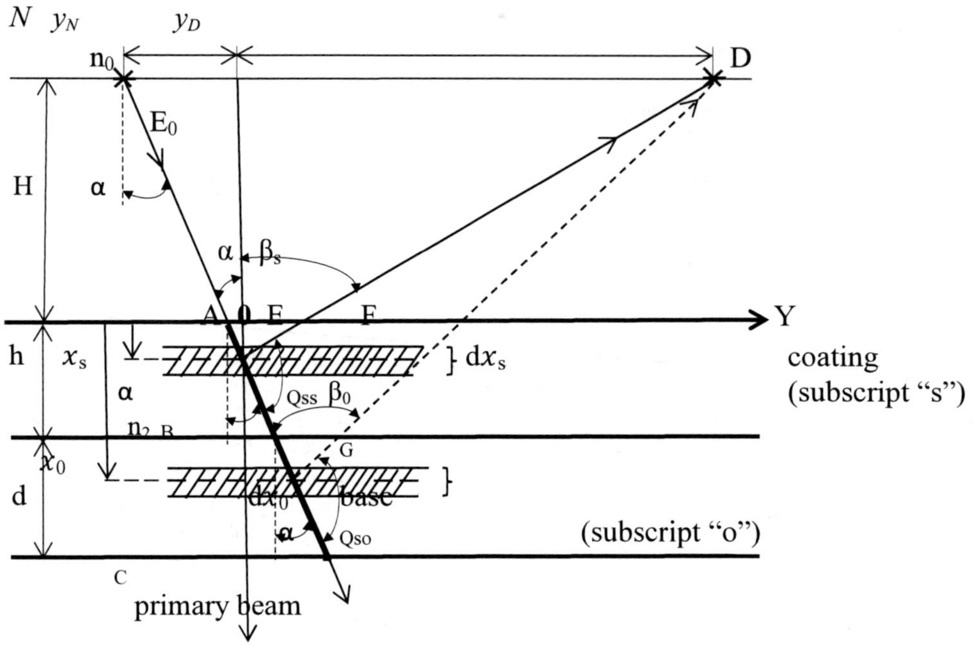

Figure 1. Diagram for calculation of fluorescence radiation and scattered radiation fluxes.

N is the source of radiation (monochromatic, energy E_0); n_0 is the number of photons per unit of time emitted by the source into a solid angle 4π (intensity); D is the detector; H is the distance from the "source – detector" base to the object; h is the thickness of the coating; d is the thickness of the base; α is the incidence angle of the primary beam; β_s, β_0 are the emergence angles of the secondary radiation from the coating and the base, respectively; Qss, Qso are the scattering angles of the secondary radiation on the detector from the coating and the base, respectively; x_s, x_0 are the coordinates of the points of interaction between the primary beam in the coating / the base, respectively; AB is the line of interaction between the primary beam and the coating; BC is the line of interaction between the primary beam and the base.

2. Monochromatic Source of Isotope Am^{241}

The radiation from source N (assume the flux to be "narrow") is incident on the object at angle α. Primary radiation quanta pass through the coating with thickness *h* along line AB where they are attenuated. Then, they get into the base with thickness *d* where they are scattered on line BC. The total secondary flux in detector D is composed of fluorescent quanta (only k-fluorescence is taken into account) from elements of the coating, Compton-scattered quanta in elements of the coating, Compton-scattered quanta in the base, with allowance for attenuation of the primary flux and the secondary flux caused by the coating and other factors. We assume that fluorescent quanta from the element of coating dx_s at point x_s escape into the detector at an angle β_s to the normal.

Radiation source *N* and detector *D* are at distance *H* from the test object in positions y_N and y_D. Let us designate the surface of the object exposed to the primary beam as *S*. We also assume that the linear dimensions of coating h and area *S* are small as compared to *H*. The interaction between the primary beam and the coating is assumed to occur along line AB, the length of which is determined as $\frac{h}{\cos\alpha}$. The intensity of the primary quantum flux at point x_s of the coating is determined by the geometric divergence and attenuation on the way from point A to the interaction point:

$$n\,(x) = \frac{n_o}{4\pi\left(\frac{H}{\cos\alpha} + \frac{x}{\cos\alpha}\right)^2} \cdot e^{-\mu_s \cdot \frac{x}{\cos\alpha}},$$

where:
μ_s is the linear attenuation factor of the primary radiation in the coating material;
$\frac{x}{\cos\alpha}$ is the path of the primary radiation in the coating to the interaction point.

Passing through the layer d x $_s$, the intensity of the primary beam is reduced by the value $dn_s\,(x_s)$:

$$dn_s\,(x_s) = -\,\mu_s \cdot \frac{1}{\cos\alpha} \cdot n_s(x_s)\,dx_s,$$

provided that h < *H*.

In this expression, let us designate the intensity of the primary quantum flux at the entry to the product as n_1. If we consider the radiation source to be a point source,

$$n_1 = \frac{n_o}{4\pi\left(\frac{H}{\cos\alpha}\right)^2}.$$

Then

$$dn_s\,(x_s) = -\,n_1 \cdot \frac{\mu_\rho^s \cdot \rho_s}{\cos\alpha} \cdot e^{-\mu_\rho^s \cdot \rho_s \cdot \frac{x_s}{\cos\alpha}}\,dx_s,$$

where:
μ_ρ^s is the mass attenuation factor of the primary beam in the coating material;
ρ_s is the density of the coating material.

The total number of quanta in the primary beam absorbed in the coating on line AB is determined by integrating $dn_s\,(x_s)$ throughout the thickness of coating h.

$$\Delta n_s\,(h) = n_1 \cdot \frac{\mu_\rho^s \cdot \rho_s}{\cos\alpha} \cdot \left(1 - e^{-\mu_\rho^s \cdot \rho_s \cdot \frac{h}{\cos\alpha}}\right).$$

The relative share of quanta in the primary beam absorbed in the coating will be:

$$\delta n_s\,(h) = \frac{\Delta n_s\,(h)}{n_1} = \frac{\mu_\rho^s \cdot \rho_s}{\cos\alpha} \cdot \left(1 - e^{-\mu_\rho^s \cdot \rho_s \cdot \frac{h}{\cos\alpha}}\right).$$

Therefore, the quantum flux of the primary beam passed through the coating, i.e. at the entry to the base, will be equal to $n_2 = n_1 - \Delta n_s\,(h)$ or:

$$n_2 = n_1 \cdot \left[1 - \frac{\mu_\rho^s \cdot \rho_s}{\cos\alpha} \cdot \left(1 - e^{-\mu_\rho^s \cdot \rho_s \cdot \frac{h}{\cos\alpha}}\right)\right].$$

Quanta of primary radiation in the coating material are subject to two processes: photoabsorption and scattering. The ratio between probabilities of these processes is given the photoabsorption section τ and the scattering section σ [10]. The total mass section of the interaction in the coating is determined as:

$$\mu_\rho^s = \tau_\rho^s + \sigma_\rho^s.$$

Hence, the intensity of the primary radiation quantum flux which has interacted with photoabsorption will be:

$$\Delta n_s\,(h)_{\mathrm{absorb}} = n_1 \cdot \frac{\tau_\rho^s \cdot \rho_s}{\cos\alpha} \cdot \left(1 - e^{-\mu_\rho^s \cdot \rho_s \cdot \frac{h}{\cos\alpha}}\right),$$

and the intensity of the primary radiation photon flux which has Compton-interacted will be:

$$\Delta n_s\,(h)_{\mathrm{Compt}} = n_1 \cdot \frac{\sigma_\rho^s \cdot \rho_s}{\cos\alpha} \cdot \left(1 - e^{-\mu_\rho^s \cdot \rho_s \cdot \frac{h}{\cos\alpha}}\right) \qquad (1)$$

The fluorescence radiation yield is associated only with photoabsorption τ. Each photoabsorbed photon of primary radiation excites one atom in the coating substance [11], and the number of the atoms that have emitted fluorescence quanta is defined by the probability ω_K of the fluorescence yield:

$$\Delta n_{fl.}^{k} = \omega_k \cdot \Delta n_s(h)_{\text{absorb.}}$$

ω_k is determined by experiment as a ratio of the fluorescent quantum flux $n_{fl.}^{k}$ to the primary quanta flux photoabsorbed in the material $\Delta n_s\ (h)_{\text{absorb}}$. The most frequently used formula Stephenson [11]. The calculated values of the fluorescence yield coefficient $\omega_K$of the elements included in the coating are shown in Table 1 [11].

If we consider only the fluorescence yield during the excitation of *K*-level of elements, doublet $K\alpha_1\alpha_2$ is excited. Its line energies $E_{K\alpha 1}$ and $E_{K\alpha 2}$ are defined by the energies of transitions from *K*-level to level L_{III} and from *K*-level to level L_{II}, respectively. The numerical values for these elements are also shown in Table 1.

3. The ratio between Intensities of Lines

The ratio between intensities of lines belonging to the same multiplet is calculated by the Burger and Dorgelo rule [11]: If we neglect splitting of the initial multiplet levels (which is permissible for *K*-level of all elements since they differ only by the spin number), the intensities of the line of multiplet $K\alpha_1\alpha_2$ will be proportional to the statistical weights of the final levels L_{III} and L_{II} respectively. Since the top level of multiplet Kα1α2 is common, then:

$$\frac{J_{\alpha 1}}{J_{\alpha 2}} = \frac{\tau_o^2\,(\alpha_1)}{\tau_o^2\,(\alpha_2)} = \frac{2}{1}.$$

The energies of lines $K\alpha_1$ and $K\alpha_2$ differ insignificantly and cannot be resolved with a scintillation spectrometer. Each doublet will be recorded in aggregate. In order to determine the energy of K-doublet, we can use the concept of effective energy defined as the sum of the weighted energies of lines $K\alpha_1$ and $K\alpha_2$. The quantitative ratio of the energy contribution of lines Kα1 and Kα2 is defined by the relative intensity of these lines. According to [11], for the elements with Z of 23 to 90, the ratio

$$\frac{J_{\alpha 1}}{J_{\alpha 2}} = 2.00$$

with a deviation of not more than ± 0.05. Therefore, this paper assumes the weight of line $K\alpha_1$ to be 2, and the weight of line $K\alpha_2$ to be 1. Then, the effective energy of the doublet will be determined as:

$$\mathrm{E}_{eff} = \frac{2}{3} \cdot \mathrm{E}_{\mathrm{K}_{\alpha 1}} + \frac{1}{3} \cdot \mathrm{E}_{\mathrm{K}_{\alpha 2}}$$

Table 1 shows the effective energies of K-doublets for elements of coatings calculated from this ratio. The weighted sum of the effective energies of K-doublets in multicomponent coatings will define the effective energy of the fluorescence radiation spectrum excited in a coating of a specific composition.

Another factor that defines the fluorescent quantum yield is the dependence on the energy of exciting radiation. The largest fluorescence yield from the K-level will be caused by the energy of primary radiation that is probably closer to the energy of the absorption K-edge E_k. But the condition $\frac{E_0}{E_k} > 1$ must be satisfied. The dependence of the excitation probability on the ratio of these energies is described by the function [11]:

$$æ\left(\frac{E_0}{E_k}\right) = \frac{\left(\frac{E_0}{E_k}\right)^3 + \frac{\sin\alpha}{\sin\beta}\cdot\frac{1}{S_k}}{\left(\frac{E_0}{E_k}\right)^4},$$

where S_k is the absorption *K*-jump.

The numerical values S_k for elements of the coating are also presented in Table 1.

The value of the function $æ\left(\frac{E_0}{E_k}\right)$ is maximum at the ratio of energies $\frac{E_0}{E_k} = 1$ and given by $\frac{S_k+1}{S_k}$ when angles α and β are equal. As E_k moves away from E_0 towards lower values, the excitation probability value æ decreases. If isotope Am-241 is used as a source, the excitation probability for K-line of elements decreases with decreasing the atomic number, as shown in Table 1. Hence, the full fluorescent quantum yield for elements of coatings made of RZK-6 and RZO-4 is described by the following ratio, provided that isotope Am-241 is excited by the radiation:

$$\Delta n_{fl.}^{k} = æ \cdot \omega_k \cdot n_1 \cdot \frac{\tau_p^S \cdot \rho_S}{\cos\alpha} \cdot \left(1 - e^{-\mu_p^S \cdot \rho_S \cdot \frac{h}{\cos\alpha}}\right).$$

Taking into account that the total number of Compton-scattered quanta in elements of radiation protection coatings is determined by ratio (1), we can determine the ratio of the flux intensity of Compton scattering to the total flux of the characteristic emission excited in the elements:

$$\delta_S = \frac{\Delta n_S\,(h)_{\text{Compt}}}{\Delta n_{fl}^{k}}, \qquad (2)$$

Table 1. Energy of K-absorption edge (E_k), effective energy lines K – Series (E_{eff}), the quantum efficiency of the fluorescence yield K – lines (ω_K), the probability of fluorescence excitation (æ), the ratio of the flux intensity of Compton scattering to the total flux of the characteristic emission excited in the elements included in RZK-6 (rare-earth concentrate) and RZO-4 (rare-earth oxides) when used as the primary radiation isotope Am -241.

Z	Element		E_k, keV	E_{eff}, keV	ω_K, %	Absorption jump S_k, relative units	æ, relative units	δ_S, relative units
11	Sodium	Na	1.073	1.043	0.035	15.2	0.017	19469
14	Silicon	Si	1.847	1.744	0.04	13.2	0.030	1533
15	Phosphorus	P	2.144	2.019	0.055	12.6	0.033	764
20	Calcium	Ca	4.037	3.700	0.165	10.7	0.067	43,2
26	Iron	Fe	7.112	6.412	0.32	9.0	0.12	4,54
39	Yttrium	Y	17.038	14.970	0.68	7.0	0.29	0,231
48	Cadmium	Cd	26.711	23.165	0.76	6.0	0.46	0,066
50	Tin	Sn	29.200	25.258	0.75	5.9	0.50	0,059
57	Lanthanum	La	38.925	33.417	0.69	5.4	0.68	0,026
58	Cerium	Ce	40.443	34.177	0.68	5.3	0.71	0,024
59	Praseodymium	Pr	41.991	35.910	0.66	5.3	0.74	0,021
60	Neodymium	Nd	43.569	37.236	0.66	5.2	0.78	0,0196
62	Samarium	Sm	46. 834	39.950	0.64	5.1	0.86	0,0161
63	Europium	Eu	48.519	41.355	0.62	5.0	0.89	0,0149
64	Gadolinium	Gd	50.239	42.823	0.61	5.0	0.89	0,0144
65	Terbium	Tb	51.996	44.279	0.60	4.9	0.94	0,0130
66	Dysprosium	Dy	53.788	45.850	0.59	4.9	1.00	0,0116
67	Holmium	Ho	55.618	47.308	0.56	4.8	1.00	0,0112
68	Erbium	Er	57.486	48.855	0.55	4.7	1.00	0,0108
70	Ytterbium	Yb	61.332	52.094	0.51	4.7	-	-
71	Lutetium	Lu	63.316	53.769	0.51	4.6	-	-
82	Lead	Pb	88.004	74.340	-	4.1	-	-
83	Bismuth	Bi	90.526	76.454	-	4.1	-	-

4. Conclusion

The analysis shows that the source of Am - 241 effectively excites K - line elements to erbium (Z = 68). K - Ytterbium element line (Z = 70), lutetium (Z = 71), lead (Z = 82), bismuth (Z = 83) are not excited. Therefore, for their account to be used L - lines, respectively: 10.50 keV; 10.87 keV; 15.85 keV; 16.39 keV. The contribution to the total flow of the scattered radiation is less than 23%, since the yttrium (Z = 39). Thus, to control the thickness (areal density) of radiation - coatings can be effectively used fluorescent method with the isotope Am - 241 as the primary radiation source.

References

[1] Kutaitseva O N 2001 *Radiation Protection Polystyrene Coatings* Thesis of Candidate of Technical Sciences 05.23.05 defended in 2001
[2] Debalchuk V F, Glazkova N A, Kirillov V N, Chevordaev V M, Bulavina N L *Patent 2054439* Russian Federation MPK C08L9/00, C08K3/22, C08L9/00, C08L23:34, G21F1/10 Rubber Mix for Production of X-ray Protective Rubber No. 92003093/04; appl. 30.10.1992; publ. 20.02.96
[3] GOST R 51532-99 (MEK 61331-1-94) 2001 *Means of Protection against X-ray Radiation in Medical Diagnostics. Part 1: Determining the Attenuating Properties of Materials* (Moscow Gosstandart Rossii: Publishing House of Standards)
[4] *Product RZO-4 (with amendments No. 1, 2). Specifications* 1985 *TU 48-4-345-84* (Moscow Gosstandart)
[5] *Concentrates of Rare-Earth Elements of Medium Group* (with amendments 1–4) 1977 Specifications: TU 48-4-382-76 (Moscow Gosstandart)
[6] *Technical Description for Radioisotope Instrument MT-01,* National Research Tomsk Polytechnic University Certificate RU.E.38.007 A. No 14441 No 24536-03 in the State Register of Measuring Instruments.
[7] *Radioisotope Coating Thickness Meter (ITP)* 5705 1972 Technical Description and User Manual 1972 5705-00-00TO/VNIIAchermet (Moscow)
[8] *Betamikrometr-2 Radioisotope Coating Thickness Meter* 1973 Technical Description and User Manual EI620030TO/VNIIRT (Moscow)
[9] *Albedo Coating Thickness Meter ATP -2* 1975 Technical Description and User Manual APSh 5.178.002 TO (Tomsk)
[10] Storm E and Israel H 1973 *Photon cross sections from 0.001 to 100 MeV for Elements 1 through 100* (Moscow ATOMIZDAT)
[11] Blokhin M A 1953 *X-Ray Physics* (Moscow, Gostekhteorizdat)

Journal of Physics: Conference Series **671** (2016) 012039 doi:10.1088/1742-6596/671/1/012039

Converter Circuit «Voltage-Voltage» Investigation for Power Calibrator

Yu M Fomichev[1], S V Silushkin[1], T S Mylnikova[2]

[1]Associate Professor, Institute of Cybernetics, Tomsk Polytechnic University, Russia, Tomsk
[2]Senior Teacher, Institute of Physics and Technology, Tomsk Polytechnic University, Russia, Tomsk
E-mail: slavasv@mail.ru

Abstract. The paper presents possible circuits to construct a voltage-voltage converter for the calibrator of fictitious power. The problems of circuit solutions were experimentally identified, and the ways of their elimination were found. One of the main problems of convectors is to provide small harmonic distortions and additional phase shift. The use of deep negative instantaneous value feedback helps to provide the desired level of nonlinear distortions and to reduce the phase shift. Corrective circuits are used to ensure the stability of the transducer at greater depths of the feedback; the half-period average value or rms value feedback is used to ensure the stability and accuracy of conversion. However, the accuracy of the power calibrator can be upgraded and its work for various types of loads can be ensured by means of application follower circuit with modern electronic components which are also discussed in the paper.

1.Introduction

Modern power calibrators and measuring system [1–4] comprise several main modules that perform various functions, such as generation of the reference signal, signal conversion, power amplification, etc. A simplified functional diagram of the power calibrator is shown in figure 1, and a complete diagram is considered by the authors in [5].

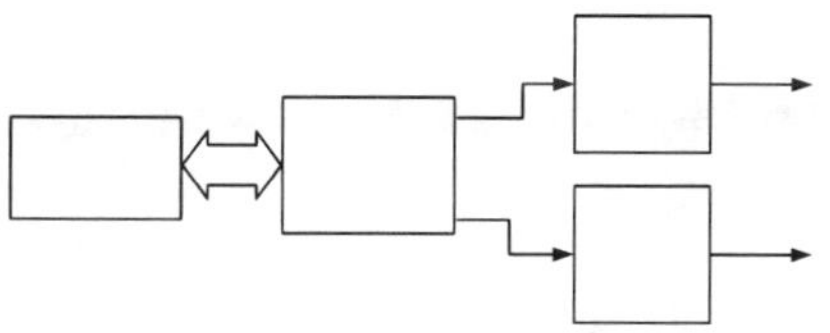

Figure 1. A simplified functional diagram of the fictitious power source.

A digital signal synthezer (DSS) generates two sinusoidal voltages with a given amplitude and phase. The required output power $P_{out} = U{\cdot}I{\cdot}\cos\varphi$ is provided by the "voltage-voltage" converter (VVC) and "voltage-current" converter (VCC). Precise setting of the required values of voltage amplitudes, current and the phase shift between them at the input of the converters is determined by the metrological characteristics of the DSG and can be currently achieved with a sufficiently high accuracy (accuracy within 0.02 ... 0.05 %), and increment in the phase shift can be provided in the range of 0.01° and a 0.02°. The calibration relative error of the power calibrator output power

$$\frac{\Delta P_{out}}{P_{out}} = \frac{\Delta U}{U} + \frac{\Delta I}{I} + \left[1 - \frac{\cos(\varphi \pm \Delta\varphi)}{\cos\varphi}\right]$$

is determined by the VVC conversion absolute error (ΔU), the VCC voltage-to-current conversion absolute error (ΔI) and additional phase shift ($\Delta\varphi$) caused by the converters.

In addition to these requirements, VVC and VCC should provide the required output power:

– for VVC, the rms voltage value is 250...260 V, and the required power is determined by the meter consumption equal to up to 10 VA according to GOST R-52323 [6], i.e. in group n meters verification $P_{out} > 10 \cdot n$VA;

– for VCC, the load is largely determined by the connecting wires and, as the experiment has shown, a voltage of 2...3 V is to be applied to the VCC output, i.e. for 60 A current, $P_{out} = 120\ldots180$ W. According to the requirements in the standard [6], the permissible distortion of output signals and does not exceed 0.2 %.

Thus, the problems occurring during converter design are as follows:

– energy characteristics of the generated signals (voltage, current, power) are to be provided;

– adequate accuracy and stability of the conversion is required;

– small non-linear distortion and minimal additional phase shift of the generated signals is to be ensured.

In addition, for group calibration of the shunt static electricity meters, galvanic separation of current and voltage channels is to be considered. In this case, the output transformer is to be used in the VVC.

The stability and accuracy of conversion can be achieved using the half-period average value or rms value feedback, however, instantaneous value deep negative feedback (NFB) is to be used to provide low distortion and additional phase shift. Another problem to be solved is the stability of the converter operation.

The paper considers several diagrams for VVC used in various measuring instruments, the diagrams for converters based on modern electronics are suggested.

2. Study of VVC diagrams

Figure 2 shows the structural VVC circuits used in power calibrators of the calibration setup "Vector".

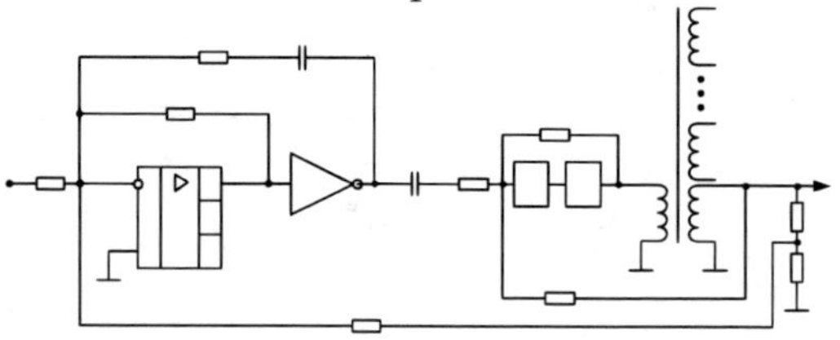

Figure 2. VVC circuits of the power calibrator "Vector".

A preamplifier with a transmission coefficient of the $K = 10$ is performed on DA_1. The positive feedback (PFB) is effected through inverter (1) and circuit elements (R_3, C_1). The voltage amplifier (VA) is designed on discrete elements and allows the output voltage of 60 V peak value. The power amplifier (PA), the follower in powerful field-effect transistors working in AB class provides power output of 100 VA. Deep voltage NFB for the VA and PA is obtained through R_4 and R_5 (DC negative feedback is 100%), which provides an insignificant zero offset at the PA output. TP_1 is a step-up output transformer which has a primary winding (W_1) and six output windings made of six bundled wires that ensure equal number of winding turns and the same transformer ratio relative to the primary winding, i.e.:

$$\frac{W_2}{W_1} = \frac{W_3}{W_1} = \ldots = \frac{W_7}{W_1} = 6\,.$$

One of the output windings is connected to the common wire, and is used for local (R_4, R_6) and total (R_1, R_7, R_8, R_9) NFB. The use of this transformer can solve two problems:

– to get six galvanically isolated voltages with a given level of up to 360 V and permissible load capacity per each winding of up to 10 VA, the switching being made from the second (W_2) winding;

– to ensure the required metrological characteristics at all outputs under similar types of loading.

The required precision of the setup, stability, low distortion and small phase shift of the output voltage are provided by deep feedback. To increase the total FB depth, local positive FB is introduced in the preamplifier via inverter (1) and circuit elements (C_1, R_3).

To calculate the FB depth, we use a directed graph of the system (figure 3).

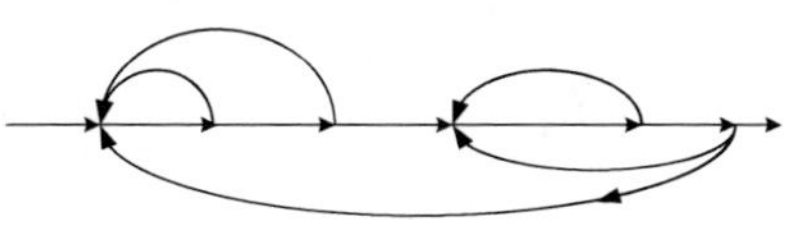

Figure 3. Directed graph of the VVC diagram of the power calibrator "Vector".

If we take the conventional values $\beta_1 = \beta_2$ and $\gamma_1 = 1$, $\gamma_2 = 1$, the VVC transmission coefficient will be determined by the expression:

$$K_{\text{VVC}} = \frac{K_1 \cdot K_2 \cdot n}{1 + K_2 \cdot \beta_3 + K_2 \cdot \beta_4 \cdot n + K_1 \cdot K_2 \cdot \beta_5 \cdot \beta_6 \cdot n},$$

where K_1 is the transmission coefficient of the input operational amplifier.

When $K_1 = 10^4$, we get the following equality

$$K_{\text{VVC}} \approx \frac{1}{\beta_5 \cdot \beta_6},$$

i.e. the accuracy and stability of K_{VVC} is determined by the resistors of the total feedback circuit. The feedback depth amounts to several thousand that allows for additional phase shift of less than 0.02 %.

For this feedback depth, when the loop includes the transformer and capacitive load, the problem of its stability is to be solved. In the research, different corrective circuits have been tested. Satisfactory results were obtained when using the circuit shown in figure 4 and gain and phase of this amplifier presented in figure 5.

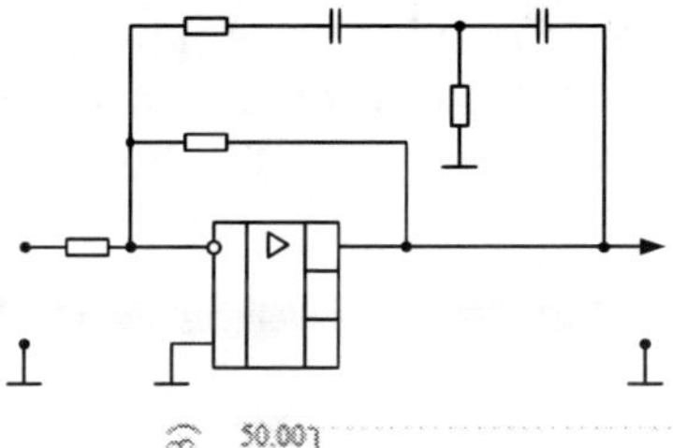

Figure 4. Example of the VVC corrective circuit.

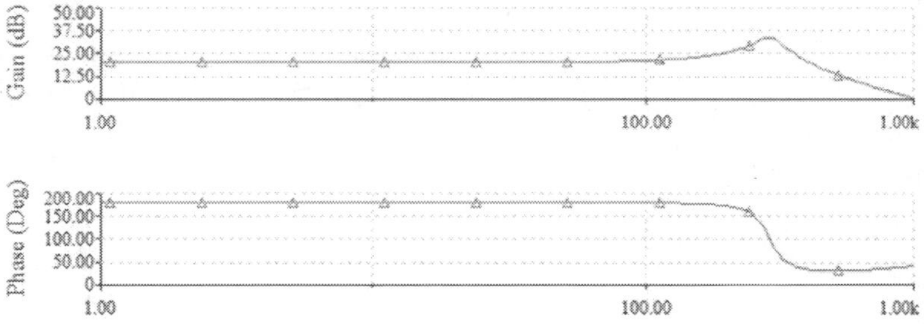

Figure 5. Gain and phase of the VVC corrective circuit.

From the graphs (figure 5), it can be seen that the amplifier operates in the frequency range of 45...65 Hz with an additional phase shift less than 0.02°. From 200 Hz, the phase shift decreases sharply by 150°. As frequency increases, the phase shift slowly grows, but the gain remains less than unity. This circuit design can be applied in the VVC input stage in the "Vector" calibrator (figure 2), and it provides stability of the system.

For group calibration of static meters, another problem has been revealed: the feed circuits of most static meters are made using half-wave rectification; therefore, the consumed current is pulsed. The input voltage circuits of the meters RIM 586.01 and SOEB-2PK/1 have been studied. Figure 6 shows the diagram of the consumed currents of the power sources for these meters.

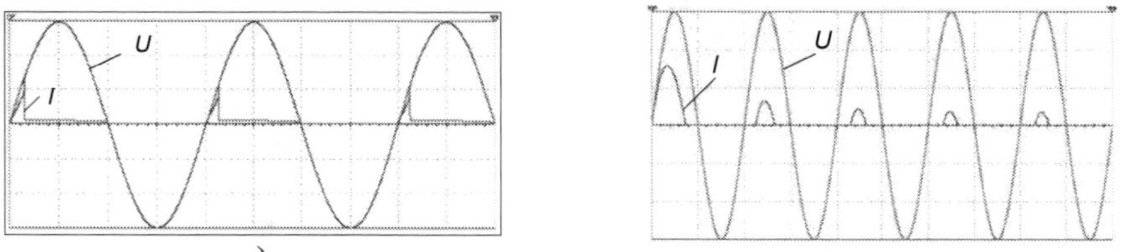

a) b)

Figure 6. Diagrams of the consumed currents: IM 586.01 (a) and SOEB-PK2/1 (b).

As can be seen from the graphs (figure 6), the initial drain is several times higher than the steady-state value, which leads to protection operation or causes excitement. To avoid these problems, the slew rate of the input and, consequently, the VVC output voltage were reduced programmatically.

3. Implementation of VVC on state of the art electronic components (transformerless VVC)

Increase in the operating frequency band for VVC and VCC up to 0.5...2 kHz is required to determine additional error of the meter caused by a number of different factors (harmonics in the voltage and current circuits and subharmonics in AC circuits) according to the requirements in GOST R-52323 [6]. To solve the problem for VVC, the output step-up transformer is to be excluded from the circuit.

A number of currently produced high voltage op amps provide permissible voltage of ±(400...450) V, for example, PA89 [7], PA91 [8], PA93 [9], PA94 [10] and PA97 [11]. These op amps can be used to make transformerless VVC, but unfortunately, the output power of the op amp is not sufficient to carry out group verification of meters, i.e. a powerful follower is required. The task is complicated by lack of powerful field-effect transistors with a p-type channel (as well as p-n-p-type bipolar transistors) for permissible voltage of $U_{DS} = 800\ldots900$ V. The designers have to find the solution using transistors IRF740 [12] with n-type channel only. Figure 7 shows one of the alternatives of this solution.

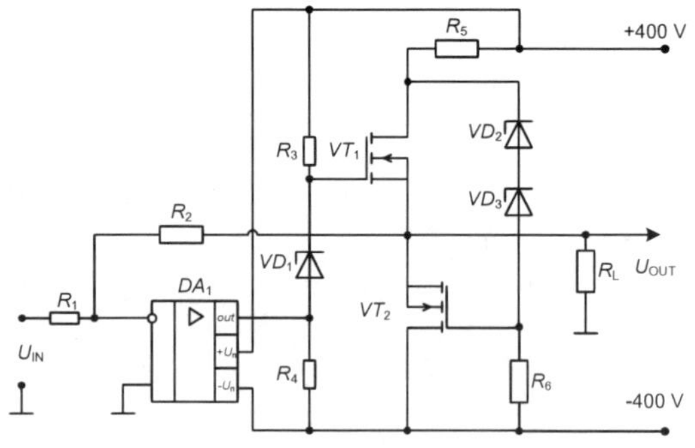

Figure 7. Circuits of the follower on single-type field-effect transistors with n-type channel.

The follower shown in figure 7 is a circuit with controlled dynamic load. The transmission coefficient of the follower is

$$K = \frac{R_L \cdot S_1 \cdot (1 + S_1 \cdot R_5)}{1 + R_L \cdot S_1 \cdot (1 + S_2 \cdot R_5)}$$

where S_1, and S_2 are the slope of the VT_1 and VT_2 transistors.

The condition of the push-pull circuit is $R_5 = 1/S_2$. In practice, the resistance R_5 is to be increased by 2...3 times to compensate for the voltage drop across the dynamic resistance of the stabilitrons VD_2 and VD_3. The studies have shown that this VVC performs well for the capacitive load as well. The disadvantage is that this follower works in class A only, hence, the efficiency = 38...45 %.

Another possible option is to make a follower on transistors of different types via series connection of two transistors IRFU 9310 [13] to p-type channel to provide allowable voltage U_{DS}. Figure 8 shows a diagram of this follower.

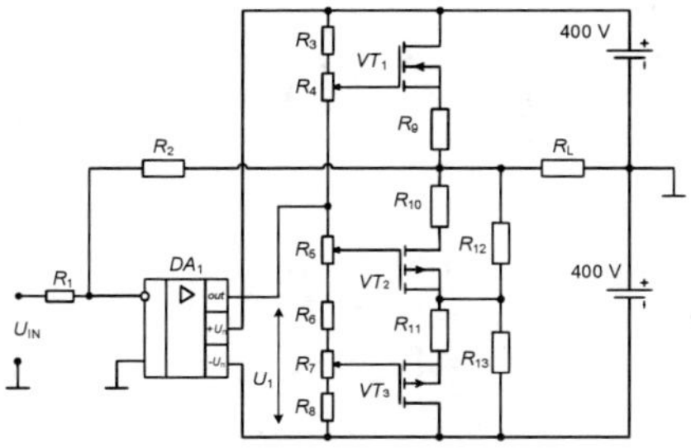

Figure 8. Circuits of the follower on different types of field-effect transistors.

Setting conditions $(R_3 + R_4) = (R_5 + R_6 + R_7 + R_8)$ and $(R_5 + R_6) = (R_7 + R_8)$ provide the voltage supplied to the VT_3 gate equal to $½ \cdot U_1$. In addition, half of the output voltage is to be supplied to the VT_3 source, which is carried out via the divider $R_{12} = R_{13}$. In order not to lose the power, high resistance is chosen.

The study of the circuit (figure 8) showed that it works well in class "AB" for active and capacitive load and is stable for standard op amp corrective circuits. Its setting is to be started with the follower, i.e. equal primary currents of the transistors are set with potentiometers, and then the op amp is connected and the overall feedback is closed. Figure 9 shows the gain and phase of this VVC for 40 gain and active-capacitive load (1 Ohm, 4 μf).

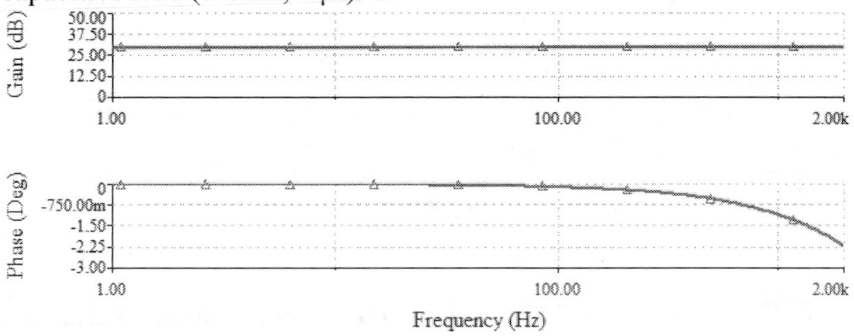

Figure 9. VVC gain and phase with active-capacitive load.

The graphs indicate that an additional phase shift up to 100 Hz is insufficient. Increase in the phase shift at higher frequencies (–2.25 degrees at 2 kHz) has a negligible effect on the overall measurement error.

It can be concluded that for group verification of static meters with this VVC design, a transformer is to be used (that is, galvanic decoupling unit); however, in this case, it is not included in the feedback loop and does not affect the system stability.

4. Conclusion

Testing of the above VVC diagrams used in various devices which require high output voltage (hundreds of volts), for example, power calibrators revealed different problems:

– additional phase shift when expanding the operating frequency;
– the stability of the diagrams in terms of deep NFB.

These problems require search not only for circuit solutions, but new electronics.

The circuit of the VVC suggested by the authors in figures 7 and 8:

– allow for a small additional phase shift, and thus improve the calibrator accuracy class;
– allow exclusion of the output transformer from the loop of the overall FB, which increases the system stability;
– work effectively not only for active, but also for resistive-capacitive load, therefore, they can be used to verify different types of meters;
– the diagram in figure8 can work in class "AB", which increases the VVC efficiency.

References

[1] Toth E, Franco A M R, Debatin R M (2005) Power and energy reference system, applying dual-channel sampling *IEEE Transactions on Instrumentation and Measurement* **54**(1) 404–408
[2] Svensson S and Ryder K E (1995) A measuring system for the calibration of power analyzers *IEEE Transactions on Instrumentation and Measurement* **44**(2) 316–317
[3] Svensson S 1999 *Power Measurement Techniques for Nonsinusoidal Conditions* (Chalmers tekniska högsk)
[4] Franco A M R, Tóth E, Debatin R M, Prada R (2001) Development of a power analyzer *11th IMEKO TC4 Symposium on Trends in Electrical Measurements and Instrumentation and 6th IMEKO TC4 Workshop on ADC Modelling and Testing 2001* 168–172
[5] Fomichev Yu M, Silushkin S V (2013) Investigation of possibilities to improve the Accuracy class of the Power Calibrator *The electronic scientific journal «Siberian Journal of Science»* **4** (10) 85–92 URL: http://sjs.tpu.ru/journal/article/view/817/575
[6] GOST 31819.22-2012 (IEC 62053-22: 2003) 2013 *Alternating-current measuring instruments. Particular requirements. P. 22. Static active energy meters of accuracy classes 0,2S and 0,5S. Intr. from 2014-01-01*(M.: Standartinform)
[7] PA89 Datasheet (PDF) – Cirrus Logic http://pdf1.alldatasheet.com/datasheet-pdf/view/304435/CIRRUS/PA89.html
[8] PA91 Datasheet (PDF) – Cirrus Logic http://pdf1.alldatasheet.com/datasheet-pdf/view/140209/APEX/PA91.html
[9] PA93 Datasheet (PDF) – Cirrus Logic http://pdf1.alldatasheet.com/datasheet-pdf/view/140210/APEX/PA93.html
[10] PA94 Datasheet (PDF) – Cirrus Logic http://pdf1.alldatasheet.com/datasheet-pdf/view/140211/APEX/PA94.html
[11] PA97 Datasheet (PDF) – Cirrus Logic http://pdf1.alldatasheet.com/datasheet-pdf/view/140213/APEX/PA97.html
[12] IRF740 Datasheet (PDF) – STMicroelectronics http://pdf1.alldatasheet.com/datasheet-pdf/view/22400/STMICROELECTRONICS/IRF740.html
[13] IRFU9310 Datasheet (PDF) – International Rectifier http://pdf1.alldatasheet.com/datasheet-pdf/view/68604/IRF/IRFU9310.html

Thermal and Mechanical Characteristics of Polymer Composites Based on Epoxy Resin, Aluminium Nanopowders and Boric Acid

O B Nazarenko[1], T V Melnikova[2] and P M Visakh[3]

[1]Professor of Ecology and Basic Safety Department, Tomsk Polytechnic University, Tomsk, Russia
[2]Student of Ecology and Basic Safety Department, Tomsk Polytechnic University, Tomsk, Russia
[3]Postdoc of Ecology and Basic Safety Department, Tomsk Polytechnic University, Tomsk, Russia
E-mail: obnaz@mail.ru

Abstract. The epoxy polymers are characterized by low thermal stability and high flammability. Nanoparticles are considered to be effective fillers of polymer composites for improving their thermal and functional properties. In this work, the epoxy composites were prepared using epoxy resin ED-20, polyethylene polyamine as a hardener, aluminum nanopowder and boric acid fine powder as flame-retardant filler. The thermal characteristics of the obtained samples were studied using thermogravimetric analysis and differential scanning calorimetry. The mechanical characteristics of epoxy composites were also studied. It was found that an addition of all fillers enhances the thermal stability and mechanical characteristics of the epoxy composites. The best thermal stability showed the epoxy composite filled with boric acid. The highest flexural properties showed the epoxy composite based on the combination of boric acid and aluminum nanopowder.

1. Introduction

Epoxy resins are thermosetting polymers which are widely used in industry as adhesives, surface coatings, construction materials due to such properties as high adhesion, chemical resistance, low shrinkage at curing, good mechanical properties [1, 2]. In the case of bonding the structural elements of the heating apparatus and electronic devices it is necessary to provide heat removal which requires the improving the thermal conductivity of adhesives. It has been shown in many works that the metal powders addition in the adhesives such as aluminum, copper, silver, etc. can improve mechanical, thermal and electrical properties of polymeric composites [3–7]. At the same time, the epoxy polymers are characterized by low thermal stability and high flammability.

To enhance both the functional and flammability properties of the epoxy polymers, nanoparticle fillers are considered to be promising [8–11]. The mechanism of thermal stability enhancement in polymer nanocomposites is associated with barrier effect, restriction of molecular motions, radical trapping, formation of protective char layer, sorption and catalytic effects [12]. Due to high specific

[1] To whom any correspondence should be addressed.

surface area of nanofillers and high surface-volume ratio, the interfacial area between the polymer and the nanofillers is greatly increased, and improvement of thermal, mechanical properties and flame retardancy of polymer nanocomposites can be achieved at very low fillers concentration compared to microfillers. However, nanoparticles alone cannot provide sufficient flame retardant effect, and it is recommended to use nanoparticles in combination with other flame retardants [12–14].

Boron compounds, including boric acid, are known as effective flame retardant additives [15, 16]. As a rule, boric acid is used in cellulosic products and coatings: wood, plywood, particle board, wood fibre, paper, cotton. At the heating, boric acid releases water in the endothermic process that reduces polymer temperature and degradation. It was found in our previous work [17] that the introduction of boric acid as a filler into epoxy resin by 1; 5; 10 % of its weight has a positive effect on the thermal stability of the composites. An increase of the temperature on 40.8 °C at 50 % weight loss and a decrease of 20.4 % weight loss at 600 °C were observed for the sample with a filler content of 10 wt. % in comparison with the pure epoxy polymer.

The aim of this work was to study the effect of aluminum nanopowder alone and in combination with boric acid on the thermal stability and mechanical strength of the epoxy composites.

2. Experimental

The materials used in this study were epoxy-diane resin ED-20 as polymer matrix, polyethylene polyamine (PEPA) as hardener (ZAO Contact, Dzerzhinsk, Russia), nanosized powder of aluminum (Al NP) and microsized powder of boric acid (BA) (Tulafarm, Tula, Russia) as fillers.

Al NP was produced by electrical explosion of wires [18, 19] in Tomsk Polytechnic University (Tomsk, Russia). The hydrogen was used as working ambient in the discharge chamber. Air-passivation was carried out by Ar+0.1 vol. % air directly after production with a view to protect nanoparticles against oxidation [19]. The properties of Al NP were studied in detail in [20]. The metal content in Al NP was 77 %. The crystalline aluminum oxide phase was not detected by X-ray diffraction analysis, but the presence of the oxide-hydroxide functional groups on the particles surface was confirmed by Fourier transform infrared spectrometry [20]. Besides, on the particles surface of Al NP water and atmospheric gases are adsorbed.

Boric acid was milled in a mortar before the sample preparation. According to the electronic microscopy study about 45 % of particles of BA had the size less than 40 μm [17].

The preparation of epoxy composites was performed as follows. The surface of fillers was modified by a hardener to improve the adhesion of the particles with the epoxy resin. Then, the required amount of fillers was added into epoxy resin. The epoxy resin with the filler was mixed by hand for 5 min at room temperature. After that PEPA was added into mixture and mixed again for 3 min. The ratio of epoxy resin and hardener was 10:1 by weight. The obtained mixtures were cured in the silicone molds at room temperature for 24 h. The formulations of the prepared epoxy composites are given in table 1. The images of the prepared samples are shown in figure 1.

Table 1. The formulations of the epoxy composites.

No.	Sample	ED-20 (wt. %)	Al NP (wt. %)	BA (wt. %)
1	E0	100	0	0
2	EA	100	5	0
3	EAB	100	5	10
4	EB	100	0	10

The obtained epoxy composites were studied by thermogravimetric analysis (TGA) and differential scanning calorimetry (DSC) using a thermogravimetric analyzer SDT Q600 (TA Instruments, USA). The thermal behavior of the epoxy composites was studied under non-isothermal linear heating (10 °C/min) in air atmosphere from room temperature to 900 °C.

To determine the mechanical properties of the composites a GOTECH AI-7000m universal testing machine (GOTECH Testing Machines Inc., Taiwan) was used. The mechanical test was carried out according to ISO 178:2010. The samples had the form of a rectangular beam with the size 40×16×2.5 mm. At least three samples of every composition were tested by three-point bending method. The cross-head rate was 2 mm/min.

a) b) c) d)

Figure 1. Images of the prepared samples: neat epoxy polymer E0 (a); Al NP filled sample EA (b); Al NP and boron acid filled sample EAB (c); boron acid filled sample EB (d).

3. Results and discussion

The results of thermal analysis for the neat epoxy polymer and epoxy composites are shown in figures 2 and 3. It can be seen that the degradation of the samples occurred in three stages. The first two stages are considered to lead to production of a stable at lower temperatures carbonaceous residue [21, 22]. The third stage is related to oxidation of remaining carbonaceous char.

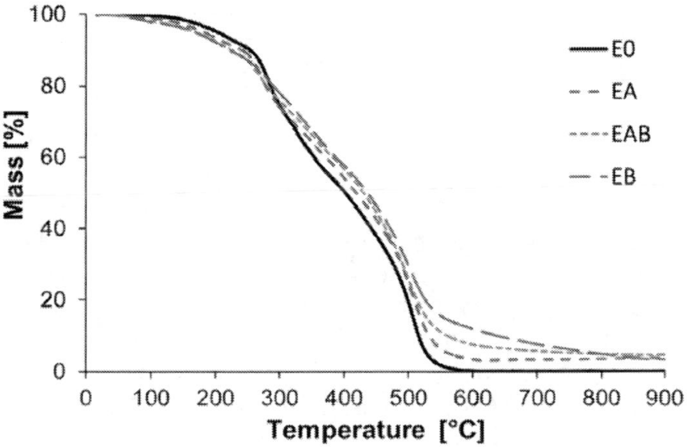

Figure 2. TGA curves of the epoxy composites.

It can be seen from figure 2 that addition of the fillers to the epoxy resin leads to increased mass loss at the heating to 280 °C because the fillers release water in this temperature range. The temperature at 5 % mass loss ($T_{5\%}$) decreases for the filled samples compared to that of the neat epoxy polymer. Evaporated water reduces the polymer temperature that leads to increasing the thermal stability of the composites at temperature above 280 °C.

The data of thermal analysis show that the introduction of Al NP has a positive effect on the thermal stability of the epoxy composites. The temperature at 50 % mass loss ($T_{50\%}$) is 419 °C and it is higher by 18 °C for the sample filled with Al NP than that of the neat epoxy polymer. The residue of the sample EA at 600 °C is 3.4 %, while the neat epoxy polymer decomposes completely.

Boric acid liberates water in an endothermic reaction during heating from 70 to 350 °C which leads to the dilution of combustible gases and provides heat sink. The temperature $T_{50\%}$ is 436 °C and higher by 28 °C for the sample filled with BA than for the neat epoxy polymer. The residue of the sample EB at 600 °C is 11.9 %.

In the case of the combination of Al NP and BA the temperature $T_{50\%}$ increases by 29 °C in comparison with the neat epoxy polymer, and the residual mass was 7.6 %.

DSC curves (figure 3) show three exothermic peaks for all samples. For the neat epoxy polymer E0 the first exothermic peak has maximum (T_{max1}) at 281 °C, the second peak has maximum (T_{max2}) at 410 °C and the third peak maximum (T_{max3}) at 506 °C. The introduction of all fillers results in shift of the exothermic peaks maxima T_{max1} and T_{max3} of the composites to lower temperature. However the shifting effect in T_{max3} for Al NP filled sample is insignificant. The peak maximum T_{max2} of the filled samples shifts to a higher temperature compared to that of the neat epoxy polymer.

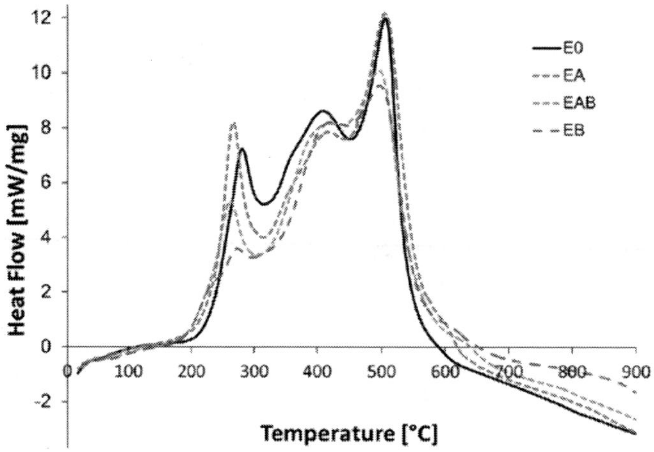

Figure 3. DSC curves of the epoxy composites.

The results of the mechanical tests are shown in table 2. The enhancement in flexural modulus is observed for all filled samples compared to that of the neat epoxy polymer. The flexural modulus for the composite filled with Al NP is 2.2 times higher than that of the neat epoxy polymer. The maximum improvement in flexural modulus is observed for the sample EAB. In the case of combination of Al NP with BA the flexural modulus is 2.4 higher than that of the neat epoxy polymer.

Table 2. The results of mechanical test.

Details	E0	EA	EAB	EB
Maximum Load [MPa]	36.1	50.9	36.5	29.0
Flexural Modulus [MPa]	569.3	1248.1	1380.2	901.5
Maximum stress at break [kgf]	8.9	11.9	11.1	8.7

The incorporation of nanosized aluminum particles in the epoxy resin provides the formation of finely crystalline structure of the polymer having fewer defects and leads to a restriction of molecular motions that result in enhancement of the thermal stability and mechanical properties. In addition, the presence of oxide hydroxide compounds on the aluminum nanoparticles surface and a high value of thermal conductivity of aluminum lead to the mentioned effects.

4. Conclusion

In the present paper, the epoxy composites filled with aluminum nanopowders, boric acid and aluminum nanopowders/boric acid were produced. The thermal stability of the epoxy composites was investigated by thermogravimetric analysis and differential scanning calorimetry. The mechanical test was carried out by three-point bending method.

The addition of all studied fillers increased the thermal stability and mechanical properties of the epoxy composites compared to the neat epoxy polymer. The epoxy composite filled with boric acid showed the best thermal stability. However, the combination of boric acid and aluminum nanopowders resulted in the highest flexural properties. The obtained results can be useful in developing the epoxy based composites with enhanced thermal stability and mechanical properties.

Acknowledgements
The authors are thankful to the Scientific and Analyzing Centre of Tomsk Polytechnic University for the providing the TGA and DSC measurements, and the Department of Technology of Organic Substances and Polymer Materials for the providing the mechanical test.

References

[1] May C A 1988 *Introduction to Epoxy Resins Epoxy Resins, Chemistry and Technology* 2nd ed C A May (New York: Marcel Dekker, Inc.)
[2] Lee H and Newill K 1982 *Handbook of Epoxy Resins* (New York: McGraw-Hill)
[3] Kilik R, Davies R and Darwish S M H (1989) Thermal Conductivity of Adhesive Filled with Metal Powders *Int. J. Adhes. Adhes.* **9**(4) 219–223
[4] Kilik R and Davies R (1989) Mechanical Properties of Adhesive Filled with Metal Powders I*nt. J. Adhes. Adhes.* **9**(4) 224–228
[5] Kahraman R, Sunar M and Yilbas B (2008) Influence of Adhesive Thickness and Filler Content on the Mechanical Performance of Aluminum Single-Lap Joints Bonded with Aluminum Powder Filled Epoxy Adhesive *J. Mater. Process. Tech.* **205** 183–189
[6] Kavak N (2013) Investigation of Effect to Mechanical Strength of Additive Powder Type into Adhesive *Int. J. Adv. Mater. Manuf. Charact.* **3** 53–55
[7] Fu Y-X, He Z-X, Mo D-C and Lu S-S (2014) Thermal Conductivity Enhancement with Different Fillers for Epoxy Resin Adhesives *Appl. Therm. Eng.* **66** 493–498
[8] Kashiwagi T, Du F, Douglas J F, Winey K I, Harris R H and Shields J R (2005) Nanoparticle Networks Reduce the Flammability of Polymer Nanocomposites *Nat. Mater.* **4** 928–933
[9] Morgan A B and Gilman J W (2013) An Overview of Flame Retardancy of Polymeric Materials: Application, Technology, and Future Directions *Fire Mater.* **37** 259–279
[10] Dasari A, Yu Z Z, Cai G P and Mai Y W (2013) Recent Developments in the Fire Retardancy of Polymeric Materials *Prog. Polym. Sci.* **38** 1357–1387
[11] Nazarenko O B, Amelkovich Y A, Ilyin A P and Sechin A I (2014) Prospects of Using Nanopowders as Flame Retardant Additives *Adv. Mat. Res.* **872** 123–127
[12] Pielichowski K, Leszczynska A and Njuguna J (2010) Mechanism of Thermal Stability Enhancement in Polymer Nanocomposites *Optimization of Polymer Nanocomposite Properties* ed V Mittal 195–210
[13] Camino G, Tartaglione G, Frrache A, Manferti C and Costa G (2005) Thermal and combustion behavior of layered silicate – epoxy nanocomposires *Polym. Degrad. Stabil.* **90** 354–362
[14] Toldy A, Anna P, Csontos I, Szabo A and Marosi Gy (2007) Intrinsically Flame Retardant Epoxy Resin – Fire Performance and Background – Part I *Polym. Degrad. Stabil.* **92** 2223–2230
[15] Awoyemi L and Westermark U (2005) Effects of Borate Impregnation on the Response of Wood Strength to Heat Treatment *Wood Sci. Technol.* **39** 484–491
[16] Shen K K, Kochesfahani S H, Jouffret F (2009) Boron Based Flame Retardants and Flame Retardancy *Fire Retardancy of Polymeric Materials* ed C A Wilkie and A B Morgan 2nd ed. (New York: CRC Press, Taylor & Francis Group) 207–237
[17] Amelkovich Y A, Nazarenko O B and Melnikova T V (2014) The Estimation of Influence of the Filler on the Thermal Stability of Epoxy Composites *Testing. Diagnostics* **13** 46–50
[18] Gromov A, Nazarenko O, Il'in A, Pautova Y and Tikhonov D (2014) Electroexplosive

Nanometals *Metal Nanopowders Production, Characterization, and Energetic Applications* ed A Gromov and U Teipel (Weinheim: Wiley-VCH Verlag GmbH & Co. KGaA) 67–78

[19] Kwon Y S, Gromov A A, Ilyin A P and Rim G H (2003) Passivation Process for Superfine Aluminum Powders Obtained by Electrical Explosion of Wires *Appl. Surf. Sci.* **211**(1–4) 57–67

[20] Nazarenko O B, Amelkovich Y A and Sechin A I (2014) Characterization of Aluminum Nanopowders after Long-Term Storage *Appl. Surf. Sci.* **321** 475–480

[21] Kandola B K, Biswas B, Price D and Horrocks A R (2010) Studies on the Effect of Different Levels of Toughener and Flame Retardants on Thermal Stability of Epoxy Resin *Polym. Degrad. Stab.* **95** 144–152

[22] Rallini M, Natali M, M. Monti, Kenny J M and Torre L (2014) Effect of Alumina Nanoparticles on the Thermal Properties of Carbon Fibre-Reinforced Composites *Fire Mater.* **38** 339–355

Journal of Physics: Conference Series **671** (2016) 012041 doi:10.1088/1742-6596/671/1/012041

Effect of Boric Acid on Volatile Products of Thermooxidative Degradation of Epoxy Polymers

O B Nazarenko[1], P B Bukhareva[2], T V Melnikova[3] and P M Visakh[4]

[1]Professor of Ecology and Basic Safety Department, Tomsk Polytechnic University, Tomsk, Russia
[2] Student of Ecology and Basic Safety Department, Tomsk Polytechnic University, Tomsk, Russia
[2] Student of Ecology and Basic Safety Department, Tomsk Polytechnic University, Tomsk, Russia
[3] Postdoc of Ecology and Basic Safety Department, Tomsk Polytechnic University, Tomsk, Russia
E-mail: obnaz@mail.ru

Abstract. The polymeric materials are characterized by high flammability. The use of flame retardants in order to reduce the flammability of polymers can lead to the formation of toxic gaseous products under fire conditions. In this work we studied the effect of boric acid on the volatile products of thermooxidative degradation of epoxy polymers. The comparative investigations were carried out on the samples of the unfilled epoxy resin and epoxy resin filled with a boric acid at percentage 10 wt. %. The analysis of the volatile decomposition products and thermal stability of the samples under heating in an oxidizing medium was performed using a thermal mass-spectrometric analysis. It is found that the incorporation of boric acid into the polymer matrix increases the thermal stability of epoxy composites and leads to a reduction in the 2–2.7 times of toxic gaseous products.

1. Introduction

Polymeric materials are widely used in industry and everyday life due to such properties as low density, high strength, chemical stability, ease of processing etc. A significant disadvantage of polymeric materials and products made of them is high flammability. Burning of polymeric materials in fire leads to environmental pollution by combustion products; it causes human injury and deaths as well as entails huge economic losses [1–4]. Thermal decomposition and combustion of polymers is accompanied by the release of heat, smoke and toxic gases such as carbon monoxide, carbon dioxide, formaldehyde, water, methane, ethylene, propylene, and others [5, 6]. Among the toxic gases carbon monoxide is considered the major cause of fire deaths.

Decrease in flammability and combustibility of polymers is an urgent problem. One of the methods of this problem decision is the incorporation of flame retardants into the polymers. The use of conventional halogenated flame retardants (bromine and chlorine) can lead to highly dangerous and persistent toxicants, having also carcinogenic action, such as phosgene, cyanogen compounds, dioxins. Therefore, current studies aimed at finding and developing environmentally friendly flame retardants [7–10].

[1] To whom any correspondence should be addressed.

Epoxy resins are thermosetting polymers which dominate in composite industry. They are widely used in electronic, aerospace, automotive industries and other structural applications. Epoxy resins possess good mechanical and electrical insulating properties, excellent adhesion, high chemical and corrosion resistance. They are applied as paint, coating, adhesive, matrices for composites. With the increasing epoxy resin market, it is of great importance to develop epoxy systems based on environmentally friendly flame retardants and study their thermal characteristics and volatile products under fire conditions. The evolved gases analysis by means of thermogravimetry, FTIR and mass spectrometry during thermal degradation of epoxy resin revealed that the main volatile product are carbon monoxide CO, carbon dioxide CO_2, water H_2O, formaldehyde CH_2O, a number of hydrocarbons [11–13].

The aim of this study was to investigate the influence of boric acid used as filler for reducing flammability, on gaseous products of thermal oxidative degradation of epoxy composites by means of a thermal analyzer coupled with a mass spectrometer. The dominant volatile products such as CO, CO_2, CH_2O were selected for the analysis and their evolution is discussed in this study.

2. Experimental

The samples of epoxy polymers were prepared from an epoxy resin ED-20. The curing of the epoxy resin was carried out using polyethylene polyamine (PEPA). The fine powder of boric acid was used as filler. In this study we investigated the thermal characteristics and the volatile products of two samples: 1) the sample E0 was the unfilled cured epoxy resin; 2) the sample EB10 was the epoxy resin filled with boric acid of concentration 10 wt. %. The studied samples were grinded to powder in mortar for all the experiments.

Analysis of gaseous products released during heating of the samples, was performed using a quadrupole mass spectrometer QMS 403 Aëolos coupled with a simultaneous thermal analyzer STA 449C Jupiter (Netzsch, Germany). Approximately 5 mg of the samples were placed in an Al_2O_3 crucible. The samples were heated from room temperature up to 800 °C at a heating rate of 10 °C/min in air with a gas flow rate of 50 mL/min. For mass spectrometry analysis, mass-to-charge ratio (m/z) values from 10 to 100 were collected and m/z intensities of interest (28, 30 and 44) were monitored with time.

3. Results and discussion

Figure 1 shows the evolution of the volatile products with mass-to-charge ratio m/z = 28; 30 and as a function of temperature. The ions with m/z = 28 and 44 correspond to carbon monoxide $[CO]^+$ and carbon dioxide $[CO_2]^+$, respectively. The value m/z = 30 indicates the release of ions of formaldehyde $[CH_2O]^+$. The main volatile gas released during thermooxidative process is carbon monoxide.

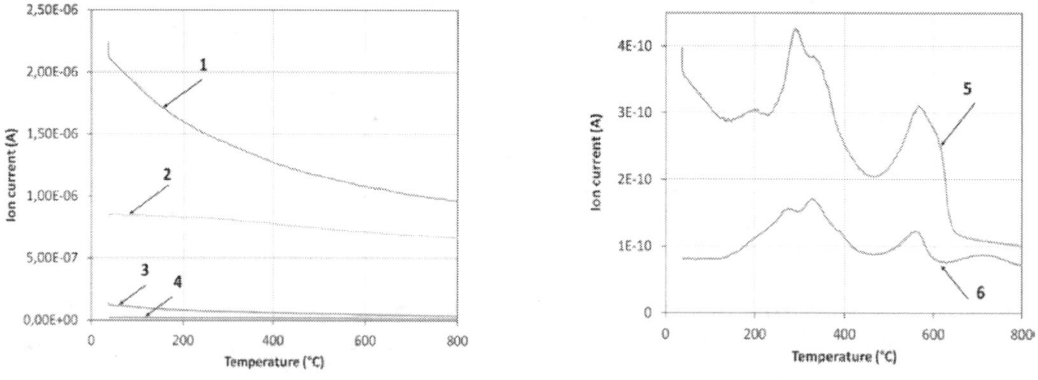

a) b)

Figure 1. Dependence of ions current on temperature a) m/z = 28 (1 and 2) and 44 (3 and 4); b) m/z = 30 (5 and 6): 1, 3, 5 – unfilled epoxy polymer; 2, 4, 6 – epoxy polymer filled with 10 wt. % boric acid.

The analysis of experimental data indicates that the incorporation of boric acid in the epoxy matrix as a filler decreases the intensity of the emission of volatile decomposition products of the epoxy samples at the heating. For the molecular ions $[CO]^+$ and $[CO_2]^+$ intense background signals are observed at mass thermograms. Nevertheless, the intensity of signals for the filled sample EB10 is less in 2–2.5 times than for the unfilled sample.

The analysis of the intensity of the ion current for the molecular ion of formaldehyde $[CH_2O]^+$ revealed that its emission during thermooxidative destruction of the samples occurs in three stages. Table 1 shows the results of analysis.

According to the data of table 1 the incorporation of boric acid in the epoxy matrix resulted in a decrease of the maximum temperature of the first peak of the ion current curve by 16 °C and the third peak by 8 °C. A decrease of the ion current over the entire temperature in 2.3–2.7 times is observed.

Table 1. The intensity of the ion current for the gaseous decomposition product with the mass number m/z = 30

Sample	Stage 1		Stage 2		Stage 3	
	T_{max} (°C)	I (A)	T_{max} (°C)	I (A)	T_{max} (°C)	I (A)
E0	288	$4.26 \cdot 10^{-10}$	330	$3.86 \cdot 10^{-10}$	568	$3.11 \cdot 10^{-10}$
EB10	272	$1.57 \cdot 10^{-10}$	330	$1.7 \cdot 10^{-10}$	560	$1.23 \cdot 10^{-10}$

Figure 2 shows the dependence of mass changes (TG) and heat release (DSC) during heating of the samples on the temperature. The results of thermal analysis of the studied samples are shown in table 2.

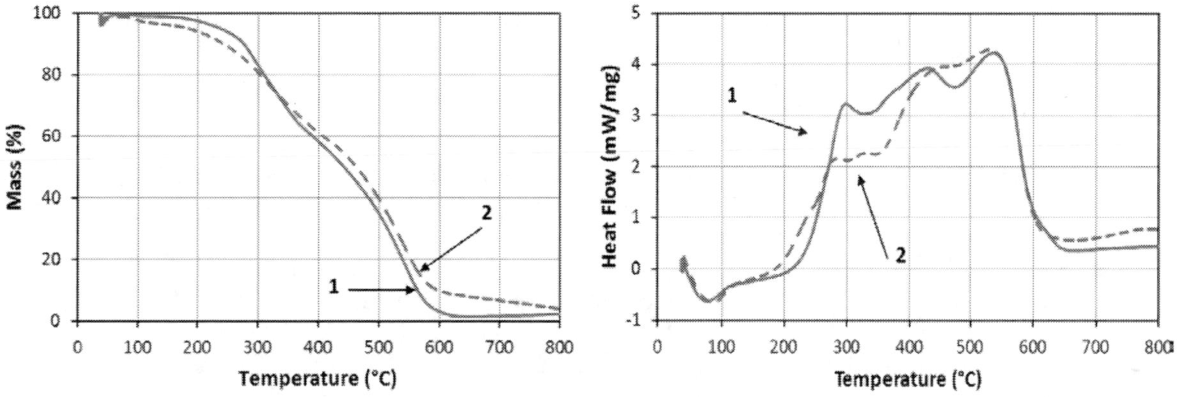

a) b)

Figure 2. TG (a) and DSC (b) curves of the epoxy composites: 1 – unfilled epoxy polymer; 2 – epoxy polymer filled with 10 wt. % boric acid.

Table 2. The results of thermal analysis

Sample	Stage 1		Stage 2		Stage 3		Residue at 600 °C (%)
	Temperature range (°C)	T_{max} (°C)	Temperature range (°C)	T_{max} (°C)	Temperature range (°C)	T_{max} (°C)	
E0	236–328	298	328–473	429	473–603	535	3.0
EB10	196–345	284	345–467	430	467–600	530	9.8

The process of thermooxidative degradation occurs in three stages with heat release in each of them. The greatest amount of heat is released upon heating the samples on the second and third stages, which correspond to the oxidation of carbon residue. According to the thermal analysis data, the onset temperature of degradation (corresponding to the 5 % of mass loss) for the filled sample EB10 is lower by 40 °C than that for the unfilled epoxy sample. The residual mass of the filled sample EB10 at 600 °C is 9.8%, which is higher by 6.8% than for the unfilled sample E0. The increased residue yield of the sample EB10 compared with the unfilled sample E0 also indicates the reducing amount of the gaseous products. The total specific heat effect for the sample E0 is 15112 J/g, and for the sample EB10 is 13086 J/g. The thermal effect of the first stage for the filled sample is less in 1.5 times than for the unfilled epoxy resin.

These results are explained by the fact that at the heating of the filled sample endothermic decomposition of boric acid occurs with release and evaporation of water and the consequent formation metaboric acid HBO_3 and boron oxide B_2O_3 [14]. This contributes to cooling of the polymer surface and dilution of volatile products of the polymer degradation. The formation of non-flammable boron oxide layer reduces the flame spread of polymer surface.

4. Conclusion

We have investigated the effect of fine powder of boric acid used as filler in the epoxy resin, on gaseous products formed upon heating of the epoxy polymers in air up to 800 °C. The thermooxidative degradation characteristics and the volatile products were studied by means of thermogravimetry and differential scanning calorimetry coupled with mass spectrometer. The incorporation of boric acid into epoxy resin resulted in increased yield of the residue after thermooxidative degradation and reducing the amount of volatile products. The analysis of the gaseous products of thermal oxidative degradation indicates a decrease in 2–2.4 times of toxic gaseous products such as carbon monoxide CO, carbon dioxide CO_2, formaldehyde CH_2O.

Acknowledgements

The authors are thankful to the Scientific and Analyzing Centre of Tomsk Polytechnic University for the providing the TGA and DSC measurements, and the Department of Silicate and Nanomaterials Technology of Tomsk Polytechnic University for the providing the mass spectrometry measurements.

References

[1] Nelson G L, Morgan A B and Wilkie C A (2009) Fire Retardancy in 2009 *ACS Symposium Series* **103** 1–7

[2] Hull T R and Kandola B K (2009) *Fire Retardancy of Polymers: New Strategies and Mechanisms* (Cambridge: RSC Publishing)

[3] Irvine D J, McCluskey J A and Robinson I M (2000) Fire Hazards and Some Common Polymers *Polym. Degrad. Stabil.* **67** 383–396

[4] Mouritz A P and Gibson A G 2006 *Fire Properties of Polymer Composite Materials* (Dordrecht: Springer Netherlands)

[5] Morikawa T (1988) Evolution of Toxic Gases from Burning Polymeric Materials in a Small-Scale Box Model *Fire and Mater.* **12**(2) 43–49

[6] Horrocks A R and Price D 2001 *Fire Retardant Materials* (Cambridge: Woodhead Publishing Limited)

[7] Papaspyrides C D and Kiliaris P 2014 *Polymer Green Flame Retardants: A comprehensive Guide to Additives and Their Applications* (Oxford: Elsevier)

[8] Zaikov G E and Lomakin S M 2002 Ecological Issue of Polymer Flame Retardancy *J. Appl. Polym. Sci.* **86**(10) 2449–2462

[9] Betts K S (2008) New Thinking on Flame Retardants *Environ. Health Perspect.* **116** (5) A210–A213

[10] Segev O, Kushmaro A and Brenner A (2009) Environmental Impact of Flame Retardants

(Persistence and Biodegradability) *Int. J. Environ. Res. Public Health* **6** (2) 478–491

[11] Ahamad T and Alshehri S M (2013) Thermal Degradation and Evolved Gas Analysis of Epoxy (DGEBA)/novolac Resin Blends (ENB) During Pyrolysis and Combustion *J. Therm. Anal. Calorim.* **111** (1) 445–451

[12] Lee L H (1965) Mechanism of Thermal Degradation of Phenolic Condensation Polymers. II. Thermal Stability and Degradation Schemes of Epoxy Resin *J. Polym. Sci. A* **3**(3) 859–882

[13] Mouritz A P, Mathys Z and Gibson A G (2006) Heat Release of Polymer Composites in Fire *Composites Part A* **37** (7) 1040–1054

[14] Shen K K, Kochesfahani S H and Jouffret F 2009 Boron Based Flame Retardants and Flame Retardancy *Fire Retardancy of Polymeric Materials* 2nd ed C A Wilkie and A B Morgan (Baco Raton: CRC Press)

Monitoring of Carrying Cable in the Well by Electric Drive of Winch at the Logging Works

I G Odnokopylov[1]and V V Gneushev[2] and I A Larioshina[3]

[1]Associate Professor, Institute of Power Engineering, Tomsk Polytechnic University, Russia, Tomsk,
[2]Senior Lecturer, Institute of Power Engineering, Tomsk Polytechnic University, Russia, Tomsk,
[3]Assistant, Tomsk State University, Russia, Tomsk
E-mail: odivan@yandex.ru

Abstract. Emergency situations during logging operations are considered. The necessity of monitoring of the carrying cable in the well was shown, especially at the jet perforation and seismic researches of wells. The way of monitoring of logging cable and geophysical probe by means of the electric drive of tripping works of the logging winch is offered. This method allows timely to identify the wedges of geophysical equipment and the tension of the cable in well without interfering into construction of logging installation by means of algorithmic processing of sensors of electric drive. Research was conducted on the simulation model; these results indirectly confirm the possibility of using of electric drive for monitoring of downhole equipment.

1. Introduction

The logging is the most common type of geophysical research of well. Logging represents detailed research of structure of cut of well by means of descent and rise of geophysical probe in it. Tripping works are conducted with using of the logging winch [1, 2]. The logging winch is used at carrying out perforation of well and at seismic researches of wells and crust, thus perforators fall on carrying logging cable.

In the process of construction, exploration and exploitation of oil and gas wells the emergencies occur associated with breakage of the logging cable, the fall of downhole equipment, etc. [3].

Very often these accidents occur during cumulative perforation and seismic researches of wells. The source of seismic fluctuations is lowered into the well at the seismic exploration; it makes a series of explosions there [4]. There is a need to carry out the rescue and recovery operations because of complications at conducting of logging operations; this causes by downtime and additional costs. Basic operations at the elimination of such accidents are the fishing operations which follow the preparatory works (shutoff of well; determining of location of break and fall of downhole equipment and its condition; milling of emergency tool, etc.).

For monitoring of the condition of logging cable and downhole equipment the logging installations can be fitted with different systems of control and registration of parameters, such as the tension of the cable, the depth and speed of movement of the geophysical probe, etc. The main part of similar registrars is intended for collecting and initial processing data received from the downhole equipment. It allows increasing efficiency of carrying out geophysical works. However, the imposition of additional methods of monitoring the condition of the logging cable and geophysical probe are the urgent task, because the cable breaks with consequent loss of downhole equipment are the common problems [3]. This problem is especially relevant at the seismic researches, when the gas-dynamic source of fluctuations is used (*GDSF*).

In this paper we propose to use the electric drive of hoisting winch as a monitoring tool. Currently a large part of the logging winch is equipped with an asynchronous frequency-controlled drive, which already contains the current sensors, voltage and encoder mounted on the reel of winch. This set of sensors allows making the additional monitoring of the logging cable. The proposed method does not require additional investments and changes in the design of the winch; it requires only a software algorithm for processing information, which you can lay into microprocessor-based control system of electric drive.

2. Research object and method

Functional scheme of logging winch is shown in figure 1. The scheme works as follows. Electrical transducer 1 receives power from three-phase AC power. The converter supplies the power to the induction motor 2 after the command for the movement of the operator. The motor rotates the gear 3 and the reel 4. The cable 5 is placed (wound) on the reel by using the cable layer 7. The geophysical probe 9 is fixed on a cable and it is lowered into the well 8. The position sensor (encoder) 6 is mounted on the shaft of the reel. Information from this sensor is fed to the frequency converter 1.

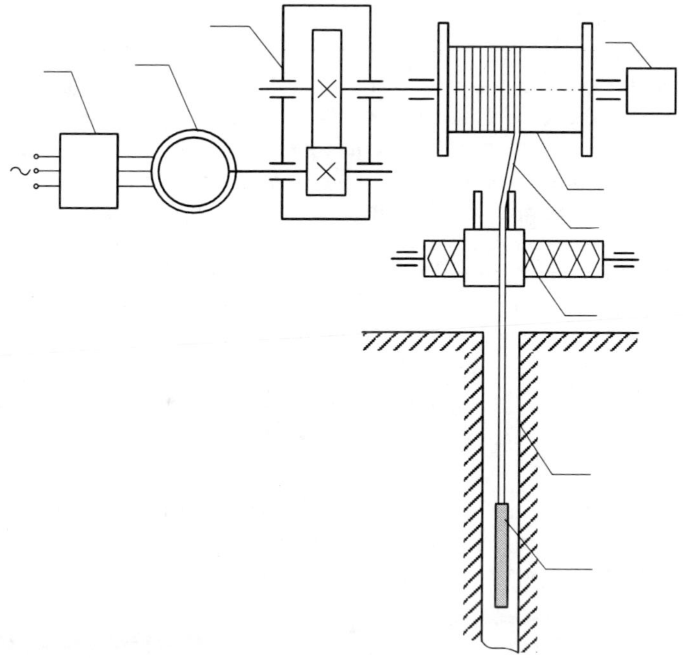

Figure 1. Functional scheme of logging winch: 1 – electric transformer; 2 – electric motor; 3 – reducer; 4 – reel; 5 – cable; 6 – position sensor; 7 – cable layer; 8 – well; 9 – geophysical probe.

The possibility of monitoring of logging cable and geophysical probe during tripping works on the logging installation is shown below. Electric AC drive typically contains three of the current sensor in each phase, three voltage sensor and encoder for feedback [5, 6]. The rise of geophysical probe with continuous recording of parameters is produced using a special above-ground electrical at conducting of geophysical research of well at a low speed. At rising of probe the subwedges and jams are possible at which the probe stops the movement. Thus the motor continues to rotate, the tension of the cable occurs. Indirect definition of load of the motor by means of current sensors can serve as alternative to the cable tension sensor. Phase currents of the motor are increased during the jamming of probe; it corresponds to increase of load, the subwedge of probe occurs. Timely shutdown of process of tripping works can save the geophysical equipment.

Deformation of the logging cable can be identified at the qualitative processing of the marks from position sensor. Descent and rising of probe have to happen evenly. The manifestation of vibrations, disruption or acceleration can be interpreted as the appearance of deformation and damage in logging cable. For detection of such faults it is need to install the position sensor as close as possible to the output link of the mechanism in this case on the reel and not on the motor.

The possibility of monitoring of logging cable and geophysical probe during seismic researches on the logging installation is shown below [7, 8]. The gas-dynamic source of fluctuations is used as a probe. The electric drive is switches-off after lowering of *GDSF* to certain depth and the series of explosions are produces with simultaneous registration of seismic fluctuations on the earth's surface. The probability of breakage of the logging cable is especially great due to shock fluctuations of the device *GDSF*. In this case, the use of electric drive as the tool allows not only to monitor the situation inside the well, but to soften the shock loads. Some researches of shock loadings are considered in [9, 10].

Alternative options at various executions of electric drives are shown below.

It is necessary to use a vector control system of the motor with feedback on speed with using of system "the frequency converter – the asynchronous motor" as the drive of the rise mechanism. Thus the zero speed of set is established at the set depth. The motor makes work with retention of a cable and the device in a motionless state. The fluctuations of device are at the explosion. In this case the electric drive is a damper; the coiling and uncoiling of cable are depending on the applied load on the motor during fluctuations of the device. Thus the resource of a cable increases; the probability of breakage of geophysical probe decreases. The coefficient of damping will depend on settings of regulators of current and speed of microprocessor control system.

It is necessary to enable a dynamic braking at carrying out the blasting operations with using of system "thyristor voltage regulator – the asynchronous motor" as the drive of the rise mechanism. It is possible to achieve a motionless condition of mechanical part due to use of a direct current. The value of DC current must be sufficient to hold the cable and the device in a stationary state, but the motor should not overheat. It can achieve by selection of accurate angle of system of pulse and phase control. This angle is formed automatically by the encoder on the motor shaft or reel. The electric drive is a damper at the explosion. In certain degree the processes will be similar to the process of electric drive with system "the frequency converter – the asynchronous motor".

3. Results and considerations

Researches of simulation model with help of the software MatLab were conducted for check of possibility of monitoring of logging cable and geophysical probe. The block diagram of the mathematical model of the electric drive of winch is presented on figure 2 [11].

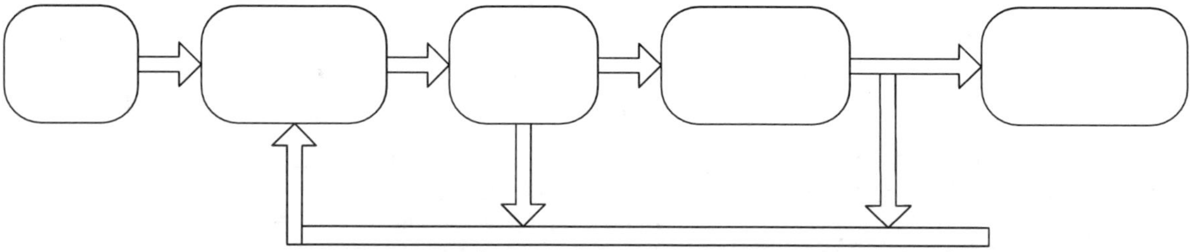

Figure 2. Block diagram of the mathematical model of the electric drive winches.

The simulation model includes a source of electric energy (three-phase alternating current network), the electric converter which transforms the variable voltage of one frequency to tension of other frequency, the asynchronous motor and two-mass mechanical system. The mechanical system consists of a first mass: motor, reducer, chain transmission, reel, etc. The device and a cable are the second mass, which is lowered into the well. The cable is the elastic element. Cable length, stiffness, and other mechanical parameters vary depending on the position of the cargo. The model allows dynamically monitoring the start processes, the stop of logging winch, reset and pounce of load, which may correspond to the real process of subwedge of the device.

Electrical converter receives the signals from the current sensors of the motor and the speed sensor. The speed sensor can be installed on the motor shaft and the shaft of reel, which is preferable from the point of view of diagnosis of subwedge of device in the well. In the electric converter the control system provides the necessary voltage to the motor depending on the condition of sensors and a set algorithm of control. In turn the motor sets in motion the device in a well.

Figure 3 shows the transient processes of the electric drive of logging winch at the jamming of the device during lifting of the device from the well.

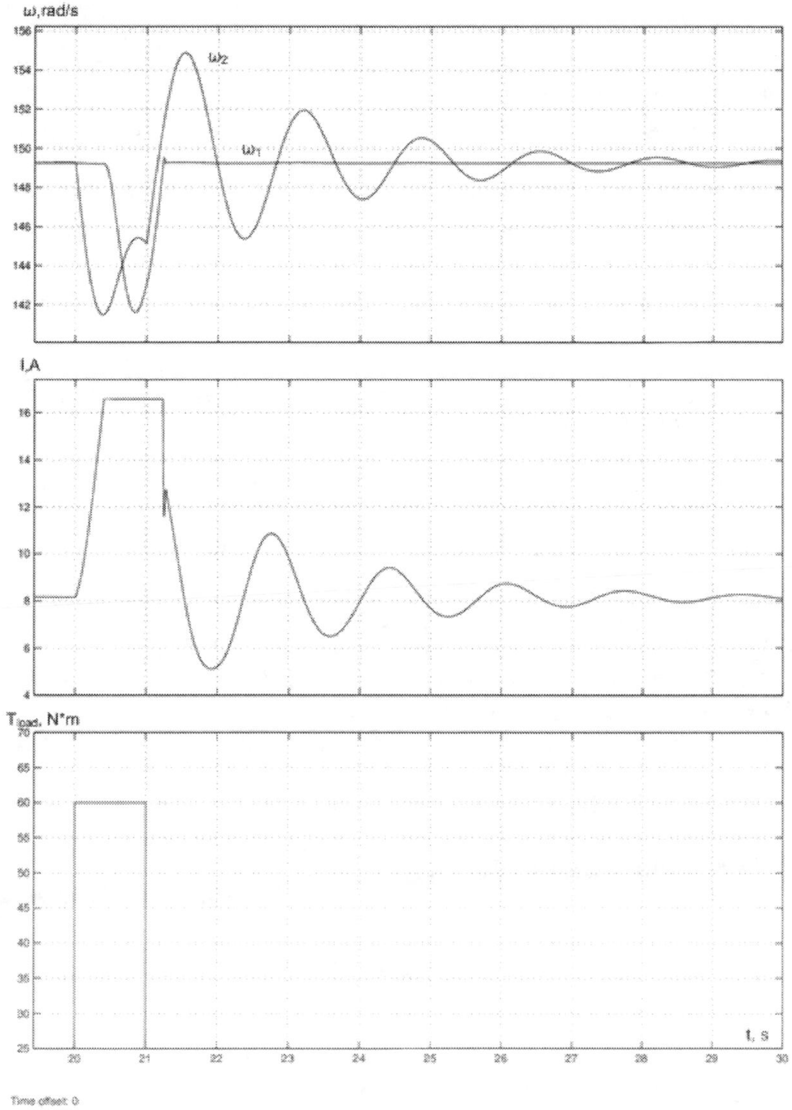

Figure 3. Transient processes of electric drive of logging winch at subwedge of device in well.

The notation adopted In figure: ω_1 – frequency of rotation of the motor (the first mass), ω_2 – frequency of rotation of the second mass, I – stator current of the asynchronous motor, T_{load} –torque of load. Jamming of device is simulated with the load increase, i.e. with increasing of

T_{load}. In this case T_{load} increases from 25 to 60 $N{\cdot}m$ at time of 20 s. Accordingly, the speed of the device ω_2 begins to decrease. At this time the rope is pulled, the motor current is gradually increased; the motor speed does not change. After 0.5 s the motor speed decreases; the current reaches a maximum value of 16.25 A, increasing from 8 A. The contour of current is in a saturated state. At time of 21 s the load reverts to the original state; the current, speeds of motor and device are leveled after the oscillatory process in the electromechanical system.

This example shows the possibility of monitoring the process of logging operations by means of the electric drive. Subwedge of device or complete jamming of the device in the well can be diagnosed in advance if information of stator current and speed of rotation is known. Thus, there is a possibility of change of the mode of operation of electric drive and the full shutdown of system can be executed.

4. Summary

1. The way of monitoring of logging cable and geophysical probe by means of electric drive of tripping works of logging installation is offered.
2. It is shown that use of electric drive at seismic researches allows reducing shock fluctuations of the device, thus, the cable resource increases, and the probability of geophysical probe break decreases.
3. Research of imitation model was conducted, and it was indirectly confirmed the possibility of using of electric drive for monitoring of downhole equipment.

References

[1] Baird T *et al* (1998) High-Pressure, High temperature, well logging, perforating and testing, *Schlumberger Oilfield Review* **10**(2)50–67
[2] Darwin V E, Singer J M 2007 *Well Logging for Earth Scientists* (Springer Science & Business)
[3] Babin C, Sarian S (2015) Cables and Skates – Improving the Weakest Links *Oilfield Review*
[4] Kholodilov V A *et al* (2006) *Mining Industry Journal* **2** 74–76
[5] Odnokopylov I G *et al* (2015) *Fundamental Research* **2** 1392–1396
[6] Rinchen G D (2014) Monitoring and Control of a Variable Frequency Drive Using PLC and SCADA *IJRITCC* **2**(10) 3092–3098
[7] Blackburn J, Daniels J (2007) *Borehole seismic Surveys: Beyond the Vertical Profile in Oilfield Review* (Elsevier) **21**
[8] Hickman S H *et al* (2012) Scientific basis for safely shutting in the Macondo Well after the April 20, 2010 Deepwater Horizon blowout *PNAS* **109 (50)** 20268–20273
[9] Zhang C P *et al* (2013) *Applied Mechanics and Materials* **401–403** 36–40 doi: 10.4028/www.scientific.net/AMM.401–403.36
[10] Michalowski S, Cichocki W (2013) The peak dynamic loading of a winch in term of the rope flexibility *Key Engineering Materials* **542** 105–117
[11] Odnokopylov I G *et al* (2015) Load balancing of two-motor asynchronous electric drive *IEEE International Siberian Conference on Control and Communications (SIBCON-2015). Proceedings* 1–4 doi: 10.1109/SIBCON. 2015.7147249

Features of X-ray Absorption Densitometry of Large-size Objects with Variable Thickness

S Osipov[1], S Chakhlov[2], O Osipov[3], S Badin[4] and A Abashkin[5]

[1]Leading reseacher, NDT Institute, Tomsk Polytechnic University, Russia
[2]Head laboratory of technical tomography and introscopy, NDT institute, Tomsk Polytechnic University, Russia
[3]Engineer-reseacher, NDT Institute, Tomsk Polytechnic University, Russia
[4]Postgraduate, NDT Institute, Tomsk Polytechnic University, Russia
[5]Undergraduate, NDT institute, Tomsk Polytechnic University, Russia
E-mail: osip1809@rambler.ru

Abstract. Features of formation and processing of the primary radiometric signals in the digital high-energy X-ray absorption densitometers for the homogeneous objects with variable thickness are examined. The densitometer's equation based on the polynomial approximation of the object's ray thickness dependence from its mass thickness is proposed. Guidance to select the capacity of the analog-digital converter is given. There is one example of the densitometer's equation coefficients calculation to examine the carbon, aluminum and steel wares with the mass density from 15 to 80 g/cm^2. It was shown that disagreement of the experimental and estimated values of the ray thickness for the similar mass thicknesses of the testing object is conditioned by the scattered radiation. On the high-energy digital radiography set with the X-ray source – the betatron MIB-4.5/9 the accuracy of the experimental estimation of the density was within 0.0086 g/cm^3 for the steel ware thickness from 25 to 100 mm.

1. Introduction

The problem of calculation of density and values related with density is typical for many branches of science, industry, construction engineering and it is solved efficiently by methods based on the gamma radiation attenuation measurement [1, 2]. Currently the digital radiography methods with X-ray sources are widely using to measure the density [3–5]. There are some factors [6–7] whose essentially affect value of the systematic inaccuracy of the testing objects (TO) parameters calculation for X-ray or gamma radiation absorption. During examination of TO with variable thickness the most important is the effect of X-ray radiation beam hardening which influence is reduced by few modes. The X-ray computational tomography uses the compensating filters [8], but this method is hard-to-use. The dual energy method (DEM) allows to calculate the density and the material's effective atomic number of TO at the same time [9]. The spectrometric method [10] is based on the energy spectrum recording. This method allows to minimize the analyzed effect influence on the parameter's estimation accuracy by the X-ray absorption method. The spectrometric version of the counting X-ray registration mode is differed by low performance. The analyzed effect compensation method based on the preliminary calibration is most productive and simple in implementation. Until now, there is a lack of discussion about features of the generation and processing of primary radiometric signals for the digital X-ray absorption densitometers of the large-size homogeneous objects with variable thickness.

2. Generation and processing of the radiometric information in X-ray absorption densitometers

The simplified geometry of the X-ray absorption densitometer is shown on figure 1. As example of the TO fragment there is selected a rectangular parallelepiped with dimensions $H\times A\times B$.

Initial data. The point source of high-energy X-ray radiation with energy spectrum $f(E,E_0)$, where E_0 is the maximal energy is located on distance F from scintillation detector. The axis of symmetry of the bar-shaped scintillation detector is directed to the radiation source, lateral dimensions are $a\times b$, the thickness is h. The geometric parameters of the examination setup are satisfied the following conditions $F >> h >$ (a and b). The slit collimator has length L, thickness D and gap width d. The TO thickness H is varied in the range from H_{min} to H_{max}.

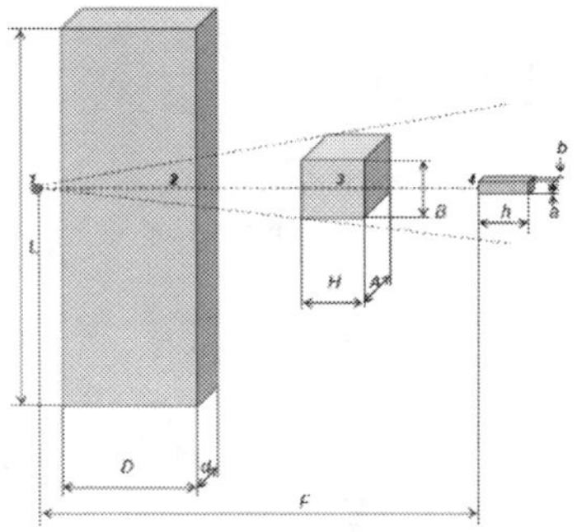

Figure 1. Geometry of X-ray absorption densitometer: 1 – radiation source; 2 – slit collimator; 3 – testing object; 4 – scintillator.

2.1. Generation, digitization and calibration of the measuring signals

The generation of the analog radiometric signal. The energy lost in scintillation detector by the X-ray photons is converted to the optical radiation. The optical radiation is transformed to the current by the photo-detector. The analog signal value on the radiometric detector output is depended from TO thickness H and its density ρ, the maximal energy of X-ray radiation E_0, the scintillator thickness and its material. The value ρH is the product of density ρ on thickness H is called the mass density and is measured in g/cm^2. The analog signal on the radiometric detector output $J(E_0,\rho H)$ equals to sum the signal $I(E_0,\rho H)$ defined by absorbed energy of the registered photons and by the own detector noises p

$$J(E_0,\rho H) = I(E_0,\rho H) + p\,. \tag{1}$$

The level of the own detector noise p can be estimated without X-ray radiation or with absolute detector protection from photon radiation. Just second approach confirms the correctness of $p=J(E_0,\infty)$.

Besides the own detector's noises level p it is interesting the output signal of the radiometric detector produced without the TO as provided by (1) and this signal logically can be denoted $J(E_0,0)$.

Digital signal generation. The output detector signal is digitized by the analog-digital converter (ADC). The main ADC feature is its capacity K. The sampling rate Δ is defined by the ADC capacity K and by the variation interval of the digitized signal (min$J$, max$J$). Evidently that the analog signal value is maximal for H=0. The zero level is selected as the minimal value of the analog signal. Hence it follows that the sampling rate is defined by the expression

$$\Delta = \frac{C_k J(E_0,0)}{2^K - 1}\,, \tag{2}$$

here C_k, $C_k > 1$ is a coefficient which allow possible variations of the analog signal maximum.

The transformation of the analog signal J to the digital signal J_d is described by formula

$$J_d = \left[\frac{J}{\Delta}\right], \tag{3}$$

where $[x]$ is integer part of number x.

The primary calibration of the measuring signal. On first stage of the primary calibration two parameters are defined: the own noise level of the radiometric detector $J_d\,(E_0,\infty)$ and the signal level without the TO $J_d(E_0,0)$. The mentioned levels were estimated many times therefore we can consider that errors of the sample mean values $\overline{J_d(E_0,\infty)}$ and $\overline{J_d(E_0,0)}$ are close to zero. The second stage is the "black" calibration which reduced to subtraction value $\overline{J_d(E_0,\infty)}$ from the digital measuring signal. On third stage the "black" calibrated digital measuring signal is normalized on the "black" calibrated digital signal without the TO. On fourth stage we find the logarithm of the "black" and "white" calibrated measuring signal. The total signal transformation is described by the expression

$$Y_d(E_0,\rho H) = \ln\frac{J_d(E_0,0) - J_d(E_0,\infty)}{J_d(E_0,\rho H) - J_d(E_0,\infty)}. \quad (4)$$

The value $Y_d(E_0,\rho H)$ called the ray thickness of the TO for the X-ray radiation with maximal energy E_0. The dependence of the ray thickness Y_d from the mass thickness ρH is the backgroud for all following analysis of the X-ray absorption densitometer.

2.2. The equation of X-ray absorption densitometer

The most common case of the equation connecting the material density, the TO thickness and the TO ray thickness expresses like that

$$P(\rho H) = Y, \quad (5)$$

here P is function of the object mass thickness ρH for the fixed value E_0. For the monotone continuous function P there exist the inverse function P^{-1}. The equation to calculate density ρ has form

$$\rho = \frac{P^{-1}(Y)}{H}. \quad (6)$$

The function P^{-1} can be created by the testing results of the specially designed graded calibration sample. This calibration sample must be made from the same material as TO. As the testing results of the calibration sample there are generated a set of pairs $(\rho H_i, Y_i)$, i=1…n, here n is number of steps in the graded calibration sample. We shall examine the polynomial approximations of the function $P^{-1}(Y)$.

The systematic inaccuracy of the density estimation conditioned by the approximation error of P^{-1} by some type function is depended from the type of function G, the object mass density range, the approximation function parameters number, the number and the calibration sample step levels. The influence of the abovementioned factors on the density precision can be examined by experiment.

The criterion of the maximal pointwise deviation of the calculated density values from the nominal ones is the most reasonable

$$\Delta_\rho = \max_{\rho H_{\min} \le \rho H \le \rho H_{\max}} \left|\frac{G(Y)}{H} - \rho\right|. \quad (7)$$

The selection of approximation type $G(Y) \approx P^{-1}(Y)$ is impossible without the analysis of functions $P(\rho H)$ the experimental dependencies of TO ray thickness Y from the mass thickness ρH.

3. The dependence *P(ρH)* calculation

3.1. The formula to calculate the ray thickness of the testing object

For the fixed maximal X-ray radiation energy E_0 the formula to calculate the TO ray thickness Y with the mass thickness ρH are

$$Y = F(\rho H) = \ln\int_0^{E_0}\frac{\overline{E}_{ab}(E)}{E}f(E,E_0)\left(1-e^{-\mu_{sc}(E)h}\right)dE - \ln\int_0^{E_0}\frac{\overline{E}_{ab}(E)}{E}f(E,E_0)e^{-m(E)\rho H}\left(1-e^{-\mu_{sc}(E)h}\right)dE, \quad (8)$$

here $\overline{E}_{ab}(E)$ is the mean value of the energy loosed by the registered photon in scintillator; $m(E)$, $\mu_{sc}(E)$ are the mass and linear attenuation coefficients of the photon radiation with energy E by the TO material and by the scintillator material, correspondingly. The expression (8) is valid for the sampling rate Δ=0. The calculation for the values of the sampling rate Δ must allow the analog to digital signal conversions that based on expressions (3), (4).

The usage of the tables of the interaction cross-section of gamma radiation with material [11] and the spline-interpolation is reasonable to calculate the ray thickness by the formula (8). Note that value $\overline{E}_{ab}(E)$ is depended from photon energy, material and sizes of scintillator [12], and for the detectors with small lateral dimensions or small thickness it is necessary take into account not only the secondary photon leakage, but also the secondary electron leakage [13]. The dependencies $\overline{E}_{ab}(E)$ are defined in the first approximation by the tables [11]. The relation $\overline{E}_{ab}(E)/E$ is no other than the part of energy lost in the detector by the registered photon.

The selection of ADC capacity is need to convert the analog signal to digital ones.

3.2. *The selection of ADC capacity*

One of the component of the density estimation error is the analog to digital signal conversion error Δd. The maximal density measurement error $\Delta\rho$ is defined on design phase of the densitometer. It is logical to claim the restriction $\Delta_d << \Delta_\rho$, for example, $\Delta_d < 0,3\Delta_\rho$ for all range of the thickness H. The ADC capacity for the examined task is adequate if satisfy the condition

$$\left[\frac{J(E_0,(\rho+\Delta_d)H_{max})(2^K-1)}{C_k J(E_0,0)}\right]-\left[\frac{J(E_0,\rho H_{max})(2^K-1)}{C_k J(E_0,0)}\right]\geq 1. \tag{9}$$

Remind the $[x]$ is integer part of the number x.

Let test the feasibility of (9) for the following conditions: K=16; the steel ware with mass thickness ρH_{max}=80 g/cm^2; the steel density 7.86 g/cm^3; E_0=4.5 and 9 MeV; Δ_d=0.002 g/cm^3; C_k=1.5. For those conditions the left part of inequality (9) is equal 1 for E_0=4.5 MeV and 2 for E_0=9 MeV, that is ADC capacity is enough for the example requirements.

3.3. *The example of Y(ρH) calculation*

To analyze the function type $P^{-1}(Y)$ the calculations of dependencies $Y(\rho H)$ in respect to the objects from carbon, aluminum and iron for $0<\rho H\leq$ 80 g/cm^2 were done. The maximal energy of X-ray radiation E_0 is changed in range from 2 up 9 MeV, that is in the high X-ray energy range. The radiometric detector was done from $CdWO_4$ with thickness 45 mm. The ADC capacity was K=16. The energy spectrum $f(E,E_0)$ of the X-ray source is described by the function Schiff's [14], E_0>1 MeV.

The typical dependencies of $Y(\rho H)$ are shown on figure 2.

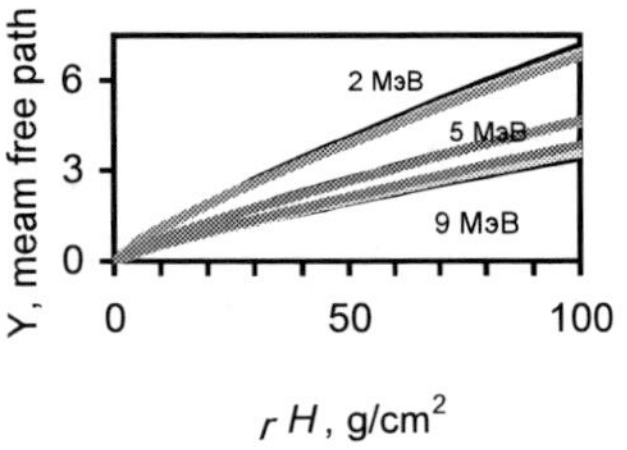

Figure 2. The typical dependencies of $Y(\rho H)$: ▬ – carbon; ▬ – aluminum; ▬ – iron.

From the analysis of the data presented on figure 2 can draw two main conclusions:

1. The functions $Y(\rho H)$ are smooth and monotone increasing in all examined range of the maximal X-ray energies.

2. There is a value E_0 for that the deviation of dependencies $Y(\rho H)$ is minimal for different materials.

The first conclusion allows use the polynomials to approximate the dependence $P^{-1}(Y)$. The second conclusion can be used to design the universal densitometers without the high accuracy requirement.

4. The study of the polynomial approximations of $P^{-1}(Y)$ function

Let analyze the approximation of table values $(\rho H_i, Y_i)$, i=1...n of the function $P^{-1}(Y)$ by the polynomial power of k. In this case the function $P^{-1}(Y)$ is described by the expression

$$P^{-1}(Y)\approx\sum_{j=0}^{k}a_jY^j, \tag{9}$$

here a_j, j=0...k are the polynomial coefficients.

The coefficients a_j, j=0…k of the polynomial regression to approximate the dependence $P^{-1}(Y)$ are founded by the least square method (LSM)

$$\min_{a_0,a_1,...,a_k} \sum_i^n \left(\sum_{j=0}^k a_j Y_i^j - \rho H_i \right)^2 \quad (10)$$

After use the polynomial approximation of the function $P^{-1}(Y)$ the equation (6) reduce to

$$\rho = \frac{\sum_{j=0}^k a_j Y^j}{H}. \quad (11)$$

Calculation results were analyzed to estimate the approximation accuracy $P^{-1}(Y)$ by polynomial power of k=3. To do that we compute the polynomial coefficients a_i, i=0…k (11) by the LSM (10) and the approximation accuracy was estimated for the TO mass thickness $15<\rho H\leq 80$ g/cm^2. The analysis of calculated data allows to draw a conclusion about the applicability of the polynomial approximation to build the densitometer equation by the X-ray absorption method for the maximal X-ray energies under study and the mass densities of the TO from 15 to 80 g/cm^2. The polynomial power of 3 or more can be recommended. The systematic inaccuracy of the density due to the lack of approximation is not exceed for carbon 0.0015 g/cm^3, for aluminum 0.0028 g/cm^3 and for iron 0.0127 g/cm^3. The indicated inaccuracies are satisfactory for most practical applications.

The abovementioned conclusions must be verified experimentally.

5. The experiment

5.1. The experimental estimation of dependence Y(ρH) for the steel

The experimental verification of the dependence $Y(\rho H)$ was carried out on the digital radiography system with the high energy X-ray source – the portable impulse betatron MIB−4/9. The detector characteristics are specified in the description of the calculation example. The geometry of measurement is closed to the layout described on figure 1. The narrow X-ray beam is shaped by the slit collimator. The width of the collimator gap equals d=5 mm. The collimator thickness is D=300 mm. The distance from the radiation source to the TO is about 3000 mm. The dependencies $Y(\rho H)$ were estimated for the maximal X-ray energies E_0=4.5 and 9 MeV. The steel objects with densities ρ=7.85 g/cm^3 were examined. Figure 3 shows the experimental dependencies $Y(\rho H)$ for the steel fragments of the TO and the maximal X-ray energies E_0=4.5 and 9 MeV. The selection of the TO material was due to the high quality of the rolled iron. To the experimental results on figure 3 were added the dependence $Y(\rho H)$ calculated by the formula (8). The analysis of the results presented on figure 3 can draw a conclusion about underestimation of the ray thickness values of the steel fragments in the experiment. We suggest a hypothesis that the most likely reason of this underestimation is an insufficient collimation of the X-ray source.

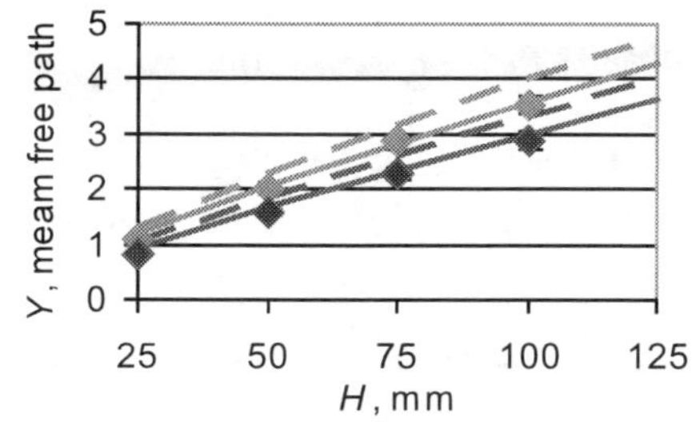

Figure 3. Typical dependencies of $Y(\rho H)$ for steel: E_0=4.5 MeV ▬ ▬ is the calculation result; ♦ is the experimental result; ▬▬ is the calculation result with correction for the scattered radiation; E_0=9 MeV ▬ ▬ is the calculation result; ♦ is the experimental result; ▬▬ is the calculation result with correction for the scattered radiation.

To validate the abovementioned hypothesis we made a series of calculations of the X-ray energy accumulation coefficient B_E by Monte Carlo method for the measurement geometry of the high energy digital radiography system of the Tomsk Polytechnic University (see figure 1). The initial data for calculation B_E are: F=4200 mm; D=400 mm; d=5 mm; A=200 mm; B=200 mm; H is a parameter variable from 25 to 125 mm; a=5 mm; b=6 mm; h=45 mm. Accounting the scattering leads to transformation of the calculated dependency $Y(\rho H)$ to $Y^*(\rho H)$

$$Y^*(\rho H) = Y(\rho H) - \ln B_E(\rho H) . \qquad (12)$$

Figure 3 shows the dependencies $Y^*(\rho H)$ produced from the primary calculation dependencies by the formula (12). The comparison of the calculated curve $Y^*(\rho H)$ and the experimental values of the ray thickness of the TO fragments validates the hypothesis.

5.2. Experimental equations for X-ray transmission densitometer

The table 1 includes the coefficients of the polynomial power of three, approximated the experimental dependencies (11) for the steel TO fragments for the maximal X-ray high energy radiation E_0=4.5 and 9 MeV. Also the table has values of the limiting approximation error Δ_ρ.

Table 1. The experimental values of equation (11) coefficients a_0, a_1, a_2, a_3 and Δ_ρ for the steel

E_0, MeV	Parameters				
	a_0	a_1	a_2	a_3	Δ_ρ
4,5	–5.707	25.57	–3.216	0.764	0.0008
9	–3.184	28.35	–1.694	0.5581	0.0086

The data analysis for table 3 draws a conclusion about applicability of the equation (11) with four parameters for the practical estimation of the TO material density by the X-ray absorption method.

6. Conclusion

The results of the theoretical and experimental researches allow draw a conclusion about possibility to measure the material density by the X-ray high energy method with the systematic error 0.01 g/cm^3. This method can be used to examine the large-size testing objects with variable thickness. It was shown that the visible shift of theoretical and experimental dependencies of the TO ray thickness from the mass thickness is due to the insufficient collimation of the high energy X-ray source.

Acknowledgements
The authors acknowledge the financial support from The Ministry of Education and Science of the Russian Federation in part of the science program and also the RFBR (Grant 13–08–98027).

References

[1] Medhat M E (2012) *Ann Nucl Energy* **40** 53–59
[2] Kumara W A S, Halvorsen B M and Melaaen M C (2010) *Int J Multiphase Flow* **36** 467–480
[3] Lechner T, Sandin Y and Kliger R (2013) *Int J Archit Herit.* **7** 416–433
[4] Bateni A, Parvin N and Ahmadi M (2011) *Powder Metall.* **54** 533–536
[5] Heindel T J (2011) *J Fluids Eng.* **133 (074001)** 1–16
[6] Nedavnii O I, Osipov S P and Nedavnii I O (2000) *Russ J Nondestr Test.* **36** 526–530
[7] Hanus R, Zych M, Petryka L and Świsulski D (2014) *Math Probl Eng.* **2014** 1–10
[8] Nedavnii O I, Osipov S P and Sidulenko O A (2002) *Russ J Nondestr Test.* **38** 261–265
[9] Gorshkov V A (2014) Features of Dual-Energy X-ray Densitometry of Multi-Component Objects *Testing. Diagnostics* **10** 25–30
[10] Zavialkin F M, Kvasnitsa M S and Osipov S P (1985) *Sov J Nondestr Test.* **21** 275–279
[11] Gamma data for elements. Department of nuclear power plant, ABBN laboratory. URL www.ippe.ru/podr/abbn/libr/groupkon.php
[12] Zav'yalkin F M and Osipov S P (1985) *Sov J At Energy* **59** 842–846
[13] Zav'yalkin F M and Osipov S P (1986) *Sov J At Energy* **60** 186–188
[14] Schiff L I (1951) *Phys Rev.* **83** 252–253

Energy Characteristics of Electric Drive of Oscillatory Motion at the Shock-Free Start

L A Payuk[1], N A Voronina[1] and O V Galtseva[2]

[1]Senior Lecturer, Institute of Power Engineering, Tomsk Polytechnic University, Tomsk, Russian
[2]Associate Professor, NDT Institute, Tomsk Polytechnic University, Tomsk, Russian
E-mail: lubapa81@mail.ru

Abstract. The peculiarity of operation of oscillatory electric drive with doubly-fed motor at realization of algorithm of shock-free start was considered. The method of evaluating of energy parameters of such motors operating at the resonance mode with учетом of quantitative and qualitative components of the process of energy conversion by an electric motor (doubly-fed motor) is proposed. The calculation of energy characteristics of electric drive of this type was made, the results of calculation show an increase of energy efficiency (an average, generalized efficiency coefficient η_E increased by 10 %), which confirms the practical significance of the research object.

1. Introduction

The modern concept of analysis and design of electric drives of oscillatory motion (*EDOM*) is based on the principles of energy and resource efficiency, and this is reflected in methods of analysis and synthesis. *EDOM* widely used in modern industry: from vibration systems for inspection of electrical equipment to the ventilation systems and human life [1–4]. Oscillations measured from a few microns to meters [1, 5]. In each case it is necessary to consider the difference between these types of drives from industrial. For example, the nonsinusoidal supply voltage, the harmonic composition of the currents in the windings of the stator and rotor of the control motor, the double frequency in the electromagnetic torque. Consequently, the generally accepted criteria of efficiency of electric drives for this type of drive are not suitable, as it does not take into account their specificity.

It is therefore proposed to use the method of evaluating the energy characteristics of controlled *EDOM* on the base on serial asynchronous motor by scheme of doubly-fed motor (*DFM*). Its oscillatory mode is excited by periodic soft reverse [6]. A distinctive feature of this method is an account of harmonic content of supply voltage, and, as a consequence, of resulting electromagnetic torque, angular velocity of rotation and law of motion of movable element of motor of drive with allowance of distortion and asymmetry.

It is known that the highest power consumption at operating of any electric drive happens at its start due to the occurrence of inrush currents and torque, which may differ from the nominal value by several times. Therefore, we proposed a method of providing of shock-free start by current and torque of *EDOM* with *DFM*. According to [7, 8] the phase correction of supply voltages reduces the magnitude of shock value of stator current on 30 % and the electromagnetic torque – in to 3 times.

Journal of Physics: Conference Series **671** (2016) 012044 doi:10.1088/1742-6596/671/1/012044

2. Research object and method

The analysis of free components of inrush currents showed that the following condition is necessary to suppress the shock values of current of *DFM* [3]:

$$\begin{aligned} &\omega_1 \cos(\alpha`) - \alpha_1 \sin(\alpha`) = 0; \\ &\alpha_3 \cos(\beta`) + \omega_2 \sin(\beta`) = 0. \end{aligned} \tag{1}$$

The formation of supply voltage of windings of excitation (*WE*) and control (*WC*) of stator (*S*) and rotor (*R*) of executive engine (doubly-fed motor) occurs in the following laws, respectively:

$$\begin{aligned} &U_{WES} = U_m \gamma_1 \sin(\omega_1 t + \alpha`); \\ &U_{WCS} = U_m \gamma_2 \sin(\omega_1 t + \beta` + \Omega t); \\ &U_{WER} = U_m \gamma_3 \sin(\omega_1 t + \gamma)\cos(\chi) - U_m \gamma_4 \sin(\omega_1 t + \varphi + \Omega t)\sin(\chi); \\ &U_{WCR} = U_m \gamma_3 \sin(\omega_1 t + \varphi + \Omega t)\cos(\chi) + U_m \gamma_4 \sin(\omega_1 t + \gamma)\sin(\chi), \end{aligned} \tag{2}$$

where U_m – peak values of supply voltage of windings of stator and rotor; α, β, γ, φ – initial phases of voltage on motor windings (*rad*); ω_1, ω_2 – angular supply frequencies of stator and rotor $\omega_2=\omega_1+\Omega$, γ_i – coefficients of signals.

Expressions for starting currents and torque shall not give here; we note only conditions of shock-free start by current and torque, based on the conditions of Table 1 (α_i – attenuation coefficients of free components of stator and rotor currents of *DFM*, which were previously found as the roots of characteristic equations upon condition of retardation of shaft of executive motor) [3].

Table 1. Conditions for calculation of shock-free start by current and torque for *EDOM*.

Shock-free start of *EDOM* with *DFM*			
by current		by torque	
$\alpha` = \text{arctg}\left(\frac{\omega_1}{\alpha_1}\right)$;	$\beta` = \text{arctg}\left(\frac{-\omega_2}{\alpha_3}\right)$	$\alpha` = \text{arctg}\left(\frac{\omega_1}{\alpha_2}\right)$;	$\beta` = \text{arctg}\left(\frac{-\omega_2}{\alpha_4}\right)$

Valuation of indicators of quality of designed system is produced by method of discrete synthesis at different frequency nutrition of *DFM* of *EDOM*. The energy quality indicators can be divided into two groups: first group determines the quality of process of electromechanical energy conversion of *EDOM*; the second group determines the comprehensive quantitative assessment of process of electromechanical energy conversion efficiency by motor and *EDOM*.

The consumption or generation of reactive power occurs at the work of electromechanical energy converters in dynamic modes except consumption of active power [6]. In this case, the total power consumption of motor is characterized by the power of distortion and asymmetry. Therefore power factor (K_P), shear factor (K_S), factor of distortion (K_D) and asymmetry (K_A) characterize the ratios between active and reactive powers and power of distortion and asymmetry. Feature of this approach to evaluating of energy characteristics allow evaluate the quality of energy consumption of *EDOM*.

Power factor K_P is determined by the formula:

$$K_{Pi} = \frac{P_i}{S_{iAV}} = \frac{P_i}{\sqrt{P_i^2 + Q_{iAV}^2 + T_{iAV}^2}}, \tag{3}$$

where P_i, Q_{iAV}, T_{iAV} – active power, average values of reactive and total power of distortion of *i* phase for the period.

The shear factor K_S describes the relationship between active and reactive power of phase and it is determined by the formula:

$$K_{Si} = \sqrt{\frac{P_i^2}{P_i^2 + Q_{iAV}^2}}. \quad (4)$$

The distortion factor K_D determines the degree of distortion of power consumption and it is calculated by the expression:

$$K_{Di} = \sqrt{\frac{P_i^2 + Q_{iAV}^2}{P_i^2 + Q_{iAV}^2 + T_{iAV}^2}}. \quad (5)$$

The factor of asymmetry is calculated according to the expression:

$$K_A = \frac{S}{S_M}. \quad (6)$$

The generalized efficiency coefficient refers to the quantitative assessment criteria and it determines the useful portion of the consumed power diverted from the network:

$$\eta_E = \frac{P_{\text{mech}}}{S} = \sqrt{\frac{P_{\text{mec,a}}^2 + P_{\text{mec,r}}^2 + P_{\text{mec,d}}^2}{S^2}}, \quad (7)$$

where $P_{mec,a}$, $P_{mec,r}$, $P_{mec,d}$ – active, reactive mechanical power and power of distortion of drive motor, respectively; S – total conditional power consumption of motor.

The mechanical power of the electric motor in an oscillatory mode of operation includes power distortion $P_{mec,d}$, active $P_{mec,a}$, and reactive $P_{mec,r}$ components. Any of these components of mechanical power are involved in useful work depending on the target of drive and nature of load. The generalized efficiency coefficient considers the qualitative (K_P) and quantitative (η) aspects of process. Therefore its maximum value is necessary to retain during operation of machine in the oscillatory mode (for providing of resonant mode).

This expression is used for determination of mechanical power of *DFM* at the feeding of windings of stator and rotor by currents of frequencies ω_1 and ω_2 at the first approximation:

$$P_{mec} = \frac{M_{EL.T}\omega_m \sin(\varphi - \alpha)}{2} \cdot [1 + \cos 2(\Omega t + \alpha)] + \frac{M_{EL.T}\omega_m \cos(\varphi - \alpha)}{2} \cdot \sin 2(\Omega t + \alpha), \quad (8)$$

where $\Omega = \omega_2 - \omega_1$ – oscillation frequency of movable element of the executive motor; α, φ – initial phases of the first harmonic components of the motion law and electromagnetic torque of *DFM*.

Analytical expressions for determination of electromagnetic torque and angular speed of rotation were determined previously [7, 9]. We give final versions for their determination. Electromagnetic torque of motor in expression (8) defined by formula:

$$M_{EL.T} = k\left[M_{ST} + M_{DE} \cdot \frac{d\chi}{dt}\right],$$

where k – coefficient of proportionality, which takes into account the specificity of motor; M_{ST}, M_{DE} – starting and damping electromagnetic moments of motor.

Angular speed of rotation ω_m in expression (8) defined by formula:

$$\omega_m = \int\left(\frac{1}{J_\Sigma}\left(L_m\left(i_{\beta s}i_{\alpha r} - i_{\alpha s}i_{\beta r}\right) - R_g\omega - C_m^{-1}\chi - M_c\right)\right)dt,$$

where R_g – coefficient of damping load torque; C_m^{-1} – coefficient of positional load torque; M_c – torque of static load on the motor shaft, χ – coordinate of movable element of motor; J_Σ – total inertia moment, which consists of the inertia moment of motor J_m and load J_l.

Instant consumption power of electric circuit AC at different frequency feeding is defined as

$$P_n(t) = U_n(t) \cdot i_n(t), \tag{9}$$

where n takes on the values 1, 2, 3, 4, which correspond to the phase currents and voltages.

Next we substitute the expressions (1) and the expressions for the currents of stator and rotor [7, 8] in (9):

$$\begin{gathered} i_1(t) = i_{\alpha s}(t);\ i_2(t) = i_{\beta s}(t);\ i_3(t) = i_{\alpha r}(t);\ i_4(t) = i_{\beta r}(t); \\ U_1(t) = U_{\alpha s}(t);\ U_2(t) = U_{\beta s}(t);\ U_3(t) = U_{\alpha r}(t);\ U_4(t) = U_{\beta r}(t). \end{gathered} \tag{10}$$

Instant values of phase currents for specified functions of regulating are defined by the formula (10); for example, the expression of instant power of phase α_s can be written in the form:

$$P_{\alpha s}(t) = P_1[1 - \cos 2(\omega_1 t + \alpha)] - Q_1 \sin 2(\omega_1 t + \alpha) - T_1 \ . \tag{11}$$

The total active and reactive power along the axes α and β are defined as

$$P_\alpha = P_{\alpha s} + P_{\alpha r}\ ;\ Q_\alpha = Q_{\alpha s} + Q_{\alpha r}\ ;\ P_\beta = P_{\beta s} + P_{\beta r}\ ;\ Q_\beta = Q_{\beta s} + Q_{\beta r}$$

Functional scheme of electric drive of oscillatory motion with potentially - phase modulation and ability of realization of shock - free start is presented in figure 1.

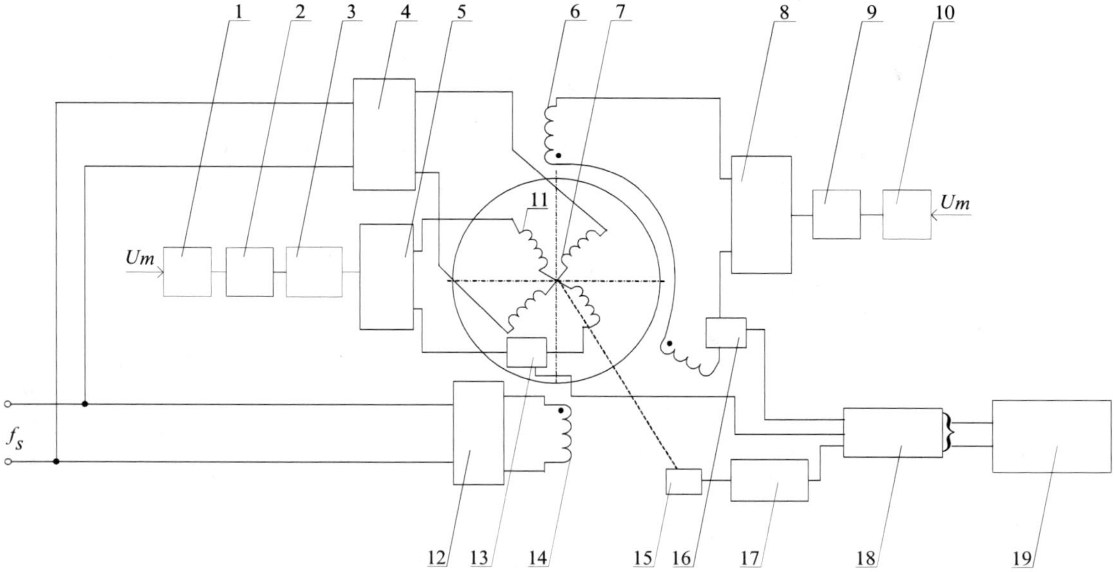

Figure 1. Functional scheme of *EDOM* with potentially - phase modulation: 1 – driving generator 2; 2 – modulator of frequency 2; 3 – phase - shifting device; 4 – linear autotransformer 2; 5 – autonomous voltage inverter 2; 6 – control winding of stator; 7 – excitation winding of rotor; 8 – autonomous voltage inverter 1; 9 – modulator of frequency 1; 10 – driving generator 1; 11 – control winding of rotor; 12 – linear autotransformer 1; 13 – current sensor 2; 14 – excitation winding of stator; 15 – speed sensor; 16 – current sensor 1; 17 – differentiator (integrator); 18 – sensor unit; 19 – personal computer.

EDOM (figure 1) consists of control system of stator and rotor of DFM realized on the basis of the asynchronous motor with slip-ring rotor, sensor unit (18) with system for converting analog signal to digital, IBM – compatible personal computer with built-in board with input-output data, providing feedback of speed and current. Control systems of stator and rotor of *DFM* consists of frequency converters with DC links, which specify the modulation frequency of linear autotransformers (4, 12), signal amplification unit and autonomous voltage inverters (5, 8), switched - on from stator and rotor side with the carrier frequency $f_1 = 50$ *Hz* and the modulation frequency $f_2 = 55$ *Hz*.

The signals from the frequency modulators (2, 9) are fed on autonomous voltage inverters (5, 8) through the windings of stator and rotor respectively, signals U_m come on them from driving generators (1, 10). The signal from the phase-shifting device (3) comes to the input (5); this signal provides realization of algorithm of shock - free start of *DFM* in the composition of *EDOM*.

Electric machine of alternating current of type *MTN* - 011-6*U*1with the following parameters was used as *DFM*: P_n=1400 *W*, nominal voltage U_n =220 *V*, nominal current of stator I_n=3.35 *A*, η=76 %, $cos\ \varphi_н$ = 0.78, nominal frequency of rotation n_n=1250 *rev/min* [8–9]. At calculation of energy indicators the following values for the components of the load were adopted (all values are given in relative units): at shock-free start by torque R_g=1.35, C_m^{-1}=1.3, M_c= 0.2, J_Σ=1.25; at shock-free start by current R_g=0.54, C_m^{-1}=0.45, M_c=0.08, J_Σ= 1.25.

Results of calculation of energy indicators of *EDOM* (expressions (3) – (7)) are shown in Table 2.

Table 2. Calculated energy indicators of the electric drive of oscillatory motion.

Indicators	Shock - free start by current of *EDOM*	Shock - free start by torque of *EDOM*	Standard start of *EDOM*
K_P	0.5	0.53	0.56
K_S	0.55	0.5	0.64
K_D	0.35	0.37	0.4
K_A	0.45	0.4	0.5
η_E, %	73.5	77	65

4. Summary

1. Analysis of the energy characteristics of *EDOM* at the realization of shock - free start by current and torque has shown that the generalized efficiency coefficient η_E increased by 10 % at the use of this method an average; power factor K_P decreased by 0.4 relative units; shear factor K_S decreased by 0.8 relative units; factor of distortion K_D decreased by 0.035 relative units; factor of asymmetry K_A decreased by 0.055 relative units. The harmonic composition of the supply voltage improves and affects on final consumption of active and reactive power.

2. The largest increase of the generalized efficiency coefficient η_E (by 12 %) and the largest decrease of the factor of asymmetry (by 0.1 relative units) occur at the realization of shock - free start by torque. This suggests that the process of energy consumption occurs with the best of quantitative and qualitative indicators at the realization of algorithm of shock - free start by torque.

3. The dependences (3) – (7) allow one to accurately assess the energy characteristics of the electric drive of oscillatory motion (of any periodic movement) and to get an adequate result and to analyze the efficiency of the designed electric drives.

References

[1] Poddar G and Ranganathan V T (2004) *IEEE Transactions on Industrial Electronics* **51**(6) 1329–1337
[2] Shimoda K *et al* 2004 *Vibrating linear actuator* Patent US 6777895
[3] Senulis A *et al* (2007) *Elektronika ir Elektrotechnika* **4** 63–66
[4] Payuk L *et al* (2015) *Applied Mechanics and Materials* **792** 379–385
[5] Li W and Zhi-xin W (2007) *J. Shanghai Jiaotong Univ. Sci.* **12 (2)** 271–282
[6] Klatt F W (2013) *IEEE Electric Ship Technologies Symposium(ESTS 2013)* 209–215
[7] Novotny D W and Schmitz N L (1962) *AIEE Great Lakes District Meeting* 652–657
[8] Poddar G and Ranganathan V T (2004) *IEEE Transactions On Industrial Electronics* **51**(5) 1089–1096
[9] Aristov A V, Voronina N A (2013) *Bulletin of the Tomsk Polytechnic University* **322 (4)** 116–120

Mathematical Simulation of Contaminant Flow in Closed Reservoir

V M Agranat[1], A M Goudov[2], V A Perminov[3]

[1]President, Applied computational fluid dynamics analysis, Thornhill, Ontario, Canada
[2] Dean of faculty, Kemerovo state university, Kemerovo, Russia
[3] Professor, Tomsk polytechnic university, Tomsk, Russia
E-mail: perminov@tpu.ru

Abstract. A mathematical model of the propagation in flooded mine lightweight contaminant due to allocation of groundwater is considered. Mathematical model was based on an analysis of experimental data and using concept and methods from reactive media mechanics. The boundary-value problem is solved numerically using the finite volume method. The distribution of fields of velocities and concentration of impurity particles in a flooded mine have been obtained at different times. These results can be used to analyze mining water treatment process due to environment and evaluate its further possible improvements.

1. Introduction

Water body pollution by mining and quarry waters is a typical problem for Kuzbass (Kemerovo region, Russia) and many other mining regions [1–3]. Mining waters usually contain particles of coal dust, clay, calcium compounds, magnesium, oil products, etc. Light substances (which density is less than water density) such as oil products accumulate on water surface while other particles remain suspended or sediment gradually. The problem of mining water treatment by pumping into abandoned mines and further use of it after precipitation of impurities (for heavy particles) or impurity floating up (for light particles) is of great interest. This methodology was presented in [4]. The model in this paper is based on laminar flow. In fact, estimates indicate that for a considered area is turbulent. The model which is being reported here is complex, and includes particles. The paper considers fluid flow containing impurity particles in a flooded mine.

2. Physical and mathematical setting

To analyze float impurities distribution a square form mine is under consideration. It has a ledge at the top (shown in figure 1).

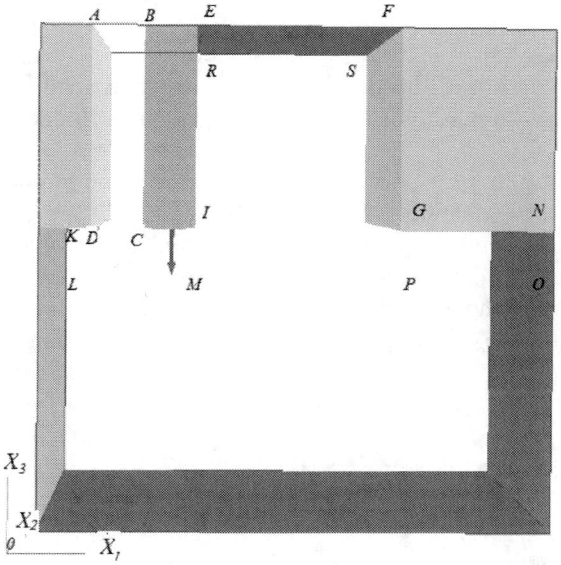

Figure 1. Computational domain scheme.

Mathematically the following differential equation system for turbulent flow should be solved in the following way:

$$\frac{\partial \rho}{\partial t}+\frac{\partial}{\partial x_j}(\rho u_j)=0 \tag{1}$$

$$\frac{\partial}{\partial t}(\rho u_i)+\frac{\partial}{\partial x_j}(\rho u_i u_j)=-\frac{\partial p}{\partial x_i}+\frac{\partial}{\partial x_j}(-\overline{\rho u_i' u_j'})-\rho S C_d u_i|\vec{u}|-\rho g_i, \tag{2}$$

$$\rho(\frac{\partial Y_k}{\partial t}+u_1\frac{\partial Y_k}{\partial x_1}+(u_3-u_{3k})\frac{\partial Y_k}{\partial x_3})=\frac{\partial}{\partial x_j}(-\overline{\rho Y_k' u_j'}), \tag{3}$$

$$\frac{\partial}{\partial t}(\rho k)+\frac{\partial}{\partial x_i}(u_i\rho k)=\frac{\partial}{\partial x_i}\left[\left(\frac{\mu_t}{\sigma_k}+\mu\right)\frac{\partial k}{\partial x_i}\right]-\mu_t\left(\frac{\partial u_i}{\partial x_j}+\frac{\partial u_j}{\partial u_j}\right)\frac{\partial u_i}{\partial x_j}-\beta\rho g_i\frac{\mu_t}{\Pr}\frac{\partial T}{\partial x_i}-\rho\varepsilon, \tag{4}$$

$$\frac{\partial}{\partial t}(\rho\varepsilon)+\frac{\partial}{\partial x_i}(u_i\rho\varepsilon)=\frac{\partial}{\partial x_i}\left[\left(\frac{\mu_t}{\sigma_\varepsilon}+\mu\right)\frac{\partial \varepsilon}{\partial x_i}\right]+C_1\frac{\varepsilon}{k}(G_k+G_B)-C_2\rho\frac{\varepsilon^2}{k}, \tag{5}$$

$$p=\rho R_0 T\sum_k\frac{Y_k}{M_k},\ \vec{g}=(0,g),\ u_{3k}=\frac{g d_k^{\ 2}}{18\nu}(\frac{\rho_k}{\rho}-1). \tag{6}$$

Here, t, x_i – time and spatial coordinates ($i=1,\ 3$); u_i – velocity vector projection on the corresponding axis of Cartesian reference system, p – pressure; g – gravitational acceleration, R_0 – absolute gas constant, M_k – molecular weight of k – component, ρ – density of fluid and particles mixture, ν – kinematic viscosity coefficient, D_t – turbulent diffusion coefficient, d_k, ρ_k, u_{3k} - diameter, density and velocity of particle settling [5], Y_k – mass concentrations of k – component ($k=1$ – water, 2 – solid particles); $\mu_t=\rho C_\mu k^2/\varepsilon$ - coefficient of turbulent viscosity, $k=\overline{u_i' u_i'}/2$ - turbulent kinetic energy ; ε – its dissipation, C_μ, σ_k, σ_ε, C_1, C_2 – empirical constants, and G_k, G_B -turbulence caused by forced convection and natural convection.

3. Numerical solution and results

Based on mathematical formulation of the problems (1) – (6) numerical calculations were made to determine the pattern of float impurity distribution process in a flooded mine with the help of **PHOENICS** [6–7].

Vector fields of velocity and impurity distribution at different time moments were obtained as the result of numerical integration of equation system (1) - (5). Side walls are considered not to influence the impurity distribution process and fluid flow. Thus the problem is solved in the two-dimensional domain X_1OX_3. A mine (length –10 meters horizontally, height-3 meters) is under consideration (figure 1.). Underground water without any impurity enters the domain. Impurity concentration equals 1 inside the domain. Particles size is $d_k = 5\cdot 10^{-5}$ m. Impurity particles density is 500 kgs/m^3 that is two times less than water density. The velocity of groundwater inflow from the upper layers is 0.1 m/s. Figure 2 shows the distribution of impurity concentration at different time moments.

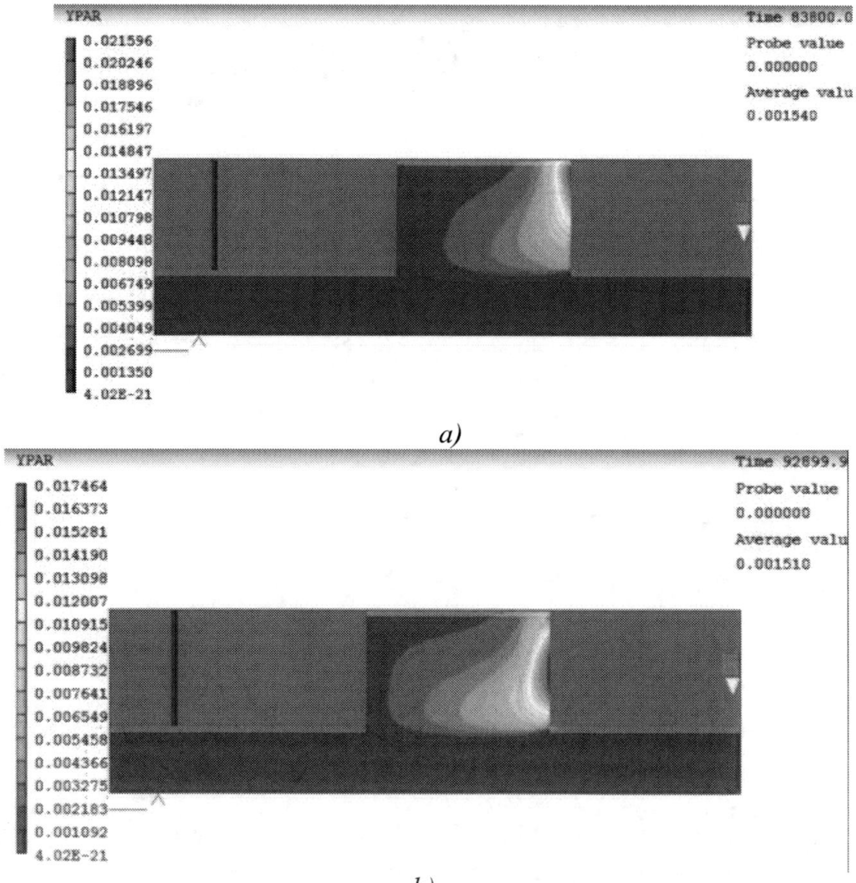

Figure 2. Impurity distribution: *a)* t=83800 sec, *b)* t=92899 sec.

Figure 2 shows, that as time goes some part of the impurities transferred by a flow leaves a mine, while the remaining part is moved to a dirt collector by a vortex. In case impurity particles density is halved (as little as 250 kgs/m^3) the distributions of impurity are numerically calculated at different time moments (figures 3–5). These figures show that the flow becomes stable and impurities accumulate in the upper part of the domain as time goes. It happens faster compared with the previous case because the particles density is two times less.

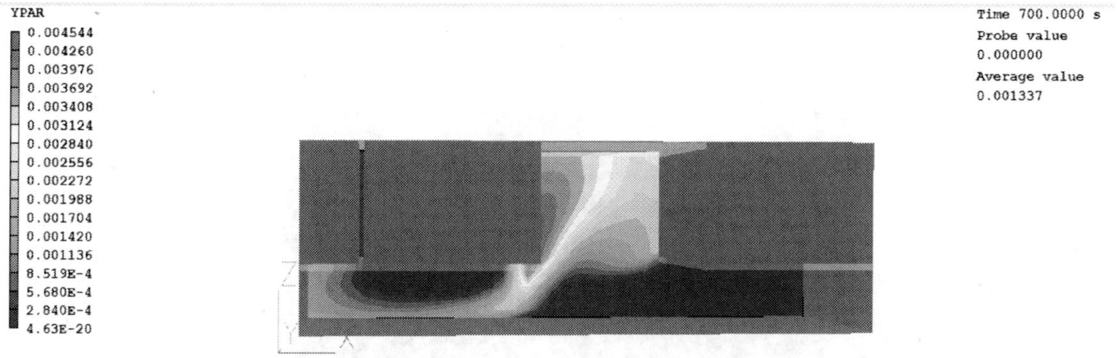

Figure 3. Distribution of impurity which density is 250 kgs/m^3 (t=700 sec).

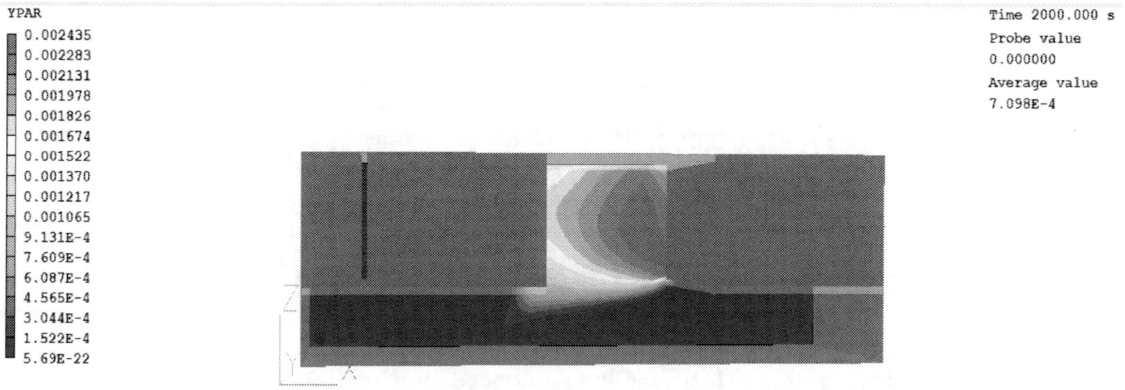

Figure 4. Distribution of impurity which density is 250 kgs/m^3 (t=2000 sec).

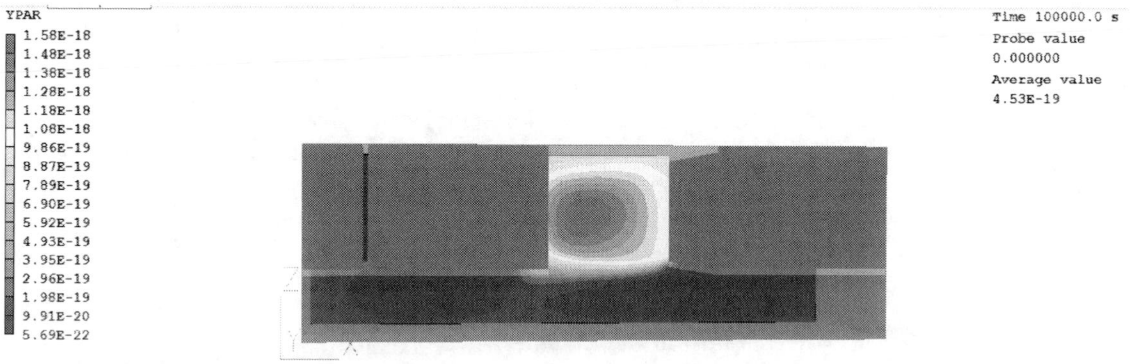

Figure 5. Distribution of impurity which density is 250 kgs/m^3 (t=10000 sec).

Later the inflow velocity was halved as compared with the previous case (0.05 m/s). When compare the results of impurity distribution (figures 6, 7) with the previous case at the same moment of time (figures 4, 5 correspondingly) it is obvious that the flow becomes stable and impurity concentration increases faster in the upper part of the domain.

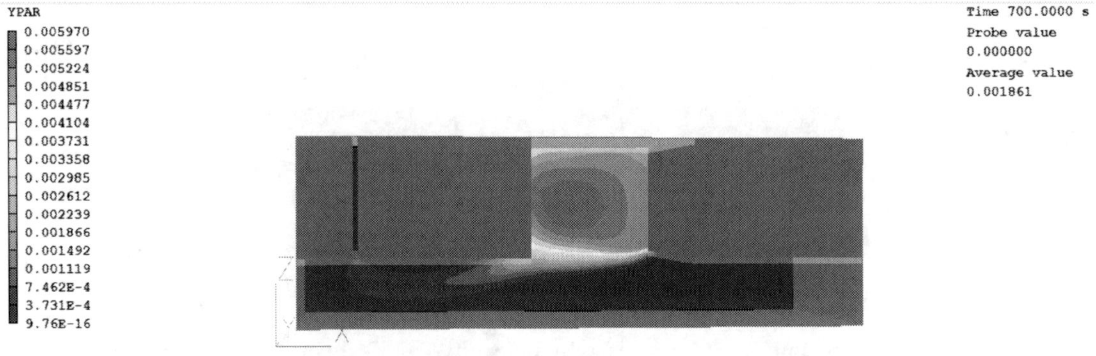

Figure 6. Impurity distribution (t=700 sec).

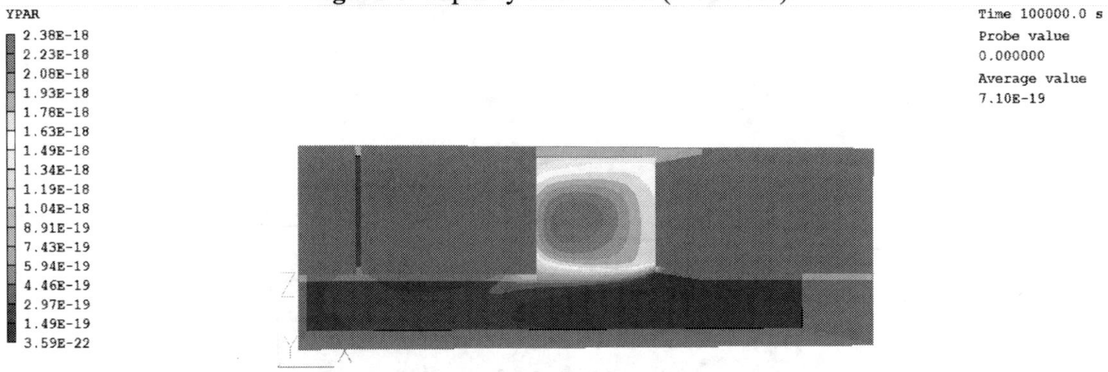

Figure 7. Impurity distribution in a flooded mine (t=10000 sec).

Based on the results of the numerical calculations, the way velocity of underground water inflow and parameters of impurity particles influence mining water treatment can be identified. The mathematical model presented in this paper can be used to analyze mining water treatment process due to environment and evaluate its further possible improvements.

The research is based on the state task № 2014/64, the state project "Scientific researches organization". The results of the numerical calculations and problem formulation will be used by the educational resources information portal that offers students, postgraduate student and academic researches different educational services.

References

[1] Morris B L, Lawrence A R L, Chilton P J C, Adams B, Calow R C and Klinck B A 2003 *Groundwater and its susceptibility to degradation: a global assessment of the problem and options for management* (United Nations Environment Programme, Nairobi, Kenya)

[2] Gundula Prokop 2004 *Ground Water Management in Mining Areas 2004 Proceedings of the 2nd IMAGE-TRAIN Advanced Study Course* (Umweltbundesamt)

[3] Adams R, Younger P L (2001) *Ground Water 39*(2) 249–261

[4] Zakharov Y N, Potapov V P, Schastlivcev E L and Chiruykina A V (2009) *Vestnik NGU (Seriya "Informacionnye tekhnologii")* **7**(4) 66–72

[5] Hinderliter P M *et al* 2010 *Particle and Fibre Toxicology* **7** Article number 36

[6] Patankar S V 1981 *Numerical Heat Transfer and Fluid Flow* (New York: Hemisphere Publishing Corporation)

[7] PHOENICS On-Line Information System: www.cham.co.uk/ChmSupport/polis.php

Choice of Parameters and Stability of Nonlinear Vibration Isolation Device

R M Mustafina[1], I A Plotnikov[2], I V Plotnikova[3], O N Tchaikovskaya[4]

[1]Associate Professor, S. Toraighyrov Pavlodar State University, Pavlodar, Kazakhstan,
[2]Associate Professor, Institute of Power Engineering, Tomsk Polytechnic University, Tomsk, Russia
[3]Associate Professor, NDT Institute, Tomsk Polytechnic University, Tomsk, Russia
[4]Professor, National Research Tomsk State University, Tomsk, Russia
E-mail: inna@tpu.ru

Abstract. Work of active vibration isolation devices with single-mass electromagnetic suspension taking into account of real characteristics of the voltage regulators is described. The analytical researches are carried out; the areas of stability of work of nonlinear vibration isolation device are defined.

1. Introduction

Reliable and accurate information on device operation is required to control and monitor current processes [1]. Vibration protection devices are widely used in engineering [2–3]. They are used to make systems with high performance characteristics. These devices are non-linear since the power characteristics of electromagnets and characteristics of the system for electromagnetic suspension stabilization are non-linear. The parameters of an active vibration isolation device should be chosen so that to ensure effective oscillation damping at a given frequency band of external disturbances and sufficient stability margin [4].

Non-linear modes of active vibration isolation devices with single-mass electromagnetic suspension [5] were analysed for objective parameters of the voltage regulator. Protected against vibrations, the object of mass m is suspended in the field of the electromagnet mounted on the basis producing oscillations. The vibrations (deviations) of the body suspended along the electromagnet, an absolute coordinate which determines the position of an object of mass m with respect to the ground. The voltage and current of the electromagnet in a position of the suspended body static equilibrium the variable components of the voltage and current of the electromagnet.

The parameters of the active vibration isolation devices should be chosen to ensure effective damping of oscillations at a given frequency band of external disturbances and sufficient stability margin. The stability margin can be provided by increasing the overall gain, however, the voltage of the inverter to convey the supply voltage for the electromagnet reaches the saturation level (figure 1), i.e. vibration isolation device is essentially nonlinear. In figure 2 is the feedback voltage controlling the voltage at the input of the nonlinear element; is the converter output voltage.

Analyze the operation of an active vibration isolation device considering that the converter to form voltage is non-linear [6]. Determine the effect of the feedback non-linearity on the performance of the tested device. For the vibration isolation device with a one-mass electromagnetic suspension the control voltage is formed as follows:

$$U_{cont} = \alpha\delta + \beta\delta',$$

where α and β are the coefficients of the control circuit to change the air gap δ and its velocity.

When studying the nonlinear active vibration isolation device, use the method of harmonic linearization to analyze the modes of the system operation affected by harmonic effects. At the input of the tested object, the wave signal

$$U_{cont} = A \sin \omega t. \tag{1}$$

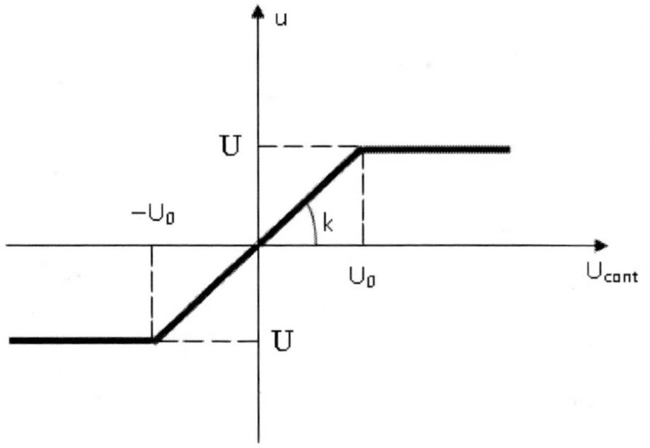

Figure 1. Plot of the voltage of the feedback U_{cont} versus the voltage at the converter output U.

2. Research object and method

Due to the electromagnetic properties of the filter, we take into account only the first harmonic of the nonlinear voltage at the converter output and using the harmonic linearization method; substitute the saturation characteristic with ratio

$$U = q(A)\, U_{cont.}$$

Coefficients $q(A)$ of the harmonic linearization for the nonlinearity have the form [4]

$$q(A) = \text{k with } U_0 < A < U_0, \tag{2}$$

$$q(A) = \frac{2k}{\pi}\left[\arcsin\frac{U_0}{A} + \frac{U_0}{A}\sqrt{1-\frac{U_0^2}{A^2}}\right] \quad \text{with } A \ge U_0, \tag{3}$$

i.e. in the saturated region, the converter is replaced by the linear element with gain $q(A)$ which decreases as the amplitude grows.

The analysis of expressions (2) and (3) shows that the nonlinearity in the feedback circuit causes decrease in equivalent coefficient q, which depends on amplitude A, in the saturation region, i.e. the effect of the nonlinear feedback on the vibration oscillation properties of the electromagnetic suspension are to be studied [6].

The linearized equations of the electromagnet winding and its traction in the operator form are

$$\begin{cases} U(s) = (R + sL_0)I(s) - a\Delta(s)S, \\ F(s) = aI(s) - b\Delta(s) \end{cases} \tag{4}$$

where R is the resistance of the electromagnet winding; L_0 is the inductance corresponding to the equilibrium point for the body of mass m suspended in the electromagnet field; $a = 2K\frac{I_0}{\delta_0^2}$ is the coefficient; K is the constant determined under the static equilibrium condition; δ_0 is the air gap in static equilibrium; $b = 2K\frac{I_0^2}{\delta_0^3}$ is the coefficient; a and b are the coefficients of linearization of the equations nearby the point of equilibrium conditions.

$$U_{cont} = (\alpha + \beta\ s)\Delta(s). \qquad (5)$$

According to the harmonic linearization method the characteristic in figure 2 is replaced by the following expression:

$$u(s) = q(A)u_{cont}(s), \qquad (6)$$

where $q(A)$ is determined according to expressions (2) and (3). Expanding equations (4–6) with the equation of the mechanical system motion

$$mX(s)s^2 = -F(s),$$

find the system transfer function

$$W(s) = \frac{X(s)}{Y(s)} = \frac{\left[q(A)\alpha\beta + a^2 - bL_0\right] s + q(A)a\alpha - bR}{mL_0 s^3 + Rms^2 + \left[a\beta\ q(A) + a^2 - bL_0\right] s + q(A)a\alpha - bR}.$$

The characteristic equation of the harmonically linearized system will be written in the form

$$mL_0 s^3 + Rms^2 + \left[a\beta q(A) + a^2 - bL_0\right] s + q(A)a\alpha - Rb = 0, \qquad (7)$$

or

$$a_3 s^3 + a_2 s^2 + a_1 s + a_0 . = 0.$$

According to the Hurwitz stability criterion, the following conditions for the third-order systems must be meet:

$$a_3 > 0,\ a_2 > 0,\ a_1 > 0,\ a_0 > 0,\ a_2 a_1 > a_3 a_0,$$

that is

$$\alpha > \frac{Rb}{aq(A)};\ \beta > \frac{bL_0 - a^2}{aq(A)};\ \beta > \frac{L_0\ \alpha\ q(A) - aR}{Rq(A)}. \qquad (8)$$

Since the linearization coefficient $q(A)$ are restrained by the following limits:

$$0 < q \leq k,$$

The stability conditions (8) for the tested system will take the form

$$\alpha > \frac{Rb}{ak};\ \beta > \frac{bL_0 - a^2}{ak};\ \beta > \frac{L_0\ \alpha\ k - aR}{Rk}.$$

Replace s by $j\omega$ in equation (7)

$$\left[q(A)a\ \alpha\ - Rb - \omega^2 Rm\right] + j\omega\ \left[a\beta\ q(A) + a^2 - bL_0 - \omega^2 mL_0\right] = 0, \qquad (9)$$

extracting the real and imaginary components in equation (9), we obtain the equation to determine periodic solutions (1) for the input value of the voltage converter

$$\begin{cases} X(A,\omega) = q(A)a\alpha - Rb - \omega^2 Rm = 0, \\ Y(A,\omega) = (a\beta\ q(A) + a^2 - bL_0 - \omega^2\ mL_0)\omega = 0. \end{cases} \qquad (10)$$

Solve equation system (10) to determine frequency ω of the periodic solution through the system parameters and find the expression to determine equivalent gain $q(A)$

$$\omega = \frac{b(\alpha\ L_0 - R\beta) - a^2\alpha}{m(\beta\ R - \alpha\ L_0)}, \qquad q(A) = \frac{aR}{(\alpha\ L_0 - R\beta)}. \qquad (11)$$

Substitute relation (2), (3) into (11) to obtain the formula relating the amplitude of periodic solution A with the system parameters

$$\frac{2k}{\pi}(\arcsin\frac{U_0}{A}+\frac{U_0}{A}\sqrt{1-\frac{U_0^2}{A^2}})=\frac{aR}{\alpha\ L_0-R\beta}. \qquad (12)$$

To determine the value of critical gain k_{cr}, which determines the system stability limit and the region of auto-oscillations, in formula (12), take $A = U_0$ and obtain

$$k_{cr}=\frac{aR}{\alpha\ L_0-R\beta},$$

i.e. oscillations occur when $k \geq k_{cr}$.

Determine the conditions of the linear system stability limit calculating in the system of equations (10) $q(A) = k$:

$$\begin{cases} ka\alpha - Rb - \omega^2 Rm = 0, \\ ka\beta + a^2 - bL_0 - \omega^2 mL_0 = 0. \end{cases}$$

$$k_{LG}=\frac{aR}{\alpha L_0-R\beta}. \qquad (13)$$

That is $k_{LG} = k_{cr}$ and considering this, the stability region limit is determined by the following equations:

$$\alpha=\frac{Rb}{ak_{cr}};\ \beta=\frac{L_0\alpha\ k_{cr}-aR}{Rk_{cr}}.$$

The value of k_{cr} is determined by relation (13) through substituting the linear system limit values α, β.

3 Results and considerations

Hence, we can conclude that if $k < k_{cr} = k_{LG}$, the active vibration isolation device is stable in both a linear response and a nonlinear (figure 2).

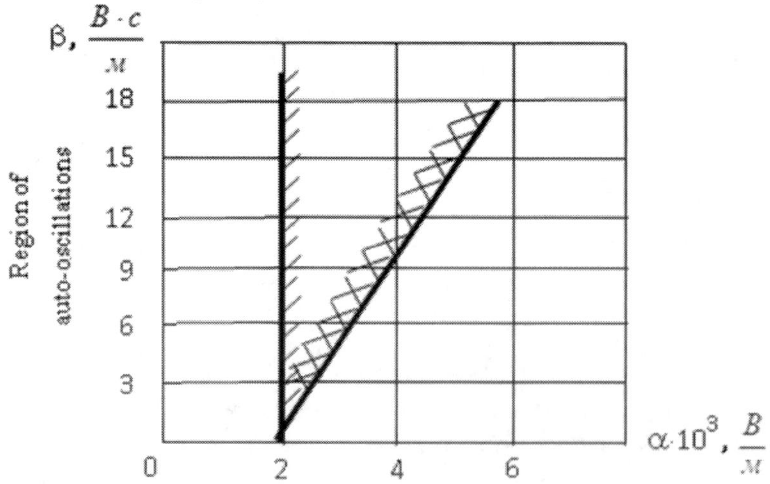

Figure 2. A graph of the stability domain of work.

Vibration When $k \geq k_{cr}$ the nonlinear device operates as a self-oscillating mode at a certain amplitude, whereas in the linear system under this condition ($k \geq k_{LG.}$) divergent oscillations would occur, i.e. the non-linear element introduced extends the stable region of the device, the device has a nonlinear vibration isolation compared to a linear high stability (figure 3).The controlled active vibration isolation devices with variable structure consist of several fixed structures which switching is made by the logical elements (*LE*) on the basis of algorithm of switching.

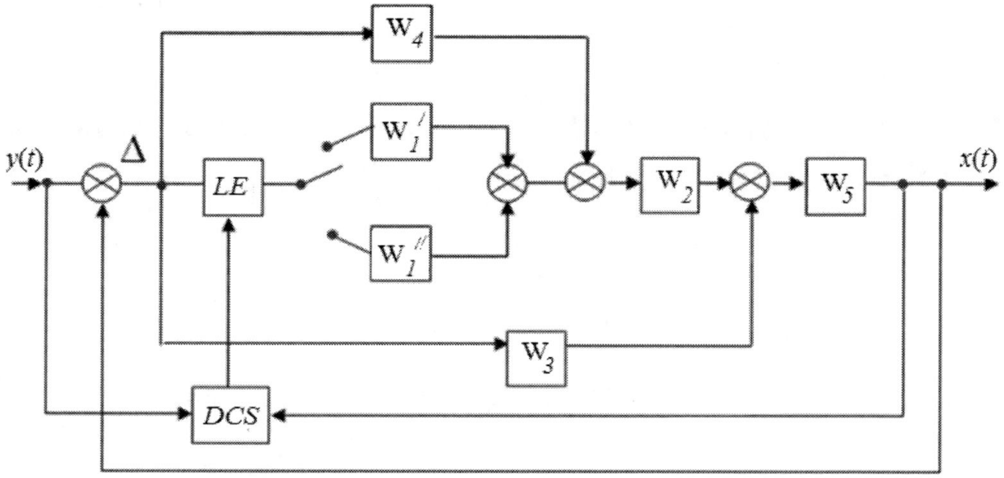

Figure 3. Algorithm of formation of information.

On figure 3 the block diagram with stepwise changing of rigidity and damping is; it is reached by automatic change of coefficients of strengthening of circuit of control of electromagnet current. The algorithm is formed based on the information about the value of external disturbance and condition of the object of vibration protection in the device of changing of structure (*DCS*). On figure 3 *y(t)* – harmonic oscillations of basement of the active vibration isolation device (external disturbance), *x(t)* – oscillations (deviations) of suspended body along the axis of the electromagnet, the absolute coordinate that defines the position of the object of vibroprotection relative to the earth. Variable parameters are the stiffness, mass, damping, etc.

4. Summary

Use in active vibration isolation device of the electromagnetic suspension with movable core allows refusing of use of accelerometers, which are relatively complex devices. In addition, the range of sensitivity is often limited, while the electromagnetic suspension with movable core provides the vibroprotection in almost the entire frequency range of external disturbances.

Thus it is established that the active vibration isolation device using the elastic forces of the electromagnetic suspension with variable structure is effectively at almost any value of frequency of the external disturbing force and the introduction of a nonlinear converter for forming voltage extends the range of stable work of the device.

References

[1] Vasendina E, Plotnikova I, Red'ko L, Zyablova (2015) Study of types of defects in wood chipboard production *IOP Conf. Series: Materials Science and Engineering* **81** 012070 1–7 doi:10.1088/1757-899X/81/1/012070
[2] Bezbah I J *et al* (2012) *Electronic Scientific and Technical Journal* **9** 317–324
[3] Melik-Shakhnazarov V A *et al* (2012) New design of active vibroprotection devices *Technical Physics Letters* **38**(3) 283–285 doi: 10.1134/S1063785012030303
[4] Kolovsky M Z 1976 *Automatic control of vibration isolation systems* (Moscow, Nauka)
[5] Fominova O V, Chernishev V I (2002) *Proceedings of the XXIX Summer School «Advanced Problems in Mechanics»* **1** 68–80
[6] Trofimov A I *et al* 1997 *Methods of the theory of automatic control, oriented to the use of computers* (Moscow, Energoatomizdat)

Journal of Physics: Conference Series **671** (2016) 012047 doi:10.1088/1742-6596/671/1/012047

Usage Autocorrelation Function in the Capacity of Indicator Shape of the Signal in Acoustic Emission Testing of Intricate Castings

Artem Popkov

Research student, Siberian State Transport University, Novosibirsk, Russia
E-mail: zabagy@gmail.com

Abstract. The article contains information about acoustic emission signals analysing using autocorrelation function. Operation factors were analysed, such as shape of signal, the origins time and carrier frequency. The purpose of work is estimating the validity of correlations methods analysing signals. Acoustic emission signal consist of different types of waves, which propagate on different trajectories in object of control. Acoustic emission signal is amplitude-, phase- and frequency-modeling signal. It was described by carrier frequency at a given point of time. Period of signal make up 12.5 microseconds and carrier frequency make up 80 kHz for analysing signal. Usage autocorrelation function like indicator the origin time of acoustic emission signal raises validity localization of emitters.

1. Introduction

There are many pilofacturing holes and imperfections in structure of freight bogie. Stochastic signals with complex shape result of acoustic emission testing of freight bogies. Coordinates of emitters and transmitters of acoustic emissions have an overriding influence on the operation factors of signals.

Topicality: algorithms and methods of processing signal are very significant part of acoustic-emissions testing. Perfection of methods and algorithms raise validity of result testing.

The purpose of work is estimating the validity of correlations methods analysing signals.

2. About autocorrelation function

The analysis objects are frequency, origins time and shape parameters of acoustic emission signals. Autocorrelation coefficient is on basis of calculating correlation coefficient by Pirson:

$$R_{j,l} = \frac{\sum_{k=0}^{w}\left(x_{k+j}-\bar{x}_1\right)\left(x_{k+j+l}-\bar{x}_2\right)}{\sqrt{\sum_{k=0}^{w}\left(x_{k+j}-\bar{x}_1\right)^2 \sum_{k=0}^{w}\left(x_{k+j+l}-\bar{x}_2\right)^2}},$$

where x – discrete acoustic emission signal, mV; j – sample of signal, w – summation interval; l – shift interval [1].

Function of autocorrelation was calculated for interval 1 and 2 with shift l (figure 1). Results of acoustic emission testing freight bogies were used in this work. It was received using the regular complex of acoustic emission control – digital acoustic emission diagnostic system SCAD 16.03 (Siberian Aeronautical Research Institute, Siberian State Transport University, Novosibirsk, Russia) with gain factor 2000, discreteness ADC 0.5 microsecond – in West-Siberian railways repair car sheds.

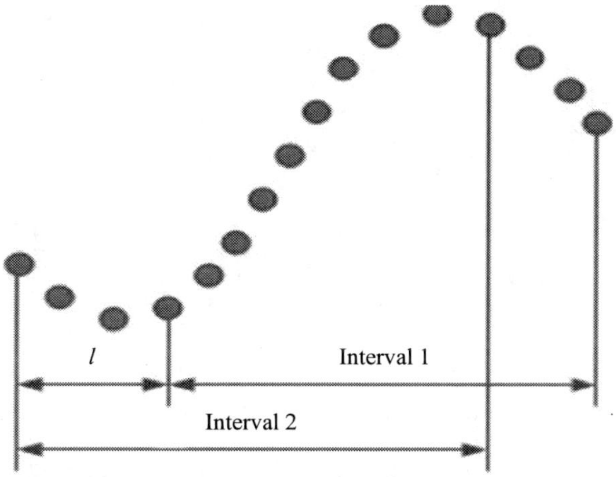

Figure 1. Scheme of summation intervals for calculating autocorrelation coefficient.

3. Determination of the operation factors

Modulus of autocorrelation function is approximately zero (below 0.2) before the origins time of signal. It happens on account of gears noises. Low-amplitude wave (below 1 mV) increase significantly autocorrelation coefficient (more 0.8). This pattern makes for increasing sensitivity of retrieval the origins time of signal with non-correlation gears noises [2].

Acoustic emission signal consist of different types of waves, which propagate on different trajectories in object of control [3]. Therefore shape of signal a lot vary with time (figure 2). Vertical lines mark turning points, which agree with confines of different parts in signal, on the diagram (figure 3).

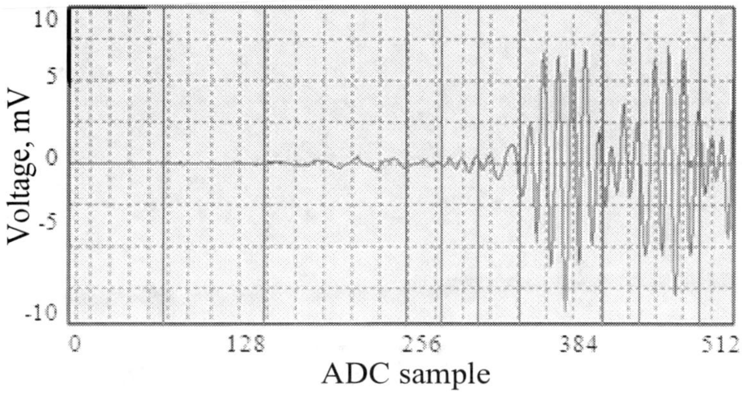

Figure 2. Acoustic emission signal.

Acoustic emission signal is amplitude-, phase- and frequency-modeling signal. It was described by carrier frequency at a given point of time. Evidently, autocorrelation function of interval 1 and interval 2 with shift l is periodic function.

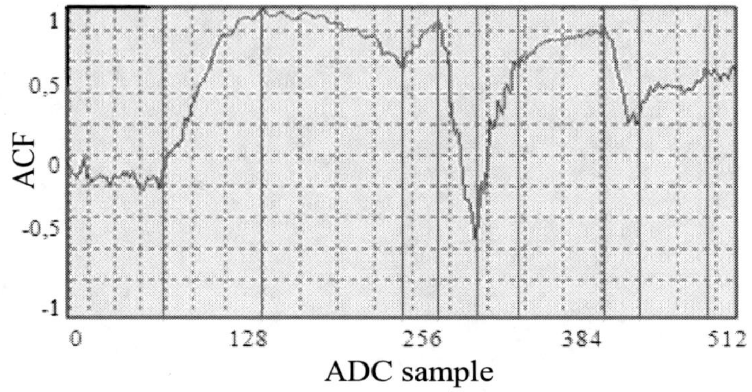

Figure 3. Autocorrelation coefficient time relation with w =64 and l = 26.

Autocorrelation coefficients with various shifts $l > 0$ (figure 4) were calculated for searching typical period of signal. Period of signal (with ADC simple) was determined according to $l > 0$ by which autocorrelation coefficient is maximal. If minimal autocorrelation coefficient fits half-period shift, then period was determined correctly [4]. Period of signal make up 12.5 microseconds and carrier frequency make up 80 kHz for diagram in figure 3. It fits the specifications of the transmitter.

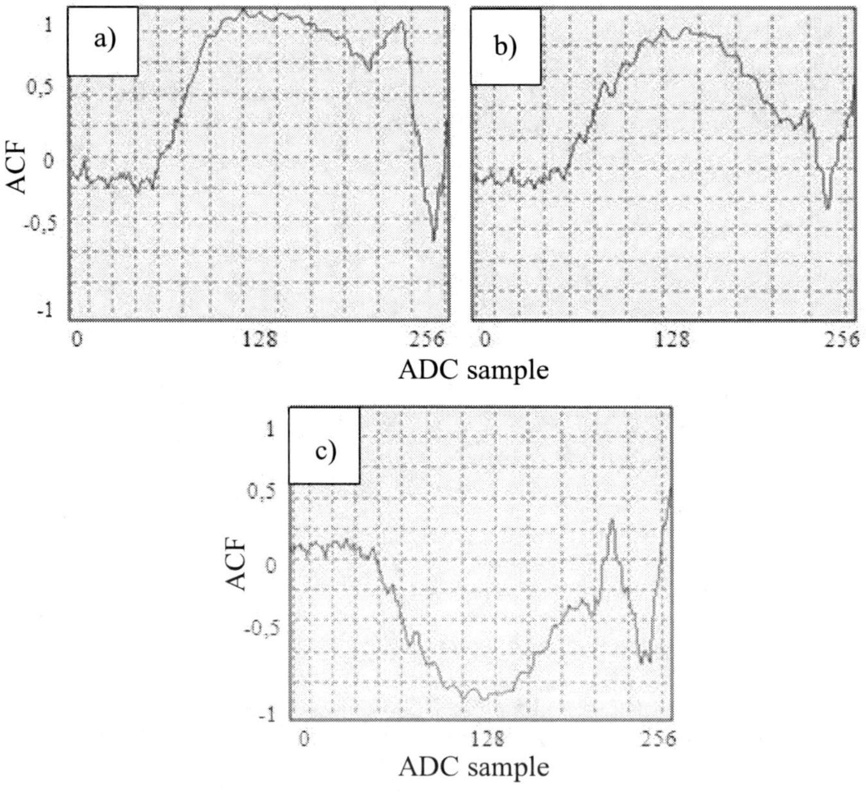

Figure 4. Autocorrelation coefficient time relation with w = 64 and l = 26 (a), l = 51 (b), l = 39 (c).

Threshold-moving-window method isn't effective. Inadmissible error may appear when the origins time of signal t_l was determined (figure 5). Usage autocorrelation function in the capacity of origin time of signal raise localization validity. Acoustic emission signal is superposition of stochastic non-correlated noise and impact part of signal [5]. Correlation coefficient more than 0.7 fits close coupling of values. Increase of coefficient autocorrelation indicates the origin time t_2 of signal (figure 5).

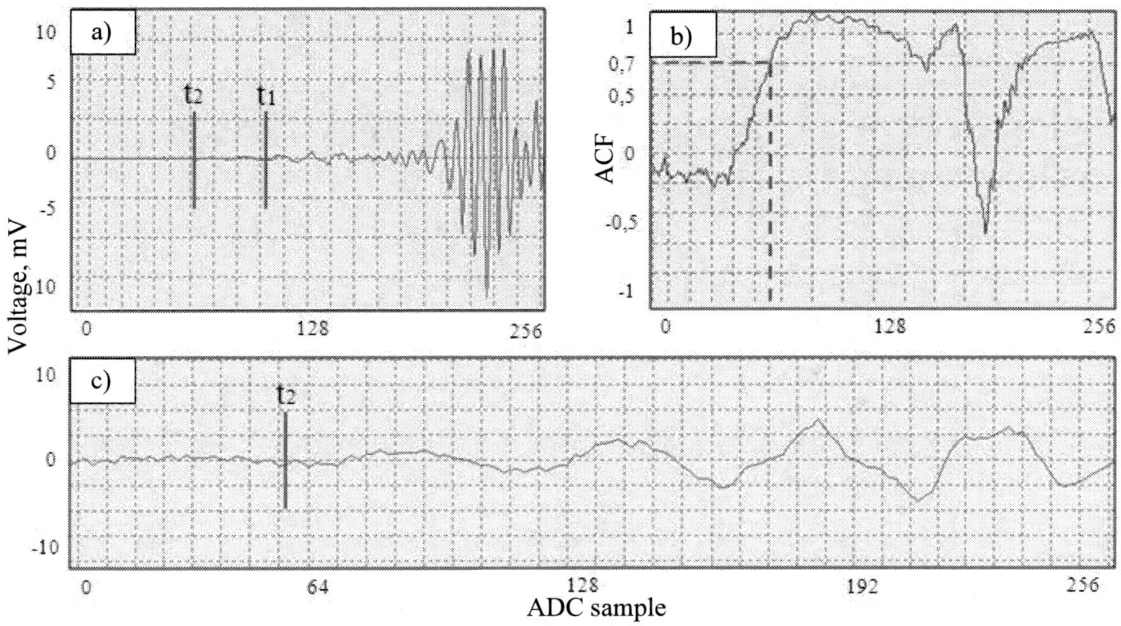

Figure 5. Indicating the origin time of signal: a) acoustic emission signal, b) autocorrelation function, c) autocorrelation in zoom.

4. Summary

In autocorrelation analysis basic operation factor is displacement of intervals. Intervals displacement is indicator which permits to calculate local value of carrier frequency. It is equal to 80 kHz. This value fits the specifications of the transmitter. Using autocorrelation analysis permits to select parts of signal with different form, which are justified different kind of oscillation in object of control.

Usage autocorrelation function like indicator shape of acoustic emission signal raises validity clusterization of emitters by dint of differentiation parts of pulse. Usage autocorrelation function like indicator the origin time of acoustic emission signal raises validity localization of emitters.

References

[1] Allen R L, Mills D W 2004 *Signal Analysis: Time, Frequency, Scale, And Structure* (NJ: IEEE Press)

[2] Ifeachor E C, Jervis B W 2002 *Digital Signal Processing: A Practical Approach (2e)* (UK: Pearson Hall)

[3] Vereskun V D, Stepanova L N, Beher S A, Bobrov A L, Kabanov S I, Lebedev E Ju (2013) Acoustic emission control of box-shaped solebar of cart of the freight car at repair in depot *Testing. Diagnostics* (1) 67—71

[4] Brandt A 2011 *Noise and Vibration Analysis. Signal Analysis and Experimental Procedures* (UK: John Wiley & Sons Ltd.)

[5] Ono K (2011) Acoustic Emission in Materials Research – A Preview *Journal of Acoustic Emission* **29** 284–308

Development of a Calculation Methodology for the Ventilation on a Besis of a Mobile Unit

A Sechin[1], A Popov[2] and O Antonevich[3]

[1] PhD, National research Tomsk Polytechnic University, Tomsk, Russia
[2] Associate Professor, National research Tomsk Polytechnic University, Tomsk, Russia
[3] Postgraduate, National research Tomsk Polytechnic University, Tomsk, Russia
E-mail: LandBio@mail.ru

Abstract: An algorithm for the analysis of safety and efficiency of the processes, which are located inside the mobile unit are developed. It follows from the calculations that the safe concentration of combustible material in the space of industrial premises is about 3.69%. Automation systems must be focused on this value. The time of occurrence of the maximum permissible concentration of pollutant was determined and amounted to 160 seconds. It is shown that the ventilation rate of 0.5 would be sufficient for functioning of the object.

1.Introduction

Technological processes, occurring in the territory of the mobile units, require both safety and operation efficiency. The analysis order and reviwering the situation development is urgent because the operation stability of the facility and staff safety health depend on the successful application of the developed actions. Emergencies, related to the ignition of flammable gases in the mobile unit, may occure in case of nonobcervance of general hygiene requirements for the air of working area, as well as fire and explosion safety. In most cases, these accidents are accompanied by high material costs for the companies involved in operation of the facility and the significant environmental damage [1,2].

The aim of this work is to develop methods (algorithm) of choice of ventilation for mobile units, which spaces are used as an industrial premises.

2.Analysis of the object

The object, for which the algorithm was developed, is a mobile unit, located in a standard 20-foot container, that allows to provide transportability, autonomy, the ability to model different operating conditions in a confined space.

Traditionally container is insulated with 10 cm styrofoam layer, floor strengthening is made by metal beams, the door is cut through for the access to the maintenance of working capacities with feedstock.

The container is lined with plasterboard inside, equipped with electrical panels and means of fire safety. Power outlets and lighting are set.

All metal parts are made of stainless steel in order to avoid premature corrosion of the metal, because operation temperature of the object varies from -30 ° C to +70 °C. Instalation design is optimized to maximize the use of final products: pipes, fittings.

3. Analysis of the facility safety

Installation, that is located in a mobile unit, is under overpressure of 0.04 atmospheres. Based on the fact that the gas is promptly removed from the gas tank - maximum pressure in the system does not exceed 1.04 atmospheres.

Part of methane obtained after purification is used for the gas electric generator disposed in the container, providing the work of all electrical systems and serving as a source of heat. The power of generator is 1000 watts. Burning of 4 m^3 of methane in power generator provides 12 kW of power and 24 kW of heat.

The main hazard in the operation of the facility is methane gas. Table 1 summarizes the main characteristics of methane [3].

Table 1. Main characteristics of methane

Indicator	**methane**
Chemical formula	CH_4
The density of the gas phase under NC*, kg/m^3	0.717
Net calorific value under NC *, MJ/m^3	35.76
The lowest concentration limit of inflammation of mixture with air under NC* , ^{0}C	5.0
Weight explosive limit, %	5.0...15.0
MPC in the air of working area, mg/m^3	7000
MPC in atmospheric air, mg/m^3	50

NC * – under normal conditions ($t = 0\ ^0$C; $p = 101.3$ kPa)
MPC – maximum permissible concentration
Fire extinguishing means: inert gases

It follows from the analysis of the results that inflammation area in air is 5,28-14,1% vol., concentration limit of inflammation (reduced to 25 °C): lowest 5.28, top. 15% vol.

We are interested in the lowest concentration limit of inflammation - about 5.28%.

Measures to ensure fire and explosion safety processes were developed according to the requirements of State Standard 12.1.004 and State Standard 12.1.010.

According to these documents to ensure fire and explosion safety of production processes, processing, storage and transportation of substances and materials it is necessary to use data rates of fire and explosion hazard of substances and materials with a safety factor $\varphi g\ no \leq 0.9\ (\varphi n - 0.7\ R)$

where φg safety - a safe concentration of the flammable substances,% vol. (gm-3); φn - the lower flammability limit of a mixture of a fuel with air,% vol. (gm-3); R – reproducibility of the method of determining the fire danger rating at 95% confidence level.

Calculations show that a safe concentration of the flammable substance is about 3.69%.

4. Definition of the time to reach maximum permissible concentration (MPC) or explosive concentrations of pollutant from the material balance

When production area has no forced ventilation or ventilation is not reliable (there are no redundant fans, electric power is not carried out by two independent feeders), the most favorable conditions for the formation of explosive concentrations in case of damages and accidents at production facilities and pipelines are created.

In this case, the amount of flammable material coming out of the equipment during the time interval $d\tau$, must be equal to the increment of the amount of flammable material in the air volume over the same period of time $d\tau$, or

$$q d\tau = V dC \quad (1)$$

where q – amount of substance emitted to the outside in the unit of time; V – volume of the room; dC – increment of the concentration of flammable material during $d\tau$.

$$\tau = \frac{C_{ЗАГ}}{q}V \tag{2}$$

Figure 1 shows a graph of a linear relationship - the result of the calculation using the equation (2), which clearly shows that the increase in the concentration of the pollutant is uniform over time. The graph shows that the time to reach a dangerous situation, if the exhaust velocity of methane in the volume space is 160 seconds. The rate of inflow of methane in the areas was taken from statistically average datas - 0.00072 g / s. The time of hazard situation was determined based on the fact that the lowest flammable limit was taken (5.28 vol.%), which was converted based on the safety factor (0.32%). Finally we got the concentration (3.69% vol.), which was correlated with the dependence graph, thereby the occurrence time of a dangerous situation was determined. The obtained time plays an important role in the choice of technological equipment of ventilation systems, which the operation of automation systems will focus on.

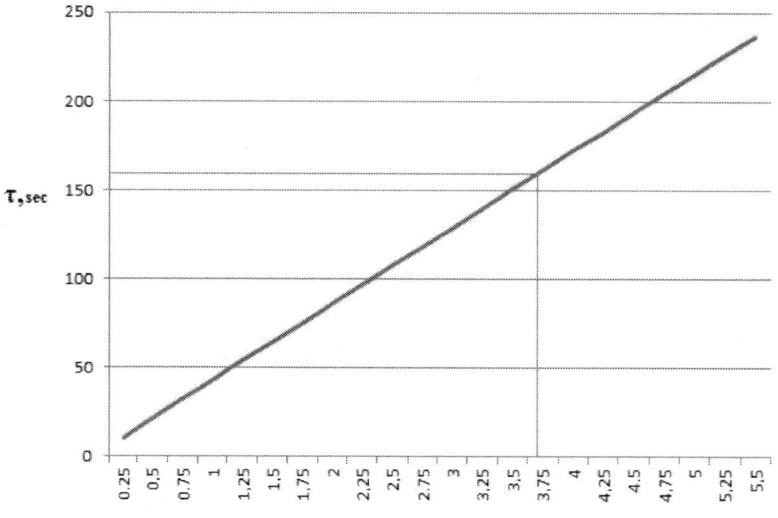

Figure 1. Graph of a linear relationship.

During the calm weather, general ventilation, which is carried by the deflector will not be effective. In this case, when the concentration of the pollutant reaches its limit, the start of emergency ventilation must be operated automatically.

Figure 2 shows the results of further calculations, taking into account the ventilation rate in the room. Analyzing the chart we see that the ventilation rate of 0.5 would be sufficient for the safe operation of the facility. This value is important in the selection of ventilation systems, because it shows the required volume of air to be disposed from the room.

In case of guaranteed ventilation rate a part of an outgoing vapors or gases in the accident are continuously withdrawn from the production area to the outside, as a result the increase in the concentration to dangerous limits will relatively slow down.

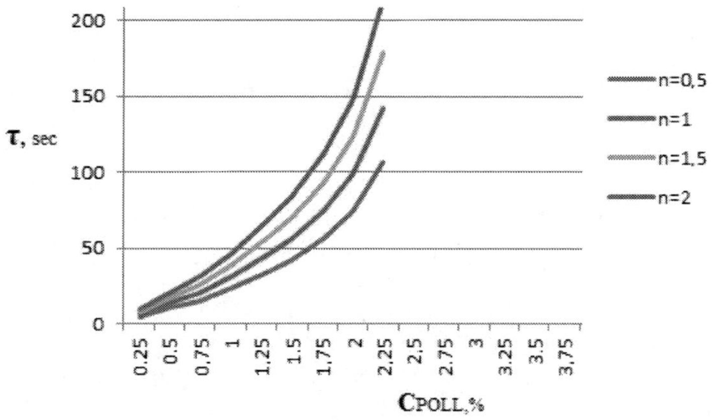

Figure 2. The rise time of the concentration of flammable vapors and gases in the volume of the mobile unit based on the ventilation rate.

In significant damages and accidents devices or pipelines with flammable vapors and gases in the premises an explosive concentration may form even when the operation of ventilation (including emergency), and so quickly that it would be impossible to implement the necessary emergency actions manually [1, 2, 4–6].

The inclusion of the ventilation system is performed when a safe concentration of a fuel (methane) is exceeded, which is about 3.69%.

5. Development of the algorithm of research for the ventilation system of the object

Figure 3 provides an algorithm for the research of the object on the basis of the mobile unit.

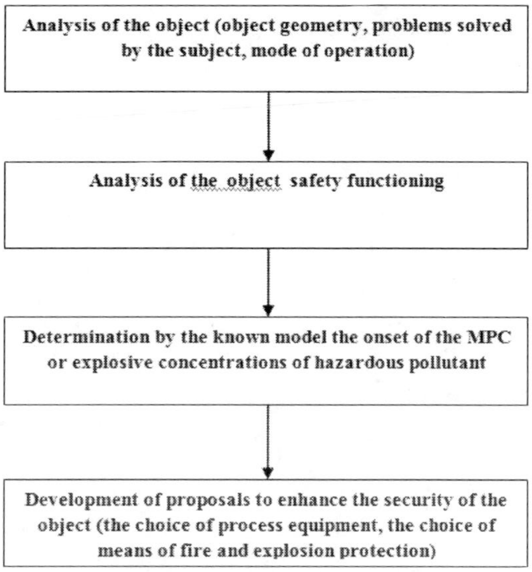

Figure 3. The research algorithm for the object on the basis of the mobile unit.

6. Conclusion

In the result of the research the algorithm to analyze the safety and efficiency of the processes, which are located inside the mobile unit was developed.

It follows from the calculations that the safe concentration of combustible material in the space of technological premises is about 3.69%. Automation systems must focus on this data.

The time of occurrence of the maximum permissible concentration of pollutant was determined and amounted 160 sec. It was shown that the ventilation rate of 0.5 would be sufficient for the investigated object.

Reference:

[1] Krasnih B A, Martyniuk V F, Sergienko T S, Sorokin A A, Feoktistov A A, Nechaev A S 2003 *An analysis of accidents and incidents on objects of gas supervision* (Moscow, OOO "Hazard Analysis")
[2] Avanesov V S, Aleksandrov A B, Alexandrov A I *et al.* 2002 *Analysis of accidents and fatalities in the oil and gas complex of Russia* (Moscow, OOO "Hazard Analysis")
[3] Monahov V T 2007 *Fire hazard substances. The analysis and prediction. Appendix 2. Profiles of fire danger of substances and materials* (Moscow, FGU VNIIPO of EMERCOM of Russia)
[4] Zerkalov D V, Lutz T E 2012 *Industrial Safety. Occupational Safety Standards* [Electronic resource] (Kiev, Osnova)
[5] Electronic resource: http://www.climatspb.ru/?node_id=824
[6] Electronic resource: http://www.refbzd.ru/viewreferat-1818-2.html

Problems of Automatic Test of Insulation in Cable Production

V V Red'ko[1], A P Leonov[2], L A Red'ko[3] and V A Bolgova[4]

[1] Associate Professor at the Tomsk Polytechnic University, Tomsk, Russia
[2] Associate Professor at the Tomsk Polytechnic University, Tomsk, Russia
[3] Associate Professor at the Tomsk Polytechnic University, Tomsk, Russia
[4] Post-graduate student of the Tomsk Polytechnic University, Tomsk, Russia
E-mail: sales@redwill.ru

Abstract. The article presents a qualitative and quantitative assessment of cable products insulation defects that can be reliably detected by means of the electrosparking control during the cable production process. The performance potential of technological control is evaluated: the limit of reliable detection of defective places in insulation taking into account the technical capabilities of modern control devices is marked.

1. Introduction

Electrosparking control of cable products insulation is regulated by a number of national and international standards [1–5] and it has been widely used in the cable industry since the 50s of the 20th century. Despite of the great world experience of its application, there are a number of unresolved issues, one of which is the lack of a unified approach to the qualitative and quantitative assessment of detected defects [6–8]. Experts in this field have different opinions about the control possibilities. There is a widespread belief that the electrosparking defect detector may detect furthermore the micro-defects in the insulation layer. Moreover, currently a large number of control devices of various types and manufacturers is applied and their defect detectors are based on different principles of operation [9, 10]. This often leads to the control reliability decreasing and production costs increasing. This article presents the results of theoretical and experimental studies aimed at addressing this issue.

2. Theory

In order to analyze the defect detection possibility it is proposed to classify the factors of the electric field and the insulation defective area interaction on the three basic physical signs [11–13].

The first and basic sign: a defect presence leads to a reduction of the insulation area dielectric strength, which in turn will leads to an electrical breakdown during the electrosparking control.

The second sign: an insulation defect consists in the geometry or the material properties changing. Both of them inevitably lead to a change in capacitance and dielectric losses in the defective area. Consequently, defects can be revealed by controlling the capacitance and / or dielectric losses of cable products insulation.

The third sign is the level of partial discharge in the insulation defects under its internal layers. The partial discharge high level indicates the presence of porosity, air pockets in insulation; violation of the solidity and homogeneity.

In this paper the defect detecting possibility upon an electrical breakdown is studied on the example of the main cable insulation materials such as PVC plasticate, polyethylene and rubber.

The main types of manufacturing defects in the cable products insulation are shown in the figure 1.

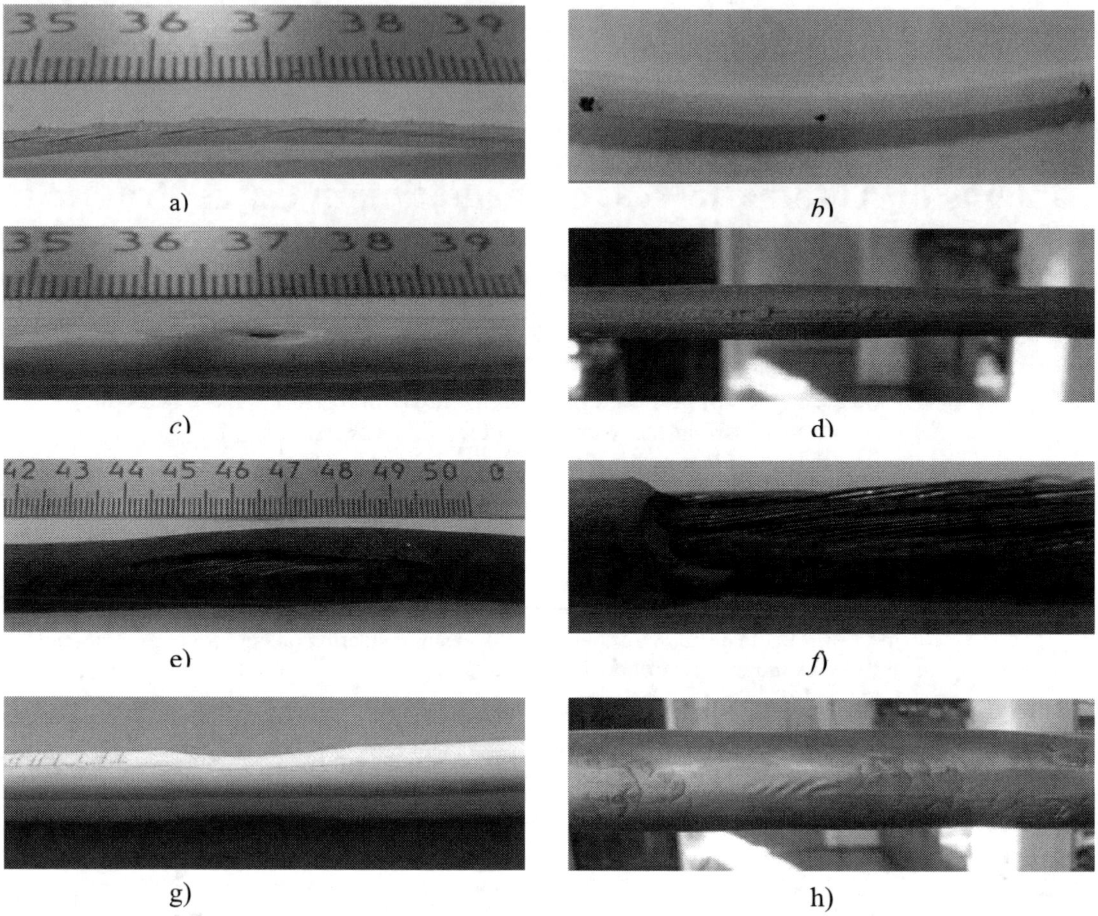

Figure 1. The main types of cable products insulation defects, insulation "slice" – a; impurity inclusion – *b*; local rupture of insulation and bumps – *c, d, e*; insulation material adhering on the cable cores – *f*, insulation porosity – *g*; outer diameter inequality, buildup along the insulation surface – *h*.

The physical principle discussed control type lets to detect only those types of defects, in which a spark discharge takes place in a short time finding the cable insulation area in the control zone (about 0.001..0.1 sec). It is shown by a long-term practice that when voltage levels selected for different thicknesses of insulation made of rubber or plasticate, in accordance with the test assessment references, the spark discharge occurs in 100% of cases only for through defects or defects close to it. It can be passing through the insulation cracks, holes, "stripping" and conducting inclusions. The question arises: is it possible to detect not only through defects by electrosparking control (air pockets, thinnings, cuts, undulations, buildups, etc.)? It is obvious that there is a minimum insulation thickness in which an electrical breakdown will occur during the control. Thickness of breaking defective insulation areas depends on the design of controlled cable products insulation material, technological features and control modes. The above factors influence estimation and examples of calculations of maximum thickness breaking defective insulation areas are given in the article.

3. Experimental part

The electric field distribution in the insulation defect such as a "crack" is shown in the figure 2.

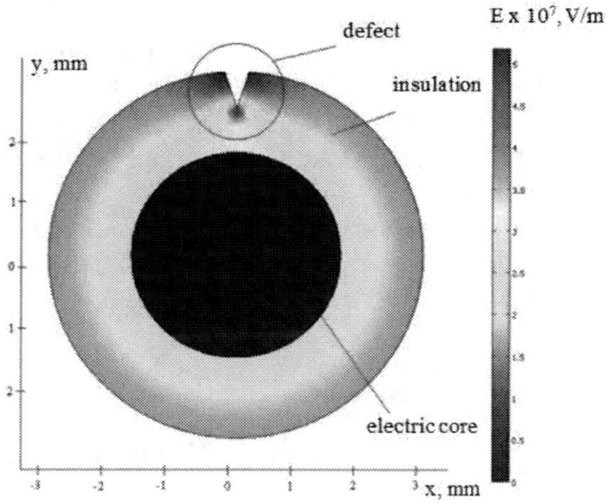

Figure 2. The electric field distribution in the insulation defect such as a "crack".

The basis for the breakdown of non-through defects under the electrosparking control is a heat balance violation in insulation.

As a result of studies a thermal model of the defective area has been proposed. The model main point consists in the defective area initial state and the control conditions comprehensive analysis. In consequence of local overheating a thermal breakdown occurs [14]. The main functional connections describing the process are given below.

$$h_d = \frac{U_{test}}{E_d(h, T, tg\delta)}, \tag{1}$$

$$\Delta T / t_{test} = \frac{U^2_{test}\omega C_{def} tg\delta}{cm}, \tag{2}$$

In the given equations E_d is an insulation dielectric strength determined as a function of the insulation radial thickness h, the actual temperature T, and the dielectric losses $tg\delta$.

It is determined that depending on the control conditions and the material properties the insulation heating rate is equal to $(0.3 \div 25)^\circ$C/sec. The results of the defective areas overheating rate calculation are shown in the Table 1.

Table 1. Overheating rates of defective areas.

Insulation material, cable product brand	h_d/h, mm/mm	$\Delta T/\Delta t$, ° C/sec
PVC plasticate, MGShV-0.35	0.14/0.62 = 0.23	13
Polyethylene, SIP-3	0.29/2.25 = 0.13	0.26
Rubber, APVR	0.33/1.2 = 0.28	25

In this calculation the initial average dielectric temperature $T_1 = 60^\circ$C, after cooling bath. The test voltage frequency $f_{test} = 1$ kHz, the test voltage application duration $t_{test} = 10$ msec.

The above information allows us to conclude that at high-speed control, at voltage application times approximately equal to several units - tens of milliseconds, insulation overheating does not exceed a few degrees and has no significant effect on the insulation dielectric strength, and therefore on the detection of defective areas with insulation thin layers probability. In case of the control at low speeds at voltage application times more than 1 second significant overheating occurs in the insulating materials with high dielectric losses. Overheating leads to decreasing dielectric strength, up to the possibility of breakdown and defect-free insulation inflammation, for example, insulation made of rubber or some brands of PVC plasticate.

The through defect minimum size is calculated (figure 3) which can be detected by means of the electrosparking control method. Obviously, there is a through defect with extremely small cross-sectional area of the conducting channel which will not be detected.

Figure 3. Example of a through defect.

An example of a through defect in the insulation made of rubber with resistance equal at least to 10^{10} Ω, 10^{14} Ω for polyvinylchloride and not less than 10^{15} Ω for polyethylene is considered. Based on the existing standards requirements [1–7] it is necessary to achieve the defects detector operation threshold at the controlling voltage $U = 3$ kV and the current through the defective area $I = 600$ μA. In this case a surface insulation resistance of not more than 5 MΩ ($R_{surf} = 5$ MΩ) must be provided. The controlling voltage $U = 3$ kV corresponds to the insulation thickness $h = 0.25$ mm. The tested material is the rubber with the lowest surface resistance. The circumference of a circular shape through defect when the desired resistance can be reached is equal to $l_c = \rho \cdot (h/R_{surf}) = 0.5$ mm. The hole diameter $d_{def} = 0.16$ mm.

The calculation of the defect smallest diameter at the electric discharge is given below. A glow discharge occurs at current values up to $I = 10^{-3}$ A. The glow discharge is characterized by a high voltage between the electrodes. The current density in the glow discharge reaches $J = 10^6$ A / m^2. Thus to achieve the defect current level $I_{def} = 600$ μA a hole diameter of at least $d_{def} = 28$ μm is required. At larger values of the controlling voltage the discharge current in defects increases and the glow discharge grows into an arc discharge. During the arc discharge the current density is in the range $J = (10^{12} \div 10^6)$ A/m^2 and characterized by a low voltage between the electrodes.

Comparing influencing factors the surface resistance and discharge in the air gap it can be concluded that in order to reliable detection of the through defects in the cable product insulation its diameter should be not less than 30 μm. The surface resistance almost has no effect to the defects detection.

4. Conclusion

Despite the widespread use of electrosparking control in the cable industry there aren't quantitative and qualitative assessments of defects detected by this type of control [1, 2, and 15].

In the above paper a method of estimating defects on three factors of the electric field and the insulation defective area interaction, such as thermal breakdown, capacitance and dielectric losses control as well as the partial discharges level control, is proposed.

The through defects or defects with the dielectric residual layer is not more than 30% of the defect-free value can be reliably detected by means of the electrosparking control. The minimum diameter of the through holes which can be detected by this control is equal to 30 μm.

References

[1] IEC 62230:2006 *Electric cables – Spark test method*

[2] British standard BS EN 50356:2002 *Method for spark testing of cables*
[3] BS 5099:2004 *Electric cables – Voltage levels for spark testing*
[4] NEMA Standards Publication WC 56-1986 (R2005) *3.0 kHz Insulation Continuity Proof Testing of Wire and Cable*
[5] NEMA Standards Publication WC 52-2005 *High-Temperature and Electronic Insulated Wire, Impulse Dielectric Testing*
[6] NEMA Standards Publication WC 53-2000 *Standard Test Methods for Extruded Dielectric Power, Control, Instrumentation, and Portable Cables for Test*
[7] NEMA Standards Publication WC 57-2004 *Standard for Control, Thermocouple Extension, and Instrumentation Cables*
[8] ASTM D149 – 09 2013 *Standard Test Method for Dielectric Breakdown Voltage and Dielectric Strength of Solid Electrical Insulating Materials at Commercial Power Frequencies*
[9] IEC 243-1 1988 *Methods of test for electric strength of solid insulating materials, Part 1: Tests at power frequencies*
[10] ANSI/NEMA ICEA T-27-581 2008 *Standard Test Methods for Extruded Dielectric Power, Control, Instrumentation, and Portable Cables for Test*
[11] ASTM D257-07 2012 *Standard Test Methods for DC Resistance or Conductance of Insulating Materials*
[12] Chursin Y A, Fedorov E M (2015) Methods of resolution enhancement of laser diameter measuring instruments *Optics & Laser Technology* **67** 86–92
[13] Leonov A P, Red'ko V V, Red'ko L A and Soldatenko E Y (2015) Determination of enamel insulation corona resistance by high-frequency modulated pulses *IOP Conference Series: Materials Science and Engineering* **81** Article number 012094
[14] Razevig D V 1976 *Tekhnika vysokikh napryazheniy [Equipment of high tension]* (Moscow: Energy)
[15] Holodniy S D 1991 *Metody ispytaniy i diagnostiki kabeley i provodov [Methods of tests and diagnostics of cables and wires]* (Moscow: Energoatomizdat)

Automation of the Technological Process to Produce Building Frame-Monolithic Modules Based on Fluoranhydrite

J Fedorchuk [1], M Sadenova[2] and O Rusina[3]

[1] PhD, National research Tomsk Polytechnic University, Tomsk, Russia
[2] Associate Professor, Vostochno-Kazakhstan state University S. Amanzholov, Ust-Kamenogorsk, Kazakhstan
[3] Postgraduate, National research Tomsk Polytechnic University, Tomsk, Russia
E-mail: olgarusina@tpu.ru

Abstract. The paper first proposes the automation of the technological process to produce building frame-monolithic modules from production wastes, namely technogenic anhydrite and fluoranhydrite. A functional diagram of the process automation is developed, the devices to perform control and maintenance with account of the production characteristics are chosen.

1. Description of the technological process to produce frame-monolithic modules based on technogenic fluoranhydrite

Fluoranhydrite is one of the most common wastes in the chemical industry. Scientists of TPU have proposed a new method for manufacturing of fluoroanhydrite-based building frame-monolithic modules.

The production scheme for anhydrite frame-monolithic modules (FMM) is shown in figure 1.

Anhydrous astringent (AA) produced from technogenic anhydrite unification is delivered to anhydrite astringent feed hopper 2 via the pneumatic pipeline through cyclone 1. From feed hopper 2, it is delivered to skip 4 via metering screw 3. The undersized fraction (less than 20 mm) of ash and slag is dumped into feed hopper 5 and delivered to skip 4 with metering screw 6. From skip 4, bulk materials are loaded into concrete mixer (CM) 7. A measured amount of water flows from tank 10 through batcher 9 to CM 7. To provide disposability of this technology, after cyclone 1, dust-loaded air is delivered to water absorber 8 (water is absorbent). After filling hopper 2, water circulating is stopped through absorber 8 and the pulp anhydrite suspended in water) is delivered to CM 7 through batcher 9. After mixing, anhydrite-slag concrete is delivered to the inter-former space of the preassembled removable reusable formwork around the metal frame on the ground where the module is produced 11. The module frame is made on the ground for module metal frame production 13. The frame is made the rolled steel pre-delivered to warehouse 12.

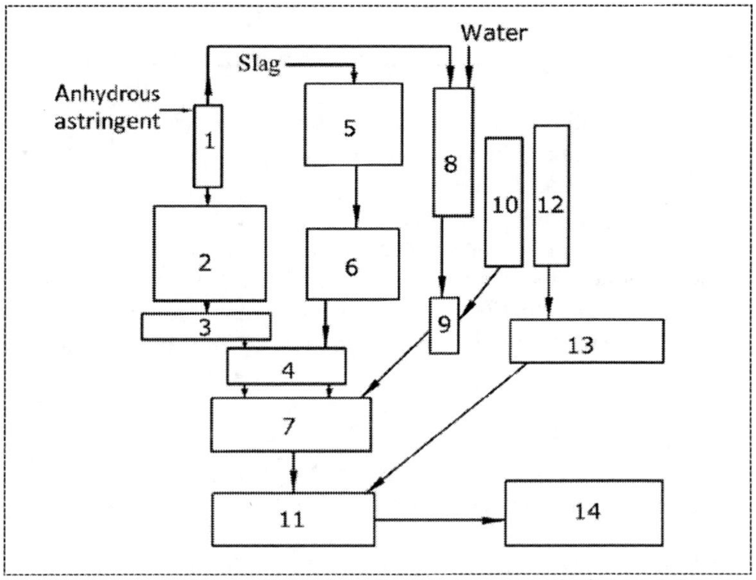

Figure 1. Production scheme for anhydrite frame-monolithic module production: cyclone (1); anhydrous astringent feed hopper (2); anhydrous astringent metering screw (3); skip (4); ash and slag feed hopper (5); ash and slag metering screw (6); concrete-mixer (7); absorber (8); batcher (9); tank with water (10); ground for frame-monolithic module manufacturing (11); warehouse to store rolled steel and removable metal forms (12); ground for module metal frame manufacturing (13); warehouse for semi-finished-products –frame-monolithic modules(14).

To meet the requirements of the construction code in the production process, all stages of the process are to be controlled. For this purpose, first of all, the parameters that facilitate the start-up, adjustment and normal technological process are to be controlled. These are input and output parameters, and if they change, the object may undergo disturbing influence.

The parameters for frame-monolithic module manufacturing are as follows:

- anhydrous astringent weight in hopper No. 1;
- undersized slag weight;
- level of water in tank No. 2;
- water flow 5;
- anhydrous astringent flow rate;
- rotation of the concrete mixer shaft;
- mass level in the concrete mixer;
- the air pressure magnitude in the compressor receiver;
- vertical and horizontal position of the metal channels to weld the module frame;
- mass level in the inter-framer space;
- mass density in the inter-framer space.

Quality control of the wall material must be carried out through controlling the amount of water-soluble calcium sulfate in the technogenic anhydrite that provides binding properties of the anhydrite astringent on the one hand. On the other hand, it is necessary to control the amount of each component of the mortar within a single loading of the concrete mixer [1–7].

2. Selection and justification of the functional scheme for automated production of the anhydrite FMM

To be reliable, the hardware and software system should comprise at least three levels. The lower level contains sensors and actuators. The middle level includes a master controller. The operator's workstation with a personal computer is located at the top level.

The process of manufacturing frame-monolithic premise modules requires accurate dosing of basic components for the specified productivity, and accuracy in geometric spatial dimensions. Therefore, a functional scheme of the automation system should comprise a frequency converter, a contactor, a normalizing summing amplifier, an induction motor, a weight gauge, a metering screw, a feed hopper, limit switches, a laser level gauge, a radioisotope densimeter, a betatron, a manometer, a level gauge and a flow meter.

3. Selection of sensors and elements for the measuring system

To control the weight of the anhydrous astringent and undersized slag in the hopper, it is necessary to install at least three weight sensors, considering the weight of the hopper and the equipment fixed on it.

The normalizing device should comprise at least three inputs for the sensors capable of accumulating the input signals.

The limit switches VK-200 and the snap-action limit switches VK-300 are taken for the operation in the electrical circuits to control AC voltage up to 660 V, a frequency of 50–60 Hz, and DC voltage up to 440 V under the influence of the controlling stops (cams) at certain path points of the controlled object. These switches are resistant to splashes and spills of accidental water or oil [8].

BURKERT S030 flow meter is suitable for measuring the volume flow rate.

The radio-wave (radar) level gauges intended for a variety of process tanks (hoppers, reservoirs, silos), stationary objects, and for sending and receiving information using other technical means of automatic control systems are suggested for non-contact continuous measurement of the bulk and liquid levels.

The gamma densimeters, which provide the density measurements in the range of 600–2500 kg/m3 with an error of no more than 2–2.5%, are suitable for measuring density.

The air pressure in the compressor receiver is measured by the manometer connected to the electrical relay circuit to control the compressor motor.

The geometric dimension control of each of the produced module frames is carried out using a laser level gauge, for example, BOSCH PLL 360 SET with a tripod: diameter working range is 20 m, measurement accuracy is 0.4 mm/m, laser beam projection is linear, number of beams is 2, laser beam alignment is automatic.

The betatrons, induction electron accelerators, are used to control the steel welded joints with a thickness of 100–900 mm. The betatron is a transformer with a primary winding fed by high voltage current, with a frequency of 50 Hz or higher.

Thus, implementation of the developed automatic system with the proposed equipment will enable the control of anhydrite frame-monolithic modules production to produce the construction products of high quality [8], and will allow the automation of the building construction in the construction industry.

References

[1] Fedorchuk Yu M, Abramov Yu P, Velichko N S, Gorbunov Yu M, Kurin N P, Kobzar' Yu F, Morev V V, Ogurtsov A V, Seredenko V A, Turaev N S *The method to neutralize fluorine-anhydride waste in anhydrous hydrogen fluoride production.* Certificate of authorship No. 1570216 USSR / No. 4405569/26, 13.01.1988, publ. 27.01.2000. Bul. No. 3

[2] Fedorchuk Yu M *The method to produce anhydrous astringen*t Patent RF No. 2277515, 01.04.2002

[3] Fedorchuk Yu M, Zykov V M, Zykova N S, Tsygankova T S *Construction mix and the method of its preparation* Patent RF No. 2266877 No. 2004119704, 28.06.2004, publ. 27.12.2005. Bul. No. 36
[4] Nedavnij O N, Fedorchuk Yu M, Belenkov S B (1993) Fluorine-anhydride based coating *News of higher educational institutions. Construction* (3) 56–57
[5] Fedorchuk Yu M, Vereschagin V I, Dyachenko A N, Komarov O G, Lazarchuk V V, Tomash Yu Ya Slag stone production line. Utility model certificate 27307 No. 2002107806/20, 01.04.2002, publ. 20.01.2003. Bul. No. 15
[6] Fedorchuk Yu M 2003 I*ndustrial anhydride, its properties and application* (Tomsk: TSU Publishing House)
[7] Lapshenkov G I, Polotskij L M 1982 *Automatic chemical production. Theory, calculation and design of automation systems* (Moscow, Chemistry)
[8] The development of the automation system to dose and mix bulk materials [electronic resource]. Information on: http://knowledge.allbest.ru/manufacture/2c0b65625a3bd78a4c53b88421306d27_0.html27_0.html

The Dosimetric Parameters Investigation of the Pulsed X-ray and Gamma Radiation Sources

S G Stuchebrov[1], I A Miloichikova[2] and X O Shilova[3]

[1] Senior Researcher, Tomsk Polytechnic University, Tomsk, Russia
[2] Engineer, Tomsk Polytechnic University, Tomsk, Russia
[3] Master student, Tomsk Polytechnic University, Tomsk, Russia
E-mail: stuchebrov@tpu.ru

Abstract. The most common type of radiation used for diagnostic purposes are X-rays. However, X-rays methods have limitations related to the radiation dose for the biological objects. It is known that the use of the pulsed emitting source synchronized with the detection equipment for internal density visualization of objects significant reduces the radiation dose to the object. In the article the analysis of the suitability of the different dosimetric equipment for the radiation dose estimation of the pulsed emitting sources is carried out. The approbation results on the pulsed X-ray generator RAP-160-5 of the dosimetry systems workability with the pulse radiation and its operation range are presented. The results of the dose field investigation of the portable betatron OB-4 are demonstrated. The depth dose distribution in the air, lead and water of the pulsed bremsstrahlung generated by betatron are shown.

1. Introduction

Recent years around the world characterized by rapid technological development for digital X-ray diagnostic devices, and as a consequence of the emergence of a variety of technical solutions based on these technologies. Today, the object's internal density visualization based on the X-ray methods is used in many fields of science, medicine and industry [1–4].

However, the development of recent decades do not allow to achieve a quantum leap in reducing the radiation dose to the object under investigation, the values of which are often very important, especially in medical diagnostic procedures. One of this problem possible solution is using the pulsed emitting source synchronized with the detection equipment.

In recent years the technological research and development for creation a radiographic equipment based on pulsed sources that meets modern requirements and reduces radiation doses for the investigated object is working hard [5–9]. In the basis of these sources can be linear accelerators, betatrons, microtrons and pulsed X-ray tubes.

The results of the development of the visualization devices based on the pulsed X-ray source and the radiation dose estimation are described in the articles [9–10]. A further object of the investigation is the development of the tomographic system based on the pulsed bremsstrahlung generated by betatron.

For the investigations the gamma beam equability, the dose rate spatial distribution, the depth dose distribution in the different areas it is necessary to select the dosimetric equipment capable of working with the pulsed radiation of the corresponding intensity and radiation spectrum. The problem consists in that the pulsed source emits portions; consequently, the dose rate in the impulse is essentially bigger than the average values. In addition, the pulse-repetition rate is high enough and the profile of the beam intensity is fast-changing. Thus, specific requirement for response rate of the recording equipments are arisen in these conditions [10]. The applied dosimetric equipment should have quick response and be capable of working with the sufficiently high dose rates.

The main objective of the research was to analyse the suitability of the different dosimetric equipment for the radiation dose estimation of the pulsed emitting source and to choose the most appropriate ones. In the first phase the dosimetric equipments were investigated by means of pulsed X-ray tube. As a result the suitable dosemeters and its operation range of the energies and the intensities of radiation were selected. The next phase was to estimate the radiation dose of the pulsed bremsstrahlung generated by betatron OB-4. The depth dose distribution in the air, lead and water of the pulsed bremsstrahlung were obtained. The spatial profile of the dose rate of the pulsed bremsstrahlung generated by betatron was measured, the gamma beam equability was estimated. The suitability of using the compact pulse betatron OB-4 as a source of bremsstrahlung for visualization purposes was analysed.

2. Materials and methods

2.1. Emitting source

In the experiment the pulsed X-ray generator RAP-160-5 (portable radiological apparatus) and the portable betatron OB-4 were used as a source of emission. These devices are producing in the small innovation business in the Tomsk Polytechnic University.

The general quantities of the pulsed X-ray generator RAP-160-5 are: the anode voltage varies from 40 to 160 kV; the anode current varies from 0.4 to 5 mA; the focal spot size is 1.2×1.2 mm; the angular divergence of a beam is 40°; the frequency of radiation impulse varies from 60 to 700 Hz; the duration of one pulse is about 140 μs; pulse form is close to squared shape. The X-ray generator can be synchronized with the other devices [11].

The portable betatron OB-4 is used as a pulsed source of bremsstrahlung. The material of the target is tungsten (0.6 cm thick). The general quantities of the portable betatron OB-4 are: the maximum kinetic energy of the accelerated particles is 4.0 MeV; the frequency of radiation impulse is 400 Hz; the duration of one pulse is about 15 μs [12].

2.2. Dosimetric equipment

The main problem associated with the dosimetry of the pulsed radiation is the response rate of the dosimeters. This problem can be solved by using the storage type of the detectors. The solid thermoluminescent dosimeters DTL-02 were used as dosimetric equipment for initial estimation of the radiation doses. The thermoluminescent material of the detectors is LiF: Mg, Ti. The detectors annealing temperature is 400°C. The dosimetric complex based on the thermoluminescent dosimeters DTL-02 is designed for the personal dosimetry [13]. The dosimeter-radiometer DKS-96 equipped with a detection unit BDMG-96 was used for radiation dose estimation of the pulsed X-ray [14].

The universal dosimeter for radiation therapy and diagnostic radiology UNIDOS E equipped with a PTW soft X-ray plane-parallel ionization chamber type 23342 and with a PTW Farmer chamber type 30013 was used in the experiments [15]. The PTW soft X-ray plane-parallel ionization chamber type 23342 (the sensitive volume is 0.02 cm^3) is used for absolute dosimetry of soft X-ray beams [16]. The PTW Farmer cylindrical ionization chamber type 30013 (the sensitive volume is 0.6 cm^3) is used for absolute photon and electron dosimetry. This chamber is waterproof and can be used in water or in solid-state phantoms [17].

2.3. Experimental setup

In the experiment using the pulsed X-ray generator RAP-160-5 between the X-ray generator and the dosimetric equipment an aluminum filter was installed (half-value layer) to reduce the contribution of soft X-rays of the spectrum, which did not correspond to the dosimeters operating modes.

In the experiment using the portable betatron OB-4 the irradiation source was placed in lead dome with the output window. The detector was positioned on the radiation axis opposite the output window.

For obtain the depth dose distribution in the lead of the bremsstrahlung generated by the accelerator the lead plates with the different thickness (0.3, 0.6, 0.8, 5.0 cm) was used.

For obtain the depth dose distribution in the water of the bremsstrahlung generated by the accelerator the water phantom 41023 for horizontal beams was used. The external phantom dimensions are approximately 30 cm × 30 cm × 30 cm. The entrance window in one of the walls has the thickness of 3 mm (the walls material is PMMA) and the size of 150 mm × 150 mm [18].

2.4. Radiation doses measurement technique of the pulsed X-ray source

One of the aims of the development is to find out the radiation doses estimation methods of the pulsed X-ray sources.

In the experiment the distance between the X-ray source focus and the detector varied from 40 cm to 60 cm. The radiation was produced by the following parameters of the X-ray tube: the anode voltage was 70 kV; the anode current varied from 0.6 to 3.5 mA.

In the experiment the thermoluminescent dosimeters DTL-02, the dosimeter-radiometer DKS-96 equipped with a detection unit BDMG-96, the dosimeter UNIDOS E equipped with an ionization chamber type 23342 were used for the dose measurements of the pulsed X-ray generator RAP-160-5.

In the next part of the experiment the universal dosimeter UNIDOS E equipped with the plane-parallel ionization chamber type 23342 was used for the measurements of the dose rates spatial distribution of the pulsed X-ray source.

2.5. Radiation dose measurement technique of the pulsed bremsstrahlung of the betatron

Another aim of the development is to use the compact pulse betatron OB-4 as a source of bremsstrahlung for visualization purposes. Therefore, it is necessary to develop the radiation dose estimation methods of the pulsed bremsstrahlung generated by the accelerator. The measurements were carried out in the three different arias: in the air, in the lead and in the water.

The irradiation source was placed in lead dome with the output window. The detector was positioned on the radiation axis opposite the output window. The thermoluminescent dosimeter DTL-02 was used for the dose measurements in the air of the pulsed bremsstrahlung generated by betatron.

In the second part of the experiment the dose and the dose rate measurements in the three different arias: in the air, in the lead and in the water were carried out by the universal dosimeter UNIDOS E equipped with the cylindrical ionization chamber type 30013.

In the experiment in the air the distance between the radiation source and the detectors was varied from 30 to 100 cm in increments of 10 cm. For obtain the depth dose distribution in the lead of the bremsstrahlung generated by the accelerator the lead plates were arranged on the radiation axis close by the output window, the detector was positioned opposite the lead plates. For obtain the depth dose distribution in the water of the bremsstrahlung generated by the accelerator the water phantom 41023 for horizontal beams was used.

2.6. The dose rate profile measurements technique of the pulsed bremsstrahlung of the betatron

For obtain better visualization results the dose rate spatial distribution of the pulsed bremsstrahlung of the betatron should be determined.

The universal dosimeter UNIDOS E equipped with the cylindrical ionization chamber type 30013 was used for the measurements of the dose rates spatial distribution of the pulsed bremsstrahlung generated by betatron. The measurements were carried out in the plane coinciding with the emitter axis. The angular divergence of a beam was scanned in the range of 70° in increments of 5°. The distance between the radiation source and the detectors was 100 cm. The scanning area is defined by the lead dome side dimension.

3. Results and discussions

The figure 1 shows the dependence between the dose rate of the pulsed X-ray beam and the anode current at the distance between the X-ray source focus and the detector equal to 50 cm (the anode voltage equal to 70 kV). The results show that the dose rate measurements obtained with the help of the solid thermoluminescent detectors DTL-02, the plane-parallel ionization chamber type 23342, the dosimeter-radiometer DKS-96 equipped with a detection unit BDMG-96 are in a good agreement and can be used for dosimetry measurements of the pulsed X-ray generator RAP-160-5.

In the figure 2 an example of the dose rate spatial distribution of the pulsed X-ray source at the 30 cm distance are shown. The dose rate measurement for the different X-ray tube parameters were averaged and normalized to 100%.

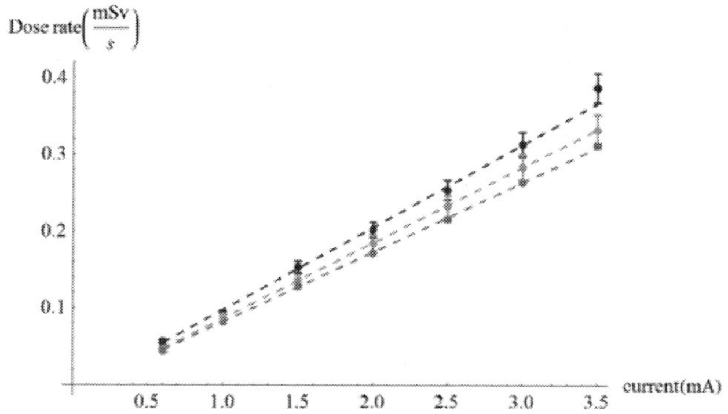

Figure 1. The dependence between the dose rate of the pulsed X-ray beam and the anode current:
● – the thermoluminescent dosimeters DTL-02;
● – the plane-parallel ionization chamber type 23342;
● – the dosimeter-radiometer DKS-96.

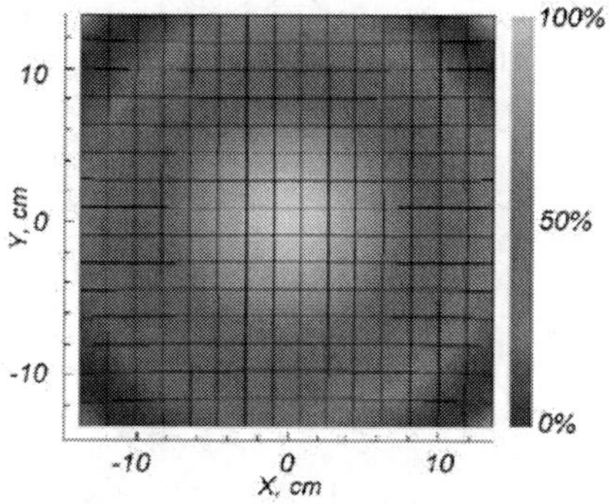

Figure 2. The dose rate spatial distribution of the pulsed X-ray source at the 30 cm distance. Array pitch is 2°.

The obtained data (figure 2) of the dose rate spatial distribution of the pulsed X-ray source show that at the distance between the X-ray source focus and the detector equal to to 30 cm – 20×20 cm for obtain better visualization results, because the dose rate spatial distribution is the most stable in this area. The obtained results show the dose rate dramatic decrease outside angular divergence of a beam (40°).

In the figure 3 the depth dose distribution in the air of the pulsed bremsstrahlung generated by betatron depend on the distance between the accelerator and the detector using the solid thermoluminescent dosimeters DTL-02 and the cylindrical ionization chamber type 30013 are presented.

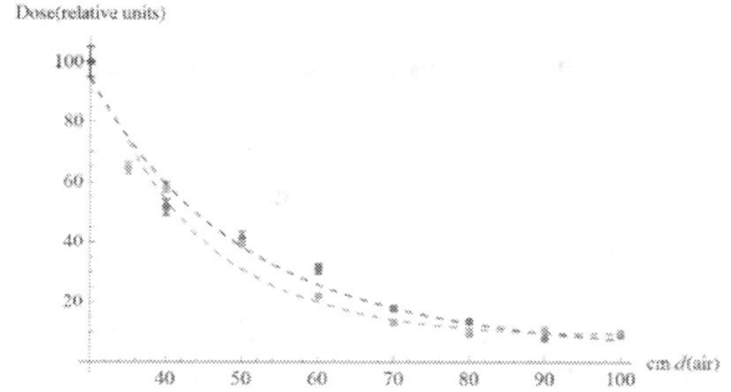

Figure 3. The depth dose distribution in the air of the pulsed bremsstrahlung generated by betatron OB-4:
● – DTL-02; ●– chamber 30013.

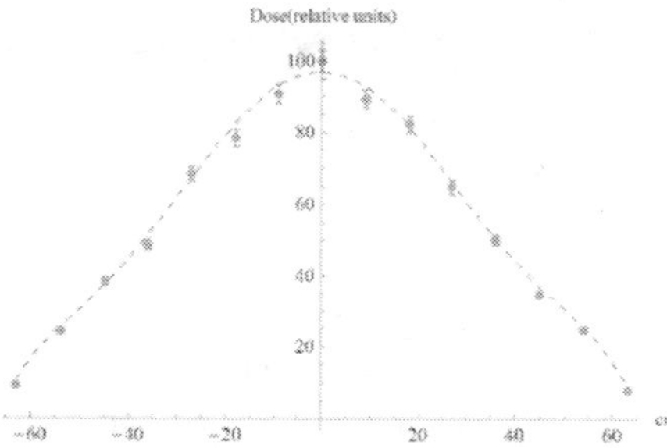

Figure 4. The dose rate profile of the pulsed bremsstrahlung generated by betatron OB-4:
● – chamber 30013.

It is shown (figure 3) that the results obtained with the help of the different types of the dosimeters are in a good agreement with each other.

In the figure 4 the dose rate profile of the pulsed bremsstrahlung generated by betatron are shown. The measurements were carried out in the plane coinciding with the emitter axis. The scanning area is defined by the lead dome side dimension. At the distance between the radiation source and the detector equal to 100 cm the angular variation equal to 5° is equivalent to 9 cm on a plane. The obtained data (figure 4) show the sufficient equability of the accelerator beam, which indicates the suitability of the emitting source for the visualization purpose.

The depth dose distribution in the water and in the lead of the pulsed bremsstrahlung generated by betatron depend on the distance between the accelerator and the detector using the cylindrical ionization chamber type 30013 are presented in the figure 5 and the figure 6 correspondingly. The obtained data (figure 5 and figure 6) show that the bremsstrahlung generated by betatron can be used for the visualization purposes of the objects in the nondestructive and medical examinations.

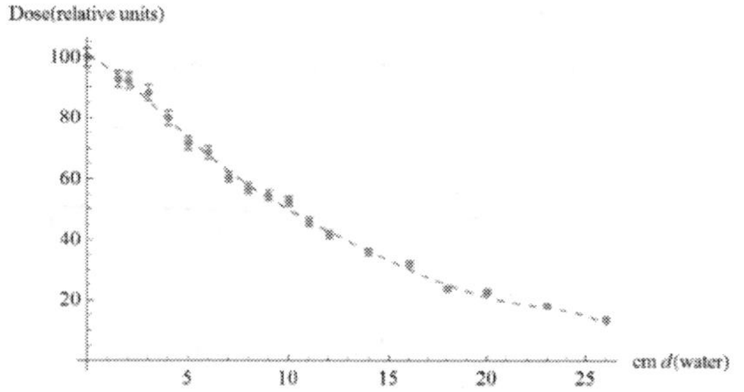

Figure 5. The depth dose distribution of the pulsed bremsstrahlung generated by betatron OB-4 in the water.

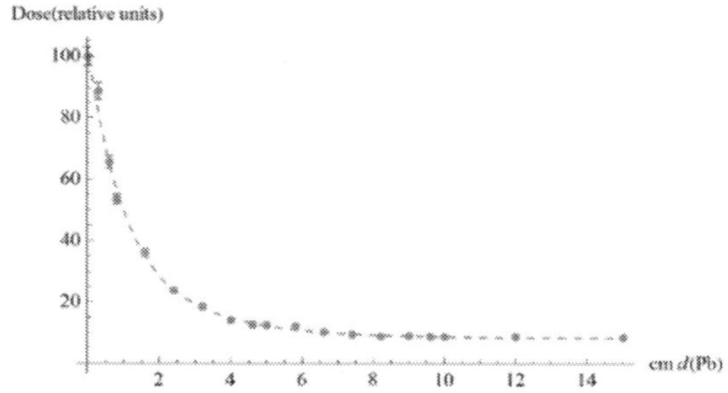

Figure 5. The depth dose distribution of the pulsed bremsstrahlung generated by betatron OB-4 in the lead.

4. Summary

The experimental results show the advisability of using the solid thermoluminescent dosimeters DTL-02, the dosimeter-radiometer DKS-96 equipped with a detection unit BDMG-96 and the universal dosimeter UNIDOS E equipped with a PTW soft X-ray plane-parallel ionization chamber type 23342 for the dosimetric parameters measurements of the pulsed X-ray generator RAP-160-5 under different radiation conditions. The obtained date show the advisability of using the solid thermoluminescent dosimeters DTL-02 and the universal dosimeter UNIDOS E equipped with a PTW Farmer chamber type 30013 for the dosimetric parameters measurements of the pulsed bremsstrahlung generated by electron accelerator with energy of 4 MeV. However, it should be noted that these dosimeters are effective at work with different energy and intensity of radiation.

After the dosimetric equipments selection the radiation dose of the pulsed bremsstrahlung generated by betatron OB-4 was estimated. The depth dose distributions in the lead and in the water of the pulsed bremsstrahlung were obtained. The profile of the depth dose distribution in the lead has an enough linear character for the adsorber thickness up to 3 cm, which indicates the suitability of the accelerator in the nondestructive testing goals. The profile of the depth dose distribution in the water has a linear character and demonstrates the suitability of radiation for visualization biological objects with thicknesses up to 25 cm, which is typical of the human body size.

The spatial profile of the dose rate of the pulsed bremsstrahlung generated by betatron and the depth dose distribution in the air was measured. The obtained data show the sufficient equability of the accelerator beam, which indicates the suitability of the emitting source for the visualization purpose. The obtained data show that the size of object under investigation should be about 50 cm in the plane perpendicular to the emitter axis for obtain better visualization results.

Acknowledgements
This work was financially supported by the Ministry of Education and Science of the Russian Federation in part of the science program (N 5.1485.2015).

References

[1] Choynzonov E L and Musabaeva L I 2006 *The intraoperative electron and gamma-ray teletherapy of the malignant neoplasms* (Tomsk: NTL) [in Russian]
[2] Cherepennikov Yu and Gogolev A (2014) Device for X-ray spectral absorption analysis with use of acoustic monochromator *J. Phys.: Conf. Ser.* **517** Article number 012037
[3] Kalender W A 2014 Dose in X-ray computed tomography *Phys. Med. Biol.* **59**(3) R129–R150
[4] Gogolev A, Stuchebrov S, Wagner A, Cherepennikov Yu and Potylitsyn A (2012) Acoustic "pumping effect" for quartz monochromators *J. Phys.: Conf. Ser.* **357**(1) Article number 012031
[5] Nakajima K(2008) Compact X-ray sources: Towards a table-top free-electron laser *Nature physics* **4**(2) 92–93
[6] Hampai D, Marchitto L, Dabagov S B, Allocca L, Alfuso S and Innocenti L (2013) Desktop X-ray tomography for low contrast samples *Nuclear Instruments and Methods in Physics Research Section B: Beam Interactions with Materials and Atoms* **309** 264–267
[7] Wagner W, Azadegan B, Sobiella M, Steiner J, Zeil K and Pawelke J 2008 An intense channeling radiation source *Nuclear Instruments and Methods in Physics Research Section B: Beam Interactions with Materials and Atoms* **266**(2) 327–334
[8] Cherepennikov Y and Gogolev A (2012) Method to reduce radiation exposure in the medical X-ray diagnostics *Proceedings - 2012 7th International Forum on Strategic Technology, IFOST 2012* Article number 6357737
[9] Stuchebrov S, Batranin A, Verigin D and Wagner A (2014) Setups for tomographic imaging with submillimeter spatial resolution *J. of Phys.: Conf. Ser.* **517**(1) Article number 012046
[10] Stuchebrov S, Batranin A, Verigin D, Lukyanenko Y, Siniagina M and Wagner A (2014) Estimation of radiation doses in X-ray visualization of biological objects *Advanced Materials Research* **880** 53–56
[11] Operations manual 2008 *Portable X-ray device for the industrial utilization* (Tomsk: Photon) [in Russian]
[12] Technical specification 2010 *Inductive cyclic accelerator O*B (Tomsk: Betta Plyus) [in Russian]
[13] Technical specification ZhBIT2.805.006RJe 2012 Thermoluminescent dosimeters DTL-02 (Moscow: Doza) [in Russian]
[14] Operations manual TE1.415313.003RJe 2012 Dosimeter-radiometer DKS-96 (Moscow: Doza) [in Russian]
[15] Information on http://www.ptw.de/unidos_e_dosemeter_ad0.html?&cId=3965
[16] Instruction manual 2013 *Soft X-ray chamber Type 23342* (Freiburg: PTW)
[17] User Manual 2013 *Ionization Chamber Type 30010, 30011, 30012, 30013* (Freiburg: PTW)
[18] Information on http://www.ptw.de/water_phantom_horizontal_beams20.html

Reflective Array with Controlled Focusing for Radiotomographic Application

S E Shipilov[1], A I Eremeev[2], V P Yakubov[3]

[1]Candidate of Physico-mathematical Sciences, associate professor the department of radiophysics, Tomsk State University, Tomsk, Russia
[2]Master student, the department of radiophysics, Tomsk State University, Tomsk, Russia
[3]Doctor of Physico-mathematical Sciences, professor the departtment of radiophysics, Tomsk State University, Tomsk, Russia
E-mail: s.shipilov@gmail.com

Abstract. It's considered the principle possibility of creation the managed reflectors for formulation of given field distribution in the focus area. Reflectors change the reflect ratio in dependence of the external control. The proposed theoretical modeling of such controlled focused device which provides focuse to a specific point in a given distribution of the reflectors. On the basis of numerical simulation it's considered the application of this approach for the solution of the problem of radiotomography.

1. Introduction

Due to the development of computing technologies enabled one to apply the radiotomography techniques for remote non-destructive testing and the diagnostics of the internal structure of radio-semitransparent media and the shape recovery of radio opaque objects. The task of radiotomography involves the transformation of the data obtained from the multiangle scanning of target objects into the 3D visualization [1]. The electronically and electromechanically scanned arrays are developed to obtain the multiangle projections of the wave field. However, the design and development of such arrays are complicated and expensive task, as it requires the development and control of numerous microwave transceiver channels.

Nowadays, the ways for simplification and cheapening the arrays with steerable pattern are looked for. It is known that two-dimensional lenses in form of the Fresnel zone plates focus radiation on a certain point. The electronic scanning can be performed in a certain angle sector with the change in geometrical arrangement of zone rings [2–6]. In these conditions, the positions of the radiator and receiver remains constant. In this paper, the solution of the problem by using the reflective array model where the reflection coefficient of every element changed by the external control is proposed.

2. Calculation of Reflection Coefficient Distribution

Consider that the externally controlled array element could have the reflection coefficient 0 and 1. Then, the recovery of the object shape using the 2D array stacked of such reflectors is split into two tasks. The first task is to calculate the reflectors distribution with the coefficient 0 and 1 for the focusing the reflected radiation on a certain point provided that the transmitter position is defined. The second task consists in calculating the field sequentially scattered by the test object and predefined distribution of reflectors. This field is detected by the receiver.

Consider the task scheme presented in figure 1, in detail. Let a transmitted spherical wave be propagated in a free space defined in the Cartesian coordinate system. The transmitter (1) is placed at Z axis and at the h height from the reflective array (2). The test object (3) is placed above the array.

The solution of the first task reduces to the calculating the shape of the Fresnel zones for the predefined position of the transmitter and the focusing point. When the focusing point lies on the Z axis at a certain height, the Fresnel zones have the shape of a circle. Focusing points with the random positions are the ellipse-shaped which centers don't coincide with the coordinate origin.

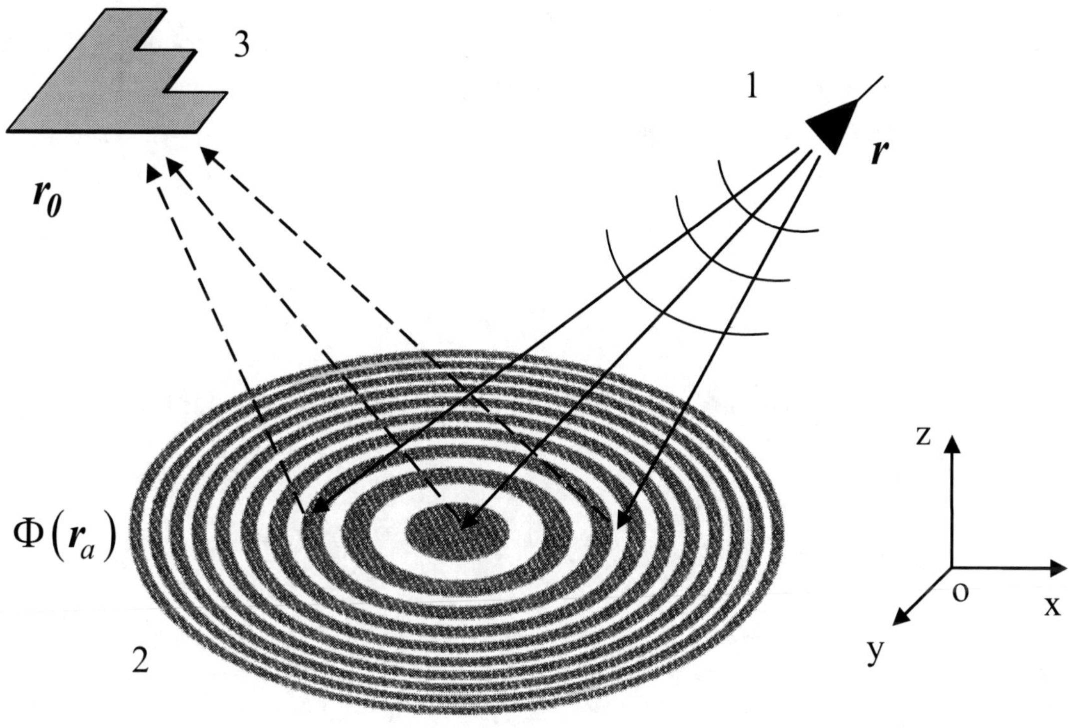

Figure 1. Task scheme.

Set the focusing point coordinates are (x_0, y_0, z_0). Then, the Fresnel zones boundaries are defined by the following equation:

$$\sqrt{z_0^2 + (x - x_0)^2 + (y - y_0)^2} + \sqrt{h^2 + x^2 + y^2} = n\lambda\left(1 + \frac{1}{4}\right),$$

where n is the number of the Fresnel zone boundary, λ is the wavelength of the sounding signal.

Consider the example when $y_0 = 0$. Then, the equation for the Fresnel zones boundaries takes the following form:

$$\frac{(x - x_c)^2}{A^2} + \frac{y^2}{B^2} = 1, \quad (1)$$

where A and B are a major and a minor semiaxis of an ellipse, respectively. The semiaxis are calculated as follows:

$$A^2 = \frac{\left(1-\frac{x_0^2}{d^2}\right)^2}{\left(\frac{qx_0}{d}\right)^2 + \left(q^2-h^2\right)\left(1-\frac{x_0^2}{d^2}\right)}; \; B^2 = \frac{1-\frac{x_0^2}{d^2}}{\left(\frac{qx_0}{d}\right)^2 + \left(q^2-h^2\right)\left(1-\frac{x_0^2}{d^2}\right)}.$$

where

$$d = n\lambda\left(1+\frac{1}{4}\right), \; q = \frac{z_0{}^2 + x_0{}^2 - d^2 - h^2}{2d}, \; x_c = \frac{qx_0}{x_0^2 - d^2}.$$

The value 1 is assigned to the even zones and 0 is assigned to the odd zones. Then, the reflection coefficient distribution $\Phi(r_a)$ is obtained (figure 1). This reflection coefficient distribution provides the radiation focusing on the point $(x_0, 0, z_0)$.

This is a general case when $y_0 \neq 0$. One can reduce it to the equation (1) with the following substitution

$$x_0 \to \operatorname{sgn}(x_0)\sqrt{x_0{}^2 + y_0{}^2}, \text{ where } \operatorname{sgn}(x_0) = \begin{cases} 1, & x_0 > 0; \\ 0, & x_0 = 0; \\ -1, & x_0 < 0. \end{cases},$$

and the following recalculation of $\Phi(r_a)$ from one coordinate system to another one using the rotation matrix

$$A_\varphi = \begin{pmatrix} \cos(\varphi) & \sin(\varphi) & 0 \\ -\sin(\varphi) & \cos(\varphi) & 0 \\ 0 & 0 & 1 \end{pmatrix}, \text{ where } \varphi = \operatorname{arctg}\left(\frac{y_0}{x_0}\right).$$

Therefore, we obtained the solution for the first task, the distribution of reflectors, which provide the radiation focusing on the predefined point.

Consider the solution of the second task. The field in the test object surface at the height z_0 after the plate reflection is written as

$$E1(r_0) = \int G(r_a - r_0)\Phi(r_a)G(r_a - r)dr_a.$$

Here, $G(r)$ is the Green's function of a point-source radiator. When radiation reflected from the object, the field at the reflector plane is written as follows:

$$E2(r_a) = \int E1(r_0)\gamma(r_0)G(r_0 - r_a)dr_0,$$

where $\gamma(r_0)$ is the function describing the geometry of the test object. The function is equal to 1 inside the object boundaries and it is equal to 0 out of the boundaries. The field in the receiver plane at the *h* height is

$$E3(r) = \int E2(r_a)\Phi(r_a)G(r_a - r_0)dr.$$

Obviously the best focusing is achieved when the receiver point coincides with the transmitter point.

Thus, the solutions of both tasks were achieved. The equation describing the reflection coefficient distribution in the reflective array to provide the radiation focusing on the certain point is obtained . This results will be applied in modeling the reflective array and taking the radio image of a test object.

It should be noted that the data on the field detected in the receiver recorded during the focusing on all points in the scanned volume is the radiotomogram of the test space.

3. Numerical Simulation

A flat array of elementary reflectors 20×20 cm was used in numerical simulation. The size of each reflector was of 0.5 cm. The coordinates of the transmitter and the receiver were $(0,0,20)$. The focusing point coordinates were $(0,0,30)$. The reflection coefficient distribution $\Phi(r_a)$ for the frequency 24 GHz was calculated (figure 2). Figure 3 presents the radiation focused by the array on the test object plane. The field distribution in the receiver is presented in figure 4. The maximum value of the scattered field coincides with the coordinates of the transmitter.

Figure 2. Fresnel zone distribution.

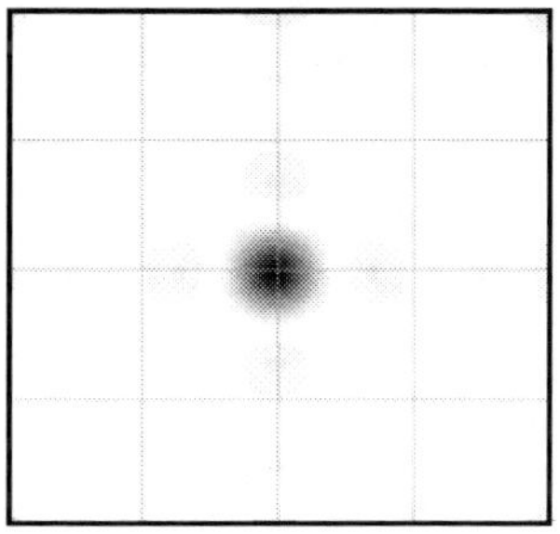

Figure 3. Field focusing on the object plane.

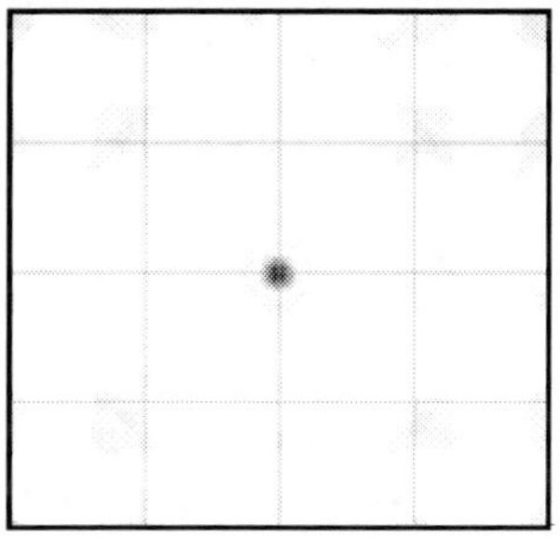

Figure 4. Field distribution on the receiver plane.

The reflection coefficient distribution for the focusing point $(5, 2, 30)$ is shown in figure 5. Radiation focused by the array in the test object plane and the field distribution for that focusing point in the receiver plane are shown in figures 4 and 5, respectively. As the figures show, the field in the object plane was focused in the same way as with the focusing point at the *oz* axis. The situation with the field distribution at the receiver pane is quite a bit different, as noise increases with the shift of the focusing point from the *oz* axis.

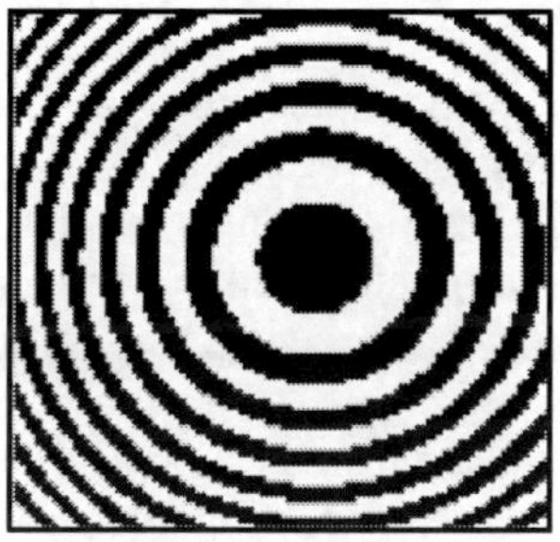

Figure 5. Fresnel zone distribution.

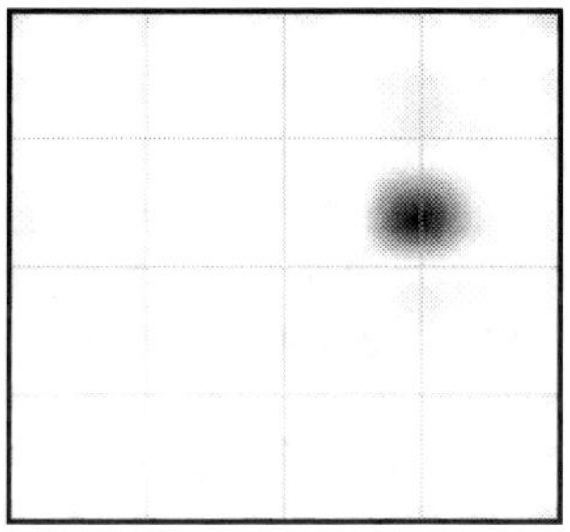

Figure 6. Field focusing on the object plane.

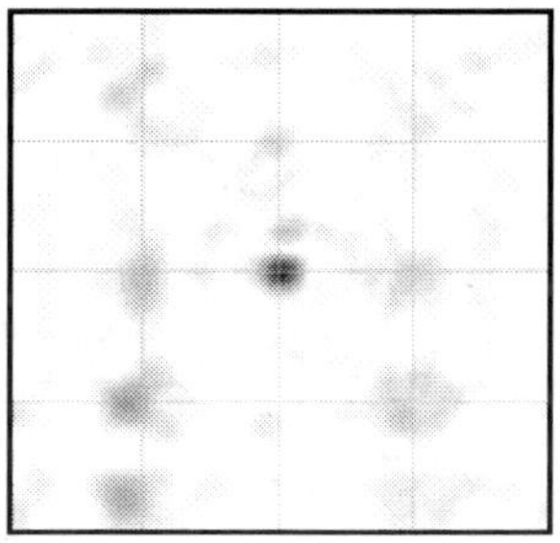

Figure 7. Field distribution on the receiver plane.

So, the best focusing is provided when the position of the transmitter coincides with the receiver position.

As a test object, a flat object in a shape of a stepped triangle was used. Each step was of 3 cm. The object was placed at the height of 10 cm above the reflective array. Figure 9 illustrates the radio image of the test object above the reflective array 10×10 cm in size. The result for the reflective array 20×20 cm in size is presented in figure 10.

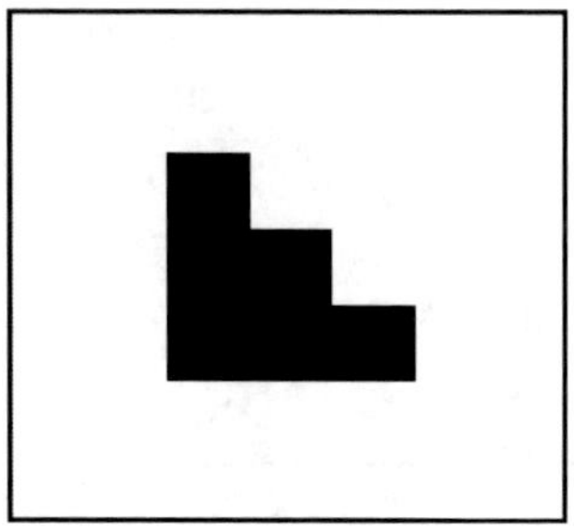

Figure 8. Test object.

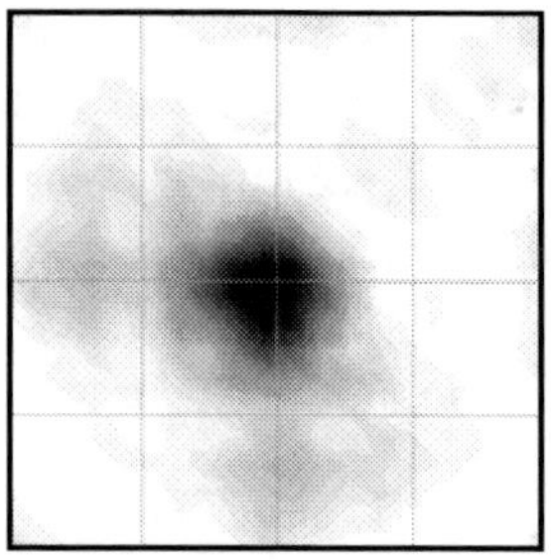

Figure 9. Radio image of the test object above the 10×10 cm array.

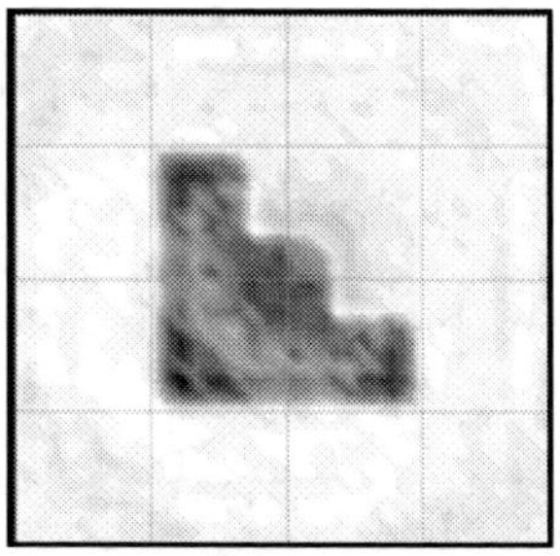

Figure10. Radio image of the test object above the 20×20 cm array.

The triangle used as the test object is plainly seen in the radio image. Using the reflective array with larger aperture provided the radio image of higher sharpness and with a lower artifact level.

4. Conclusion

The conducted research demonstrated the using the 2D reflective array with controlled reflection coefficient to focus the monochromatic radiation for radiotomographic application. The numerical simulations with a test object in a form of a stepped triangle approved the recovery of the test object's shape.

The research was supported by the Russian Federation Ministry of Education and Science as part of state job no. 3.694.2014/K

References

[1] Yakubov V P, Shipilov S E, Sukhanov D Ya, Klokov A V 2014 *Radiovolnovaya tomographiya: dostizheniya i perspectivy* (Tomsk: NTL)

[2] Leon G, Herran L, Munoz M, Las-Heras F, Hao Y (2014) Millimeter-wave offset Fresnel zone plate lenses characterization *Progress In Electromagnetics Research, – Proceedings* 125-131

[3] Hajian M, Vree G, Ligthart L (2003) Electromagnetic Analysis of beam-scanning antenna at millimetr-wave band based on photoconductivity using Fresnel-zone-plate technique *IEEE Antennas and propagation magazine* **45**(5) 13–25

[4] Webb G W, Rose S C, Sanchez M S, Osterwalder J M (1998) Experiments on an optically controled 2-D scanning antenna *Antenna Application Symposium. Monticello, Proceedings* 35–50

[5] Minin I V, Minin O V (2004) Scanning properties of diffractive element forming the axial-symmetric diffraction limited wave beam *Computer optics* **26** 65–67 (In Russian)

[6] Ksenofontov Yu G (2014) The Fresnel antenna design methodology applicable for satellite mobile television system *Vestnik gosudarstvennogo universiteta morskogo i retchnogo flota imenie S.O. Makarova* **3** 9–14 (In Russian)

Investigation Converter Circuit «Voltage-Current» for Power Calibrator

Yu M Fomichev[1], S V Silushkin[2], I A Larioshina[3]

[1]Associate Professor, Institute of Cybernetics, Tomsk Polytechnic University, Russia, Tomsk,
[2]Associate Professor, Institute of Non-Destructive Testing, Tomsk Polytechnic University, Russia, Tomsk,
[3]Assistant, Tomsk State University, Russia, Tomsk
E-mail: slavasv@mail.ru

Abstract. The paper presents alternative circuits for voltage-current converters to be used in the calibrator of fictitious power. The experimental studies have revealed a number of problems related to the stability of the system in deep feedback and zero level stabilization of the amplifier. The circuit solutions given in the article allow elimination of these problems and improve the accuracy of calibrator current calibration. For example, correction/corrective circuits are used to ensure the stability of the converter at deep depths of the feedback, and operational amplifier based circuit solution and compensation condition are proposed to reduce the additional phase shift. To improve the accuracy of the calibration current values specified by the calibrator we propose to connect the feedback circuit to the measuring current transformer. However, further improvement of the accuracy class of the power calibrator is impossible without modern electronic components.

1. Introduction

Measuring instruments of parameters electrical energy require mandatory verification, which may be performed using automated measurement system [1–4], including the fictitious power calibrator (FPC). This construction of measuring instruments for verification is necessary for the generation of signals arbitrary waveform when the working techniques connected with electrical nets with non-linear distortions [5–7].

In [8], the authors noted that the FPC should contain a voltage-to-current converter (VCC), which would provide the current within a range from 10 mA to 50 (100) A [9]. According to the requirements of normative documents [9, 10], the following metrological characteristics of the calibrator are normalized:

– current calibration error is less than 0.1%;
– instability is not more than 0.05% per 5 minutes;
– distortion is less than 0.2%;
– phase shift between the input voltage and the output current is not more than 0.01°.

The main tasks that need to be addressed:

– to obtain the required values of the VCC output current;
– to provide high metrological characteristics of the converter;
– maintain the VCC stability in deep negative current feedback (NFB) under virtual inductive load.

The article discusses several VCC circuits and the results obtained.

2. Study of power amplifiers circuits for current calibrator

To obtain the required values of the output currents for the power calibrator, a power amplifier (PA) can be implemented in the following three circuits:

- transformerless PA;
- PA with the output transformer;
- combined PA.

Figure 1 shows a circuit for possible implementation of the transformerless PA with the output stage which is a push-pull follower of AB class on power field-effect transistors (IRF740 [11], IRFPF40 [12]). The circuit is powered from two nonstabilized power supplies E_3 and E_4 with earthed middle point (the value of the sources is selected based on the desired value of the load voltage).

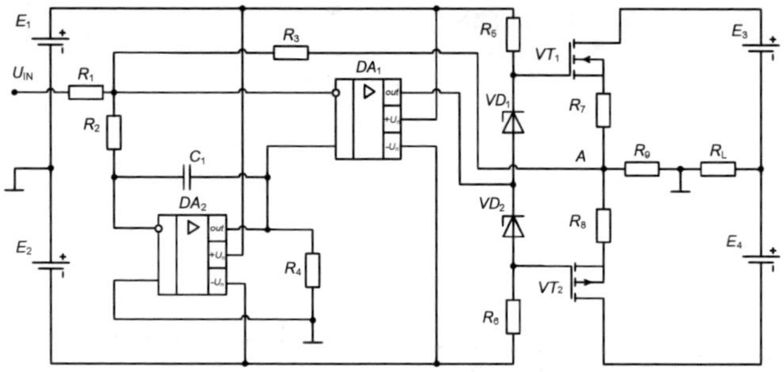

Figure 1. Transformerless PA circuit.

The preamplifier in the circuit is an operational amplifier (op amp) powered from stabilized power supplies E_1 and E_2 with the middle point connected to earth. Resistance R_9 is included between the repeater output (*point A*) and earth. The load is switched between earth and the middle point of the E_3 and E_4 power supplies. The overall current NFB in PA is applied via R_1 and R_3 resistors recordered.

An important condition of the circuit operation is zero DC voltage on R_9, since $I_L = I_9$, then if $U_{R9} = 0$, $I_L = 0$ as well. The problem can be solved through additional feedback application to PA via the integrator using op amps with low offset voltage (for example, OP 177 [13]). The disadvantage of this implementation is that the signal passing through the integrator causes additional phase shift, which is not admissible.

We investigated the possibility of introducing NFB through op amp DA_2 (Fig. 1). In this case, the input signal of the amplifier (input voltage between op amp DA_1) does not affect the phase shift since its value is very small. Figure 2 shows the graph of this amplifier with llowance for the offset voltage of the operational amplifiers U_{offset1} (DA_1), U_{offset2} (DA_2) U_{offset3} (output stage offset voltage).

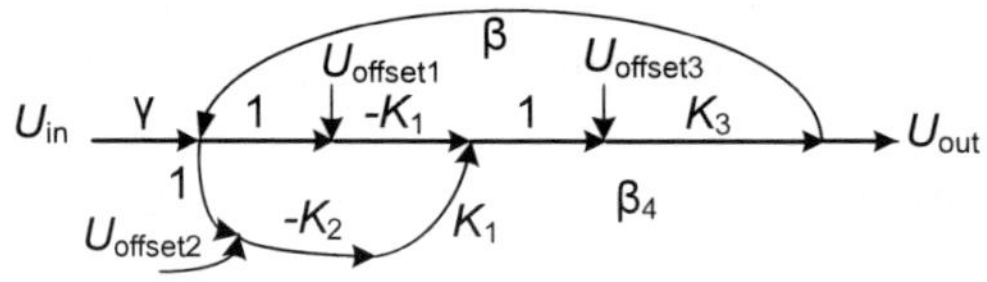

Figure 2. Graph of the transformerless power amplifier.

In the graph $\gamma = \dfrac{R_3}{R_1 + R_3}$; $\beta = \dfrac{R_1}{R_1 + R_3}$; $-K_1$ and $+K_1$ are the transmission coefficients through the inverting and non-inverting DA_1 inputs, respectively; $-K_2$ and $+K_3$ are transmission coefficients for DA_2 and the output stage. Write the following equation (figure 2):

$$U_{\text{out}} = -\frac{\gamma}{\beta}\cdot U_{\text{in}} - \frac{U_{\text{offset1}}}{\beta\cdot(1+K_2)} - \frac{U_{\text{offset2}}}{\beta} + \frac{U_{\text{offset3}}}{\beta\cdot K_1\cdot(1+K_2)}.$$

Thus, zero offset at point A is determined by operational amplifier DA_2 and this determines the choice. The PA using high-power operation amplifiers, such as PA-05 [14] and PA-07 [15] can be implemented in a similar way.

The fictitious power calibrator "Vector" developed at the Department of Computer Measuring Systems and Metrology, Tomsk Polytechnic University, uses the combined solution: in the circuit shown in figure 1, a single PA working up to 10 A is used without a transformer, and in the range of 10...50 A, the output transformer is used.

In VCC, to obtain high precision calibration of the generated current, one more NFB loop is introduced, in which the current-to-voltage converter is used as a current transformer. Figure 3 shows a VCC circuit diagram, where PA is a power amplifier, DA_1 is a pre-amplifier, T_1 is a current transformer with a transformation coefficient of n_1.

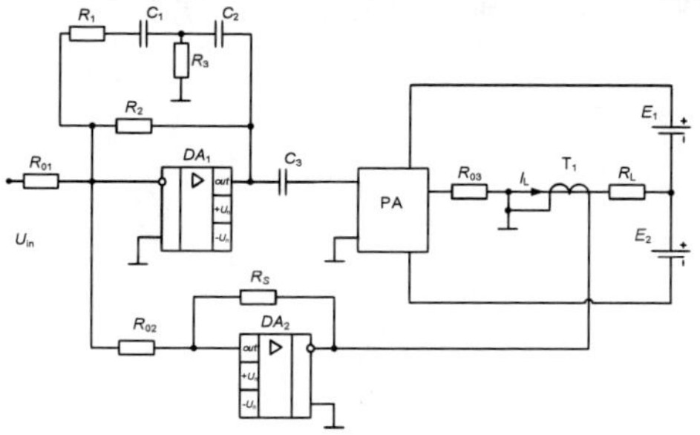

Figure 3. Circuit diagram of VCC with measuring current transformer.

The current-to-voltage converter is performed on op amp DA_2. For this design, the relation between the output current and input voltage is described by the expression

$$I_L = U_{in} \cdot \frac{R_{02}}{R_{01}} \cdot \frac{n}{R_s}, \tag{1}$$

i.e. the ratio R_{02} / R_{01} can be chosen to set the required coefficient for conversion of the input voltage into the load current I_L.

For high-ampere load currents, the output transformer T_2 with a conversion rate of n_2 is switched figure 4.

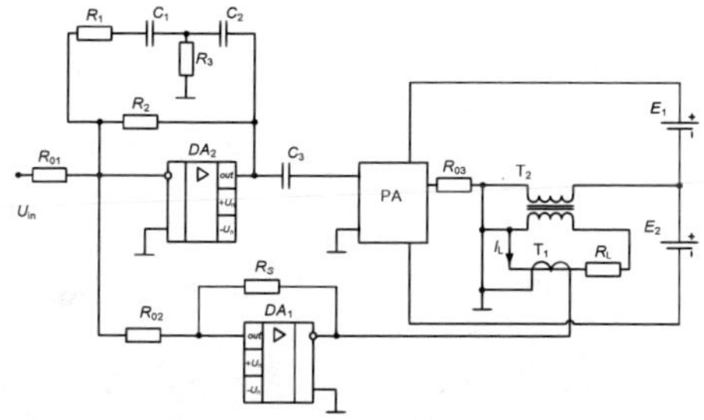

Figire 4. The VCC circuit with the output transformer.

In the circuit shown in figure 4, the load current is determined by relation (1), but the PA works with the output currents $I_{PA} = I_L / n_2$.

As can be seen from expression (1), the properties of the current transformer (linearity, phase shift) unambiguously determine the calibration accuracy of the VCC output current. When using the current transformer as an output current-to-voltage converter (figure 5), the phase shift φ between E_2 and I_1 according to [16] is equal to

$$\varphi = \psi + \alpha,$$

where ψ is determined by the properties of the current transformer core, and angle α is the inductance of the secondary winding (L_2), its active resistance I_1 and load R_L

$$\alpha = arctg \frac{R_i + R_L}{\omega \cdot L_2}.$$

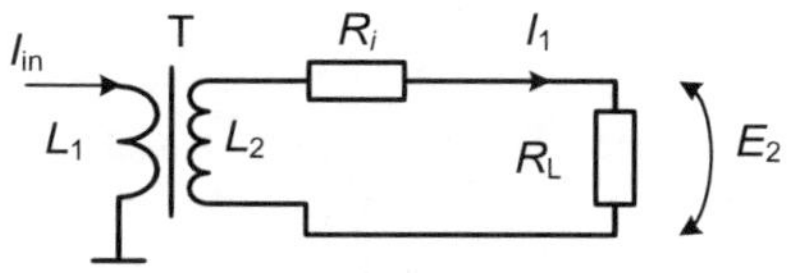

Figure 5. The circuit of the output current-to-voltage converter with current transformer.

Angle α can be considerably reduced using the circuit designs, for example, the circuit shown in figure 6.

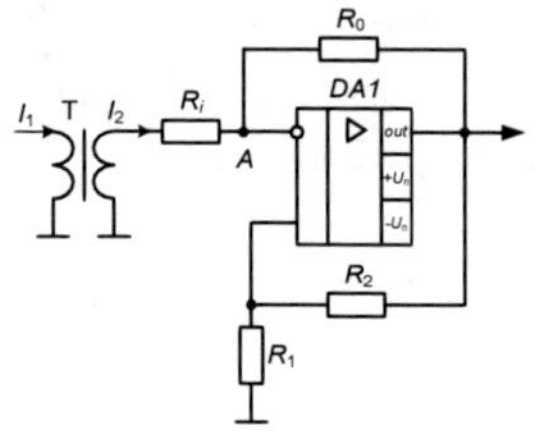

Figure 6. The circuit for reducing the phase shift caused by the inductance of the transformer secondary winding, its active resistance and load.

The input resistance at point A is

$$R_A = \frac{R_0}{1+K}, \tag{2}$$

where K is op amp gain covered by positive feedback (PFB) through R_1 and R_2

$$K = \frac{K_0}{1-K_0 \cdot \beta}, \quad \beta = \frac{R_1}{R_1 + R_2},$$

where β is the transfer coefficient of the PFB circuit; K_0 is the op amp gain.
Provided $K_0\beta >> 1$, we obtain

$$K = -\frac{1}{\beta},$$

and expression (2) takes the form:

$$R_A = \frac{R_0}{1-\frac{1}{\beta}} = -R_0 \cdot \frac{R_1}{R_2}. \tag{3}$$

From (3), we can write the compensation condition

$$R_i = R_0 \cdot \frac{R_1}{R_2}. \tag{4}$$

Thus, using compensation condition (4), angle α can be reduced to zero. When using current transformer cores made of amorphous alloys, the obtainable angles ψ are of units of minutes that provides the measurement error within 0.1% [17, 18].

A number of current transformers (TO2, TO3-TM, TO3-AS, NRC3-XQ1005 and NRC03A) have been studied. The experimental circuit is shown in figure 7. The tested transformers were loaded with resistance of 100 Ohm (type S2-29, Class 01). The linearity and phase shift were checked by comparing the readings of the precision current transformer (PRISMA-TT/1) and the reference shunt (developed by VNIIM).

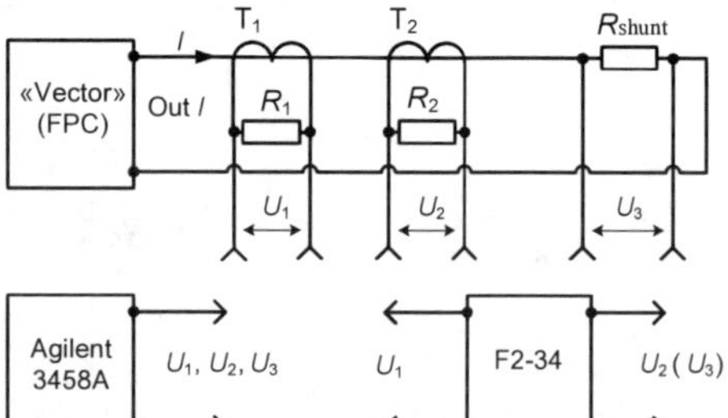

Figure 7. Diagram to study the performance of measuring current transformers.

To measure the voltage, the multimeter Agilent-3458A was used, the phase shift was measured by the phase meter F2-34 at currents of 10 A, 20 A, 30 A and 50 A. The experimental results showed that the tested current transformers have good linearity and low phase shift (with respect to the precision instruments does not exceed 0.01°...0.02°), i.e. their use allows development of VCC with the designed parameters.

It was noted earlier that while designing VCC, the problem to tackle is the stability of the system with very deep NFB. The studies have shown that in case of current calibrator, satisfactory results are obtained using a circuit with the correction circuit (figure 8). The inclusion of this circuit in the pre-amplifier circuit is shown in figures 3 and 4, and gain and phase are shown in figure 9. Hence, the selected resistances R_2 and R_3 can eliminate parasitic oscillations.

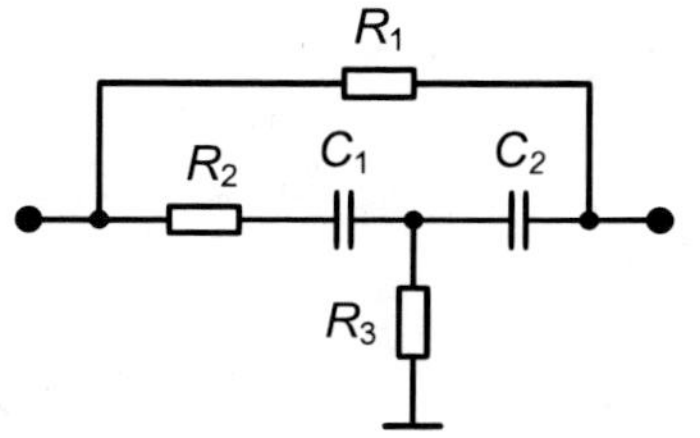

Figure 8. Correction circuit.

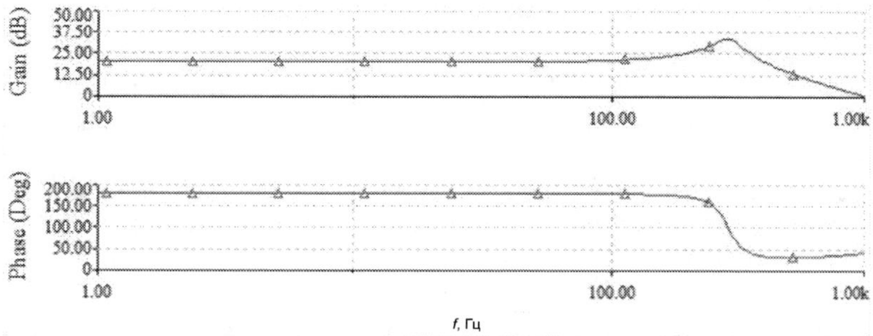

Figure 9. Gain and phase of the correction circuit.

3. Conclusion

Development and implementation of precision voltage-to-current converters faces the problems related to the design, as well as the selection of electronic components (operational amplifiers, transistors, transformers, etc).

The proposed circuit solutions allow elimination of a number of problems such as:

– the frequency correction circuit provides stability of the system with deep negative current feedback;

– variant of compensation for the phase shift of the measuring current transformer reduces the conversion error;

– application of modern operational amplifiers reduces the zero offset.

Thus, the found circuit designs and state of the art electronic components can improve the accuracy of the converters included in the power calibrator and, consequently, improve the calibrator accuracy class.

Reference

[1] Toth E, Franco A M R, Debatin R M (2005)Power and energy reference system, applying dual-channel sampling *IEEE Transactions on Instrumentation and Measurement* **54**(1) 404–408

[2] Svensson S and Ryder K E (1995) A measuring system for the calibration of power analyzers *IEEE Transactions on Instrumentation and Measurement* **44**(2) 316–317

[3] Franco A M R, Tóth E, Debatin R M, Prada R (2001) Development of a power analyzer *11th IMEKO TC4 Symposium on Trends in Electrical Measurements and Instrumentation and 6th IMEKO TC4 Workshop on ADC Modelling and Testing 2001* 168–172

[4] Iljazi I, Arsov L *et al* (2012) Calibration of a virtual instrument for power quality monitoring http://www.icrepq.com/icrepq'12/370-iljazi.pdf

[5] Zhang X-B, Li Y-H and Cui X-M (2014) Active power measurement based on multiwavelet transforms *Mathematical Problems in Engineering* **2014** Article number 905108

[6] Svensson S 1999 Power Measurement Techniques for Nonsinusoidal Conditions (Chalmers tekniska högsk)
[7] Emanuel A E (1990) Powers in non-sinusoidal situations-a review of definitions and physical meaning *IEEE Transactions on Power Delivery* **5**(3) 1377–1389
[8] Fomichev Yu M, Silushkin S V (2013) Investigation of possibilities to improve the Accuracy class of the Power Calibrator *The electronic scientific journal «Siberian Journal of Science»* **4** (10) 85–92 URL: http://sjs.tpu.ru/journal/article/view/817/575
[9] GOST 31819.21-2012 (IEC 62053-21:2003) 2013 *Alternating-current measuring instruments. Particular requirements. P. 21. Static active energy meters of accuracy classes 1 and 2. Intr. from 2014-01-01* (M.: Standartinform)
[10] GOST 31819.22-2012 (IEC 62053-22: 2003) 2013 *Alternating-current measuring instruments. Particular requirements. P. 22. Static active energy meters of accuracy classes 0.2S and 0.5S. Intr. from 2014-01-01*(M.: Standartinform)
[11] IRF740 Datasheet (PDF) – STMicroelectronics http://pdf1.alldatasheet.com/datasheet-pdf/view/22400/STMICROELECTRONICS/IRF740.html
[12] IRFPF40 Datasheet (PDF) – Vishay Siliconix http://pdf1.alldatasheet.com/datasheet-pdf/view/251481/VISHAY/IRFPF40.html
[13] OP177 Datasheet (PDF) – Analog Devices http://pdf1.alldatasheet.com/datasheet-pdf/view/174958/AD/OP177.html
[14] PA05 Datasheet (PDF) – Cirrus Logic http://pdf1.alldatasheet.com/datasheet-pdf/view/140181/APEX/PA05.html
[15] PA07 Datasheet (PDF) – List of Unclassifed Manufacturers http://pdf1.alldatasheet.com/datasheet-pdf/view/98111/ETC/PA07.html
[16] Afanasyev V V, Adonyev N M, Kibel V M *et al* 1989 *Current transformers. Second Edition, revised and enlarged* (L.: Energoatomizdat)
[17] Raskulov R F, Smirnov A S *Effect of the secondary load power factor on measuring transformer error* URL: http://www.cztt.ru/articles/138.html
[18] Raskulov R F, Ignatenko E V *About the advantages of transformers for commercial electricity metering* URL: http://www.cztt.ru/articles/148.html

Journal of Physics: Conference Series **671** (2016) 012054 doi:10.1088/1742-6596/671/1/012054

Simulation of Surface Oscillation of Ultrasound Sensor Based on Piezoelectric Semiconductor Transducer

A I Soldatov[1], A A Soldatov[2], M A Kostina[3], V F Tolkatchev[4]

[1]Professor, Institute of Non-Destructive Testing, Tomsk Polytechnic University, Tomsk, Russia
[2]Master of Science, Institute of Non-Destructive Testing, Tomsk Polytechnic University, Tomsk, Russia
[3]PhD student, Institute of Non-Destructive Testing,, Tomsk Polytechnic University, Tomsk, Russia
[4]Proffessor, Research Institute of Applied Mathematics, Tomsk State University, Russia
E-mail: asoldatof@mail.ru

Abstract. In the article theoretical and experimental studies of the surface deformation of piezoelectric sensors are carried out. Round membrane was used as a model of piezoelectric sensors. Experimental studies were conducted on mass-produced sensors and self-made sensor. Good match fluctuations in membrane model with fixed edges real sensors is shown.

1. Introduction

The ultrasonic testing based on excitation of piezoelectric sensors by electric pulse, which converts the electric pulse to the ultrasonic signal and passes one in the researched object. Ultrasonic signal interacts with a testing object, echo signal receive and one is analyzed [1-4]. When an electrical signal arrives to piezoelectric sensors, piezoelectric plate fluctuations occur synchronously over the entire surface, that follow from the inverse piezoelectric effect equation [5, 6]:

$$U = s\sigma + dE + \sigma T \ ,$$

where U – mechanical deformation, s – mechanical compliance, σ – mechanical stress, E –electrical field strength, d – piezoelectric constant, T – temperature.

It is considered that all changes of acoustic signal happen only in the testing object. However, when the sensor is manufactured, flat side of a piezoelectric plate fixes by a box or another construction unit [7]. Therefore, a piezoelectric plate deformation has intricate form (shape). Study of natural oscillations of piezoelectric plates were devoted to a work [8–10]. But one did not take into account the influence of degree of fixing flat side of a piezoelectric plate in the manufacture of the sensor. The purpose of this work is to study fluctuations of surface sensor in mode excitation, which must be taken into account when analyzing the reflected echo-signal.

2. Theoretical aspects of phantom head designing

The behavior of the sensor is studied during of shock excitation (this mode uses the most frequently). The analogue sensor oscillations can be fluctuations circular membranes, the edges of which either rigidly fixed or on the edge of the membrane acting elastic force proportional to the deviation of the membrane. Options considered to correspond to the boundary conditions of the problem of the membrane oscillation [11]. An equation of free oscillations of the membrane with fixed edges is:

$$U(r,t)_{tt}=c^2\,\Delta U\ , \tag{1}$$

at initial conditions:

$$U(r,t)\big|_{t=0}=A, \qquad U(r,t)_t\big|_{t=0}=0,$$

and the boundary condition:

$$U(r,t)\big|_{r=b}=0,$$

where A is the residual amplitude after shock excitation, $U(r,t)_t$ – first derivative of oscillations amplitude with time, r – current radius coordinate, b – the radius of the membrane, $U(r,t)_{tt}$ – the second derivative of oscillations amplitude with time, c –t he speed of propagation of flexural vibrations, ΔU – Laplace operator.

For the free vibrations of the membrane with elastic fastening edges only boundary conditions are changed:

$$G\cdot U(r,t)_r\big|_{r=b}=K\cdot U(r,t)\big|_{r=b} \;\Rightarrow\; U(r,t)_r+H\cdot U(r,t)\big|_{r=b}=0\,,$$

where G is the shear modulus, $G\cdot U(r,t)_r$ – elastic force, $U(r,t)_r$ – the first derivative of oscillations amplitude with radius.

Focus on task with a stiff pinning the edges of the membrane:

$$U(r,t)_{tt}=c^2\cdot[U(r,t)_{rr}+\frac{1}{r}\cdot U(r,t)_r],$$

The solution to the equation is:

$$U(r,t)=2\cdot A\cdot b\cdot\sum_{k=1}^{\infty}\frac{1}{\alpha_k^2}\cdot J_1(\alpha_k)\cdot J_0(\alpha_k\cdot\frac{r}{b})\cdot\cos(\frac{\alpha_k\cdot c}{b}\cdot t)\ , \tag{2}$$

where α_k -the positive roots of the equation $J_0(x)=0$

Table 1 shows the values of the first of six roots of the equation $J_0(x)=0$ for stiff α_k and elastic β_k pin of the membrane edge.

Table 1. The first of six values of the roots.

№ root	1	2	3	4	5	6
α_k	2.40	5.52	8.65	11.79	14.02	18.10
β_k	3.832	7.016	10.173	13.324	16.471	19.616

Using the obtained values of the roots calculations of longitudinal amplitude oscillations of the surface of the membrane were carried out. The results of the calculations for expression (2) are presented figure 1.

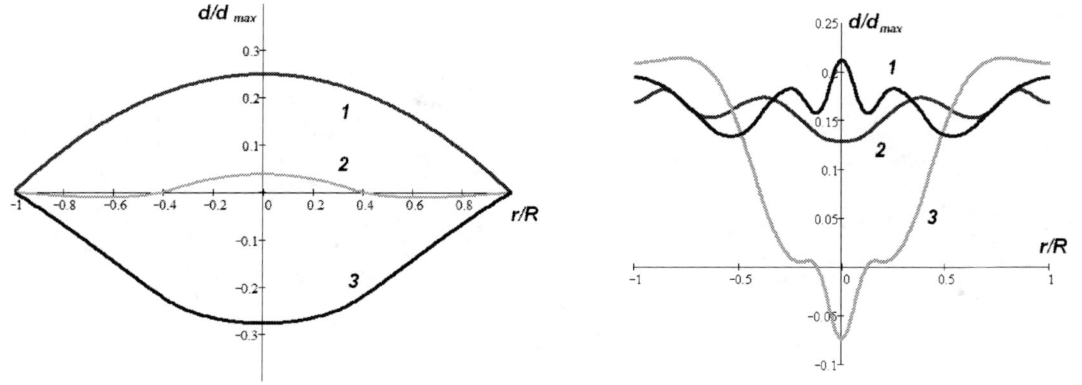

a b

Figure 1. Membrane fluctuations: a– edge stiff fixed, b– edge elastic fixed; 1–start of the excitation, 2 – on time π/2, 3– on time π.

3. Experimental research

To verify the correctness of the calculations experimental research of deformation of the surface of self-made and mass-produced sensors were carry out.

For this purpose test stand has been used [12], which allows you to move point receiver on the surface of the emitter with the specified step. Minimum scan step is 0.1 mm. The movement control is performed by a personal computer. The resulting data set was used for the reconstruction of 3D images of the surface piezoceramic sensors using algorithms described in [13–15].

The sensor P111-1, 25-K20 company IC "Fizpribor" was researched. The results are showed in figure 2 and figure 3. The surface of the emitter shifts down during the shot (figure 2a), except the outer edge which one is bonded to the sensor box. On the edge of the converter starts to bending wave after the finish of the shot. It moving from the edge to the center of the plate (figure 2b)

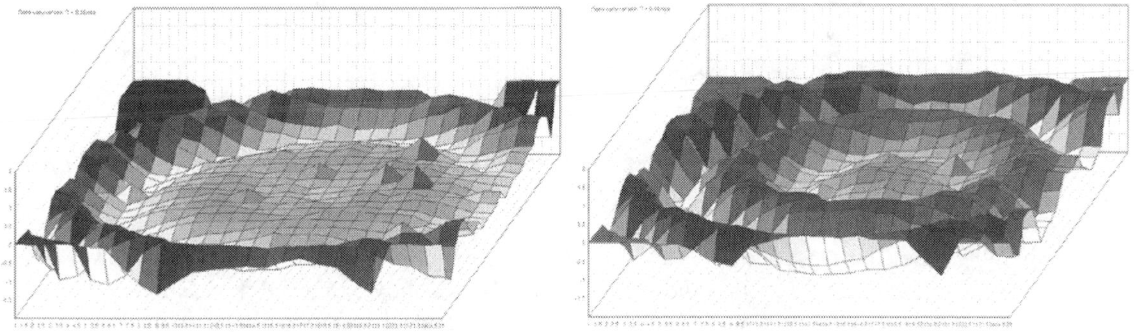

a) b)

Figure 2. 3D image deformation of the on the surface mass-produced sensor: a – the start shot, b – after the finish shot.

After that the second wave on the surface sensor is formed, which also moves to the center of the sensor (figure 2b). At this time the shot was finished. As a result, the interference of waves, coming from the edge of the emitter, at the center of the sensor turned out spike; it has much more amplitude than the amplitude of the waves coming from the edges (figures 3a and 3b).

Considering that scan step on the surface of the emitter was 1 mm, it is possible to estimate the geometric size of the spike, which had 6 mm diameter. A total diameter sensor is 25 mm. The spike in the center of the sensor turned out on time of 5 period. After that the process dies.

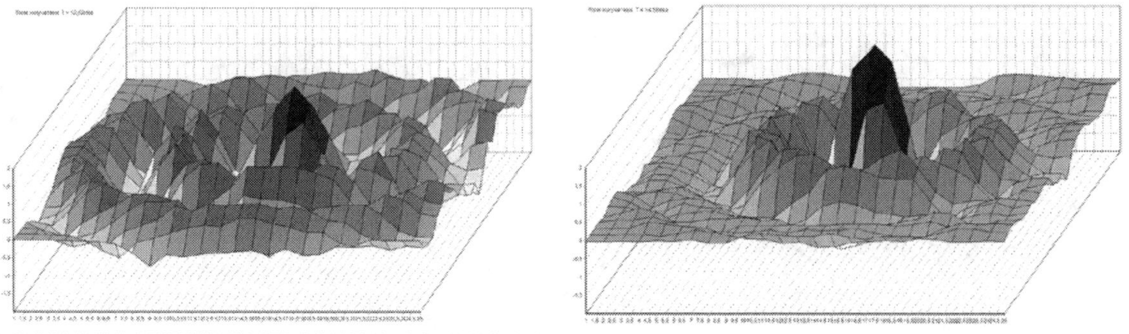

a b

Figure 3. Result of interference of flexural waves on the surface mass-produced sensor: a – on time 8π, b – on time 10π.

The results of the study of ultrasonic field on the surface emitter self-made are presented in figure 4. Constructively sensor consists of tread, a piezoelectric plate of 25 mm diameter and a dash-pot in the shape of a cone.

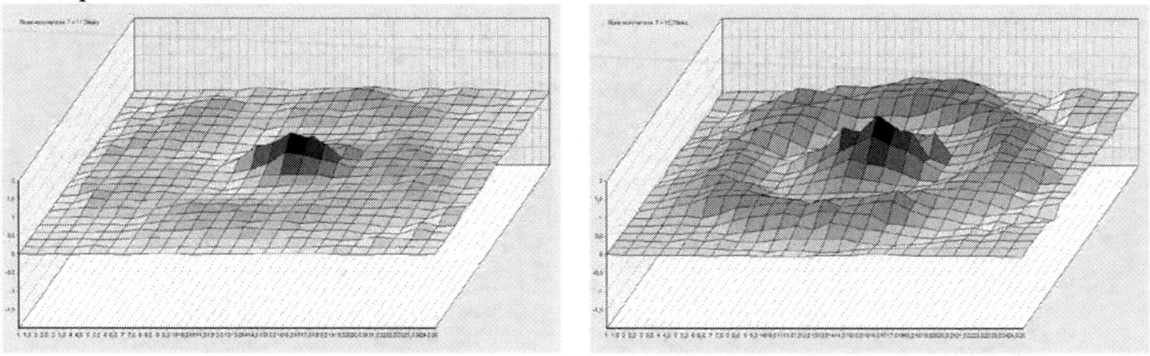

a b

Figure 4. Result of interference of flexural waves self-made sensor: a – on time 8π, b – on time 10π.

Figure 4b shows that the frequency of flexural vibrations almost one and a half times lower and central peak has a smaller amplitude than the figure 3b. Figure 4b shows that the frequency of flexural waves a half times lower and spike has a smaller amplitude than the figure 3b. This is due to the stronger damping emitter. However, the general nature of the deformation of the surface of the emitter has remained unchanged.

The figure 5 presents the results of experimental researches of self-made sensor for which piezoelectric cell of square shape was used.

Change the shape of piezoelectric cell lead to change the deformation of the surface of the sensor.

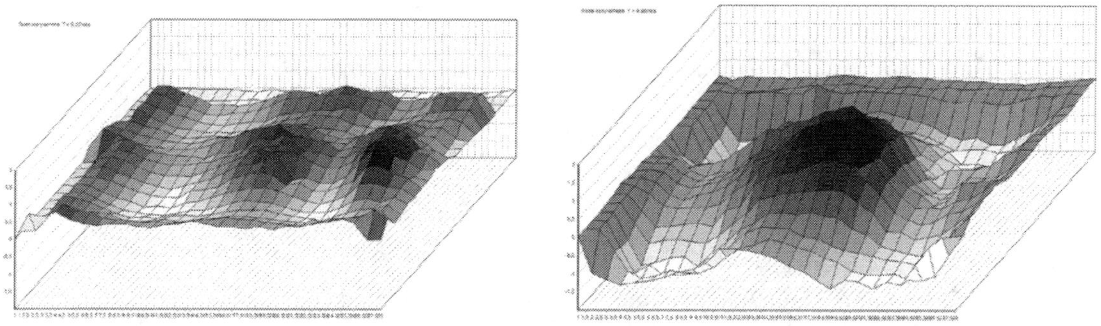

a b

Figure 5. Result of the interference of flexural waves on the surface of piezoelectric cell of the square shape: a – the spike starts to formation, b – the spike finishes to formation.

Figure 5 shows that the sensor edges are formed by two bending waves moving in mutually perpendicular directions, and the result of their interference is observed not only in the center but on the edges of the sensor at equal distance from the corners. Design of both sensors is the same, so the amplitude of deformation and frequency of flexural waves of these sensors are the same too.

4. Conclusions

Comparison of theoretical and experimental studies leads to the conclusion of the adequacy of the proposed model of the sensor with fixed flat side by a box. Using the proposed model we can predict technical features of the sensor and determine the nature of the emitted signal and analyze the reflected signal with regard to the shape and duration of the echo signal. This will allow to increase the reliable estimate of testing result and to eliminate false rejection of products.

References

[1] Birks A S, Green R E *et al* 1991 *ASNT Nondestructive Testing Handbook: Ultrasonic Testing* (American Society for Nondestructive Testing) **7** 893

[2] Josef Krautkrämer, Herbert Krautkrämer 2013 *Ultrasonic testing of materials* (Springer Science & Business Media)

[3] Drury J C 2004 *Ultrasonic Flaw Detection for Technicians* 3rd ed. (UK: Silverwing Ltd.) **1**

[4] Bolotina I, Dyakina M., Kröning M, Mohr F, Reddy K M, Soldatov A, Zhantlessov Y (2013) Ultrasonic arrays for quantitative nondestructive testing an engineering approach *Russia Journal of Nondestructive Testing* **49**(3) 145–158

[5] Ermolov I N 1986 Ultrasonic piezoelectric transducer for Nondestructive testing (Moscow, Mashinostroenie)

[6] Gautschi G (2002) Piezoelectric Sensors *Piezoelectric Sensorics: Force, Strain, Pressure, Acceleration and Acoustic Emission Sensors, Materials and Amplifiers* (Springer Berlin Heidelberg) pp 73–91 doi:10.1007/978-3-662-04732-3_5

[7] Lach M, Platte M, Ries A 1996 Piezoelectric materials for ultrasonic probes *NDTnet* **1** (09) http://www.ndt.net/article/platte2/platte2.htm

[8] Huang C-H, Ma C-C (2001) Vibration characteristics of composite piezoceramic plates at resonant frequencies: Experiments and numerical calculations *IEEE Transactions on Ultrasonics, Ferroelectrics, and Frequency Control* **48**(4) 1147–1156

[9] Ivina N F, Balabaev S M (2001) Characteristic vibrational modes of piezoceramic plates with variable thicknesses *Russian Journal of Nondestructive Testing* **37**(1) 37–43

[10] Ivina N F (2001) Analysis of inherent oscillations in round piezoceramic plates of variable thicknesses *Russian Journal of Nondestructive Testing* **37**(6) 386–393

[11] Asmar N H 2005 *Partial Differential Equations with Fourier Series and Boundary Value Problems* 2nd edition (Pearson Prentice Hall)

[12] Soldatov A, Makarov V, Sorokin P, Kvasnikov K, Selesnev A, Kroening M (2012) Dynamic imaging acoustic fields in research practice *Proceedings – 2012 7th International Forum on Strategic Technology, IFOST 2012* Article number 6357734

[13] Kvasnikov K G, Soldatov A I, Bolotina I O, Krening K M, Potapenko A A (2013) The use of geometrical acoustics for the solution of visualization problems *Russian Journal of Nondestructive Testing* **49**(11) 625–630

[14] Soldatov A I, Kozhemyak O A, Soldatov A A and Shulgina Yu V (2015) Measurement error reducing in the ultrasound time-pulse systems *IOP Conference Series: Materials Science and Engineering* **81**(11) Article number 012117 doi: 10.1088/1757-899X/81/1/012117

[15] Sorokin P V, Shulgina U V, Soldatov A A, Soldatova M A (2015) Influence of the echo shapes on the result of tomographic image *2015 International Siberian Conference on Control and Communications, SIBCON 2015 – Proceedings* Article number 7147305 doi: 10.1109/SIBCON.2015.7147305

Comparison of Three Different Methods for Pile Integrity Testing on a Cylindrical Homogeneous Polyamide Specimen

Y D Lugovtsova [1], **A I Soldatov** [2]

[1] Postgraduate, NDT Institute, Tomsk Polytechnic University, Russia
[2] Proffessor, NDT Institute, Tomsk Polytechnic University, Russia
E-mail: yevgeniya.lugovtsova@bam.de

Abstract. Three different methods for pile integrity testing are proposed to compare on a cylindrical homogeneous polyamide specimen. The methods are low strain pile integrity testing, multichannel pile integrity testing and testing with a shaker system. Since the low strain pile integrity testing is well-established and standardized method, the results from it are used as a reference for other two methods.

1. Introduction

The main method currently used for quality assurance of foundation piles is low strain pile integrity testing or PIT, which is well-established and standardized [1]. A pile head is hit with the hammer, what produces a stress wave travelling down to the pile toe and reflecting back to the pile head. The received wave is measured by an accelerometer mounted on the pile head. The length as well as the location of large flaws can be determined after the interpretation of the results [2,3]. Multichannel measurements approach was taken from geophysics, it is called "Ultraseismic" [4] or multichannel pile integrity testing [5], and the purpose of it is to improve the testing of the pile walls or piles under the structures (e.g. slabs, pile caps, etc.). As part of the PileInspect project, a shaker system instead of a hammer was offered to use, which allows not only performing repeatable measurements, but also enables a better controllability of the input signal [6, 7]. Furthermore, the use of higher frequencies may increase the resolution for the detection of smaller flaws.

Since the low strain pile integrity testing is well-established and standardized method, the results from it are used as a reference for the other two methods. All experiments were performed at Bundesanstalt für Materialforschung und -prüfung, Berlin, Germany. To compare three different methods a homogeneous material as polyamide has been chosen in the form of a 2 m long pole with 20 cm in a diameter.

2. Low strain pile integrity testing

The theory behind the PIT using a hammer was well described by Stahlmann et al. [8], where a compression stress wave induced by a hammer blow travels from the pile top downwards and reflects from the pile toe in regard with the impedance changes. The graphical representation of the results is a so-called reflectogram or a velocity-time plot. Thus, according to it one can estimate the length of the pile using the following expression [8]:

$$L = \frac{c \cdot t}{2}, \qquad (1)$$

where c is the wave velocity and t is the travel time.

The test was carried out on the polyamide pole using the PET (Pile Echo Tester) device from the PileTest company [9], which is shown in figure 1.

a) b)

Figure 1. Experimental set-up for the PIT: a) the specimen, b) PET device (figure from [9]).

The corresponding reflectogram can be seen in figure 2. The aim is to adjust the velocity so, that the maximum of the second peak is positioned exactly at the line corresponding to the length of the pole, which is 2 m in our case. Thus, the resulting wave velocity is 1850 m/s.

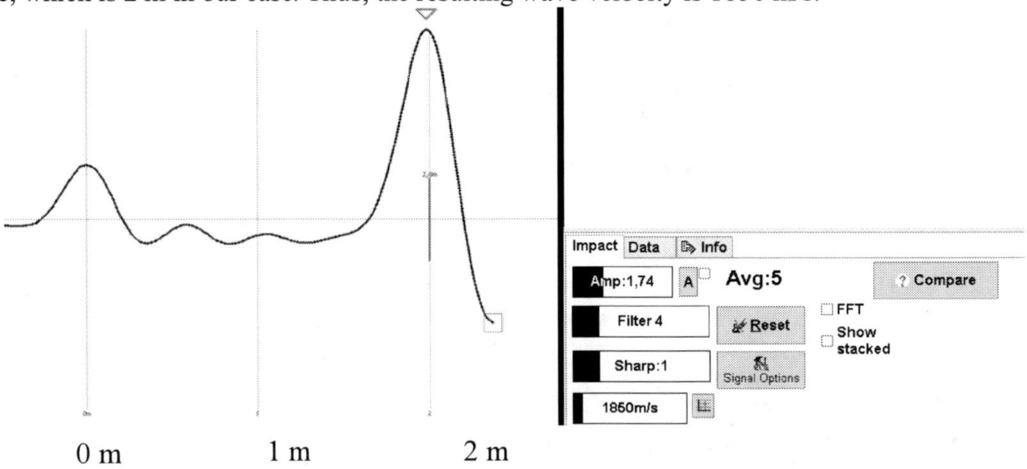

Figure 2. Results of the PIT using hammer on the polyamide pole.

3. Multichannel pile integrity testing

As it was mentioned above to improve the testing of the piles under the structure (e.g. slabs, pile caps, etc.) the multichannel pile integrity testing has been developed [4]. Basically, the classical PIT with the hammer blow was modified by increasing the amount of accelerometers, which are placed vertically along the pile shaft. The number of accelerometers depends on the accessibility of the pile, however, one can note that the method requires at least 1 to 1.8 m of exposed area to provide accurate field measurements (within 5% or better) [10].

The test was accomplished by using a hammer and 6 accelerometers, which were mounted each 20 cm, starting from the top (zero point) going down to 1 m. The signals from the accelerometers were collected by a data acquisition system (DAQ) with the sampling frequency of 51.2 kHz and sent to a field laptop. The whole system is shown in figure 3.

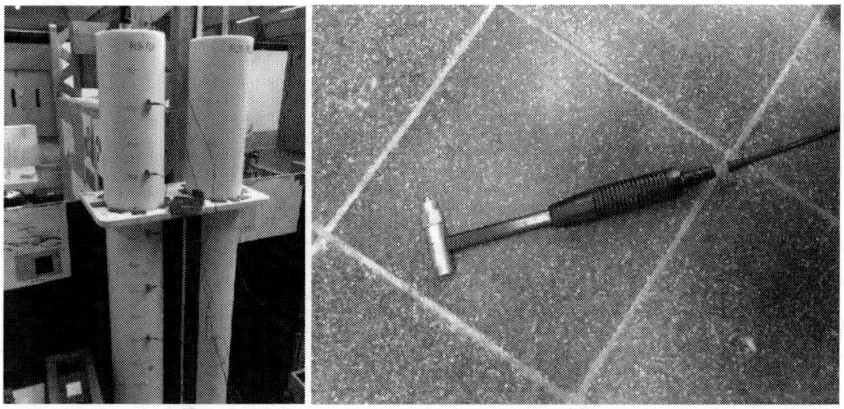

a) b)

Figure 3. Experimental set-up for the multichannel pile integrity testing: a) the specimen with the sensors, b) the testing hammer.

The wave velocity can be determined using the slope (red line in figure 4), which corresponds to the time of the first wave arrivals at each accelerometer, the further away a sensor is the more time is needed for a wave to reach it. Also, the time values can be taken from the negative peaks, which indicate the wave passing the sensor location. The velocity is calculated at the time of first arrivals at 0.2 and 1 m sensors positions using following expression:

$$c = \frac{\Delta s}{\Delta t} = \frac{1m - 0.2m}{0.4189ms - (-0.002137)ms} = 1900\frac{m}{s} \tag{2}$$

Based on the velocity, now the length of the pole can be determined using the negative peaks of the accelerometer at the zero position as follows:

$$L = \frac{c \cdot \Delta t}{2} = 1900\frac{m}{s} \cdot \frac{(2.4777ms - 0.2955ms)}{2} = 2.07m$$

This gives an error of 0.07 m, resulting in a percentage-wise error of 3.5 %. However this method allows the estimation of both the wave velocity and the length of a pile/pole, which is useful in practical applications, where the information about velocity is missing and has to be assumed.

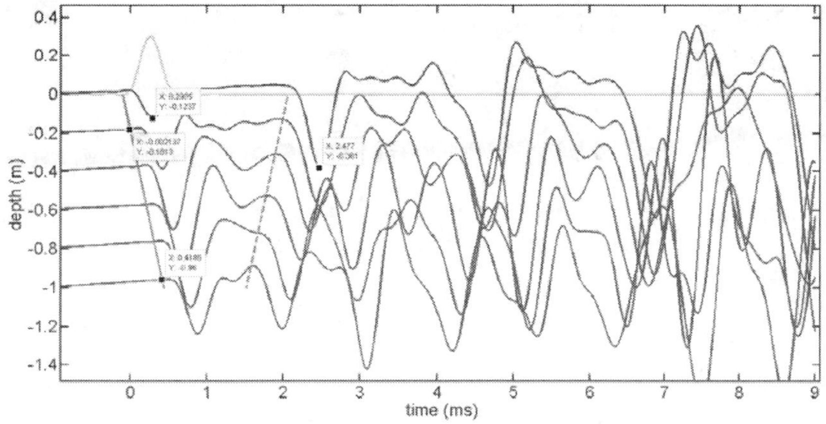

Figure 4. Result of multichannel measurement: green line – hammer blow signal; blue lines – received signals from accelerometers at corresponding depth; red line – first wave arrivals; red dashed line – the time arrivals from the toe reflection.

4. Testing with a shaker system

The idea of using the shaker (tactile transducer) as a source of the excitations was taken from Paquet and Briard (1976) [11] and suggested to use in the frame of PileInspect project, where the main motivation was not only the low cost and easy handling, but also the possibility to use new signal processing techniques on the data obtained. The theory, on which the data processing is based on, is described in [7] and it is suggested to use the impulse response function by performing the deconvolution of the recorded (output) signal with the excitation (input) signal. One of the important points to obtain the impulse response is that the deconvolution must be stabilized using the regularization factor (method after Tikhonov [12]):

$$I(\omega)=\frac{S(\omega)^{*}\cdot X(\omega)}{(S(\omega)^{*}\cdot S(\omega)+\lambda)\cdot dt}, \tag{3}$$

where $S(\omega)$ is the input of the signal's spectra, $X(\omega)$ is the recorded signal's spectra, * is complex conjugate of the spectrum, λ is the regularization factor and dt is the time increment of the measurement.

Then $I(\omega)$ is transformed into time domain to retrieve the impulse response $i(t)$.

The experiment was performed using a shaker, which was positioned on the pole top. The input signal was recorded with the low sensitive accelerometer of 10 mV/g, which was attached on the top of the shaker with some adhesive putty. The accelerometer, with the sensitivity of 100 mV/g on the top of the pole, was used to record the output signal. Figure 5 depicts the experimental set-up. The signals were generated and analysed using the software, designed earlier using LabVIEW specifically for this application. The input and output signals, as well as the result of the deconvolution using 500-1000 Hz logarithmic sweep with the length of 0.1 s are shown in figure 6.

a) b)

Figure 5. Experimental set-up for the testing with the shaker system: a) the specimen with the shaker and sensors, b) complete set-up.

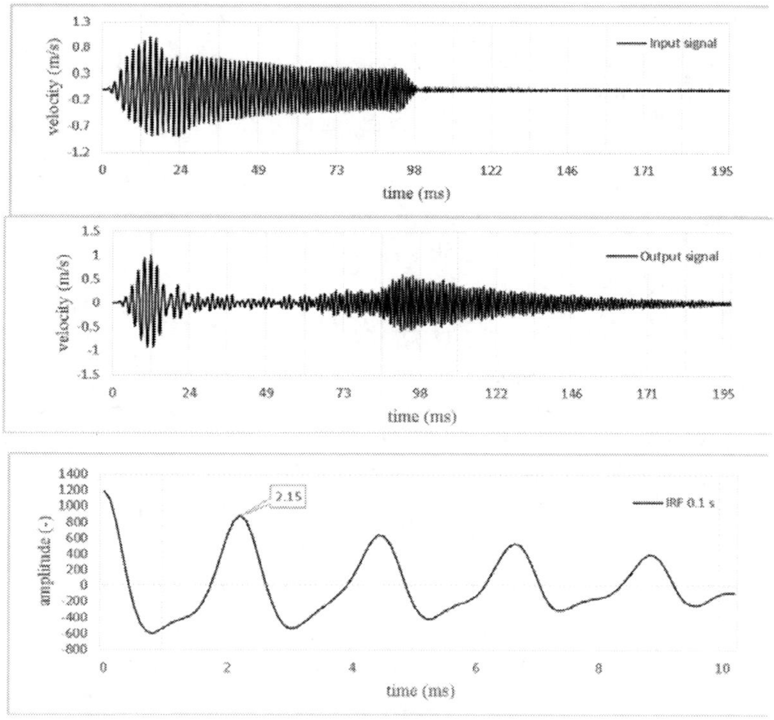

Figure 6. Results of the shaker experiment using a logarithmic sweep of 500–1000 Hz with the length of 0.1s.

Using expression (1) and the wave velocity of 1850 m/s obtained with PIT, the length of the pole can be calculated as follows:

$$L = 1850\frac{m}{s}\cdot\frac{2.15\cdot 10^{-3}s}{2} = 1.99m,$$

which gives the percentage-wise error of 0.5%.

Also, for the same frequency range different sweep lengths were tested. The results, which are depicted in figure 6 were deconvolved and processed based on exp. (3) using the same regularization factor. It can be seen that the peak of the impulse response function (IRF) for the 1 s sweep length is slightly shifted to the right, compared with other two. So, for the time of 2.15 ms the corresponding length is 2.08 m, which gives the percentage-wise error of 4%.

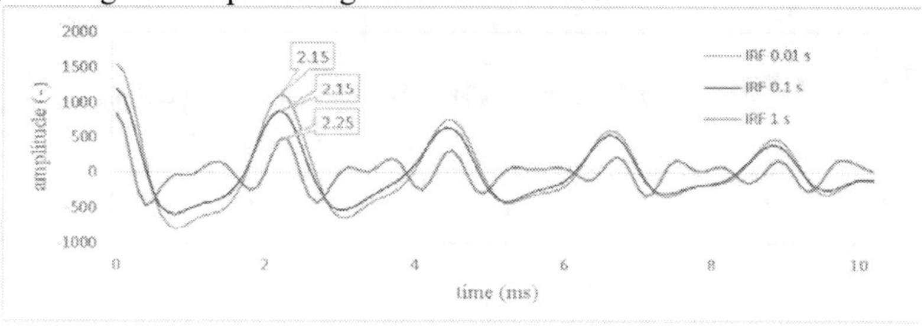

Figure 6. Comparison of the impulse response functions obtained by sweeps with different lengths.

5. Conclusion

Thus, the results show that it is possible to determine both the length of the pole and the velocity of the wave only by the multichannel pile integrity testing. For other two methods at least one parameter has to be assumed. In practical applications, for instance, after the production of new piles, when the pile length is known, the velocity and the presence of flaws can be determined. In other cases, there is no information about the velocity and the length, and these parameters have to be assumed. All results are summarized in Table 1.

Table 1. Results of velocity/length determination using three different methods.

Method	Velocity (m/s)	Length (m)
PIT	1850	2
Multichannel testing	1900	2.07
Testing with shaker	taken from PIT	1.99 (2.08)

The velocity for the testing with the shaker system was taken from PIT results, due to the fact that this method is well-established and standardized. The different sweep lengths were tested and corresponding impulse response functions were compared (figure 6), the time value has a small deviation, which may indicate that sweep length shorter that 1 s gives more accurate results in length estimation. Also, different types of filtering and algorithms for the regularisation of the deconvolution may affect the results, thus the further investigations are needed.

6. Acknowledgements

The authors would like to thank the division 8.2 “Non-destructive Damage Assessment and Environmental Measurement Methods” of Bundesanstalt für Materialforschung und –prüfung.

References

[1] ASTM D5882-01 (2013) *Standard test method for low strain impact integrity testing of deep foundations* (ASTM International, West Conshohocken)

[2] Soldatov A I, Kozhemyak O A, Soldatov A A, Shulgina Yu V (2015) Measurement error reducing in the ultrasound time-pulse systems *IOP Conference Series: Materials Science and Engineering* **81** (1) Article number 012117

[3] Shulgina Y V, Soldatov A I, Shulgin E M, Rozanova Y V, Kroning M (2014) Mathematical analysis of the echo-impulse position detection by the dual-frequency sensing method *Proceedings of 2014 International Conference on Mechanical Engineering, Automation and Control Systems, MEACS 2014* Article number 6986888

[4] Jalinoos F and Olson L D (1996) Determination of unknown depth of bridge foundations using nondestructive testing methods *Structural Materials Technology. An NDT Conference* 91–97

[5] Hardage B A (2000) Vertical Seismic Profiling: Principles, Third Edition. Handbook of Geophysical Exploration: Seismic Exploration (California, Pergamon Press)

[6] http://pileinspect-project.eu/ [Accessed 18 September 2015]

[7] Zheng H, Kappatos V, Niederleithinger E, Ertel J-P *et al* (2015) Defect detection in concrete pile using impulse response measurements with sine sweep excitations *International Symposium on Non-Destructive Testing in Civil Engineering (NDT-CE)* 9

[8] Stahlmann J, Kirsch F, Schallert M, Klingmüller O and Elmer K-H (2004) Pfahltests – modern dynamisch und/oder konservativ statisch *Kolloquium 'Bauen in Boden und Fels', Technische Akademie Esslingen* **20** 23–40

[9] http://www.piletest.com/ [Accessed 18 September 2015]

[10] http://www.olsoninstruments.com [Accessed 27 September 2015]

[11] Paquet J. and Briard M (1976) Non-destructive control of concrete piles – sonic logging and mechanical impedance method *ANN ITBTP* 337 49–80

[12] Vogel C R 2002 Computational methods for inverse problems. Frontiers in Applied Mathematics (Tom 23) (USA,Siam)

Control of Cable Insulation Quality by Changing of Electrical Capacitance Per Unit During High Voltage Testing

N S Starikova[1], V V Redko[2], G V Vavilova[3]

[1] Assistant, National Research Tomsk Polytechnic University, Tomsk, Russia
[2] Assistant professor, National Research Tomsk Polytechnic University, Tomsk, Russia
[3] Graduate student, National Research Tomsk Polytechnic University, Tomsk, Russia
E-mail: nadushasns@tpu.ru

Abstract. The paper describes the complex method of insulation quality control. It was found that electrical contact between bead chain electrodes and insulation surface can be provided by surface discharges along the entire length of the controlled zone. The pattern of electromagnetic field was developed by using Comsol Myltiphysics software.

1. Introduction

One of the parameters, which have high influence on the operational characteristics of the cable product, is a quality of the insulation (continuity, external and internal geometry, chemical compounds). During the manufacture of the cable products it is necessary to provide continuous control of this parameter for timely correction of technological regime and reduction of mass spoilage probability of the products.

Nowadays mainly two methods of control during cable products manufacturing is used: the electro capacitive method [1] and the spark method [2] of control.

1.1. Electro capacitive method of control

Under the control of electric capacity the cable is passing through the cylindrical electrode, low voltage is applied to cable to measure the electrical capacity of the controlled area. To provide electric contact between cable insulation surface and the electrode water in cooling bath is used. Value of the linear electric capacity is changed if defected section of cable insulation passes through the controlled area.

1.2. Spark method of control

The spark method of control is performed with electrode (chain-type, spring-type or brush-type) through which high voltage is applied to the insulation surface. When defected section of cable insulation passes through the controlled area electrical breakdown occurs and it is registered with the automatics [3].

Described method of control does not allow detecting all types of the defects separately. Thus, the reliability of control is low [4,5]. To increase the reliability of cable insulation control it was suggested to combine two existing methods to the one complex method. Suggested complex method consists in measuring of electric capacity per meter during a spark testing of controlled cable (figure 1).

The aim of this work is to assess the ability of reliable electrical capacity measurement during a spark testing of controlled cable.

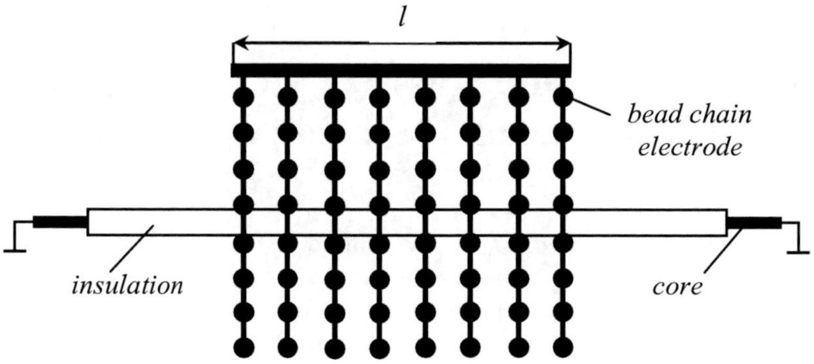

Figure 1. Theoretical model of the complex method of control.

2. Complex method of control

To provide reliable electrical capacity measurement it is necessary to apply test voltage to whole surface of controlled area, but not only to point of contact of electrode bead chain and insulation surface. It is realised in complex method due to high test voltage.

The electrode is presented as a set of chains (figure 1) therefore the electric field in control area is non-uniform. The figure 2 shows a diagram of electric field distribution in controlled zone at the initial moment of test voltage applying before appearance of ionisation processes. It may be concluded from the figure that normal component of electric field dominates over the tangential component. Predominance of the normal component leads to thermal ionisation, which eases the process of electrical discharge.

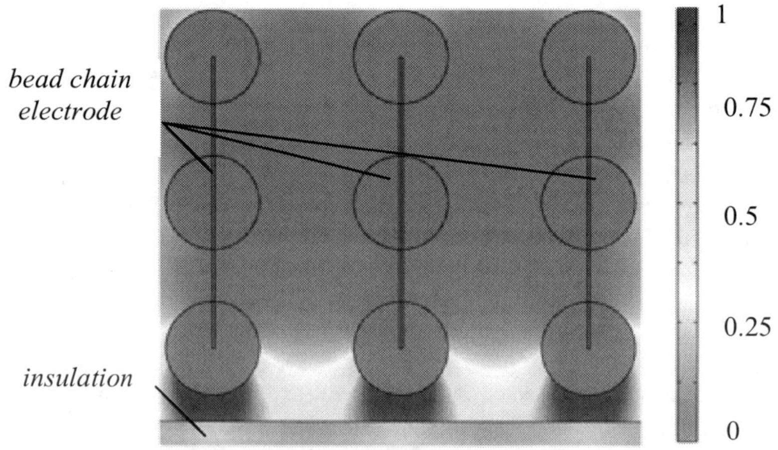

Figure 2. Electrical field distribution between bead chain electrode and cable insulation surface (maximal relative electric field intensity equals 1, minimal equals 0).

In presence electric field sliding discharge is occurred [6]. Different factors effect on the spreading of the sliding discharge. These factors are materials of cable insulation, state of insulation surface, type and value of the test voltage. Length of the sliding discharge can be calculated with empirical Tepler's formula [7]:

$$l_{sd} = k \cdot C^2 \cdot U^5 \cdot \frac{dU^{1/4}}{dt},$$

where k – empirical coefficient, C – specific surface capacity, U – voltage. Thereby, according to the Tepler formula, value of voltage and specific surface capacity have a significant influence on the length of sliding discharge spreading. To explore the test voltage distribution along an cable insulation surface the experiment was carried out.

3. Experimental part

For the experiment it was necessary to use device for applying and measuring of signal. Applying equipment was the system consisting of signal generator, power amplifier and high-voltage transformer. The measuring part consisted of oscilloscope divisor and digital oscilloscope. For applying and measuring were used ring-type electrodes. In current experiment test voltage was applied to the zero coordinate point and potential distribution along the surface of cable product was defined with measurement device.

After the experiments the dependences of the test voltage distribution along an insulation surface from test voltage parameters was obtained (figure 3).

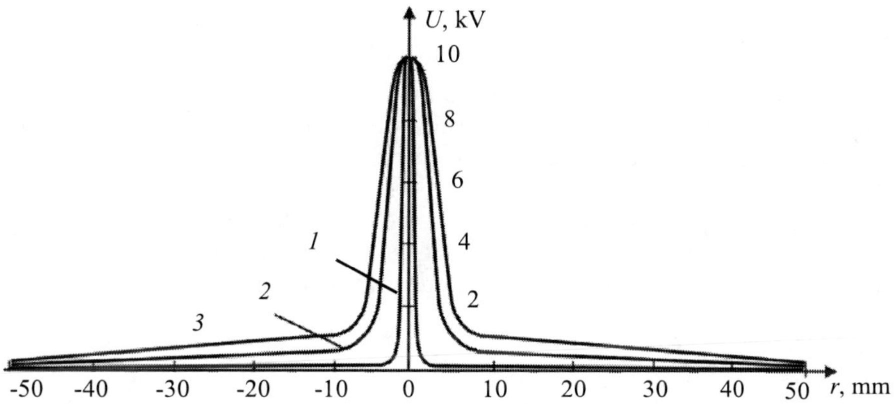

Figure 3. Distribution of the test voltage along the insulation surface with constant voltage (1), with alternating voltage with frequency 50 Hz (2), 1 kHz (3).

Dependences shown on the picture was obtained with the 10 kV test voltage value.

The obtained dependences for test voltages with values 3, 10 and 15 kV were analysed. As a result it is possible to conclude that noticeable distribution of the potential presents only with alternative voltage. It appears due to specific surface capacity of the insulation material, which is confirmed by Teplers formula. In case of direct voltage a specific surface capacity does not affect the spreading of the discharge. The discharge is similar to discharge in the air gap for this case.

Based on obtained dependences of voltage distribution, dependence of the electrode elongation from the type and value of voltage was theoretically found (figure 4).

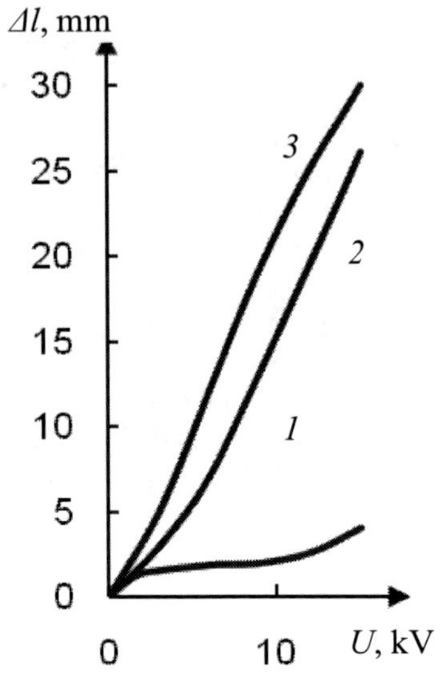

Figure 4. Dependence of electrode elongation from the value of the direct (1), and alternative voltage with 50Hz (2) and 1kHz (3) frequency).

4. Conclusion

Admitting the association of voltage distribution and the length of the sliding discharge it can be noted that value of applied alternating voltage has a significant influence on the electrode elongation, which is corresponds to the Tepler's formula. In case of direct voltage only small elongation of the electrode was found.

Analysis of the research results shows that appearance of the sliding discharge allows providing of electric contact between electrode and insulation surface not only at points of mechanical contact of electrode chains and cable insulation, but along the whole length of controlled zone using the alternating voltage.

5. Acknowledgments

This work was partly supported by the State Organization "Fund for Assistance to Small Innovative Enterprises in Science and Technology" (competition of "UMNIK-2014").

References

[1] GOST (State standart) 27893–88 1989 *Telephone cable. Test methods* (Moscow: standards Publishing House)

[2] GOST (State standart) 2990-72 1986 *Cables, wires and cords. Methods of voltage testing* (Moscow: standards Publishing House)

[3] Redko V V, Starikova N S, Redko L A and Vavilova G V (2015) Determination of sensitivity for in-process control of cable product insulation *IOP Conf. Series: Materials Science and Engineering* **81**(1) 012083

[4] Starikova N S and Redko V V (2013) Research of the reliability of cable insulation control implemented with electric capacity changing in strong electric fields *Non-destructive testing: Proc. Russian Conf.-training* 153-159

[5] Starikova N S, Redko V V and Vavilova G V (2015) Comparison of cable insulation control in weak and strong electric fields *Applied Mechanics and Materials* **756** 486–490

[6] Budenkov V A, Vazhov V F, Kuznetsov Y I, Kurtenkov G E, Lavrynovych V A, Mytnikov A V, Pichugin M T and Startseva E V 2010 *Technics of high voltage: a tutorial* (Tomsk: TPU)

[7] Skanavi G I 1958 *Physics of Dielectrics (strong electric fields)* (Moscow:State Publishing House of physical and mathematical literature)

Measurement Technique of Dose Rate Distribution of Ionization Sources with Unstable in Time Beam Parameters

S G Stuchebrov[1], I A Miloichikova[2] and I B Danilova[3]

[1] Senior Researcher, Tomsk Polytechnic University, Tomsk, Russia
[2] Engineer, Tomsk Polytechnic University, Tomsk, Russia
[3] Bachelor student, Tomsk Polytechnic University, Tomsk, Russia
E-mail: stuchebrov@tpu.ru

Abstract. The article describes a new technique for the average values of radiation dose measurement for the unstable gamma-ray sources which are used in non-destructive testing. The method is based on usage of different types of compact accumulative dosimeters. Spatially distributed position sensitive dosimetry system based on compact sensitive elements was created. Size and spatial resolution of the system of the dosimetry system are chosen taking into account sources characteristics. The proposed method has been tested on the measurement of dose distribution of several sources of X-ray and gamma-radiation based on X-ray tubes, electronic accelerator betatrons and linear electron accelerators.

1. Introduction

The tendency for increasing the research results quality of the internal structure of products in the performance of nondestructive testing in industries is specified requirements on the stability characteristics of the probing beams and correspondingly, of their emitters [1–3]. Despite this, today is often used unstable X-ray and gamma-ray sources. The basic parameters of such sources are changing while working up to several tens of percent. The causes of the parameters changing are: the external pickups on the emitting source; the sensibility to the instability of the external power supply; the heating of the device; the imperfection of individual units of the emitting source. Such systems are used, for example, for the research purposes. This systems use the experimental equipment, debugging of which is often either not possible or does not make sense [4–6].

As a consequence, the use of such devices requires the measurement and monitoring of the characteristics of the dose fields. The variability of the parameters does not allow using of conventional methods based on the data acquisition from a single dosimeter moving within the radiation field because the unstable field must be measured simultaneously in the entire volume.

To solve this problem we studied the possibility of constructing a spatially distributed dosimetry system. For this purpose were used compact storage dosimeters placed in the space in a pre-calculated points dose radiation fields generators. The position detecting elements is caused by the shape of the radiation field of the test generator and the desired spatial resolution of the dosimetry system.

The investigations were carried out on various sources of X-ray and gamma-ray radiation. The spatial distributions of the pulsed X-ray generator RAP-160-5, of the X-ray generator RAP-60-25, of the portable betatron OB-4, of the pulsed electron accelerator "ASTRA" were obtained.

The obtained results showed the suitability of spatially distributed dosimetry systems for the radiation field analysis, including the pulsed sources which have a very sharp change of the beam profile and the high intensity within individual pulses.

2. Materials and methods

2.1. Emitting source

In the investigation the pulsed X-ray generator RAP-160-5, the X-ray generator RAP-60-25, the portable betatron OB-4 and frequency pulsed electron accelerator "ASTRA" were used as an emitting source.

The basic parameters of the pulsed X-ray generator RAP-160-5 are: the anode current is 0.4 – 5 mA; the anode voltage is 40 – 160 kV; the beam angular divergence is 40º; the focal spot size is 1.2 ×1.2 mm; the frequency of radiation impulse is 60 – 700 Hz; the duration of one pulse is about 140 μs [7].

The basic parameters of the X-ray generator RAP-60-25 are: the maximum anode current is 40 mA; the anode voltage is 20 – 60 kV; the beam angular divergence is 16º [8].

The portable betatron OB-4 was used as source of bremsstrahlung. The basic parameters of the portable betatron OB-4 are: the maximum energy is 4.0 MeV; the frequency of radiation impulse is 400 Hz; the duration of one pulse is about 15 μs [9]. The material of the target is tungsten (0.6 cm thick).

In addition to the betatron as a source of the bremsstrahlung was used the frequency pulsed electron accelerator "ASTRA". The general quantities of the accelerator "ASTRA" are: the maximum energy is 450 keV; the frequency of radiation impulse is 50 Hz; the duration of one pulse is about 60 ns [10].

2.2. Dosimetric equipment

The use of different types of the storage dosimeters such as t thermoluminescent dosimeters DTL-02 and capacitor type dosimeters D-2R make it possible to increase the reliability of the experimental data, the measurement accuracy and to extend the application area of the dosimetry system using the different operating mode of the dosimeters.

The solid thermoluminescent dosimeters DTL-02 were used as dosimetric equipment. The thermoluminescent material of the detectors is LiF: Mg, Ti [11].

The individual capacitor type dosimeters D-2R designed for the exposure dose measurement of gamma radiation. Detector D-2R is a cylinder with the ionization chamber and two fluoroplastic capacitors (the charging capacitor and the measuring capacitor), located at the ends of the detector [12].

The small size of the storage type detectors affects the accuracy of their indications. To solve this problem, each of them was calibrated with the help of specialized clinical dosimetry system of high accuracy UNIDOS E equipped with a PTW soft X-ray plane-parallel ionization chamber type 23342 (the sensitive volume is 0.02 cm^3) and with a PTW Farmer chamber type 30013 (the sensitive volume is 0.6 cm^3) [13–15].

The obtained results were allowed not only to evaluate the functionability of each dosimeter, and enter to obtain the correction factors for each of them.

The dosimetric equipment specifications are listed in the Table 1.

Table 1. The dosimetric equipment specifications.

The dosimeter type	Physical principle of the radiation registration	The radiation type of registration	The energy range	The dose range
The dosimeter DTL-02	thermoluminescense	gamma radiation	15 keV – 10 MeV	20 μSv – 10 Sv
The dosimeter D-2R	ionization	gamma radiation	12 keV – 1.25 MeV	0.1 mSv – 20 mSv
The plane-parallel chamber type 23342	ionization	soft X-ray	7.5 keV – 70 keV	3 mGy – 30 Gy
The cylindrical chamber type 30013	ionization	photon	30 keV – 50 MeV	0.1 mGy – 1 Gy

3. Results and discussions

The figures 1, 2 show an examples of the dose rate spatial distribution of the pulsed X-ray source at the 30 cm distance and the dose rate spatial distribution of the portable betatron OB-4 at the 100 cm distance [16–17].

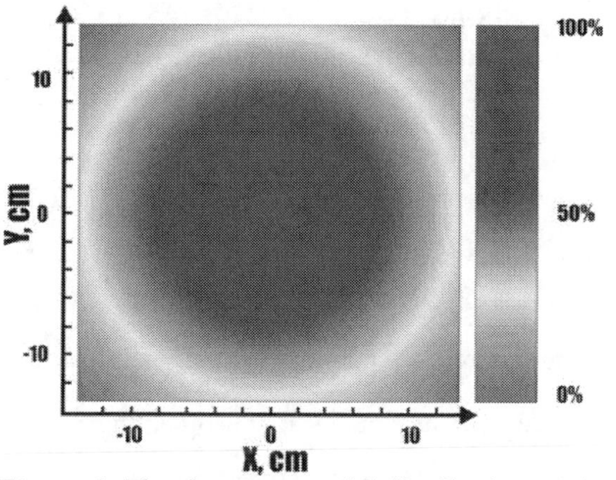

Figure 1. The dose rate spatial distribution of the pulsed X-ray source. Array pitch is 2°.

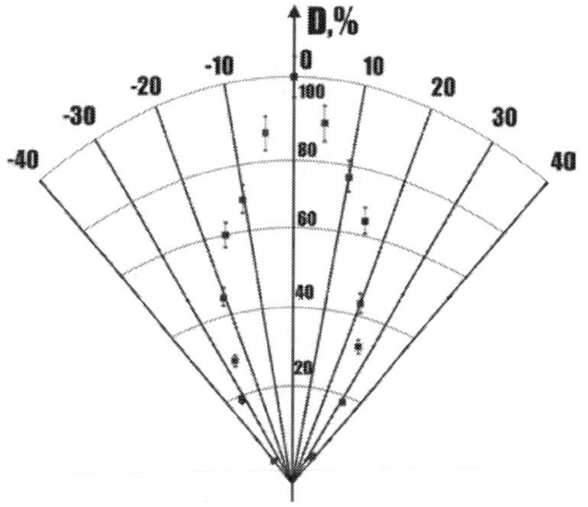

Figure 2. The dose rate profile of the bremsstrahlung generated by betatron.

To calculate the dose rate it is necessary to consider the radiation exposure. However, additional radiation dose on the detectors appearing during the emitting source in the process of stabilization is a problem. The method of calculate the dose received by dosimeter during the time of the source output mode is applied in plants which are not instantaneous get the operating mode [18].

The figure 3 shows a simplified timing diagram of the dose rate of the source at the process of stabilization.

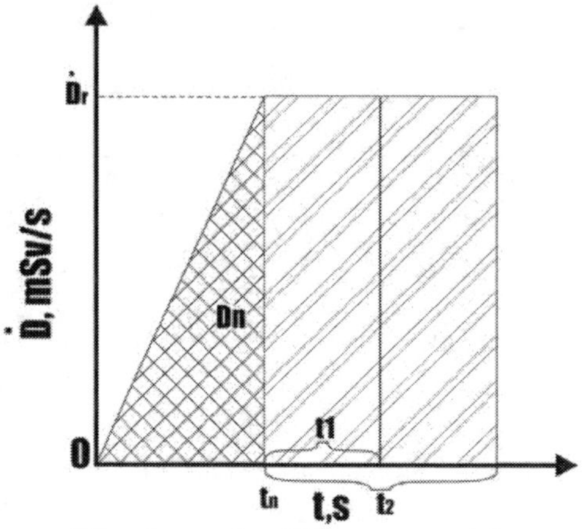

Figure 3. Timing diagram of the dose rate of the source at the process of stabilization.

As it shown in the figure 3, two different exposure times of the dosimeters were chosen. In both cases, total time consists of the output mode time and the operating time of the predetermined mode. The dose received by the detector is the area under the curve of the timing diagram of the dose rate (figure 3). Thus, from the figure 3 that the dose from the time of emitting source in the process of stabilization can be found from the following system of equations:

$$\begin{cases} D_n + t_1 \cdot \dot{D}_r = D_1 \\ D_n + t_2 \cdot \dot{D}_r = D_2 \end{cases} \Rightarrow \dot{D}_r = \frac{D_1 - D_2}{t_1 - t_2} \Rightarrow D_n = D_1 - t_1 \cdot \dot{D}_r = D_2 - t_2 \cdot \dot{D}_r,$$

where D_n – the dose from the time of the emitting source in the process of stabilization; t_1, t_2 – operating time of the source; $\dot{D}_r$ – the dose rate of the source at the operating mode; D_1, D_2 – the total dose accumulated by the dosimeter during times t_1 and t_2, correspondingly.

The proposed technique is tested on the dose rate spatial distribution measurements of the several sources of X-ray and gamma-radiation, based on the X-ray tubes, betatrons and linear electron accelerators. The small size of the dosimeters allows to achieve the high spatial measurements accuracy. The obtained correction coefficient for each of the detecting elements is increased the reliability of the measurements.

4. Summary

In this investigation using the proposed method the dose spatial distribution of the different gamma sources, as pulsed and continuous were obtained. The use of different detection elements allows significantly extending the energy range of the registered radiation: from a few keV to a few MeV. To improve the accuracy of the dosimetric system the additional calibration of each individual storage dosimeter was carried out. To improve the reliability of the results the errors associated with the dose accumulation during the time of the source output mode took into account with the help of the developed method.

Acknowledgements

This work was financially supported by the Ministry of Education and Science of the Russian Federation in part of the science program (N 5.1485.2015).

References

[1] Kurokawa C and Urushiyama A (2013) Measurements for dose distribution with a photo-stimulated luminescence dosimeter sheet *J. Phys.: Conf. Ser.* **444**(1) 012035
[2] Cherepennikov Y and Gogolev A (2014) Device for X-ray spectral absorption analysis with use of acoustic monochromator *J. Phys.: Conf. Ser.* **517**(1) Article number 012037
[3] Low D (2015) The importance of 3D dosimetry *J. Phys.: Conf. Ser.* **573**(1) Article number 012009
[4] Gogolev A, Stuchebrov S, Wagner A, Cherepennikov Yu and Potylitsyn A (2012) Acoustic "pumping effect" for quartz monochromators *J. Phys.: Conf. Ser.* **357** (1) Article number 012031
[5] Kaderka R, Schardt D, Durante M, Berger T, Ramm U, Licher J and Tessa C La (2012) Out-of field dose measurements in a water phantom using different radiotherapy modalities *Physics in Medicine and Biology* **57**(16) 5059–5074
[6] Cherepennikov Y and Gogolev A (2012) Method to reduce radiation exposure in the medical X-ray diagnostics *Proceedings - 2012 7th International Forum on Strategic Technology, IFOST 2012* Article number 6357737
[7] Operations manual 2008 *Portable X-ray device for the industrial utilization* (Tomsk: Photon) p. 34 [in Russian]
[8] Service log 2011 *X-ray generator for the technical application* (Tomsk: Photon) p.17 [in Russian]
[9] Technical specification 2010 *Inductive cyclic accelerator OB* (Tomsk: Betta Plyus) p. 15 [in Russian]
[10] Egorov I S, Kajkanov M I, Lukonin E I, Remnjov G E and Stepanov A V (2013) The Astra repetitive-pulse electron accelerator *Instruments and Experimental Techniques* **56**(5) 568–570
[11] Technical specification ZhBIT2.805.006RJe 2012 Thermoluminescent dosimeters DTL-02 (Moscow: Doza) [in Russian]
[12] Information on http://www.aes.pp.ua/RSafety/P11-02.htm
[13] Information on http://www.ptw.de/unidos_e_dosemeter_ad0.html?&cId=3965.
[14] Instruction manual 2013 Soft X-ray chamber Type 23342 (Freiburg: PTW) 15 p
[15] User Manual 2013 Ionization Chamber Type 30010, 30011, 30012, 30013 (Freiburg: PTW)
[16] Miloichikova I A, Stuchebrov S G , Krasnykh A A and Vagner A R (2015) Dose rate spatial distribution produced by the pulsed X-ray source in the radiographic examination *Adv. Mater. Res.* **1084** 598–601
[17] Miloichikova I A, Ruchjeva V A, Shuvalov E N and Stuchebrov S G (2014) Depth dose distribution of the bremsstrahlung generated by the betatron OB-4 in different environments *XXIV Russian Particle Accelerator Conference: Proceedings* 266-268
[18] Miloichikova I A, Stuchebrov S G , Krasnykh A A and Vagner A R (2015) Radiation dose measurement technique of the X-ray source in the process of stabilization *Adv. Mater. Res.* **1085** 478-481

Focal Spot Size Estimation for a 4 MeV Small-size Betatron Using Digital X-ray Detector

K V Sukharnikov[1], M M Rychkov[3], V G Gentselman[2]

[1]Post-graduate student, Institute of NDT, TPU, Tomsk, Russia
[2]Head of laboratory, Institute of NDT, TPU, Tomsk, Russia
[3]Engineer, Institute of NDT, TPU, Tomsk, Russia
E-mail: k.sukharnikov@gmail.com

Abstract. One of the main advantages of small-size betatrons is small focal spot size. However there is no conventional procedure of focal spot size measurement for betatrons. Techniques described in standards are suitable only for low- and mid-energy tubes. The task become more complex if the focal spot size should be estimated quickly using modern digital detectors. In this paper we present preliminary results of focal spot size measurement of a 4 MeV small-size betatron with two different procedures using digital x-ray detector.

1. Introduction

A betatron is a pulsed induction electron accelerator. The first operable betatron was created by Donald W. Kerst in 1940, although patents describing devices based on similar principles had been published decades before [1]. The development of betatrons until reaching the energy up to about 300 MeV was of increasing interest. Then the high energy betatons have been gradually supplanted by linacs. Nevertheless, low energy machines remain effective sources of hard X-rays used in non-destructive testing.

One of the most promising fields of application of the betatrons is high-energy X-ray computed tomography (CT). The main advantage of the betatrons in CT over the other types of hard X-rays sources is small focal spot size [2]. Obtaining the focal spot sizes within the limits of 200...300 μm is possible because of the cyclic nature of betatrons that allows multiple interaction of the accelerated electron beam with a bremsstrahlung target mounted inside the acceleration chamber. In contrast with linacs, the size of the target can be reduced with minor effect on the radiation intensity.

Focal spot size is one of the most significant parameters of the X-rays source for CT. However there is no conventional procedure of focal spot size measurement for betatrons. Techniques described in European Standard EN 12543 are applicable for x-ray tubes with voltage restricted to 500 kV. As well as Russian Standard GOST 22091.9–86 is suitable only for low- and mid-energy tubes. The task become more complex if the focal spot size should be estimated quickly using modern digital detectors.

The aim of this work is to present preliminary result of focal spot size measured of a 4 MeV small-size betatron with two different procedures using digital x-ray detector. The methods to be applied in the research are slit camera radiographic method and indirect method based on image unsharpness measurement using image quality indicator (IQI).

2. Experimental setup

The betatron (figure 1) used in the experiments is a small-size pulsed betatron MIB-4 with maximal bremsstrahlung energy 4 MeV and claimed focal spot size 0.3×3 mm. Positioning was performed with a motorized positioning systems by PI GmbH & Co. KG.

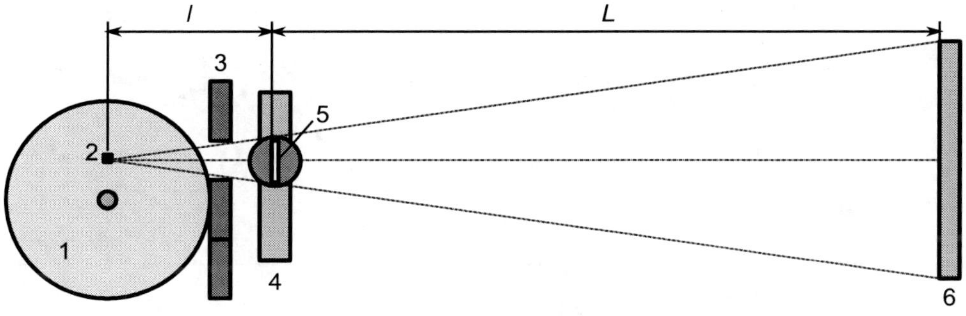

Figure 1. Experimental setup layout: 1 – betatron; 2 – tungsten target mount inside acceleration chamber; 3 – collimator; 4 – positioning system; 5 – test object; 6 – imaging system.

A digital X-ray detector Perkin Elmer XRD 1622 was used as the imaging system (figure 1). Total pixel number is 2048×2048 with pitch 200 µm. An important characteristic of the X-ray detector that must be taken into account and corrected is line spread function (LSF). Full width at half maximum (FWHM) of the X-ray beam $FWHM_I$ can be found from the image obtained using the relation [3]:

$$FWHM_R = FWHM_I + FWHM_{LSF} \quad (1)$$

Where $FWHM_R$ is FWHM of the obtained image and $FWHM_{LSF}$ is FWHM of detector's LSF. Likewise it was done in [3] the LSF of the detector system was found by differentiation of the edge spread function (ESF) obtained by contact radiography of the test object with sharp edge. Gaussian approximation of the LSF is shown on figure 2.

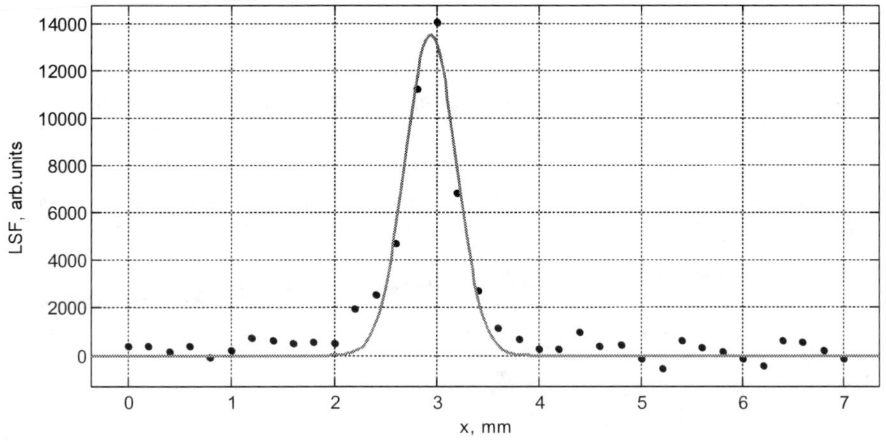

Figure 2. Gaussian approximation of the LSF. $FWHM_{LSF} = 0.6mm.$

3. Slit camera method

Focal spot size measurement using pinhole and slit cameras is based on the camera obscura principle. Figure 3 indicates the common layout of the experimental setup used for this method.

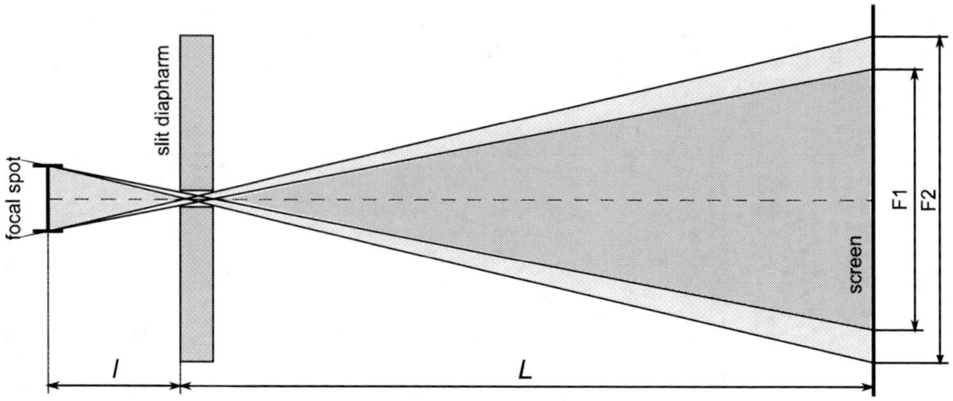

Figure 3. Slit camera experiment layout.

In the ideal case the thickness of the edges is negligible small that is practically unachievable. Taking into consideration the finite slit camera thickness, intensity distribution shown on figure 2 can be introduced. $F1$ and $F2$ can be calculated through geometrical optics approach:

$$F1=(f-d)\frac{2L+h}{2l-h}-d \tag{2}$$

$$F2=(f+d)\frac{2L-h}{2l+h}+d \tag{3}$$

Where f is focal spot size, d is slit size, h is slit camera thickness. $F1-F2$ indicates the penumbra size. Influence of the finite slit camera thickness can be corrected by using the FWHM of the X-ray beam profile.

$$f=\frac{FWHM_I}{M-1}-d\left(1-\frac{1}{M}\right) \tag{4}$$

$$M=\frac{L+l}{l} \tag{5}$$

Setting h and d equal to 0 (thickness is negligible small) expressions can be transformed to the ideal case (no penumbra is observed):

$$F1=F2=f(M-1) \tag{6}$$

Taking into consideration (2) and (3) one can figure out an expression for the edge thickness and slit size limits by solving the equation $F1-F2=0$. The solution is as follows:

$$h_{max}=\frac{2dl}{f};d_{min}=\frac{fh}{2l} \tag{7}$$

Using the expressions (7) as well as the expression for linear attenuation coefficient for the desired material of the slit camera one can evaluate boundaries for the geometrical size of the camera. Figure 4 indicates geometry of the steel diaphragm used in the experiment.

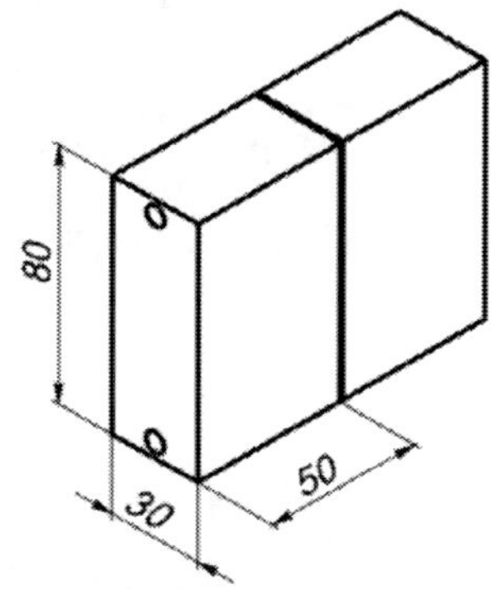

Figure 4. The camera with variable slit size used in the experiment.

The focal spot size (horizontal value) of the 4 MeV betatron is claimed to be $f = 0.3$ mm. So assuming the distance $l = 250$ mm (figure 1) and slit size $d = 0.1$ mm one can make certain that the maximal slit camera thickness does not exceed the maximal value allowed. The slit camera is chosen to be made of steel because of simplicity of manufacturing.

In article [3] slit camera made of tungsten was used to measure the focal spot size of a 5 MeV linac. In comparison to the linac the betatron has lower radiation intensity as well as smaller spot size. It diminishes the requirements for slit camera material used in our experimental setup. Also usage of the digital X-ray detector in our experimental facility (figure 1) allows enhancing contrast of the images obtained in comparison of an X-ray film. However an experiment with tungsten-made slit camera is planned.

Profile of the image obtained with 100 μm slit is shown on figure 5.

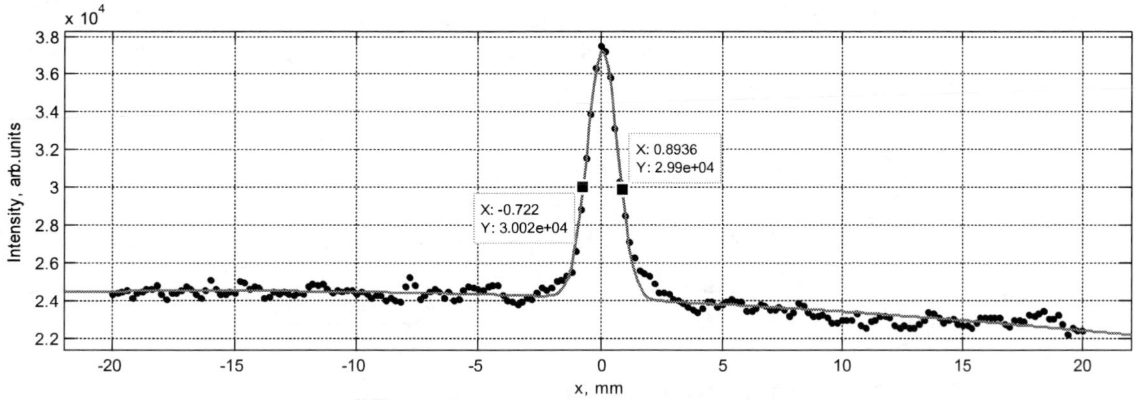

Figure 5. Slit camera image profile.

The focal spot size can be now found using (1) and (4) with following parameters: $l = 250$ mm, $L = 1000$ mm, $M = 5$, $d = 0.1$ mm. The evaluated value is $f = 0.1738$ mm.

4. Indirect measurement using IQI

The indirect method is based on relation between basic spatial resolution and focal spot size. Equation (8) taken from ASTM E 2698 [4]. It can be used for focal spot size assessment by measuring the basic spatial resolution SRb and image unsharpness U_i .

$$f = \frac{M}{M-1}\sqrt[3]{U_i^3 - \left(\frac{2.0}{M}SRb\right)^3} \tag{8}$$

One of the most widely used method for measurement of SRb and U_i uses duplex wire type IQI.

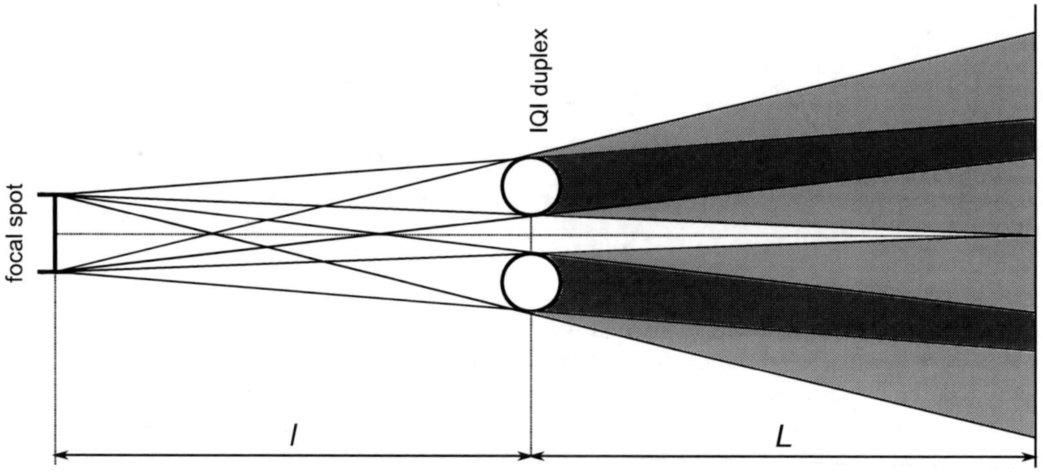

Figure 6. Duplex wire type IQI experiment layout.

The penumbra areas indicated on figure 6 merge when focal spot size f matches wire diameter and spacing *D*. Gray level profiles for various cases are shown on figure 7. Results of geometrical optics modeling provide an intuitive representation of the technique. But the relations are hardly to apply practically, by virtue of the fact that the wires are penetrable for the hard x-rays. Also probabilistic nature of interaction of radiation with matter as well as detector noises should be taken into account. For this reason a method based on spatial resolution measurement can be applied.

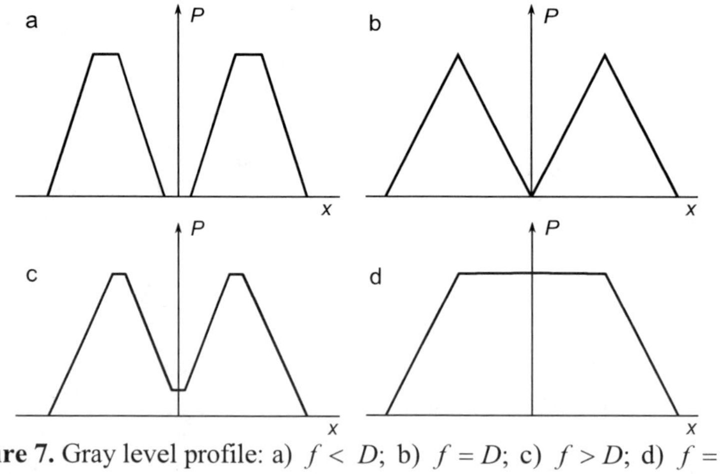

Figure 7. Gray level profile: a) $f < D$; b) $f = D$; c) $f > D$; d) $f = 2D$.

Profiles of radiograph captured on the real experiment are shown on figure 8. The experiment is based on the layout presented on figure 1 with $l = 245$ mm, $L = 1045$ mm, $M = 5.265$.

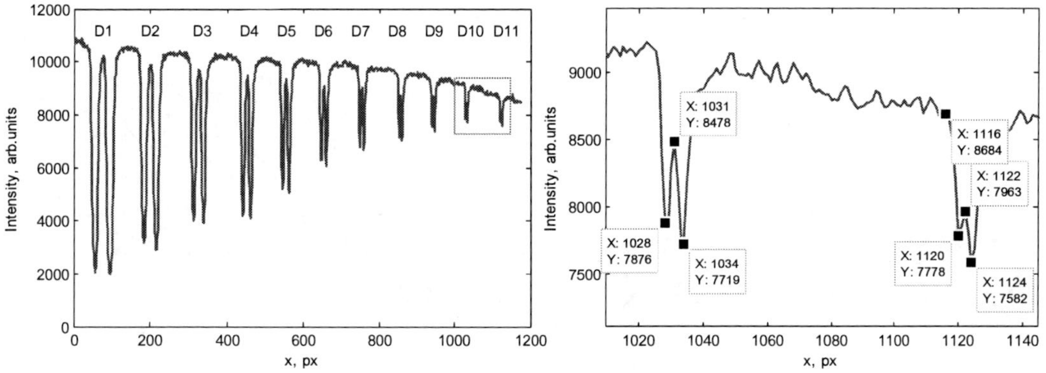

Figure 8. Radiograph profile of duplex wire IQI.

Obtained radiograph can be used to assess a spatial resolution *SRb* and image unsharpness U_i by measurement of modulation depth (dip) for neighboring IQI wires. The results are shown in table 1.

Table 1. Results of dip measurement

IQI Element No	D8	D9	D10	D11	D12
Measured dip, %	75.9	73.39	52.35	28.36	0
Maximum Unsharpness, mm	0.32	0.26	0.20	0.16	0.125
Maximum Basic Spatial Resolution, mm	0.16	0.13	0.10	0.08	0.063

Finally the focal spot size can be found from (8) using *SRb* and U_i values for D11. The evaluated value is $f = 0.1972$ mm.

5. Conclusion

Focal spot size of a 4 MeV betatron was measured using slit camera and resolution methods. The results of the experiments are very close and do not exceed claimed focal spot size 0.3 mm. However the presented results are course and have to be verified. Ways to get more accurate measurements are as follows:

- Use of the slit camera made of tungsten.
- Precise measurement of the LSF of the imaging system.
- Reiteration of the experiments using different detectors to collect statistics.
- Monte-Carlo simulation of the experiment to obtain extra data.
- Use of a betatron with a changeable bremsstrahlung target.

References

[1] Kerst D W (1946) Historical development of the betatron *Nature* **157** 90–95
[2] Vaynberg E I, Kasyanov V A, Chakhlov V L, Stein M M (2004) Experience of Using Small-Size Betatron Mib-5 in the Structure of Industrial Computed Tomograph Bt-500Xa *16th World Conference on NDT (WCNDT 2004) Proceedings* 1–5
[3] Gambaccini M, Cardarelli P, Taibi a., Franconieri a., Domenico G D, Marziani M, Barnà R C, Auditore L, Morgana E, Loria D, Trifirò A. and Trimarchi M (2011) Measurement of focal spot size in a 5.5MeV linac *Nuclear Instruments and Methods in Physics Research, Section B: Beam Interactions with Materials and Atoms* **269** 1157–1165
[4] Bavendiek K, Ewert U, Riedo A, Heike U and Zscherpel U (2012) New Measurement Methods of Focal Spot Size and Shape of X-ray Tubes in Digital Radiological Applications in Comparison to Current Standards *18th World Conf. Nondestruct. Test.* 16–20

Journal of Physics: Conference Series **671** (2016) 012059 doi:10.1088/1742-6596/671/1/012059

State of the Art and Development Trends of the Digital Radiography Systems for Cargo Inspection

V Udod[1], J Van[2], S Osipov[3], S Chakhlov[4] and A Temnik[5]

[1]Professor, Department of mathematical methods and information technologies in economics. Tomsk State University, Russia,
[2]Postgraduate student, NDT Institute, Tomsk Polytechnic University, China,
[3]Leading reseacher, NDT Institute, Tomsk Polytechnic University, Russia,
[4]Head laboratory of technical tomography and introscopy, NDT institute, Tomsk Polytechnic University, Russia,
[5]Head of section, NDT Institute, Tomsk Polytechnic University, Russia
E-mail: temnik_ak@mail.ru

Abstract. Increasing requirements for technical parameters of inspection digital radiography systems are caused by increasing incidences of terrorism, drug trafficking and explosives via variety of transport. These requirements have determined research for new technical solutions that enable to ensure the safety of passengers and cargos in real-time. The main efforts in the analyzed method of testing are aimed at the creation of new and modernization of operated now systems of digital radiography as a whole and their main components and elements in particular. The number of these main components and elements includes sources of X-ray recording systems and transformation of radiometric information as well as algorithms and software that implements these algorithms for processing, visualization and results interpretation of inspection. Recent developments of X-ray units and betatrons used for inspection of small- and large-sized objects that are made from different materials are deserve special attention. The most effective X-ray detectors are a line and a radiometric detector matrix based on various scintillators. The most promising methods among the algorithms of material identification of testing objects are dual-energy methods. The article describes various models of digital radiography systems applied in Russia and abroad to inspection of baggage, containers, vehicles and large trucks.

1. Introduction

The terrorism is one of the principal dangers for the society in XXI century. In support that we can mentioned scaled terroristic acts on transport (2001 – September 11 attacks in USA; 2009 – 2010 – Moscow Metro bombing, Nevsky Express bombing; 2011 – Domodedovo International Airport bombing) [1, 2]. As relation in Russia and aboard there is generated a whole scientific and technical area named «anti-terroristic diagnostics» to secure the passenger operations and freight service [2–6].

The systems for anti-terroristic diagnostics (inspection of cargo) solve a wide spectrum of tasks: explosive, toxic and radioactive material detection, and also all kinds of explosive devices and killing agents [2].

All types of the Custom Equipment can be divided on the following big groups [6]:

1. The technical means and systems to detect the objects by its visually perceptible forms (by outward appearance, internal structure image, thermal image and etc.) are:

– optical magnifying devices (magnifying glass), inspection mirrors, technical endoscopes, borescopes and videoscopes;

– inspection television systems;

– infrared imagers;

– inspection of radiation.

2. The technical means and systems to detect the objects by indicative of material properties (by conductivity, permeability or permittivity, density, chemical composition, presence of nonlinear electromagnetic properties, radioactive radiation and etc.) are:

– metal detectors;

– nonlinear radars;

– ultrasonic echolocation devices;

– narcotic and explosive detectors (dreyfspektrometry, gas analyzers, radiation gamma quantum and corpuscular imaging devices, kits chemicals, immunochemical diagnostic tools);

– radiation monitoring devices.

3. The technical means and systems to detect the objects by features work of mechanisms and electronic devices (by distinctive sound, vibration, electromagnetic radiation and etc.) are the radio radiation detection equipment (including the side electromagnetic radiation), acoustic and vibro-acoustic signals.

4. Integrated Inspection Systems providing combining the functions and possibilities of two or more types of inspection equipment in one system are: metal detectors with built-in sensor of the ionizing radiation, mobile robotic systems and other combined system.

As follows from numerous publications (for example, [5–10]), currently among different kinds of the inspection systems the digital radiography system (DRS) are occupied one of the dominant positions. The description of its state of the art, the practical usage and also the further development trends would be presented below.

2. The digital radiography systems for cargo inspection

The term «digital radiography» means a set of methods of non-destructive radiation testing and diagnostics for whose the radiation image of the inspected object (IO) is converted on some stage into a digital signal. Further this digital signal is stored in computer memory and redistributed to the two-dimensional array of the measured data, which can be digitally processed (contrast enhancement, scaling, dissection, smoothing etc.) and, finally reproduced on display as a half-tone image, directly perceived by the operator [11, 12].

The main efforts of the radiation inspection experts are directed to create a high-performance sources and detectors of the ionized radiation, and also to develop algorithms and software to process the results of detection of radiation. For example, Non-destructive Testing Institute of Tomsk Polytechnic University (NDT-institute TPU) with company «Foton» created the following sources of the ionized radiation [13]:

– charged particle accelerators – betatrons MIB – 7,5D and MIB – 9D (Shtok–T) used to the trucks inspection;

– X-ray machines RAP 160-2D, RAP 200-4D, RAP 300-4D used to the cars inspection.

In the same article [13] it was noted that in the inspection DRS to record the transmitted radiation most often used three base types of detectors:

– fluorescent screens with CCD camera;

– fluorescent screens with photodiode matrix;

– scintillation crystals in combination with photodiode array.

In NDT institute of TPU there was developed the original specialized software «Diada» to visualize the internal structure of IO, which was used in different DRS specifically – to detect prohibitive electronic components of printed-circuit boards.

In work [14] there is analysis of requirements for the betatron's bremsstrahlung radiation dose rate meters (monitors), designed to use in the inspection systems and to inspect large size articles. The designed monitors have the radiation resource no less 10000 hours, work in wide temperature range from – 20 to +50ºC and provide measurement of dose rates from 0.05 to 50 R/min. The geometrical sizes of one monitor for energy 9 MeV are 30×25×285 mm^3.

In some inspection tasks it is necessary to identify the material of the inspected object. It can be done to identify the IO class which is prohibited or strictly regulated to cross-border movement or to transport by air. For the material identification there are applied methods using the ionized radiation sources. The dual energy methods (DEM) are most widely used [15] and based on the evaluation of X-ray attenuation for two sources with specially selected maximal energies. There are existed also some modifications of this method: twice X-raying – for two different X-ray tube voltages (it corresponds to two effective radiation energies); single X-raying with the radiation recording by two placed one after the other (along the beam path) detectors and etc. By this method the IO components can be divided by its effective atomic number (Z_{eff}) and the color image [16] is generated by color identification palette, where:

"light-weight" materials with Z_{eff}<10 have shades of orange;

"middle-weight" materials with 10< Z_{eff}< 20 have shades of green;

"heavy-weight" materials with Z_{eff}>20 have shades of blue.

The work [15] estimates contribution of the various components to the integrated measurement error of the effective atomic number of the homogenous IO by the dual energy method for energy range below 200 keV. It was shown that the largest contribution to the measurement error of the effective atomic number is due to the maximum energy deviations in the X-ray radiation spectra.

The work [17] researches the high-energy realization of DEM. This realization is based on the IO prescan by narrow beams of the high-energy X-ray radiation with two maximal energies, on the generation of two digital radiographs, on the further separate estimation of contributions of the Compton scattering and the effect of pair production into X-ray attenuation and into the generation of the identification parameter image.

For the mass-produced portable betatron MIB 4.5/7.5 in work [18] a series of calculations to estimate the different physical factors influence on the identification quality were carried out. The following factors were considered: maximum energy deviations of the high-energy X-ray radiation from the nominal values; the radiation, scattered within IO; the radiation, scattered within construction of inspection system. Studies prove the needs to consider mentioned factors for the expection system design with material identification option by the dual high-energy method.

The work [19] describes a new algorithm to identify materials for the X-ray luggage inspection system, used in airports and strategic locations. The algorithm is based on the DEM and the spectral analysis of the digital radiographs.

In order to improve the accuracy of automatic recognition of nuclear and explosive materials in inspected objects (e.g., containers in seaports) using DEM and high-energy X-ray sources (with a maximum energy of 2.5 MeV to 9 MeV) [20] it proposed to carry out X-raying OK by three fan-beams from three sources. Thus, each beam is formed from a single radiation source and is recorded on the proper line of detectors.

The work [21] made a simulation of the container inspection system, comprising as the radiation source accelerator 4 MeV/9 MeV and 960-detectors array with the detector aperture size 6×6 mm^2. To identify the materials there was used DEM realization based on the proposal that detectors have a function of the energy discriminator, i.e. there is an electronic separation of the impulse flows on the detector's output. Developed as a result of research methodology permit to identify the following materials: tissue-equivalent plastic, Al; Fe; Ag.

The authors in [16, 22] studied the questions more informative radiographs due to the X-ray registration passed through the IO for two, three and more energy ranges. Three-energy X-ray images of the inspection objects with the effective atomic number from 7.08 to 8.07 were generated. The material recognition precision by the effective atomic number was about 5 % when using detectors with type scintillator-photodiode on base of scintillator ZnSe(Te).

The work [23] reports that «Rapiscan Systems» company developed the inspection system on base the linear accelerator for 9 MeV to detect a special nuclear material within freight containers. This system uses the two step inspection. On the first stage (main scanning) the container is checked quickly and fully by images, generated by use two independent detector's arrays: conventional main array with the high space resolution and the array with «rough» energy resolution. Those two detector arrays are used to detect high-Z objects, i.e. the objects with high atomic number (such as lead, tungsten, uranium), which can be potential protective materials or special nuclear materials. On second stage, container areas, whose were identified as high-Z, rechecked. This is accomplished by X-raying those areas by X-ray beam and by recording the secondary radiation (fast neutrons, delayed gamma rays), which generated if high-Z area contains special nuclear material.

The work [24] presents the description of experience of practical use DRS, intended for cargo container inspection to detect a special nuclear material. Reportedly, in particular, that scan speed is about 45 cm/s. The build system has a similar structure with one from [23].

The DRS on base the linear accelerator 3 MeV / 6 MeV to detect contraband in aircraft containers is described in work [25]. To accelerate the suspicious cargo detection the container radiography is doing under several directions (multi-view). The detector array with the detector aperture size 6×6 mm^2 was used. The scan speed was about 0.2 m/s. The time to acquire the container images in two orthogonal views was about 40 s.

3. Summary

1. The inspection DRS development is carried out mainly by the X-ray source and detector enhancement, and also by the enhancement of algorithm and software for digital data processing.
2. The inspection performance can be significantly increased by use two-step procedure – the main scanning all object and the subsequent rescan of selected suspicious areas.
3. To improve the inspection efficiency of DRS it is advisable use the multi-angle-raying (multi-view) with simultaneous use several radiation sources and detector arrays.
4. To improve the reliability of the unauthorized objects detection it is necessary further development as methods to improve the space resolution of DRS, and as the material identification methods (dual energy methods and its versions).

The work was supported by National Research Tomsk Polytechnic University (grant VIU INDT 66 2014).

References

[1] Kluev V V, Bobrov V T (2011) Technic diagnostics – the nation security base *Testing. Diagnostics* **5** 55–61
[2] Bobrov V T (2011) Session of the science concil for automated system diagnostics and testing Russian Academy of Science for the transpoty security problems *Testing. Diagnostics* **5** 61–67
[3] Park J S and Kim J K (2011) Calculation of effective atomic number and normal density using a source weighting method in a dual energy X-ray inspection system *J Korean Phys Soc.* **59** 2709–2713
[4] Kovalev A V (2004) Anti-terrorism and criminalistic diagnostics *Testing. Diagnostics* **2** 21–29
[5] Lebedev M B, Sidulenko O A and Udod V A (2008) State of the art analysis and development of the digital radiography system *Bulletin of the Tomsk PolytechnicUniversity* **312** 47–55
[6] Arkanov A P 2006 *Technical means of anti-terror* (Moskow, Publ. Phoenix)
[7] Spirin D O, Berdnikov Ja A and Gavrish Ju N (2010) Principles of introscopy for the large-scale cargo *Sc and Eng gazette SPbSPU* **2** 120–127
[8] Sidulenko O A, Kas'yanov V A, Kas'yanov S V and Osipov SP (2008) Research of an opportunity of application small-sized betatrons for identification of substances of the testing objects by the method of dual energy *Testing. Diagnostics* **8** 46–52
[9] Vorobeichikov S E, Udod V A, Klimenov V A and Schetinkin S A (2014) An algorithm for the

automatic detection of inclusions in an inspected object with a scanning digital X-ray imaging system (one-dimensional variant) *Russian J Nondestr Test.* **50** 359–368

[10] Bukley A A (2009) Researching and creation of portable inspection X-ray techniques and equipments of NDT. Working out of technology and its application *Testing. Diagnostics* **4** 76–80

[11] Kluev V V and Sosnin F R (1999) Modern state of digital X-ray equipment *Russian J Nondestr Test.* **37** 576–591

[12] Sosnin F R (1994) Modern methods and means of digital X-ray radiography (review) *Ind Lab.* **60** 28–34

[13] Klimenov V A, Kasianov V A, Lebedev M B, Moskalev Ju A, Temnik A K, Shtein M M and Chakhlov S V (2011) State-of-the-art and prospects for the creation of compettive systems of digital radiography (SDR) *Testing. Diagnostics* special issue 25–29

[14] Volkov V G and Shtein M M (2013) Betatron bremsstrahlung monitors for inspection of large-sise products and vehicles *Testing. Diagnostics* **9** 78–80

[15] Klimenov V A, Osipov S P and Temnik A K (2013) Identification of the substance of a test object using the dual-energy method *Russian J Nondestr Test.* **49** 642–649

[16] Ryzhikov V D, Opolonin A D, Volkov V G, Lisetskaja E K, Galkin S N and Voronkin E F (2013) Three energy digital radiography to separate materials with low effective atomic number *Vestnik NTU «KhPI»* **34** 43–51

[17] Klimenov V A, Alhimov Ju V, Shtein A M, Kasianov S V, Babikov S A, Batranin A V and Osipov S P (2013) Application and development of digital radiography methods for thechnical diagnostics of non-destructive testing and inspection *Testing. Diagnostics* **13** 31–42

[18] Osipov S P, Temnik A K and Chakhlov S V 2014 The effects of physical factors on the quality of the dual high-energy identification of the material of an inspected object *Russian J Nondestr Test.* **50** 491–498

[19] Pourghassem H, Fesharaki N J,Tahmasebi A (2012) Material detection based on GMM-based power density function estimation and fused image in dual-energy X-ray images *Proceedings 4th International Conference on Computational Intelligence and Communication Networks, CICN* **6375134** 364–368

[20] Frosio I, Borghese N A, Lissandrello F, Venturino G and Rotondo G (2011) Optimized acquisition geometry for X-ray inspection *Conference Record – IEEE Instrumentation and Measurement Technology Conference* **5944195** 300–305

[21] Gil Y, Oh Y, Cho M, Namkung W (2011) Radiography simulation on single-shot dual-spectrum X-ray for cargo inspection system *Appl Radiat Isot.* **69** 389–393

[22] Opolonin O D and Ryzhikov V D (2013) Increasing informativity of digital radiographic systems *Funct Mater.* **20** 528–533

[23] Stevenson J, Gozani T, Elsalim M, Condron C and Brown C (2011) Linac based photofission inspection system employing novel detection concepts *Nucl Instrum Methods Phys Res Sect A* **652** 124–128

[24] Miller E A, Caggiano J A, Runkle R C, White T A and Bevill A M (2011) Scatter in cargo radiography *Appl Radiat Isot.* **69** 594–603

[25] Duan X, Cheng J, Zhang L, Xing Yu, Chen Zh and Zhao Zi (2009) X-ray cargo container inspection system with few-view projection imaging *Nucl Instrum Methods Phys Res Sect A* **598** 439–444

High Prf Metal Vapor Laser Active Media For Visual And Optical Monitoring

S N Torgaev[1,2,3], M V Trigub[4,5], G S Evtushenko[6] and T G Evtushenko[7]

[1]Assistant of professor, Tomsk polytechnic university, Tomsk, Russia
[2]Research assistant, V.E. Zuev Institute of Atmospheric Optics SB RAS, Tomsk, Russia
[3]Assistant of professor, Tomsk State University, Tomsk, Russia
[4]Assistant of professor, Tomsk polytechnic university, Tomsk, Russia
[5]Research assistant, V.E. Zuev Institute of Atmospheric Optics SB RAS, Tomsk, Russia
[6]Professor, Tomsk polytechnic university, Tomsk, Russia
[7]Assistant of professor, Tomsk State University, Tomsk, Russia
E-mail:torgaev@tpu.ru

Abstract. In this paper the feasibility of using metal vapor lasers for visual and optical monitoring of fast processes is discussed. The theoretical calculations consistent with the experimental study have been performed. The possibility of visualizing objects with pulse repetition frequency of the brightness amplifier up to 60 kHz has been demonstrated. The visualization results of the corona discharge are also given.

1. Introduction

Today many scientific and technical problems are related to the necessity of real-time visual monitoring of different objects and processes shielded by the intense broadband background illumination. The examples of such processes are self-propagating high-temperature synthesis, the processes of interaction of intense energy fluxes with the surface, welding, plasma discharge, etc. Visualization of such processes is possible using active optical systems with brightness amplifiers, for example, a laser monitor [1–3].

Active optical systems based on metal vapor brightness amplifiers have a number of advantages such as high spectral brightness of radiation, high gain in a narrow spectral range (2–5 pm), operation in a pulse-periodic mode. For the visualization of fast processes it is necessary to use a high-frequency brightness amplifier. The typical pulse repetition frequency (PRF) of metal vapor amplifiers, such as copper bromide vapor amplifier, is about 10–30 kHz. The pulse repetition frequency is limited by a number of plasma-chemical processes within the active medium of the amplifier. In [4,5] the authors showed that obtaining high pulse repetition rate is possible with brightness amplifiers operating with a low input energy into the discharge. In this case, the negative impact of cumulative effects in a plasma discharge are partially leveled. Such a mode can be obtained when the amplifier is pumped by high frequency pulses of short duration (20–50 ns, with typical duration of 200–300 ns), as well as it operates in a low-current mode, i.e. when the current going through the gas discharge tube (GDT) is lower than its typical values and equals a few amperes [6].

2. Optical schemes. Calculation and experimental data

One of the features of metal vapor active media on self-terminating transitions is that they can operate effectively both as a generator and an amplifier. Moreover, operating as a brightness amplifier, along with the high gain of the active medium up to 100 dB/m, the metal vapor active media emit intense superradiance, that is 20–30% of the power of the laser.

These advantages of metal vapor active media inspired engineers to design a laser projection microscope, and later, a laser monitor [1,3,7,8]. Laser monitors can be built using two schemes: monostatic [7] and bistatic [9]. In the first one, the active medium is both a radiation source and a brightness amplifier (figure 1a). In the bistatic scheme the object is illuminated by the active medium that operates as a generator (laser), and the reflected signal is amplified by the other similar active medium (figure 1b).

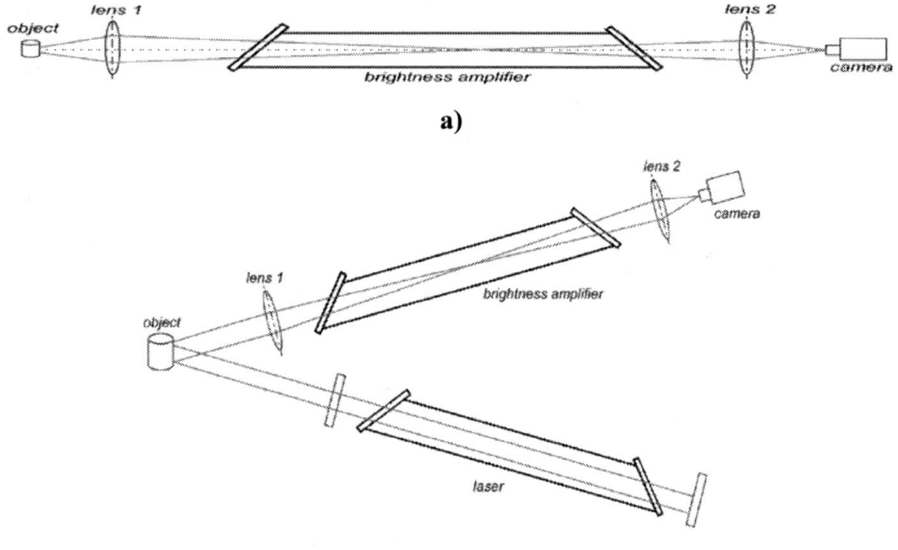

Figure 1. Laser monitor optical schemes: a) monostatic, b) bistatic.

The high repetition frequency (300 kHz, and then 400 kHz) in a copper bromide vapor laser was obtained in [10,11]. In [11] a GDT with a diameter of 0.5 cm and a length of 24 cm was utilized. Subsequently, in [12], using a GDT with a diameter of 0.7 cm and active medium length of 14 cm the pulse repetition frequency reached 700 kHz. In the works mentioned above the laser was operated at a low energy input into the discharge which was obtained when the laser active medium was pumped by high-voltage pulses of short duration. It was observed that the mode of high pulse repetition frequencies could be achieved in a GDT with a small diameter. In a GDT of a small diameter the interpulse plasma recombination is more effective due to diffusion processes. However, the use of lasers (brightness amplifiers) with a small diameter GDT in active optical systems is not effective because the field of view is reduced. Thus, the challenge was to obtain high pulse repetition frequencies (and gain), up to 100 kHz and above in GDTs with a diameter more than 1 cm [13].

In this paper the experimental data are analyzed and theoretical estimations of electrical characteristics for two GDTs are provided. The experimental conditions and the GDT parameters are given in table 1.

Table 1. Experiment and GDT parameters.

	GDT №1	GDT №2
GDT diameter, cm	2	0,7
Active medium length, cm	50	14
Pulse repetition rate, kHz	5–40	630
Neon pressure, torr	20	20-25
GDT wall temperature, K	650	550

As it was mentioned above, it is possible to obtain high pulse repetition frequency of metal vapor brightness amplifiers when they are operated at a low energy input into the discharge. However, when using such a mode the output energy decreases, which is confirmed by the modelling and experimental results. Figure 2 shows the dependence of the output energy of the copper bromide laser (at a green wavelength of 510.6 nm) on the pump pulse repetition rate in the range of 10 kHz to 40 kHz [5] for GDT № 1. These dependences were obtained both experimentally and by means of simulation.

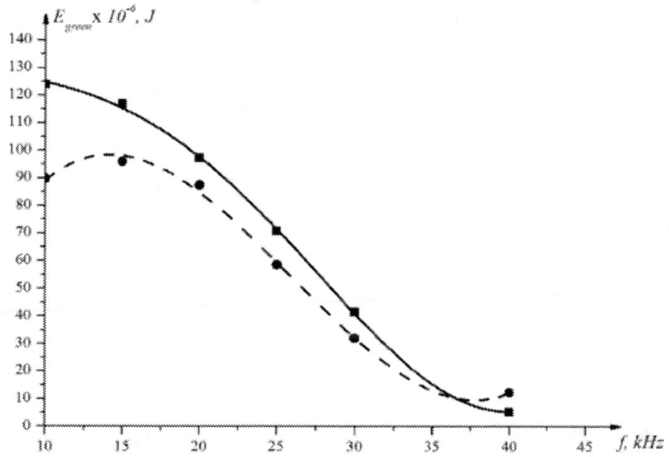

Figure 2. Dependence of the radiation energy (green wavelength of 510.6 nm) on the frequency: experimental (solid line) and calculation (dashed line) data.

In a number of practical tasks the decrease in the generation energy (superradiance) can be quite critical, for example, when visualizing objects with a high absorption coefficient at amplifier wavelengths. At the same time the energy of the radiation reflected from the object brought to the amplifier may be insufficient for the amplification, i.e. the energy of the input signal can be lower than the noise energy of the amplifier. And in this case the input signal carrying the image will not be amplified. For example, in GDT № 1 at a pulse repetition rate of 40 kHz the noise energy of the amplifier is $E_{N\_GDT№1} = P_{N\_GDT№1} \cdot \tau = 2.8 \cdot 10^{-12} J$. And the energy of the radiation (for the green line) is $E_{GDT№1} = 5 \cdot 10^{-6} J$. And when the amplifier operates at a frequency of 630 kHz (GDT №2) the noise energy of the amplifier is $E_{N\_GDT№2} = 8.7 \cdot 10^{-12} J$, and the energy of the radiation (for the green line) is $E_{GDT№2} = 0.12 \cdot 10^{-6} J$. In this case, the experimental data are given for the copper bromide vapor active element that operates in a generator mode. When it is used as a brightness amplifier it is operated in a superradiance mode (without a resonator). In this case the energy of the radiation is 5 times less approximately. Consequently, when the active medium is operated in a brightness amplifier mode the radiation energy for GDTs presented above will be approximately $E_{GDT\,№1} = 1\cdot 10^{-6} J$ and $E_{GDT\,№2} = 6.8\cdot 10^{-9} J$ for GDTs №1 and №2, respectively.

The presented estimates show that when the pulse repetition frequency increases the radiation energy decreases much faster than the energy of the noise of the amplifier but remains higher than the last one. This allows to visualize objects though the image quality will decrease. The pulse energy of the brightness amplifier noise is determined by the following equation [14]:

$$E_N = h\cdot\nu\cdot\Delta\nu\cdot\tau_{INV}\cdot\left(\frac{d^2}{\lambda\cdot L}\right)^2,$$

where τ_{INV} is the radiation pulse width, $\Delta\nu$ is the radiation bandwidth due to the Doppler broadening, ν is the frequency of the green radiation wavelength, d is a GDT diameter, L is a GDT active medium length. This ratio shows that the amplifier noise energy is not directly dependent on the pulse repetition rate. The increase in the pulse repetition rate reduces the duration of inversion (the duration of the laser pulse, τ_{INV}). However, the experimental data show that when the pump frequency varies from 10 to 40 kHz, the laser pulse duration (FWHM) is reduced by a factor of 1.5, while the radiation energy is reduced by a factor of more than 20 (figure 2). When obtaining generation in a GDT with a diameter of more than 1 cm, which is necessary for its application as an amplifier, it will be necessary to reduce the input energy which will also reduce the energy of the radiation.

Thus, as the pulse repetition frequency increases the radiation energy significantly decreases. When the active element is used as a brightness amplifier in a monostatic scheme of the laser monitor (figure 1a) for visualizing objects with a small reflectance coefficient of the corresponding wavelength, the energy of the signal brought to the amplifier can be of the same order as the noise energy. This effect leads to a considerable deterioration of the quality of the image formed by the laser monitor. For the visualization of such objects it is necessary to use a brightness amplifier with active additives, which will increase the generation energy. For the visualization of distant objects and objects with a low reflectance coefficient it is necessary to use a bistatic laser monitor scheme (figure 1b). In this scheme it is possible to significantly increase the energy of the illumination using an independent source that operates as a generator and a brightness amplifier synchronized with the latter in time.

The analysis suggests the possibility of designing a brightness amplifier with the pulse repetition rate up to 100 kHz when it is operated in a low energy input mode. At the same time the time resolution of the laser monitor will be 10^{-5} s. And the use of active additives (e.g., HBr) will increase the efficiency of laser plasma recombination during the interpulse period. This will increase the aperture of the brightness amplifier active medium and pulse repetition frequency.

3. Results of visualization

Based on the estimates, the scheme of a monostatic laser monitor that allows to obtain an image formed by a single superradiance pulse was designed. In this experiment the brightness amplifier operated at frequencies of up to 60 kHz, i.e with the time resolution ≈ 16.7 μs. For obtaining the sufficient spatial resolution, the GDT with a diameter of the active medium of 2 cm was used. A further increase in the pulse repetition frequency was limited by the power source being used. The results of visualization of the test object (metal grid) at different pulse repetition frequencies are shown in figure 3.

a) b)

Figure 3. Visualization of the metal grid using a laser monitor with different PRF: 30 kHz (a) and 60 kHz (b).

In our experiment, as the frequency increased the superradiance beam diameter decreased from 1.2 cm at 30 kHz to 0.5 cm at 60 kHz. The visualization was performed in a scheme of the laser monitor with frame-by-frame image registration. In figure 3 it is shown that when the beam diameter, and hence the width of the gain profile decreased, the field of view was reduced.

Figure 4 shows the results of the visualization of the corona discharge obtained by the laser monitor [15] at different moments of time.

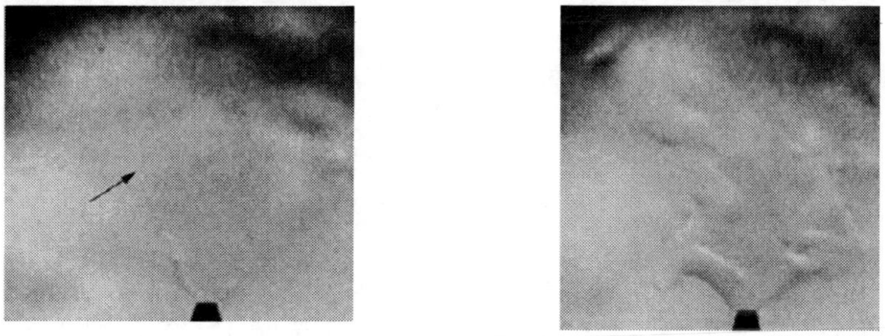

a) b)

Figure 3. Visualization of the corona discharge process development at different moments of time: 375 μs (a) и 1125 μs (b).

It can be seen that the laser monitor allows us to visualize processes that occur during the evolution of the corona discharge and to study their dynamics. In particular, using the frame-by-frame registration mode it is possible to calculate the velocity of the discharge evolution, the movement of the thermal front, etc.

4. Conclusion

Thus, it has been shown that copper bromide vapor brightness amplifiers with high PRF can be used for designing high speed laser monitors. High pulse repetition frequencies are obtained using the mode of low energy input into the discharge. The current measurements suggest that for solving a number of problems it is advisable to use a bistatic laser monitor construction which will allow to monitor the objects, remote objects as well, with a low reflectance coefficient at the operating wavelength.

The experimental results show that the use of brightness amplifiers as an integral part of the laser monitor with frame-by-frame registration makes it possible to obtain images of high quality with the time resolution of up to 20 μs.

The authors appreciate the help and comments of the senior researcher of V.E. Zuev Institute of Atmospheric Optics SB RAS D.V. Shiyanov

Acknowledgements
The work was supported by the Russian Science Foundation, Project № 14-19-00175.

References

[1] Batenin V M, Klimovskij I I, Selezneva L A (1988) A study of coal arc electrode surface during the combustion process *Doklady Earth Sciences* **303** 857–860

[2] Prokoshev V G (2009) Micro- and nanostructure and hydrodynamic instabilities induced by laser radiation on solid state surface and their diagnostic using laser and probe microscopy methods (Vladimir: dissertation on doctor habilitation degree of physical and mathematical sciences)

[3] Evtushenko G S, Trigub M V, Gubarev F A, Evtushenko T G, Torgaev S N, Shiyanov D V (2014) Laser monitor for non-destructive testing of materials and processes shielded by intensive background lighting *Review of Scientific Instruments* **85**(3) 033111 doi: 10.1063/1.4869155

[4] Torgaev S N, Boychenko A M, Evtushenko G S, Shiyanov D V (2013) Simulation of a CuBr–Ne–HBr laser with high pump pulse repetition frequencies *Russian Physics Journal* **55**(9) 1039–1045

[5] Boichenko A M, Evtushenko G S, Nekhoroshev V O, Shiyanov D V, Torgaev S N (2015) CuBr-Ne-HBr laser with a high repetition frequency of the lasing pulses at a reduced energy deposition in the discharge *Physics of Wave Phenomena* **23**(1) 1–13

[6] Fedorov V F, Fedorov K V, Shiyanov D V, Evtushenko G S (2014) Low-current discharge in a copper vapor laser *Proc. Conf. on Metal; Vapor Lasers* 84 [in Russian]

[7] Zemskov K I, Isaev A A, Kazaryan M A, Petrash G G (1974) Laser projection microscope *Quantum Electronics* **1** 14–15

[8] Kuznetsov A P, Buzhinskij R O, Gubskij K L, Savelov A S, Sarantsev S A, Terehin A N (2010) Visualization of plasma-induced processes by a projection system with a Cu-laser-based brightness amplifier *Plasma Physics Reports* **36**(5) 428–437

[9] Buzhinskij O I, Vasiliev N N, Moshkunov A I, Slivitskaya I A, Slivitsky A A (2002) Copper vapor laser application for surface monitoring of divertor and first wall in ITER *Fusion Engineering and Design* **60**(2) 141–155

[10] Evtushenko G S, Petrash G G, Sukhanov V B, Fedoeov V F (1999) CuBr-laser with a 300 kHz pulse repetition rate *Quantum Electronics* **29** 775–777

[11] Evtushenko G S, Gubarev F A, Fedoeov V F, Sukhanov V B, Zaikin S S (2008) CuBr-laser operating at 400 kHz pulse repetition rate *Bulletin of the Tomsk Polytechnic University* **312** 106–107

[12] Nekhoroshev V O, Fedorov V F, Evtushenko G S, Torgaev S N (2012) Copper bromide vapour laser with a pulse repetition rate up to 700 kHz *Quantum Electronics* **42**(10) 877–879

[13] Trigub M V, Shiyanov D V, Vlasov V V (2014) Brightness Amplifiers with PRF up to 100 kHz *International Conference of Young Specialists on Micro/Nanotechnologies and Electron Devices, EDM* 301–304

[14] Pasmanik G A, Zemskov K I, Kazaryan M A and others (1988) Optical systems with brightness amplifiers (Russia: Gorkij) [in Russian]

[15] Rybka D V, Andronikov I V, Evtushenko G S, Kozyrev A V, Kozhevnikov V Yu, Kostyrya I D, Tarasenko V F, Trigub M V, Shut'ko V Yu (2013) Corona discharge in atmospheric pressure air under a modulated voltage pulse of 10 ms voltage pulse *Atmospheric and Oceanic Optics* **26** 85–90

Analytical Approximation of Spectrum for Pulse X-ray Tubes

S Vavilov[1], G Koshkin[2], V Udod[3,4], O Fofanof[1]

[1]Associate Professor of Cybernetics, National Research Tomsk Polytechnic University, Tomsk, Russia
[2] Professor, Department of Applied Mathematics and Cybernetics, National Research Tomsk State University, Tomsk, Russia
[3] Professor, NDT Institute, National Research Tomsk Polytechnic University, Tomsk, Russia
[4] Professor, Department of Economics, National Research Tomsk State University, Tomsk, Russia
E-mail: baron@tpu.ru

Abstract. Among the main characteristics of the pulsed X-ray apparatuses the spectral energy characteristics are the most important ones: the spectral distribution of the photon energy, effective and maximum energy of quanta. Knowing the spectral characteristics of the radiation of pulse sources is very important for the practical use of them in non-destructive testing. We have attempted on the analytical approximation of the pulsed X-ray apparatuses spectra obtained in the different experimental papers. The results of the analytical approximation of energy spectrum for pulse X-ray tube are presented. Obtained formulas are adequate to experimental data and can be used by designing pulsed X-ray apparatuses.

1. Introduction

One hundred and twenty years ago, Wilhelm Rontgen made one of the greatest discoveries of modernity, which gave him the title of the first Nobel laureate in physics. X-rays were discovered – quanta with wavelengths ranging between the optical range and gamma radiation. At the phenomenological level, Rontgen studied all the basic properties of X-rays, and, in particular, made an attempt to evaluate the energy spectrum of the open radiation on the basis of a penetrating facility of radiation. In modern parlance, Rontgen used the method of filters, which nowadays is the most accurate, coupled with making use of modern pulsed X-ray detectors.

It is interesting to note that in his first experiments this Rontgen identified two main areas of application of X-rays: medicine and fault detection. And now, in literature, you can find links to pictures of hands of Roentgen's assistant and Roentgen's spouse, which show the structure of their hands. Less well known is the fact that one of the first images taken by Roentgen was a shadow image of a gun barrel. Here is how the researcher wrote: "Through one handset ... I got a pretty good picture shotgun barrels. It was clear and sharp to see all the details of bullets and internal irregularities Damascus barrels "[1].

In the late 1930s, it had been received short pulses of X-rays lasting a few tenths of microseconds marking the birth of pulsed X-ray technology. In the 1960s, it began learning to handle the nanosecond range of the X-ray pulses. They created the first small-size pulsed X-ray apparatuses (PXAs) and offered empirical methods of calculating the basic characteristics of the pulse tubes [2–4].

Among the main characteristics of the PXAs the spectral energy characteristics are the most important ones: the spectral distribution of the photon energy, effective and maximum energy of quanta, pulsed exposure dose, and power of exposure dose.

Knowing the spectral characteristics of the radiation of pulse sources is very important for the practical use of them in non-destructive testing. In one way or another all flaw characteristics – the thickness of the tested material, the focal length, exposure, sensitivity – are connected with the energy of the radiation and its distribution. Knowledge of the spectral and energy characteristics has become essential in new areas by use of the pulsed X-ray sources for inspection and customs testing, in the X-ray location and navigation.

2. Processing of experimental data

At the end of the 1960s, the making of the first home apparatuses of series "IRA" and "MIRA" has led to the extension of theoretical and experimental studies of the spectrum of pulsed devices [5–7]. In the majority of papers, the spectral distribution is calculated from the experimental dose characteristics. Note, as detectors, silicon and thermoluminescence detectors are used [8,9].

The intensity distribution of the X-ray pulse can be described by the following relation [10]:

$$F(\lambda) = \frac{(\lambda - \lambda_0)^2}{\lambda^4},$$

where $F(\lambda)$ is a density of radiation intensity; λ is a wavelength of radiation quantum; λ_0 is a short-wavelength of radiation border. The most likely wavelength quantum is determined from the relationship:

$$\lambda = 2\lambda_0 \tag{1}$$

Thus, the spectrum of pulse emitters has longer wavelengths than the spectrum of X-ray apparatuses with the sinusoidal and constant anode voltage.

The experimental results of a study of the spectrum of devices with supply voltages in the range of 100-500 kV show a number of significant deviations from these dependences. Enumerate the main reasons of such situation: the errors of detector, the influence of radiation absorption, the contribution of scattered radiation, peculiarities of the supply voltage. The effective power can become displaced to lower energies than it follows from (1) [10–12].

We have attempted on the analytical approximation of the PXA spectra obtained in the different experimental papers. We analyzed data on the PXA spectrum starting with the papers, in which were used the first serial models of the pulsed X-ray apparatuses “IRA”, “RINA”, and the like. The level of the supply voltage in these devices does not exceed 400–500 kV.

Further research we conducted with making use of “MIRA-2”. Its energy distribution $F(x)$ [5] was tabulated by values shown in table 1, where $x = 15\times(E/E_0)$, E–energy of quantum, E_0– energy of maximum quantum.

Table 1. The tabulation of energy distribution $F(x)$.

x	0.5	1	2	3	4	5	6	7	8	9	10	11	12	13	14	15
F(x)	0	1.6	4.3	7	9.5	11	11.5	12	10.7	9.6	8	6.3	5	3	1.2	0

Using a discrete set of data, i.e. table 1, it is necessary to obtain an estimate of the analytical form of the function $F(x)$.

The problem was solved for the two options: the first, on the base of all data of table 1; the second, on the base of the same table, but without the first column of data, i.e. without the pair of numbers 0.5 and 0. Such approach allows to "cut" short-wave boundary of the spectrum, for which the most likely error.

The analytical form of the function $F(x)$ was obtained using parametric and nonparametric approaches. In the case of the parametric approach, the function $F(x)$ is approximated by a polynomial of the third degree and the corresponding parameters are found by the least squares method. As a result, we got the expressions

$$F_1(x) = 0.01352x^3 - 0.5284x^2 + 5.0642x - 3.0522,$$

$$F_2(x) = 0.01546x^3 - 0.581x^2 + 5.4851x - 3.9801$$

for the first and second options, respectively.

The graphs of functions $F_1(x)$ and $F_2(x)$ are shown in figure 1.

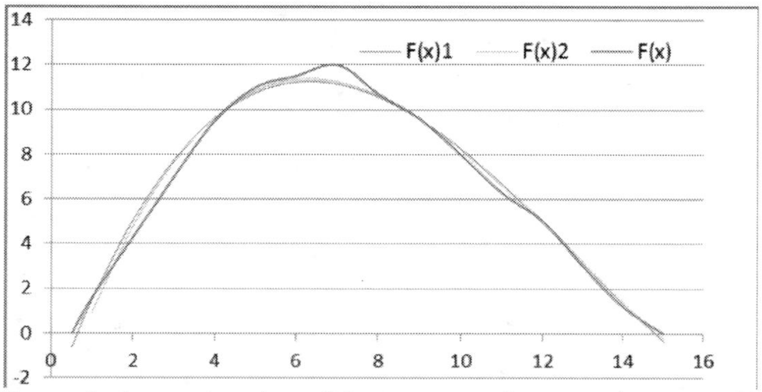

Figure 1. Approximation of the spectrum (parametric approach).

To evaluate the accuracy of the parametric models, absolute deviations of the estimates from the true values were calculated by the formulas:

$$A_1(x) = |F(x) - F_1(x)|;$$

$$A_2(x) = |F(x) - F_2(x)|.$$

The simulation results and respective deviation values are shown in table 2.

Table 2. Results of approximation and deviations.

x	$F(x)$	$F_1(x)$	$F_2(x)$	$A_1(x)$	$A_2(x)$
0.5	0	-0.651	-	0.6505	-
1	1.6	1.4971	0.9395	0.1029	0.6605
2	4.3	5.0708	4.7898	0.7708	0.4898
3	7	7.7498	7.6635	0.7498	0.6635
4	9.5	9.6154	9.6536	0.1154	0.1536
5	11	10.749	10.853	0.2514	0.1473
6	11.5	11.23	11.354	0.2695	0.1464
7	12	11.142	11.249	0.8578	0.7509
8	10.7	10.565	10.632	0.1352	0.068
9	9.6	9.5794	9.5951	0.0206	0.0049
10	8	8.267	8.231	0.267	0.231
11	6.3	6.7089	6.6326	0.4089	0.3326
12	5	4.986	4.8927	0.014	0.1073
13	3	3.1795	3.104	0.1795	0.104
14	1.2	1.3705	1.3593	0.1705	0.1593
15	0	-0.36	-0.249	0.36	0.2486

Approximation of the spectrum using a nonparametric approach was carried out by the formulas

$$F_3(x)=\frac{\sum_{i=1}^{n}F(x_i)\exp\left(-\frac{(x_i-x)^2}{0.084}\right)}{\sum_{i=1}^{n}\exp\left(-\frac{(x_i-x)^2}{0.084}\right)},\quad n=16;$$

$$F_4(x)=\frac{\sum_{i=2}^{n}F(x_i)\exp\left(-\frac{(x_i-x)^2}{0.076}\right)}{\sum_{i=2}^{n}\exp\left(-\frac{(x_i-x)^2}{0.076}\right)},\quad n=16$$

for the first and second options, respectively.

The graphs of functions $F_3(x)$ and $F_4(x)$ are shown in figure 2.

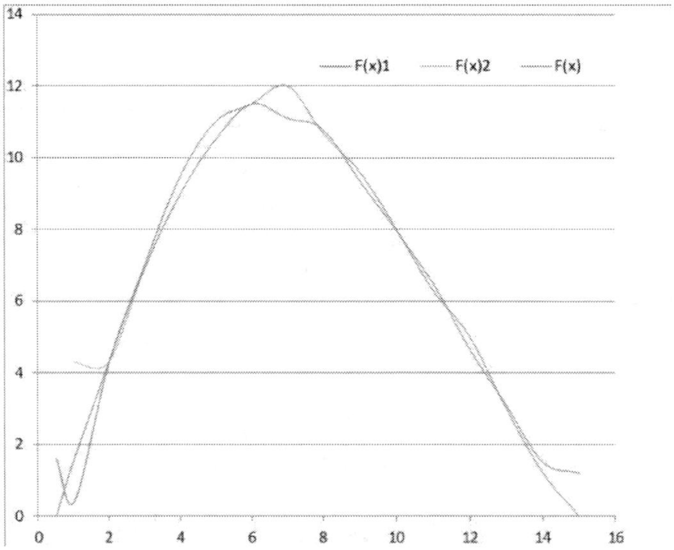

Figure 2. Approximation of the spectrum (non-parametric approach).

The accuracy of the non-parametric model is evaluated by analogy with the parametric approach using the following formulas:

$$A_3(x)=\left|F(x)-F_3(x)\right|;$$

$$A_4(x)=\left|F(x)-F_4(x)\right|.$$

The simulation results and values of the corresponding deviations are presented in table 3.

Table 3. The simulation results (non-parametric approach).

x	$F(x)$	$F_3(x)$	$F_4(x)$	$A_3(x)$	$A_4(x)$
0.5	0	1.607		1.607	-
1	1.6	0.419	4.3	1.181	2.7
2	4.3	4.248	4.3	0.052	0
3	7	6.9	6.9	0.1	0.1
4	9.5	9	9	0.5	0.5
5	11	10.5	10.5	0.5	0.5

6	11.5	11.5	11.5	0	0
7	12	11.1	11.1	0.9	0.9
8	10.7	10.8	10.8	0.1	0.1
9	9.6	9.35	9.35	0.25	0.25
10	8	7.95	7.95	0.05	0.05
11	6.3	6.5	6.5	0.2	0.2
12	5	4.65	4.65	0.35	0.35
13	3	3.1	3.1	0.1	0.1
14	1.2	1.5	1.5	0.3	0.3
15	0	1.2	1.2	1.2	1.2

As shown in tables 2 and 3, obtained analytical approximations of the spectral characteristics have rather good accuracy for practice. So, they can be used in the design of PXAs, as well as to estimate the effective atomic number of the substance, based on the dual-energy method.

References

[1] Rontgen V K 1933 *A New Kind of Rays* (Moscow, Gos. Tehniko-Theoretical. ed.)
[2] Vavilov S P 1985 *Pulsed X-ray Technique* (Moscow, Energy)
[3] Vavilov S P and Gorbunov V I 1985 *Pulsed X-ray Emission in the Inspection* (Moscow, Energoatomizdat)
[4] Senczyk D 2003 *Krajowa konferencjia badan radiograficznych "Popov in 2003"* (Poland, Poznan)
[5] Morgovsky L and Kurbatov V M (1977) *The Equipment and Methods of X-ray Analysis* **18** 175–179
[6] Tolchenov Yu M and Chepek A V (1972) *Instruments and Experimental Techniques* **4** 233–235
[7] Chakhlov V, Vavilov S P and Pushin V S (1975) *Instruments and Experimental Techniques* **2** 23–25
[8] Johnson D (1974) *Research instruments* **45**(2) 47–52
[9] Cracnell A P (1970) The determination of the photon energy spectrum of a flash X-ray tube *Radiation Effects* **2**(4) 233–242
[10] Artemov K P *et al* (2004) Pulse-periodic X-ray source (PPXS) *Pribory i Tekhnika Eksperimenta* **5** 166-167
[11] Gerasimov A I *et al* (2007) An transportable high-power pulse X-ray radiation source *Instruments and Experimental Techniques* **50**(2) 258–266
[12] Lavrinovich I V *et al* (2013) A compact pulsed X-ray source for high-speed radiography *Instruments and Experimental Techniques* **56**(3) 329–334

In-process Measuring of Capacitance Per Unit Length for Single-core Electric Wires

A E Goldshtein[1], G V Vavilova[2] and S V Mazikov[3]

[1] Professor, National Research Tomsk Polytechnic University, Tomsk, Russia
[2] SeniorLecturer, National Research Tomsk Polytechnic University, Tomsk, Russia
[3] Student, National Research Tomsk Polytechnic University, Tomsk, Russia
E-mail: wgw@tpu.ru

Abstract. The paper describes technical in-process implementation of the electrical method to measure the electrical capacitance per unit length of a single core electric wire. The basic design values of the electro-capacitive measuring transducer are determined. The impact of changes in water conductivity on measurement results is analyzed. Techniques to offset from the impact of changes in water conductivity on the results of the electrical capacitance per unit length control based on indirect electrical conductivity measurement are considered. An appropriate correction of the conversion function is made.

1. Introduction

Capacitance per unit length for electric cable and related impedance are considered to be important parameters for a number of cable products such as communication cables, radio-frequency cables and LAN-cables [1]. The cable capacitance is tested to meet the standard requirements according to GOST 27893-88 for final quality [2]. The drawback of this technique is that the cable cannot be tested along its entire length, and the quality of the product can be assessed after its manufacturing only.

This drawback can be eliminated through the in-process testing of the single core cable wire at the stage of the conductive core insulation. For this purpose, a tubular electrode is immersed in water (typically, in the extrusion bath with cooling water). During the in-process testing, the capacitance value of the capacitor is measured. The two electrical components of the capacitor are the wire core and the cooling water with the immersed tubular electrode. The water provides an electrical contact between the core sheath and the tubular measuring electrode. This technique is employed by the leading companies specializing in the production of inspection devices for the cable industry: Sikora, Zumbach [3–6].

2. Problem formulation

Figure 1 presents the design of the electro-capacitive measuring transducer (ECMT) used to implement the above testing technique.

The ECMT consists of cylindrical metal housing 1, tubular measuring electrode 2, a pair of cylindrical guard electrodes 3 and dielectric 4. The dielectric is placed between metal housing 1 and electrodes 2 and 3. Tested wire 5 continuously moves along the common axis of the three electrodes. The guard electrodes provide a uniform electric field in the adjacent ends of the measuring electrode.

The wire core and the transducer housing are earthed. The tubular electrodes are connected to alternating voltage generator $\dot{U}$ with circular frequency ω. The current in the measuring electrode electric circuit is measured using current transformer 6. The transducer and a segment of the tested wire core therein are immersed in the cooling water. The water is an aqueous solution containing salts, acids and bases.

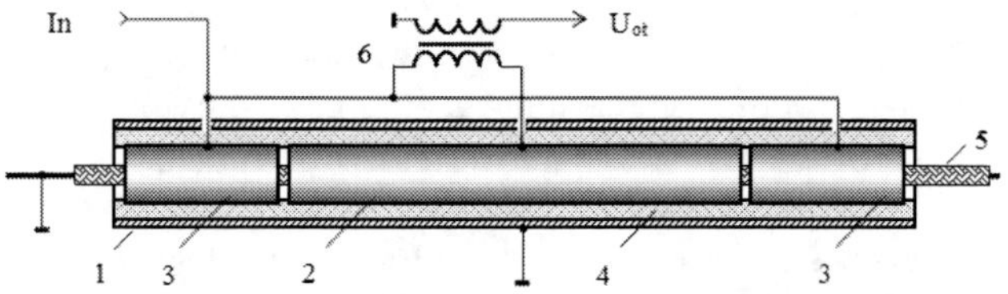

Figure 1. Design of the ECMT and switching circuit.

The purpose of the research: to study the impact of the basic ECMT design parameters and electrical conductivity of water on the conversion function of the ECMT, to select the optimal values for the ECMT design parameters to provide the maximum reliability of the method to test the capacitance per unit length of the electric wire, to develop an algorithm to offset from the impact of water conductivity on the results of capacitance per unit length measurement, and to experimentally check the effectiveness of the suggested techniques to offset.

3. Selection of the optimal design values for ECMT

The basic ECMT design values are the inner diameter of the tubular electrodes, the length of the measuring and guard electrodes, the distance (gap) between the measuring and guard electrodes, the inner diameter of the ECMT cylindrical housing.

An optimal ECMT design will provide the highest longitudinal (axial) uniformity of the electric field between the inner surface of the measuring electrode and the wire core of the tested electric wire. This minimizes the dependence of the function of the capacitance per unit length conversion into the ECMT output signal on the geometric dimensions of the wire, electrical insulation properties and changes in water conductivity.

The uniformity of the electric field is indicated through the identical values of the radial component E_r of the electric field intensity vector $\overline{E}$ and equality of the longitudinal component E_x (directed along the wire axis x) to zero for any specified value of the radius r inside the measuring electrode.

Variable β is introduced as a criterion of the electric field uniformity. Variable β is equal to the ratio of the linear capacitance in the central part of the measuring electrode (C_0) to the linear capacitance along the total length of the measuring electrode (C_1). (In the central part of the measuring electrode, the electric field is known to be uniform). For an optimal ECMT design β tends to 1. The central part of the measuring electrode, the length of which is equal to half of the total electrode length was assumed to be the region of a uniform field. The remaining part of the measuring electrode, for which the electric field uniformity was evaluated in the longitudinal direction, was assumed to be the measuring electrode.

To provide the longitudinal uniformity of the field inside the measuring electrode, the tubular electrodes are to be with a minimal inner diameter and maximal length. However, for free movement of the tested wire inside the ECMT to provide local testing, the inner diameter of the tubular electrodes is to be at least two times larger than the outer diameter of the wire sheath, and the measuring electrode length is to be of the order of (200 ... 300) mm.

To minimize the initial capacitance value of the ECMT (without the test object), the inner diameter of the cylindrical metal housing should be two times higher than the outer diameter of the tubular electrodes and an insulator is the air gap. The optimal values of the other design parameters may be determined through computer simulation of the ECMT. The 3D model of the transducer was made using COMSOL Multiphysics 3.5a.

The following parameters of the model were specified. The diameter of the wire core was 2 mm, the diameter of the wire insulation was 6 mm, the length of the measuring electrode was 200 mm, the inner diameter of the electrodes was 20 mm, the inner diameter of the housing was 40 mm, the thickness of the electrode and the housing wall was 1 mm, the amplitude of the electrode potential was 5 V, and the field frequency was 10 kHz. In the simulation, two versions of the electrical properties of water were set: distilled water (conductivity $\sigma = 10^{-4}$ S/m) and salt water ($\sigma = 0.8$ S/m).

The equipotential lines and color spectrum in figure 2 show the distribution of the electric potentials for the interaction of the ECMT electric field with the electrical wire for the case when no guard electrodes and distilled water are used.

Figure 2. Electric field created by the ECMT without guard electrodes (distilled water).

The analysis of the simulation results (figure 2) shows that in the absence of the guard electrodes "buckling" of the electric field occurs at the ends of the measuring electrode and, as a consequence, the field in these areas is highly non-uniform.

The value of the capacitance between the measuring electrode and the core wire can be determined by the mathematical expression, which identifies the relationship between the electric field energy of the capacitor W, the potential difference of its electrodes U, capacitance of the capacitor C, the electric field intensity E and the electric field induction D [7]:

$$W = \frac{CU^2}{2} = \int_V \frac{\bar{D}\bar{E}}{2} \mathrm{d}V \quad (1)$$

where V is the volume of the electric field.

From (1), we can obtain an expression to calculate the capacitance value by the numerical method using COMSOL Multiphysics: $C = \frac{1}{U^2} \int_V \bar{D}\bar{E} \mathrm{d}V$. For the case of the ECMT without guard electrodes, the calculated values are: $C_0 = 1.259 pF$; $C_1 = 1.499 pF$; $\beta = \frac{C_1}{C_0} = 1.9$.

Thus, in this case, the degree of the longitudinal (axial) uniformity of the electric field is approximately 20%, which causes a methodological error of the same order of magnitude in measuring capacitance per unit length. Distilled water used instead of salt water provides qualitatively similar pattern of the field and quantitatively similar parameters.

Radical increase in the longitudinal uniformity of the electric field along the entire length of the measuring electrode can be achieved by means of the guard electrodes.

Figure 3 shows the distribution of the electric potentials in the interaction of the ECMT electric field with the electrical wire for the case of guard electrodes and distilled water used. The length of the guard electrodes is equal to the inner diameter of the measuring electrode $R = 10$ mm, and the distance (gap) between the measuring electrode and the guard electrode is 1 mm.

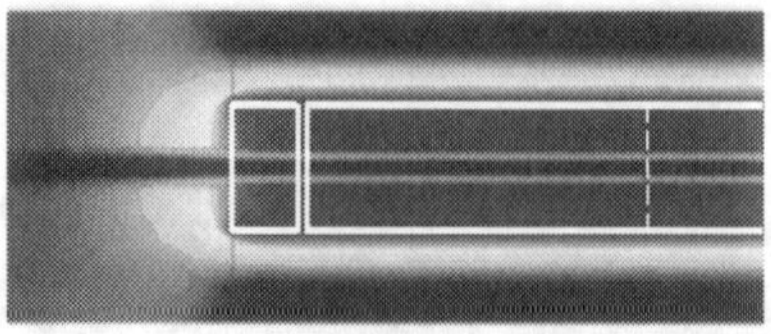

Figure 3. Electric field created by the ECMT with guard electrodes with a length of 1R and a gap of 1 mm (distilled water).

Figure 4 shows the distribution of the values of the longitudinal E_x and radial E_r spatial components of the ECMT electric field intensity vector along the longitudinal axis.

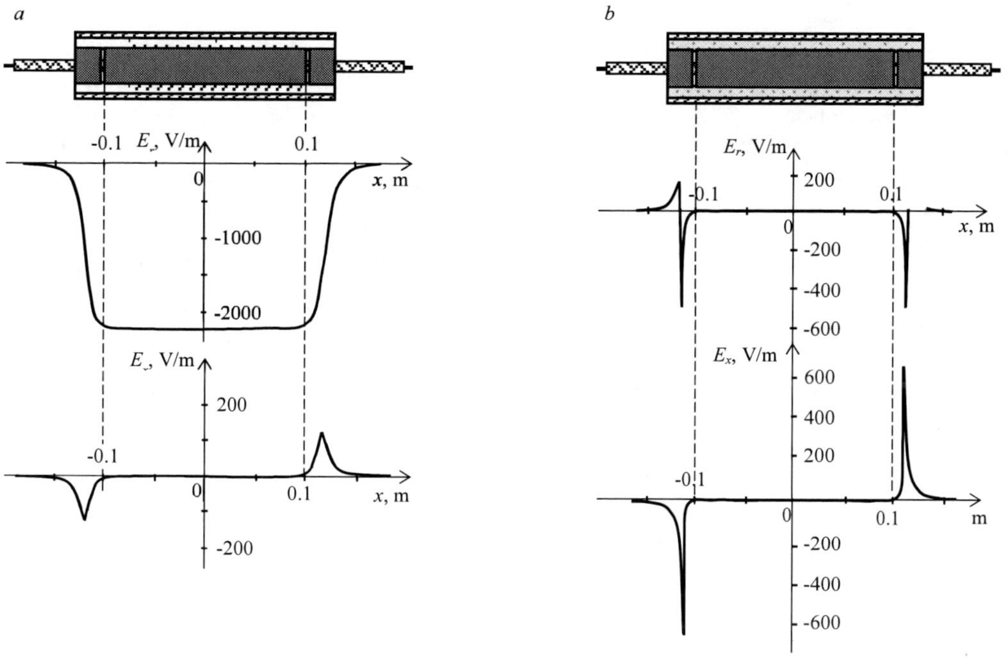

Figure 4. Distribution of the radial E_r and longitudinal E_x components of the electric field intensity vector in the middle part of the wire sheath (a) and near the inner surface of the measuring electrode (b) for ECMT with guard electrodes with a length of $1R$ and a gap of 1 mm.

The analysis of the simulation results (figure 4) shows that in case of using guard electrodes, the electric field "buckling" occurs at the distant edges of the guard electrodes. The electric field in the longitudinal direction is found to be highly uniform along the entire length of the measuring electrode. In this case, the value β is equal to 1 that corresponds to the optimal version of the ECMT design.

The developed computer model was used to study the impact of the guard electrode length on the degree of the electric field uniformity. The electric field is highly uniform in the longitudinal direction if the length of the guard electrodes is 0.5R or more. Virtually the same results were obtained for salt water.

The effect of the gap between the measuring and guard electrodes on the ECMT electric field was investigated as well. Summarizing the results on the effect of the guard electrode length and the gap between the guard electrodes on the ECMT characteristics, it can be concluded that the optimal ECMT design parameters when using both distilled and salt water are the guard electrode length equal to (0.5...1)R and the gap equal to (1...3) mm.

4. Experimental study on the impact of water conductivity on the ECMT conversion function

To study the effect of water conductivity on the results of the electric wire capacitance measurement, we used a single-core electric wire with an outer diameter of 4°mm, the capacity ranging from 160°pF/m to 460°pF/m, and similar values of the dielectric loss factor. The actual value of the wire capacitance per unit length was determined in compliance with GOST 27893-88 [2]. NaCl was solved in fresh water to change the electrical conductivity of water. The salt concentration of water λ varied from (0…4)°g/l.

For experimental studies we used the ECMT with the basic design and electrical parameters similar to those used in mathematical modeling.

Figure 5 presents the hodograph diagrams to show the dependence of the relative current $\dot{I}^*$ on the change in the wire capacitance per unit length C_p (solid line) and saline concentration/ salt concentration λ (dotted line).

The value of the current for the case of ECMT with no wire used (which corresponds to the value of the wire capacitance per unit length C_p=0) and $\lambda \to 0$ (distilled water) was taken as a fiducial value of the current.

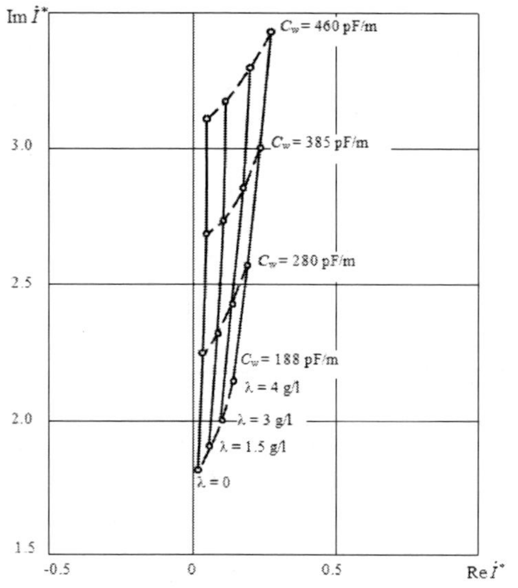

Figure 5. Hodograph diagram of the signal of the transducer (ECMT) for the change in the wire capacitance per unit length C_w and salt concentration λ.

The analysis of the dependencies (figure 6) shows that the amplitude of the current $\dot{I}^*$ increases linearly as the wire capacitance per unit length grows. Therefore, it is expedient to use the current amplitude as an informative parameter of the ECMT output signal when measuring the wire capacitance per unit length. The change in the salt concentration in the studied range changes the current amplitude in the range from 10 % for large capacitance values to 20 % for small capacitance values. Consequently, measurement of the capacitance per unit length with no account for the impact factor causes poor measurement accuracy.

5. Techniques to offset from the impact of changes in water conductivity on ECMT conversion function

The experimental results (figure 6) were used to find the function of the inversion of the measured current value into the value of the wire capacitance per unit length with the account for the impact of water salinity:

$$C_w = C_{01}(\lambda) + k_1(\lambda) \cdot I^* \qquad (2)$$

where $C_{01}(\lambda)$ is a constant component, $k_1(\lambda)$ is a proportionality factor.

Both of the functions depend on the salt concentration of water λ and they can be approximated by the second order polynomials. The coefficients of the functions are determined by the ECMT design parameters. In this paper, the techniques to offset from the impact of changes in the salt concentration of water on measurement results are investigated based of indirect measurements.

As can be seen in figure 6, as salt concentration of water increases, the phase angle φ between the vector current $\dot{I}^*$ and the imaginary axis of the complex plane monotonically grows. Consequently, $t = \frac{\operatorname{Re} \dot{I}^*}{\operatorname{Im} \dot{I}^*}$ and tg φ tend to grow. Therefore, the results of measuring the complex current components $\dot{I}^*$ can be used to offset from the impact of changes in water conductivity.

By analogy with (2), the equation of the inverse conversion of the amplitude of the current $\dot{I}^*$ into the wire capacitance per unit length is described by the linear dependence:

$$C_w = C_{02}(t) + k_2(t) \cdot I^* \quad (3)$$

where functions $C_{02}(t)$ and $k_2(t)$ are the values *t* indicating water conductivity.

Functions $C_{02}(t)$ and $k_2(t)$ as well as functions $C_{01}(\lambda)$ and $k_1(\lambda)$ can be described by the second order polynomials. The coefficients of these polynomials are determined by the ECMT design parameters and are found experimentally during the initial ECMT calibration.

A numerical experiment was carried out using conversion function (2) to assess the efficiency of the described technique to offset from the impact of changes in water conductivity on the results of the electrical capacitance per unit length testing. The experimental data was obtained for single-core electric wires with the capacity ranging from 160°pF/m to 460°pF/m and salt concentration of water varying from (0…4)°g/l.

The analysis of these results shows that without offset from the impact of changes in water conductivity the relative measurement error can reach 20 %, however, when offsetting, the values of the capacitance per unit length calculated by formula (3) for these ranges of parameters affecting the actual values differ by not more than 2.5 %. A limited range of changes affecting the parameters allows many-fold reduction in the measurement error.

6. Conclusion

Based on the results of computer simulation of the interaction of the electric field of the ECMT for capacitance per unit length meter with an electric wire, the optimal design parameters of the main elements of the ECMT are selected: the inner diameter of the tubular electrodes, the length of the measuring and guard electrodes, the distance (gap) between the measuring and guard electrodes, the inner diameter of the ECMT cylindrical housing. It is shown that the change in water conductivity significantly impacts the results of the in-process measuring of the wire capacitance per unit length. The change in water conductivity can be caused by changes in salinity of water, in which the ECMT is immersed, as well as the changes in its temperature and composition of impurities contained in water. Based on indirect measurement of water conductivity and error correction in the conversion function, the techniques to offset from the impact of these factors on measurement results are suggested. It is shown that these offset techniques allow multifold increase in the measurement accuracy.

References

[1] GOST (State Standard) 27893-88 1989*Communication Cables. Methods of Tests* (Moscow: Standards publishing house)

[2] GOST (State Standard) 11326.0-78 1978 *Radio-frequency cables. General specifications* (Moscow: Standards publishing house)

[3] Blohm W (1999) Advanced on-line measuring and analysis techniques for an economical cable production *Proceedings Intl. IWMA Conf. Economical Processing of Rod to Wire & Cable* pp 67–75

[4] CAPAC®/FFT *Capacitance measurement systems for cable production lines* (Zumbach Electronic AG, SCHWEIZ: Hauptsitz) URL:http://www.zumbach.com/pdf/Literature_DE/Catalogs/CAPAC/CAPAC_Familie_CAPA.002.0002.D.pdf

[5] Fleming P and Coleman L R 2001 A system for monitoring fluctuations in the thickness of a cable insulating sheath GB Patent 2358928

[6] Fleming P and Coleman L 2003 Capacitance monitoring systems US Patent 20030128038

[7] Govorkov V A 1968 Electric and magnetic fields (Moscow: Energy)

The Vibration Monitoring of the Gascompressor Units with Application of Hurst's Statistics

O V Zakharov[1], S V Novichkov[2], A V Kochetkov[3]

[1]Ph. Doctor, Professor of the Designing technical and technological complexes department of Yuri Gagarin State Technical University, Saratov, Russia
[2]Candidate of Sciences, Associate Professor of the Energy Department of Yuri Gagarin State Technical University, Saratov, Russia
[3]Ph. Doctor, Professor, Perm National Research Polytechnic University, Perm, Russia
E-mail: zov17@mail.ru

Abstract. The gascompressor units work in the unsteady regime which depends on many internal and external factors. For the solution of problems of control, management and forecasting the statistical assessment of a condition of object in real time according to noisy observations of one realization of stochastic process is carried out. The traditional spectral analysis on the basis of Fast Fourier Transformation in this case is inefficient if time scale is much less than duration, which is subject to the analysis. Therefore for monitoring of a condition of the equipment it is offered to use Hurst's statistics by means of which it is possible to investigate quantitatively law of development of a vibration signal and unambiguously to analyze a condition of the gascompressor units. Calculation of economic efficiency from introduction of such technology of monitoring will provide decrease on average by 2.5% of a consumption of gas and specific costs of capital repairs of the gascompressor units at the expense of increase in a total operating time by 20%.

1. Introduction

In monitoring systems and monitoring of a condition of gascompressor units the low-frequency range of a range of fluctuations of its mechanical system is mainly used. This range consists of harmonic and subharmonic oscillations of rotors of turbines of a low and high pressure which are well described by system of the usual differential equations. Application of classical approach for definition of high-frequency components leads to unjustified increase in an order of system, and determination of coefficients of the equations is difficult in view of a large number of the analyzed knots. Important part in the analysis of non-stationary processes in multifrequency signals is existence of a constant control of changes in its amplitude-frequency components. The traditional spectral analysis on the basis of Fast Fourier Transformation (FFT) is inefficient for non-stationary signals with a temporary scale much smaller duration which is subject to the analysis of realization. It is connected with averaging of power of fluctuations in the spectral analysis on all time of supervision of a signal. The most obvious way of application of FFT to the analysis of non-stationary signals will be a splitting realization into separate short sites of identical length with the subsequent application of algorithm of FFT to each of them. However this method has the restriction of accuracy of estimation of frequency of spectral peaks determined by the signal interval size. This feature limits use of BPF in the analysis of the fast-changing stochastic signals to which fluctuations of the processing equipment working in the transitional or quasistationary mode are peculiar. When using autoregression models the accuracy of a frequency scale, close to FFT, can be reached only at a considerable (more than 10000) order of model. It demands considerable computing resources and time in comparison with algorithm of FFT.

Unlike periodic components of a range which are formed at interaction of the structural blocks of gascompressor units among themselves or with a working environment, the frequency of thermoacoustic oscillations can't be described a certain value. For example, for the GT-750-6 unit, with the external combustion chamber, thermoacoustic fluctuations have the limited range of frequencies, and also area of "an attraction of fluctuations", that is the most part of fluctuations has the frequency close to frequency to the maximum amplitudes. These thermodynamic processes proceed in the combustion chamber and define process of functioning of the unit. The received results allowed to confirm existence of such communication, in the range of 2800–5300 Hz. The analysis of the thermoacoustic fluctuations allocated from vibrosignal range of the unit showed that their characteristics along with pressure and temperature of gas mixture, reflect a condition of thermodynamic processes in the engine combustion chamber. This fact was the basis for development of program software for monitoring of gascompressor unit in parameters of thermoacoustic fluctuations.

2. Problem definition

At traditional approach periodic fluctuations are analyzed, and the problem of a filtration consists at the removal of noise from the received signal. At research of thermoacoustic fluctuations the return problem is solved. Namely the useful information about thermoacoustic fluctuations is drawn by removal of periodic components of a signal. At an exception of periodic components of fluctuations there is a loss of some part of information on fluctuations because of zeroing of values of amplitudes of fluctuations. However because periodic fluctuations possess narrow frequency range, their number in a range is limited. And as the power of thermoacoustic fluctuations is distributed in the wide range of frequencies, the small part of power is the share of the deleted spectral harmonicas. Thus, losses of information are small and don't influence results of the further analysis of thermoacoustic fluctuations.

Typical examples of results of work of the offered algorithm are presented in figures 1–4. In figure 1 the site of a range of a vibrosignal of HK-12CT from the periodic component formed by the compelled fluctuations of shovels of the free turbine is represented. On peak of amplitude noticeable change of fluctuations frequency because of change of a rotor frequency of rotation. After processing of a range the same site of a range from a remote periodic component is shown in figure 2. In figure 3 the range site with periodic components against thermoacoustic fluctuations, is given in figure 4 – the same site after processing. The received results clearly demonstrate possibilities of algorithm when determining characteristics in a range the thermoacoustic of fluctuations for the purpose of their further use for monitoring of gascompressor units. For the solution of problems of control, management and forecasting the statistical assessment of a condition of object of management in real time is carried out. The statistical assessment is carried out according to noisy measurements of one realization of casual process. The gascompressor unit often works in the non-stationary mode which can be caused by such factors as change of composition of gas, pressure differences in the main pipeline, change of ambient temperature. Thus there is a loss of efficiency and is reflected in change of a gasdynamic subsystem of gascompressor unit and, as a result, amplitude and frequency characteristics of a signal. Therefore the Hurst's statistics for monitoring of a condition of GPA is offered to use by means of which it is possible to investigate quantitatively regularities of formation of a vibrosignal and unambiguously to judge a condition of gascompressor unit [1, 2].

3. Application of Hurst's statistics

Now became conventional that many stochastic processes in the nature and equipment possess fractal structure. In the fractal analysis of ranks Hurst exponent is applied to identification of existence and a type of long-term dependence [3]. Processes for which $0.5 < H < 1$ show steady behavior of a trend that is if a row increases (decreases) during the previous period, it will keep this tendency in the future. The probability of such behavior of subjects is higher, than the indicator is closer to unit. Processes for which $0 < H < 0.5$ is characteristic a sign-variable tendency. For such processes growth during the previous period is replaced by recession during the next period of supervision with probability, that bigger, than closer to zero. At $H = 0.5$ processes in which the trend is absent take place. Than Hurst exponent to 0.5 is closer, a row more noisy and the trend component is less expressed to those. Hurst's method or the analysis is nonparametric [4–6], that is he doesn't contain requirements to the law of distribution of observed casual process. This property of a method should be considered as advantage before the methods of the analysis of casual processes focused on a certain law of distribution.

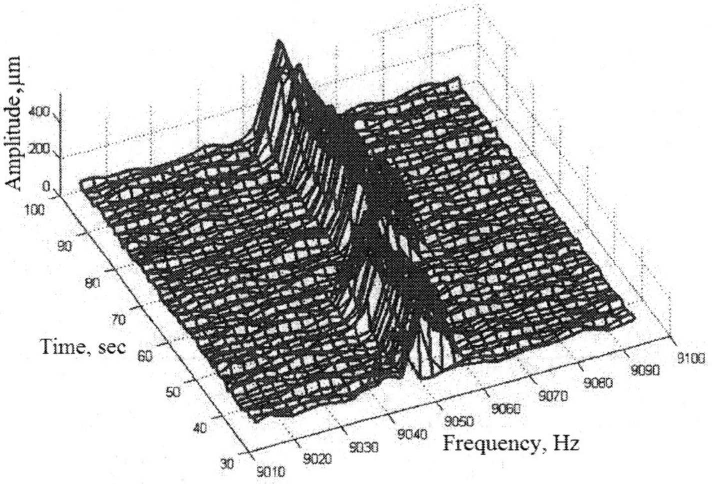

Figure 1. Initial fragment of a range of vibrofluctuations of the GPA-Z-6.3 unit.

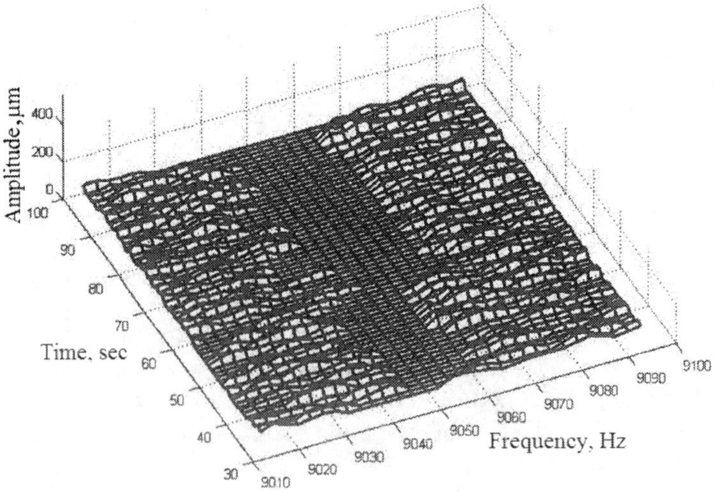

Figure 2. A fragment of a range of vibrofluctuations of the GPA-Z-6.3 unit after processing.

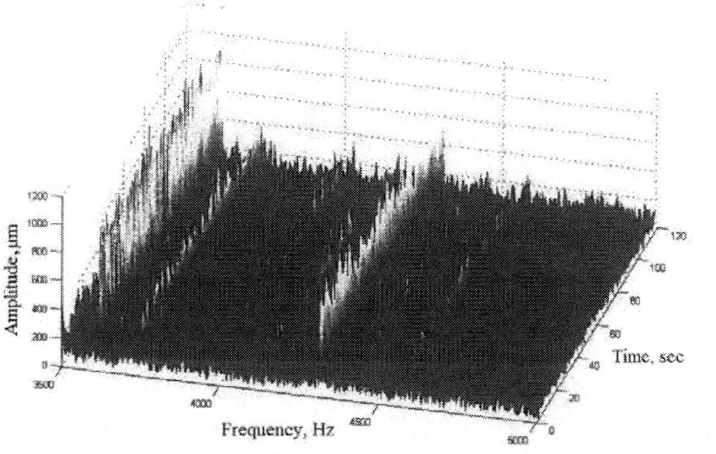

Figure 3. A fragment of a range of a signal of vibrofluctuations of the GPA-Z-6.3 unit.

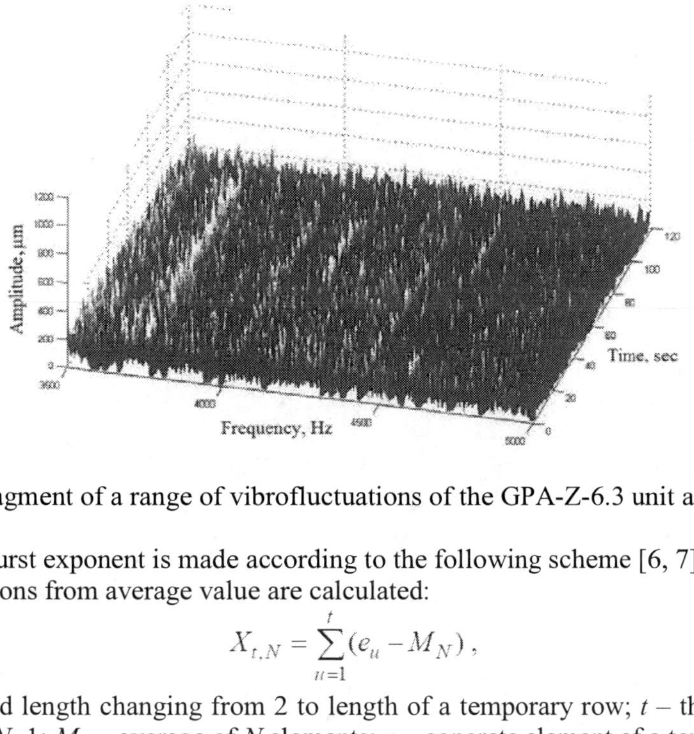

Figure 4. A fragment of a range of vibrofluctuations of the GPA-Z-6.3 unit after processing.

Calculation of Hurst exponent is made according to the following scheme [6, 7].

1. At first deviations from average value are calculated:

$$X_{t,N} = \sum_{u=1}^{t}(e_u - M_N),$$

where N – the period length changing from 2 to length of a temporary row; t – the variable changing the value from 1 to N -1; M_N – average of N elements; e – concrete element of a temporary row.

2. On each iteration we receive N – 1 values of Xt,n which we use in the following formula:

$$R := Max(X_{t,N}) - Min(X_{t,N}),$$

where R – scope of a deviation X.

3. Further we normalize scope division into a standard deviation of S which is calculated on N values.

4. We take the logarithm *R/S* and *N*, we build the schedule on the basis of the obtained data.

5. On a function graph of *log*(*R/S*) from log(*N*) we find an inclination by linear approximation

The method of statistical modeling by us found tolerance on values of statistics of Hurst. The interval between admissions sets area of operability of gascompressor unit and defines its optimum situation. By results of pilot studies strategy of decision-making in the course of monitoring at change of provision of statistics of Hurst concerning its optimum situation is created (figure 5).

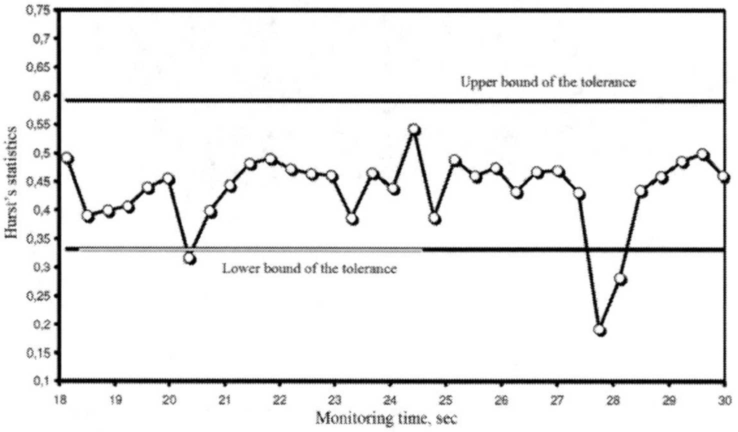

Figure 5. Monitoring of a condition of gascompressor unit with use of Hurst's statistics.

4. Summary

The received results allow to organize of conclusion of gascompressor unit in repair only when its state really displays need for carrying out repair. Then becomes possible to pass from service according to regulations to service on a state. Calculation of economic efficiency from introduction of such technology of monitoring will provide decrease on average by 2.5% of a consumption of gas and specific costs of capital repairs of gascompressor unit at the expense of increase in a total operating time by 20%.

5. Acknowledgments

The work has been performed within the framework of the State Contract No.14.579.21.0019 (unique agreement identifier #RFMEFI57914X0019) on the subject Developing the device for fetal and maternal health monitoring during pregnancy by means of control over cardiovascular parameters based on acoustic data concluded between the Ministry of Education and Science of the Russian Federation and LLC Diagnostica +.

References

[1] Feder Jens 1988 *Fractals plenum* (Press New York)

[2] Lubkov S V and Novichkov V I 2011 *Vibration of cars and mechanisms at operation of the equipment of thermal power plant* (Saratov SSTU)

[3] Bendat J S and Piersol A G 2011 *Random data: analysis and measurement procedures* (New York John Wiley & Sons Inc.)

[4] Combescure D and Lazarus A (2008) Refined finite element modelling for the vibration analysis of large rotating machines: Application to the gas turbine modular helium reactor power conversion unit *Journal of Sound and Vibration* **318**(4–5)1262–1280

[5] Abu-El-Yazied T G, Abu-El-Haggag S Y and Al-Fares F S (2009) Identification of vibrating structures with application to a steel tower *World Applied Sciences Journal* **6**(12) 1673–1674

[6] Zargar O A (2013) Different aspects of gas turbine siemens 162MW-V94.2 vibration analysis Middle-East Journal of Scientific Research **18**(4)546–556

A Device for Fetal Monitoring by Means of Control Over Cardiovascular Parameters Based on Acoustic Data

L A Khokhlova[1,2], A I Seleznev[1,2], D S Zhdanov[3,4], I Yu Zemlyakov[5,6] and E Yu Kiseleva[7,2]

[1]Engineer, LLC Diagnostica +, Tomsk, Russia
[2]Engineer, Tomsk State University, Tomsk, Russia
[3]Senior programmer, LLC Diagnostica +, Tomsk, Russia
[4]Jr. researcher, Tomsk State University, Tomsk, Russia
[5]Senior researcher, Tomsk State University, Tomsk, Russia
[6]Chief engineer, LLC Diagnostica +, Tomsk, Tomsk, Russia
[7]Deputy director for project work, LLC Diagnostica +,Tomsk, Russia
E-mail: hohlova@diagnostic.tom.ru

Abstract. The problem of monitoring fetal health is topical at the moment taking into account a reduction in the level of fertile-age women's health and changes in the concept of perinatal medicine with reconsideration of live birth criteria. Fetal heart rate monitoring is a valuable means of assessing fetal health during pregnancy. The routine clinical measurements are usually carried out by the means of ultrasound cardiotocography. Although the cardiotocography monitoring provides valuable information on the fetal health status, the high quality ultrasound devices are expensive, they are not available for home care use. The recommended number of measurement is also limited. The passive and fully non-invasive acoustic recording provides an alternative low-cost measurement method. The article describes a device for fetal and maternal health monitoring by analyzing the frequency and periodicity of heart beats by means of acoustic signal received on the maternal abdomen. Based on the usage of this device a phonocardiographic fetal telemedicine system, which will allow to reduce the antenatal fetal mortality rate significantly due to continuous monitoring over the state of fetus regardless of mother's location, can be built.

1. Introduction

Nowadays the primary trend of perinatal medicine has been changed in the Russian Federation. Delivery in favor of fetus has become a routine practice. The Russian Federation has set a goal to reduce the rate of maternal and infant mortality in no less than 2 times by 2025 [1]. One of the key criteria of outpatient perinatal institutions' work is the absence of antenatal fetal death cases [2].

A solution for reducing the number of antenatal deaths can be either long-term inpatient hospitalization of high-risk pregnant women or the use of substituting technologies.

Fetus condition in the antenatal period is controlled by means of various methods: survey, visual examination, abdominometry, fundal height measurement, assessment of the fetal position, its movement activity and heart rate as well as different ultrasound examinations. However, traditional clinical methods of examination do not always allow assessing fetus development and detecting certain potentially dangerous conditions, e.g., fetal hypoxia. Despite seeming insignificance of this condition, it is a cause of most intrauterine deaths and multiple complications at the early antenatal period [2].

Journal of Physics: Conference Series **671** (2016) 012064 doi:10.1088/1742-6596/671/1/012064

Cardiotocography is still the primary method for detecting the signs of fetal hypoxia. In this case, the fetal heart rate is registered by an ultrasound detector utilizing the Doppler effect. Nowadays a lot of various models of fetal monitoring devices are available (BabyCare, Bionet, SonicaidOne, Oxfordmedical, etc.) [3,4]. However, all of them are used at medical institutions only; therefore, they cannot guarantee a prompt response in case of sudden deteriorations in fetal health. Moreover, the use of these devices is connected with difficulties for high-risk pregnant women who live in the areas located far from medical centers.

Mobile fetal cardiotocography concepts are actively developed in order to solve these problems. The most successful option is the Sense4Baby project that currently undergoes approval procedures in the USA [5]. At the same time, the safety of ultrasound radiation used for cardiotocography is not an established fact.

Besides mobile cardiotocographs, the Monica AN24 system (Great Britain) has become popular within the segment of commercially available autonomous wearable fetal monitoring devices. The AN24 daily monitoring device is meant for registering fetal and maternal ECG, uterine contractions and fetal movement activity under the condition of pregnant woman's free movement. It is noteworthy, that in order to overcome the problem of signal to noise ratio, the Monica AN24 device depends on the patient's abdomen receiving a thorough skin preparation with a special abrading solution that removes the stratum corneum of the epidermis [6].

Phonocardiography (PCG) can be an alternative method to acquire information on the fetal cardiac activity. PCG allows recording fetal heart beats in a passive and completely non-invasive way [7]. Signals can be registered by placing an acoustic sensor on the mother's abdominal surface without using any gel, which is a necessary condition for normal operations of the devices that are used to register fetal ECGs or utilize Doppler sonography. Phonocardiographic signals potentially can be used to detect fetal cardiac pathologies, such as cardiac murmur, premature beats and dyadic/triadic atrial contractions [8]. Standard cardiotocography and other techniques do not provide such opportunities [7].

2. Description of the device for fetal monitoring

In connection with good prospects of the use of phonocardiography methods in the field of antenatal health protection, the LLC Diagnostica + team develops a device for non-invasive monitoring of the state of health of fetus and pregnant woman by analyzing the frequency and periodicity of their heart beats on the basis of acoustic data. The device will be used as part of the monitoring system, which also includes patient smartphone software, providing data on the medical institution server, and obstetrician workstation.

The hardware part of monitoring system is an mobile electronic device for reception, transmission and primary processing of acoustic data. The general structure of the hardware component of the experimental prototype is presented in figure 1.

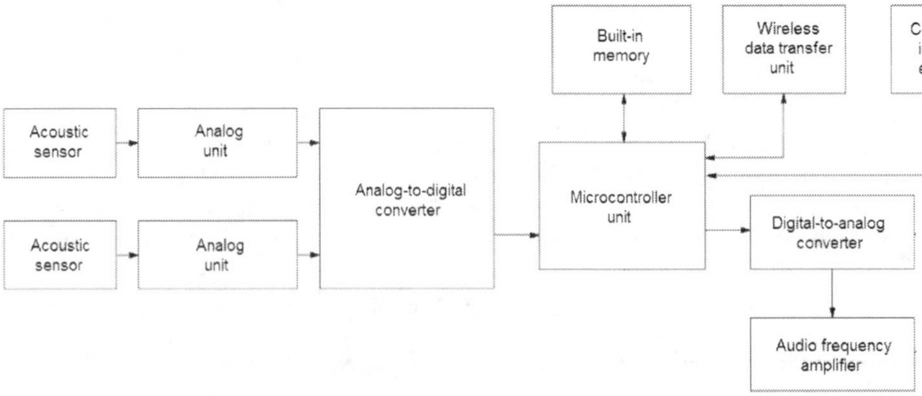

Figure 1. General structure of the fetal monitoring device.

The device works in the following way. Acoustic sensors made as compact highly-sensitive contact microphones located in the abdominal area of a pregnant woman register weak acoustic signals of fetal heart beats and convert them into electric signals. Signals are transferred to the analog unit, where they are amplified and filtered in order to reduce the impact of interferences and unwanted noises on the device.

Then the microcontroller (microcontroller unit) uses the analog-to-digital converter to transform analog signals into a digital code. These data are saved in the built-memory and shared with the software component installed on a patient's smartphone by means of the wireless data transfer unit. Control and indication elements are used to display the device status and operation mode, audio indication (digital-to-analog converter and audio amplifier) inform about dangerous conditions. Next step is the transmission of the obtained data to a medical server and workstation, where obstetrician can assess fetus health condition and monitor the fetal heart rate and, if necessary, appoint additional examinations or take actions in case of a dangerous condition.

Figure 2 presents signals acquired by means of the experimental prototype.

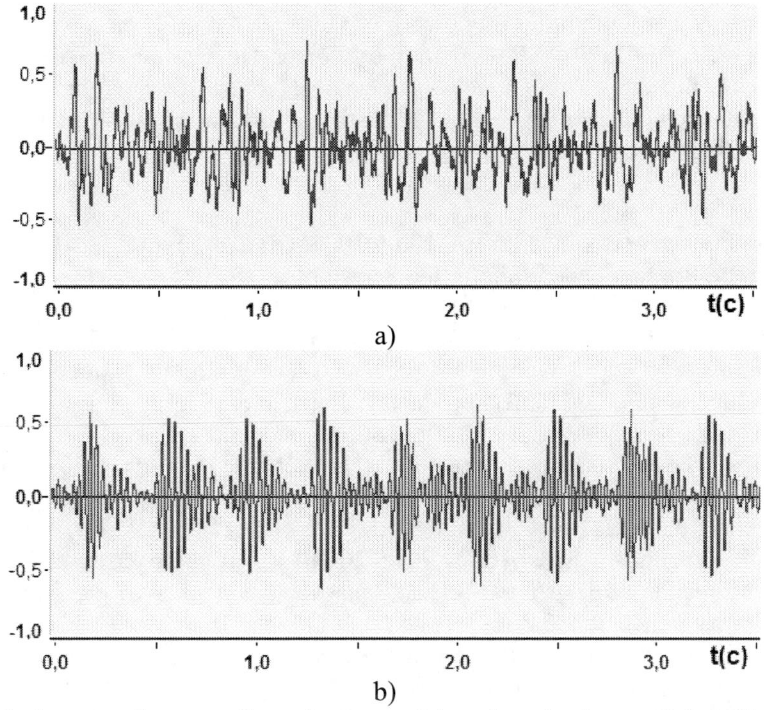

Figure 2. Fetal phonocardiogram. Gestation term: 34 weeks, a) prior to digital filtration; b) after digital filtration.

The key problem connected with the use of acoustic data is proper detection of fetal heart beats in noise-contaminated signals and further processing of these data. Noise contamination of PCG signals is connected with the following factors [7,9]: phonocardiographic signals depends on the fetal presentation; fetal movements cause acoustic noises typical of specifically its movement activity; pregnant woman's GIT sounds can overlay PCG signals; phonocardiograms contain pregnant woman's cardiac and respiratory sounds as well as external noises (figure 2a). Additional filtration and amplification of audio records allow acquiring clear fetal heart beats (figure 2b). These signals can be used for further digital extraction of the heart beats and measurement of the fetal heart rate.

3. Summary

The device under development will allow conducting dynamic monitoring over the cardiovascular system of the fetus and pregnant woman without the need for staying inside a medical institution and being under medical personnel's supervision. In addition, phonocardiography is safe due to the absence of any exposing radiation, so this technique can be used for any period of time according to obstetrician's recommendations. The device is implemented in the form of a mobile, wearable, data-transferring unit, which will allow timely informing physicians about changes in the fetal health without interfering the patient's daily life. The smartphone software will eventually provide a visual representation of the fetus health condition gathered by the wearable monitoring device. The signals acquired by means of the experimental prototype can be satisfiable for digital extraction of the heart beats after analogue pre-processing and digital filtration. However for introducing the method described into medical practice, it is necessary to solve the major task of receiving stable signals in various types of fetal presentation and mother's anthropometric data.

Acknowledgments

The work has been performed within the framework of the State Contract No.14.579.21.0019 (unique agreement identifier #RFMEFI57914X0019) on the subject Developing the device for fetal and maternal health monitoring during pregnancy by means of control over cardiovascular parameters based on acoustic data concluded between the Ministry of Education and Science of the Russian Federation and LLC Diagnostica +.

References

[1] RF Presidential Decree No.1351 as of 10.09.2007 On Approving the Demographic Policy Concept in the Russian Federation for the Period of up to 2025

[2] Frolova O G, Palenaya I I, Shuvalova M P, Sukhanova L P (2011) Regional aspects of stillbirth rates in the Russian Federation in 2008 *Obstetrics and Gynecology* **1** 105–109

[3] Bionet, Compact Fetal Doppler, 05.26.2014. [Online source]. Available at: http://www.bio2net.com/eng/products/products1_1.asp?idx=193&category=55&search_type=0&search_word=&page_size=8&page=1 [Date of reference: 05.26.2014]

[4] Oxford Medical, Fetal Heartbeat Detectors – Sonicaid®One, 05.26.2014 [Online source]. Available at: http://www.oxford-medical.ru/detectory_serdcebieniya_ploda_sonicaid_one.html [Date of reference: 05.26.2014]

[5] West Health Institute, Sense4Baby, 05.26.2014. [Online source]. Available at: http://www.westhealth.org/news/press-release-sense4baby-licenses-wireless-fetal-monitor-system [Date of reference: 05.26.2014]

[6] Adam J (2012) The future of fetal monitoring *Rev. in Obstet.Gynecol* **5**(3–4) 132–136

[7] Kovács F, Horváth C, Török M, Hosszú G (2006) Long-term phonocardiographic fetal home monitoring for telemedicine systems *Annual International Conference of the IEEE Engineering in Medicine and Biology - Proceedings* 3946–3949

[8] Varady P, Wildt L, Zoltaîn B, Achim H (2003) An advanced method in fetal phonocardiography *Computer methods and programs in biomedicine* **71**(3) 283–296

[9] Ruffo M, Cesarelli M, Jin C, Gargiulo G, McEwan A, Sullivan C, Bifulco P, Romano M, Shephard R W, Schaik A 2011 Non invasive foetal monitoring with a combined ECG–PCG system *Biomedical Engineering, Trends in Electronics, Communications and Software* (Croatia, InTech) pp 347–366

Journal of Physics: Conference Series **671** (2016) 012065 doi:10.1088/1742-6596/671/1/012065

The Capacitive Magnetic Field Sensor

D O Zyatkov[1], A V Yurchenko[2], V B Balashov[3], V I Yurchenko[4]

[1]Postgraduate student, The department of physical methods and devices of quality control, Tomsk Polytechnic University, Tomsk, Russia
[2]Doctor of Technical Sciences, professor the department of physical methods and devices of quality control, Tomsk Polytechnic University, Tomsk, Russia
[3]Engineer, Research Institute of Semiconductor Devices,Tomsk, Russia
[4]Head of department №110, Research Institute of Semiconductor Devices Tomsk, Russia
E-mail: zyatkov_do@niipp.ru

Abstract. The results of a study of sensitive element magnetic field sensor are represented in this paper. The sensor is based on the change of the capacitance with an active dielectric (ferrofluid) due to the magnitude of magnetic field. To prepare the ferrofluid magnetic particles are used, which have a followingdispersion equal to $50 < Ø \leq 56$, $45 < Ø \leq 50$, $40 < Ø \leq 45$ and $Ø \leq 40$micron of nanocrystalline alloy of brand 5BDSR. The dependence of the sensitivity of the capacitive element from the ferrofluid with different dispersion of magnetic particles is considered. The threshold of sensitivity and sensitivity of a measuring cell with ferrofluid by a magnetic field was determined. The experimental graphs of capacitance change of the magnitude of magnetic field are presented.

1.Introduction

Naturally occurring liquidsweakly interact with the magnetic field. However, the ability to control fluid by a magnetic field is attractive in quite a number of technical problems. To solve these problems magnetic fluid was invented in 1960. Magnetic fluids or ferrofluids are colloidal suspensionsof magnetic particles dispersed stably in a carrier liquid. The properties of the magnetic fluid are determined by the size magnetic particles, property of this particle, surface-active substance and carrier liquid. Parameters of ferrofluid depend on the composition of magnetic liquid.Magnetization of magnetic liquid on a magnetic fielddepends on the chemical composition of the substance and its structure. Magnetic liquids thanks to itsthe properties are perspective materials and they have different application in various areas of measurement technology [1, 2].

These fluids have very unusual physical phenomena, sometimes a unique combination of strong magnetic properties and fluidity.

Generally, magnetic fluid consist of magnetic particles of size 10 nm [2].

Existence of a uniform suspension of small particles in magnetic fluid without settling occur with magnetic particles of size 10 nm.In this case, the thermal motion is sufficient to maintain the particles in suspension [2, 3].

However, with increase of the particles size sedimentation stability due to adhesion particles is violated because they have big magnetic moments.

However, sedimentation stability is violated due to increase of the particles size. It takes place because of agglomeration of particles at the expense of the big magnetic moment. Also separation of particles occur in its gravitational force field due to difference of densities solid and fluid phases.

In order to avoid agglomeration of particles and to prevent subsidence (sedimentation) by gravitational force field due to integration of particles (formation of lumps), it is necessary to stabilize particles in the liquid carrier [4].

Surface – active substances which are adsorbed on a surface of particles, forming a thin covering with a thickness 3–4 nm are used for this purpose. The covering is enough in order to particles aren't approached each other within short distance, when the forces of interaction between particles which lead to aggregation of particles, will dominate [5, 6].

Chain aggregates due to interaction of particles are formed [7]. Average number of particles in chain aggregates is determined by the parameter of the its magnetic-dipole interaction and concentration of the magnetic fluid. In a weak field the chain aggregates due to thermal fluctuations has a "vermiform" appearance.The formation of chain aggregates is capable of leading to considerable changes of magnetic and other properties of ferrofluids which are appeared in capacity change of the condenser [8, 9].

Thus, the our researchis the creation of a capacitive sensor with ferrofluid for fixation and control of weak magnetic fields.

2. Object and methods of experimental research

Magnetic fluids in experimental investigations are used. Thesemagnetic fluids consist of a polimetilfenilsiloksan (PFMS-4) which have a following dispersions of magnetic particles: $50 < Ø \leq 56$, $45 < Ø \leq 50$, $40 < Ø \leq 45$ and $Ø \leq 40$microns. These magnetic particles were obtained by grinding of ribbonnanocrystalline alloy in an agate mortar and siftingits through 56, 50, 45 and 40 micron sieve.The concentration of magnetic powder in PFMS-4 was 15 volume percent.

Measuring cell had the form of a dielectric cylindricity vessel which made from the insulation material and provided with two plane-parallel plates with flexible leads. The distance between the plates is equal to 3 mm and them area is equal to 10 mm^2. Active dielectric (magnetic fluid) between the plates is placed. Figure 1 shows the experimental setup layout.

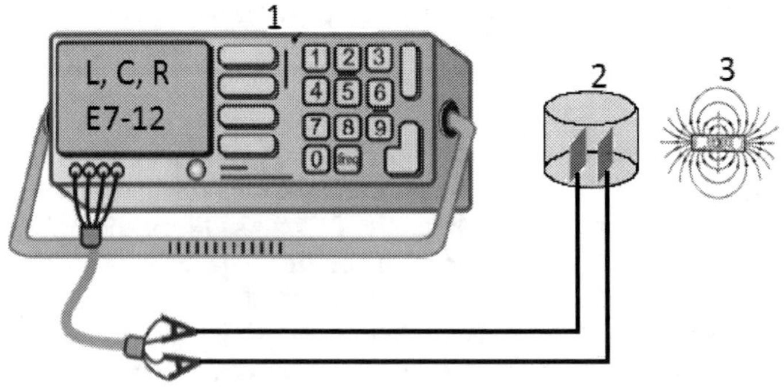

Figure 1. Experimental setup to study the magnetic fluid, where 1 – measuring device, 2 – electrical measuring cell filled with magnetic fluid, 3 – magnet.

The capacity of plane condenser consisting of two parallel plates (figure1) is determined from the formula:

$$C = \frac{\varepsilon \varepsilon_0 s}{d}$$

where$s$ – the area of electrodes, d – the distance between electrodes, ε_0 – dielectric constant, ε – the dielectric permeability of the medium filling the space between the plates.

The capacity change of measuring cell was measured by digital L, C, R meter type E7-12 with an adapter which reduces of influence of stray parameters on the frequency of the measuring signal of 1

MHz.It should be noted that the value of the experimental measuring electrical field of cell was weak enough and didn't have effect on the structural state of the emulsion in a measuring cell, that is, the structure of emulsion is formed only under applied external magnetic field. To study the effect of magnetic fluid on the electrical parameters of the measuring cell it was exposed toan external magnetic field. The measurements were made under magnetic field which is oriented parallel and perpendicular of the orientation of the measuring electrical field. The magnetic field was created by a permanent magnet.

It is should be noted that surface-active substance was not used, because the measurements were taken a couple of minutes, therefore, the magnetic fluid can be considered as a stable system and the sedimentation of particles is neglected.

3. The results of experimental studies

To determine greatest sensitivity of magnetic particles on dispersion from magnetic field, the measuring cell filled the study magnetic fluid which has different dispersion of particles was exposed to action of the external magnetic field created by a permanent magnet.The action of a magnetic field leads to the capacity change of a measuring cell due to reorganization and structuring suspension (figure 2 and figure 3).

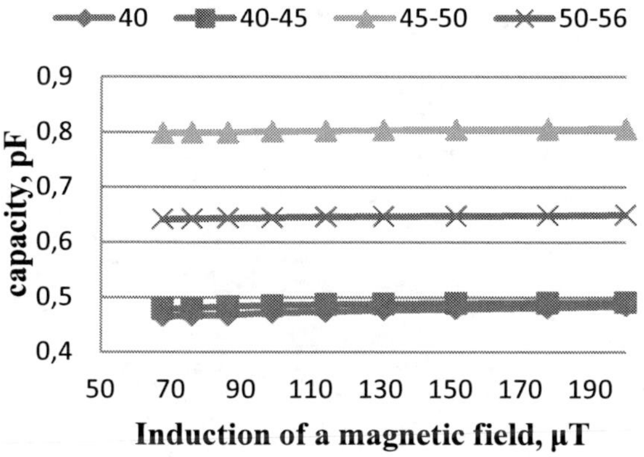

Figure 2. The capacity change of a measuring cellin the perpendicular orientations of the external magnetic and measuring electrical fields.

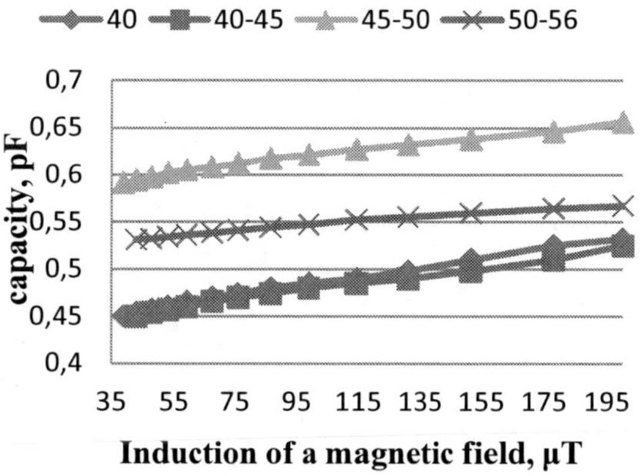

Figure 3. The capacity change of a measuring cellin the parallelorientations of the external magnetic and measuring electrical fields.

In that way, the capacity change is equated with the change of mutual positioning of particles in the fluid. The mutual ordering of particle distribution are sited along the magnetic field vector. Figure 4 and figure 5 shows the dependence of the relative capacity change of a measuring cell with magnetic particles of different dispersions: $50< \varnothing \leq 56$, $45< \varnothing \leq 50$, $40 < \varnothing \leq 45$ and $\varnothing \leq 40$ micron from the magnitude of the magnetic field in the various mutual orientations of the magnetic and electric fields.

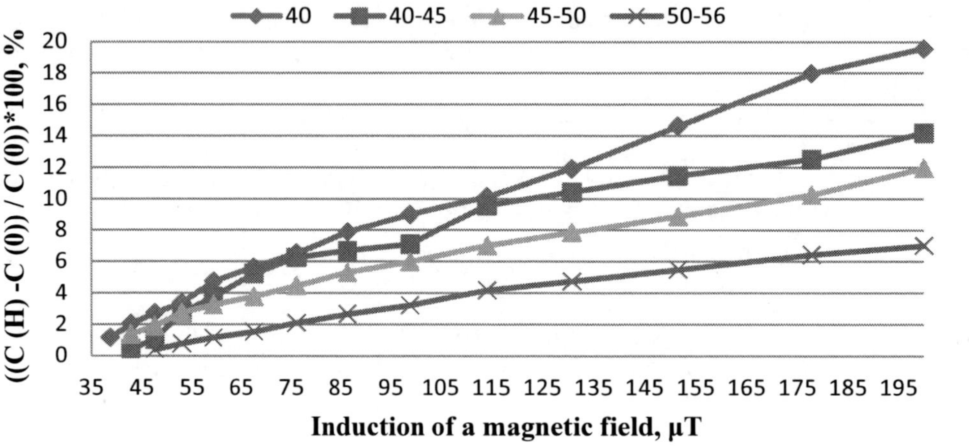

Figure 4. The relative capacity changein the parallelorientations of the external magnetic and measuring electrical fields.

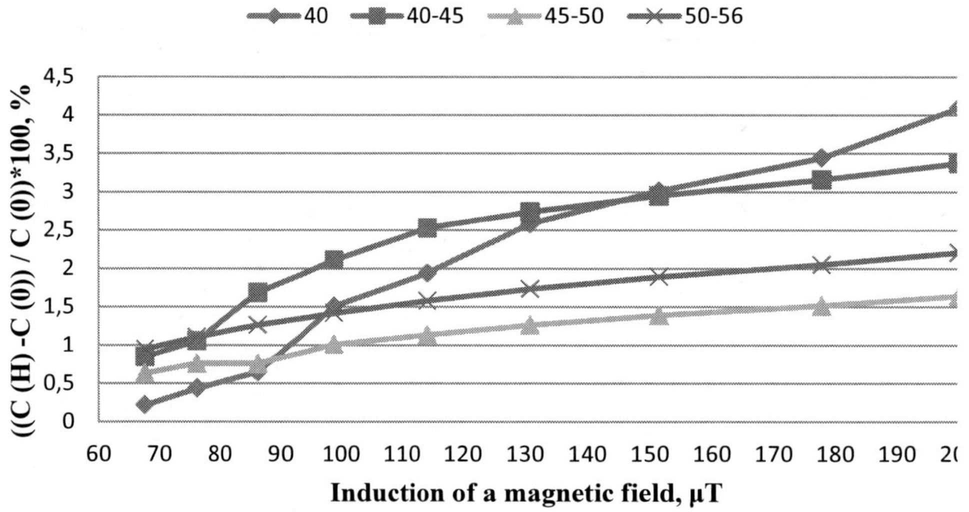

Figure 5. The relative capacity changein the perpendicular orientations of the external magnetic and measuring electrical fields.

The relative capacity change is defined from the formula:

$$\Delta C = \frac{\mathrm{C(H)} - \mathrm{C(0)}}{C(0)} * 100\%$$

where *C(H)*is the capacity of measuring cell in dependence from the magnetic field, *C(0)* – the capacity of measuring cell in case absence magnetic field.

Figure 4 shows a consistent decrease of the sensitivity threshold of a measuring cell with the magnetic fluid by decreasing the size of the magnetic particles. The threshold of sensitivity (figure 4) for particles which have a size of dispersion less 40 microns is equal to 38.7 μT. Given the linear dependence, we can assume, that further decrease of particle size will be reduced threshold of sensitivity to the magnetic field in case accounting of linear dependence.

The threshold of sensitivity is the lower-range value of the magnetic field magnitude, which generated appearance output signal.

Figure 6 the sensitivity (S) of a measuring cell with the magnetic fluid depending on the dispersity of the particles in a carrier liquid is presented.

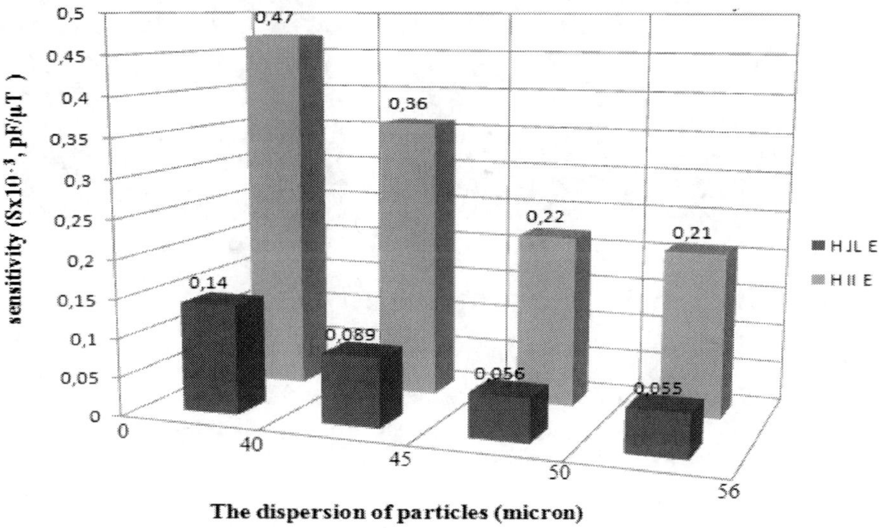

Figure 6. The sensitivity of a measuring cell with the magnetic fluid depending on the dispersity of the particles.

Figure 7 shows a photograph of the particles of dispersity: $50 < Ø \leq 56$, $45 < Ø \leq 50$, $40 < Ø \leq 45$ and $Ø \leq 40$ micron.

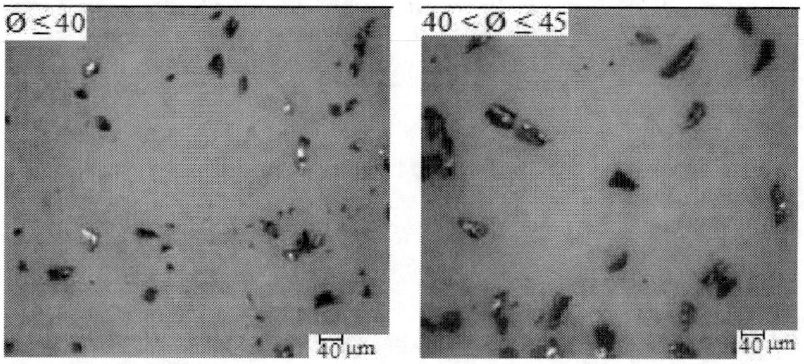

a) b)

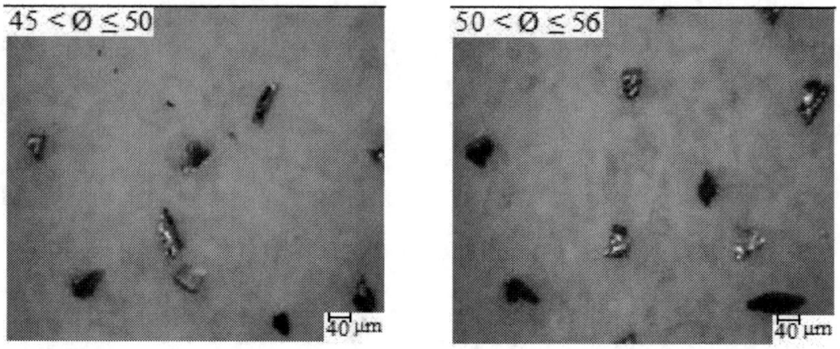

c) d)

Figure 7. The dispersion of particles: Ø ≤ 40 (a), 40 < Ø ≤ 45 (b), 45 < Ø ≤ 50 (c) and 50 < Ø ≤ 56 (d) micron.

Figure 7 shows that the particles have extended (in the form of "needles") or flat (in the form of "disks") forms and according to [10] they are easier to magnetic texturing, i.e. the ordering of the directions of the magnetic axes of the particles.

4. Conclusion

According to the results of experimental research it can be concluded that magnetic particles move in the liquid matrix orientating across the field by changing the physical characteristics of the matrix. Figure 4 shows a consistent decrease of the sensitivity threshold due to decreasing the size of the magnetic particles. The detection limit of the magnetic field by a measuring cell with the magnetic fluid in the parallel orientation of the external magnetic and measuring electrical fields is equal to 38.7 µT when use a dispersion of particles less 40 microns. The sensitivity of the measuring cell to the magnetic field is 0.47 fF/µTby the parallelorientations of the external magnetic and measuring electrical fields and 0.14 fF/µT by the perpendicular orientations of the external magnetic and measuring electrical fields.

References

[1] Taketomi S and Chikazumi S 1988 *Magnetic Fluids-principle and Application* (Tokio: Nokkan Kogyo Shinbun)
[2] Rosensweig R E 1987 Magnetic fluids *Ann. Rev. Fluid Mech.* **19** 437
[3] Popplewell J and Rosensweig R E 1996 *J. Phys. D: Appl.Phys.* **29** 2297
[4] Lopez-Lopez M T, Kuzhir P, Lacis S, Bossis G, Gonzalez-Caballero F and Duran J D G 2006 *J. Phys.: Condens. Matter* **18** S2803
[5] Elimelech M, Gregory J, Jia X and Williams R 1995 *Particle Deposition & Aggregation* (Oxford: Butterworth-Heinemann Ltd)
[6] Helgesen G, Svasand E and Skjeltorp A T (2008) *J. Phys.: Condens. Matter* **20** 204127
[7] Scherer C 2005 *J. Magn. Magn. Mater.* **289** 196
[8] Zyatkov D O, Yurchenko A V, Balashov V B, Yurchenko V I and Mehdiyev A D (2015) *Siber. J. Scien.***15** 161–167 (in Russian)
[9] Zyatkov D O, Yurchenko A V, Balashov V B and Yurchenko V I (2015) Polzunovsky vestnik **3** 161–164 (in Russian)
[10] Gubin S P, Koksharov Yu A, Khomutov G B and Yurkov G Yu (2005) *Russ. chem. rev.* **74** (6) 489–520